Springer Proceedings in Mathematics & Statistics

Volume 371

This book series features volumes composed of selected contributions from workshops and conferences in all areas of current research in mathematics and statistics, including operation research and optimization. In addition to an overall evaluation of the interest, scientific quality, and timeliness of each proposal at the hands of the publisher, individual contributions are all refereed to the high quality standards of leading journals in the field. Thus, this series provides the research community with well-edited, authoritative reports on developments in the most exciting areas of mathematical and statistical research today.

More information about this series at http://www.springer.com/series/10533

Albert N. Shiryaev · Konstantin E. Samouylov ·
Dmitry V. Kozyrev
Editors

Recent Developments in Stochastic Methods and Applications

ICSM-5, Moscow, Russia, November 23–27, 2020, Selected Contributions

Springer

Editors
Albert N. Shiryaev
Steklov Mathematical Institute of RAS
Moscow, Russia

Dmitry V. Kozyrev
Applied Probability & Informatics,
Department of Applied Probability
and Informatics
RUDN University
Moscow, Russia

Konstantin E. Samouylov
Applied Probability & Informatics,
Department of Applied Probability
and Informatics
RUDN University
Moscow, Russia

ISSN 2194-1009 ISSN 2194-1017 (electronic)
Springer Proceedings in Mathematics & Statistics
ISBN 978-3-030-83265-0 ISBN 978-3-030-83266-7 (eBook)
https://doi.org/10.1007/978-3-030-83266-7

Mathematics Subject Classification: 60Gxx, 60Hxx, 62Pxx, 60K25

This Springer imprint is published by the registered company Springer Nature Switzerland AG
The registered company address is: Gewerbestrasse 11, 6330 Cham, Switzerland

Preface

This volume contains a collection of revised selected full-text papers presented at the 5th International Conference on Stochastic Methods (ICSM-5), held in Moscow, Russia, November 23–27, 2020.

ICSM-5 is the successor to the previous conferences—the All-Russian Colloquium School on Stochastic Methods (held from 1994 till 2013) and the ICSM series—which took place in the last 5 years. ICSM-5 inherits its conferencing features configuration and the traditions initiated by the previous conferences and keeps and develops the ICSM participants' international community formed in the last several years of successful work. The aim of ICSM-5 is to unite the efforts of Russian and foreign researchers from various academic and research organizations for the development, exchange and generalization of the accumulated experience in the field of stochastic analysis and applications of stochastic modeling methods. The geography of the conference participants is traditionally wide and covers all parts of the world. In 2020, ICSM-5 gathered 110 submissions from authors from 19 different countries. From these, 87 high-quality papers in English were accepted and presented during the conference. The current volume contains 29 extended papers which were recommended by session chairs and selected by the Scientific Committee for Springer post-proceedings.

The content of this volume is related to the following subjects:

- Analytical modeling,
- Asymptotic methods and limit theorems,
- Stochastic analysis,
- Markov processes,
- Martingales,
- Insurance and financial mathematics,
- Queuing theory and stochastic networks,
- Reliability theory and risk analysis,
- Statistical methods and applications,
- Stochastic methods in computer science,

- Machine learning and data analysis,
- Probability in industry, economics and other fields.

All the papers selected for the post-proceedings volume are given in the form presented by the authors. These papers are of interest to everyone working in the field of stochastic analysis and applications of stochastic models.

February 2021

Albert N. Shiryaev
Scientific Committee Chair

Organization

ICSM-5 was jointly organized by Steklov International Mathematical Center, Steklov Mathematical Institute of RAS, Lomonosov Moscow State University (MSU), Peoples' Friendship University of Russia (RUDN University) and Don State Technical University (DSTU).

Organizing Committee

Chairmen

A. N. Shiryaev	Steklov Mathematical Institute of RAS, Russia
V. M. Filippov	RUDN University, Russia

Vice-chairmen

K. E. Samouylov	RUDN University, Russia
P. A. Yaskov	Steklov Mathematical Institute of RAS, Russia
I. V. Pavlov	Don State Technical University, Russia
V. V. Rykov	Gubkin University, and RUDN University, Russia
A. S. Holevo	Steklov Mathematical Institute of RAS, Russia

Members

A. A. Muravlev	Steklov Mathematical Institute of RAS, Russia
M. V. Zhitlukhin	Steklov Mathematical Institute of RAS, Russia
Yu. V. Gaidamaka	RUDN University, Russia
D. V. Kozyrev	RUDN University, and V.A.Trapeznikov Institute of Control Sciences of RAS, Russia
D. A. Shabanov	Lomonosov Moscow State University, Russia
E. B. Yarovaya	Lomonosov Moscow State University, Russia
S. I. Uglich	Don State Technical University, Russia

A. V. Bulinski
M. V. Zhitlukhin Steklov Mathematical Institute of RAS, Russia

Support

The conference was supported by the Ministry of Science and Higher Education of the Russian Federation (agreement no. 075-15-2019-1614).

This publication has been supported by the RUDN University Strategic Academic Leadership Program.

Contents

Probability and Statistics

On Threshold Selection Problem for Extremal Index Estimation

Igor Rodionov([✉])

Trapeznikov Institute of Control Sciences of RAS, Moscow, Russian Federation

Abstract. We study the properties of the new threshold selection method for non-parametric estimation of the extremal index of a stationary sequence proposed in [15]. The method is to apply the so-called discrepancy method based on the Cramér–von Mises–Smirnov's statistic calculated by the largest order statistics of a sample. The limit distribution of this statistic is derived if the proportion of the largest order statistics used tends to some nonzero constant. We also use the non-standard modification of the Cramér–von Mises–Smirnov's statistic to propose the goodness-of-fit test procedure of ω^2 type for distribution tails.

Keywords: Extremal index · Threshold selection · Discrepancy method · Stationary sequence · Cramér–von Mises–Smirnov's statistic · Goodness-of-fit · Distribution tail

1 Introduction

Let (X_n) be a strictly stationary sequence with common cumulative distribution function (cdf) F. Introduce the following

Definition 1 [11]. *The sequence (X_n) is said to have the extremal index $\theta \in [0, 1]$, if for each $\tau > 0$ there exists a sequence of real numbers $u_n = u_n(\tau)$ such that*

$$\lim_{n\to\infty} n(1 - F(u_n)) = \tau, \quad \lim_{n\to\infty} P(M_n \le u_n) = e^{-\theta\tau},$$

where $M_n = \max(X_1, \ldots, X_n)$.

The extremal index exists for a wide class of stationary sequences and reflects a cluster structure of an underlying sequence and its local dependence properties. The extremal index of an independent sequence is equal to 1, and the opposite is not true. In particular, if the Berman's condition holds, then a Gaussian stationary sequence has the extremal index equal to 1, [11].

We consider the problem of non-parametric estimation of the extremal index. It is important to note that absolutely all non-parametric extremal index estimators require the selection of the threshold parameter u and/or the block size

The author was partly supported by the Russian Foundation for Basic Research (grant No. 19-01-00090) and by Young Russian Mathematics Award.

A. N. Shiryaev et al. (Eds.): ICSM-5 2020, PROMS 371, pp. 3–16, 2021.
https://doi.org/10.1007/978-3-030-83266-7_1

b or other declustering parameter. The well-known blocks estimator [1] and its later modifications [19] depend of the choice of both b and u. Another classical one, the runs estimator [24], requires the selection of u and the number of consecutive observations r running below u separating two clusters. The intervals [8] and K-gaps [21] estimators require the choice of u only, whereas the sliding blocks estimators proposed in [16] and simplified in [2] depend on the block size only.

Less attention in the literature is devoted to methods of selection of the mentioned parameters, in particular, the threshold parameter u. Usually, the value of u is taken from high quantiles of an underlying sequence (X_n) or selected visually corresponding to a stability plot of values of some estimate $\hat{\theta}(u)$ with respect to u. Fukutome et al. [9] proposed the procedure of selection among pairs (u, K) for the K-gaps estimator based on Information Matrix Test.

Markovich and Rodionov [15] proposed the non-parametric tool to select one of the necessary parameters for extremal index estimation. Although the proposed method can be applied for selection of arbitrary aforementioned parameter, they focused on the selection of a threshold parameter u. The developed method is an automatic procedure of extremal index estimation in cases if it is based on estimators requiring the choice of only one parameter, in particular, the intervals and K-gaps estimators. But in [15] this procedure was established only if the proportion of the largest order statistics of a sample used vanishes as $n \to \infty$ (more precisely, see Theorem 3.3 [15]). The aim of this work is to investigate the opportunity of justification of the Markovich and Rodionov's method if the mentioned proportion tends to some positive constant c. The problem of goodness-of-fit testing of distribution tails is also studied.

2 Preliminaries

2.1 Inter-exceedance Times and Their Asymptotic Behavior

Let us discuss the properties of a stationary sequence (X_n) and its extremal index θ. Let L be the number of exceedances of level u by the sequence $(X_i)_{i=1}^n$ and $S_j(u)$ be the j-th exceedance time, that is,

$$S_j(u) = \min\{k > S_{j-1}(u) : X_k > u\}, \quad j = 1, \dots, L,$$

where $S_0 = 0$. Define *the inter-exceedance times* as

$$T_j(u) = S_{j+1}(u) - S_j(u), \quad j = 1, \dots, L-1$$

and assume its number equal to L for convenience.

Introduce the following φ-mixing condition.

Definition 2 [8]. *For real u and integers $1 \le k \le l$, let $\mathcal{F}_{k,l}(u)$ be the σ-field, generated by $\{X_i > u\}$, $k \le i \le l$. Introduce the mixing coefficients $\alpha_{n,q}(u)$,*

$$\alpha_{n,q}(u) = \max_{1 \le k \le n-q} \sup |P(B|A) - P(B)|,$$

where the supremum is taken over all sets $A \in \mathcal{F}_{1,k}(u)$ *with* $P(A) > 0$ *and* $B \in \mathcal{F}_{k+l,n}(u)$.

The next theorem states that for some sequence of levels (u_n) it holds

$$\overline{F}(u_n)T_1(u_n) \xrightarrow{d} T_\theta = \begin{cases} \eta, \text{ with probability } \theta, \\ 0, \text{ with probability } 1-\theta, \end{cases}$$

where η is exponential with mean θ^{-1}.

Theorem 1 [8]. *Let the positive integers* (r_n) *and the thresholds* (u_n), $n \geq 1$, *be such that* $r_n \to \infty$, $r_n\overline{F}(u_n) \to \tau$ *and* $P\{M_{r_n} \leq u_n\} \to exp(-\theta\tau)$ *as* $n \to \infty$ *for some* $\tau \in (0, \infty)$ *and* $\theta \in (0, 1]$. *If there are positive integers* $q_n = o(r_n)$ *such that* $\alpha_{cr_n,q_n}(u_n) = o(1)$ *for all* $c > 0$, *then for* $t > 0$

$$P\{\overline{F}(u_n)T_1(u_n) > t\} \to \theta \exp(-\theta t) =: 1 - F_\theta(t), \qquad n \to \infty. \tag{1}$$

The well-known intervals estimator of the extremal index is based on inter-exceedance times and is found via method of moments applied to the limit distribution (1). It is defined as ([8], see also [1], p. 391),

$$\hat{\theta}_n(u_n) = \begin{cases} \min(1, \hat{\theta}_n^1(u)), \text{ if } \max\{T_i(u_n),\ 1 \leq i \leq L\} \leq 2, \\ \min(1, \hat{\theta}_n^2(u)), \text{ if } \max\{T_i(u_n),\ 1 \leq i \leq L\} > 2, \end{cases} \tag{2}$$

where

$$\hat{\theta}_n^1(u_n) = \frac{2\left(\sum_{i=1}^L T_i(u_n)\right)^2}{L\sum_{i=1}^L (T_i(u_n))^2} \quad \text{and} \quad \hat{\theta}_n^2(u_n) = \frac{2\left(\sum_{i=1}^L (T_i(u_n)-1)\right)^2}{L\sum_{i=1}^L (T_i(u_n)-1)(T_i(u_n)-2)}.$$

It is known that

$$\sqrt{L}(\hat{\theta}_n(u_n) - \theta) \xrightarrow{d} N(0, \theta^3 v(\theta)), \tag{3}$$

where $v(\theta)$ is the second moment of the cluster size distribution $\{\pi(m)\}_{m\geq 1}$, [18]. Moreover, Theorem 2.4 [18] states that non-zero elements of the sequence

$$Z_i = \overline{F}(u_n)T_i(u_n), \quad i = 1, \ldots, L, \tag{4}$$

are asymptotically independent under the assumptions of Theorem 1 and some assumptions on the cluster structure of the initial stationary sequence (X_n).

To be able to use these properties, Markovich and Rodionov [15] assume that there exists a sequence $(E_i)_{i=1}^l$, $l = [\theta L]$, of independent exponentially distributed random variables with mean θ^{-1} such that

$$Z_{(L-k)} - E_{(l-k)} = o\left(\frac{1}{\sqrt{k}}\right) \tag{5}$$

uniformly for all $k \to \infty$ with $k/L \to 0$ as $L \to \infty$. This assumption is based on the following reasoning. It follows from Theorem 3.2 [18], that the limit distribution of the statistic

$$\sqrt{L}\left(\sum_{i=1}^L f(Z_i) - Ef(Z_1)\right)$$

for some class of continuous f does not depend on substitution of the set of r.v.s $\{Z_i^*\}_{i=1}^L$ with cdf F_θ instead of $\{Z_i\}_{i=1}^L$ under some regularity conditions. Moreover, for these r.v.s Theorem 2.2.1 and Lemma 2.2.3 [7] imply that if $k/L \to c,\ c \in [0, \theta)$ as $k \to \infty$, $L \to \infty$, then

$$\sqrt{k}(E_{(l-k)} - \ln(l/k)/\theta) = O_P(1).$$

2.2 Discrepancy Method

The method proposed in [15] is based on the so-called discrepancy method initially introduced in [12] and [22], see also [13], for optimal bandwidth selection in the problem of density estimation and applied at the first time for extremal index estimation in [14]. Let $\rho(\cdot, \cdot)$ be some distance on the space of probability measures, $\hat{F}_n$ be the empirical cdf of the sequence $(X_i)_{i=1}^n$ and $\{F_u, u \in U\}$ be the family of cdfs parametrized by one-dimensional parameter u. Then the optimal value of u can be found as a solution of the discrepancy equation

$$\rho(\hat{F}_n, F_u) = \delta, \tag{6}$$

where δ, the so-called discrepancy value, is defined by the choice of ρ. The statistic of the Cramér–von Mises–Smirnov goodness-of-fit test (CMS statistic)

$$\omega_n^2 = n \int_{\mathbb{R}} (\hat{F}_n(x) - F_0(x))^2 dF(x)$$

was chosen as ρ in [15], though the statistics of other goodness-of-fit tests, e.g., the Kolmogorov and Anderson-Darling tests, can be applied in the discussing problem. Then quantiles of the limit distribution of the CMS statistic can be used as δ. The choice of the parameter u as $\hat{u} = \text{argmin}_u \rho(\hat{F}_n, F_u)$ is usually not optimal.

3 Main Results

In this section we consider the problem of the discrepancy method application to choose the threshold/block-size parameter of the extremal index estimator. To simplify, let us assume that we choose the threshold parameter u. It seems that for this purpose one can take $F_u = T_{\hat{\theta}(u)}$, $\hat{F}_n$ be equal to the empirical cdf of the sequence $\{Z_i\}_{i=1}^L$ and ρ be equal to the ω^2 distance in (6). However, we cannot directly apply the discrepancy method coupling with the ω^2 distance to this problem since T_θ is not a continuous distribution and thus the limit distribution of the CMS statistic would depend on θ. To overcome this difficulty, we introduce the modification of the CMS statistic based only on the largest order statistics corresponding to $\{Z_i\}_{i=1}^L$, since, as was mentioned in Sect. 2.1, the largest elements of this sequence are continuously distributed and asymptotically independent. Thus we face the problem of goodness-of-fit testing of left-censored data and, in particular, distribution tails, see [20] for the principles of testing of distribution tails.

Let $(Y_i)_{i=1}^n$ be independent identically distributed random variables with common continuous cdf F_Y. Recall that if the hypothesis $H_0 : F_Y = F_0$ for continuous cdf F_0 holds, then the CMS statistic can be rewritten as

$$\omega_n^2 = \sum_{i=1}^{n} \left(F_0(Y_{(i)}) - \frac{i - 0.5}{n} \right)^2 + \frac{1}{12n},$$

where $Y_{(1)} \leq \ldots \leq Y_{(n)}$ are the order statistics corresponding to $(Y_i)_{i=1}^n$. It is well-known that the limit distribution of ω_n^2 (denote its cdf as A_1) under H_0 does not depend on F_0.

Goodness-of-fit procedures for various types of left- and right-censored data were proposed in a large number of works, we refer to the classical monograph [4] and recent monograph [23]. But to the best of author's knowledge, there are no works in the literature proposing the modifications of goodness-of-fit statistics for censored data having the same limit distribution as their full-sample analogues. Introduce the following modification of the CMS statistic

$$\hat{\omega}_k^2 = \sum_{i=0}^{k-1} \left(\frac{F_0(Y_{(n-i)}) - F_0(Y_{(n-k)})}{1 - F_0(Y_{(n-k)})} - \frac{k - i - 0.5}{k} \right)^2 + \frac{1}{12k}.$$

Theorem 2. *Let the hypothesis* $H_0^t : \{F_Y(x) = F_0(x) \text{ for all large } x\}$ *holds. Then there is c such that*

$$\hat{\omega}_k^2 \xrightarrow{d} \xi \sim A_1$$

as $k \to \infty$, $k/n < c$, $n \to \infty$.

Theorem 3.1, [15], is a particular case of the latter theorem for $F_0 = F_\theta$, where F_θ is defined in (1). It is worth noting that there is no necessity to require the continuity of F_0 for all real x; we need this only for all sufficiently large x. Theorem 2 allows us to propose the goodness-of-fit test for continuous distribution tail of significance level α in the following way

$$\text{if } \hat{\omega}_k^2 > a_{1-\alpha}, \text{ then reject } H_0^t,$$

where $a_{1-\alpha}$ is the $(1-\alpha)$-quantile of A_1. Additionally, one can show that this test is consistent as $k \to \infty$, $k/n \to 0$, $n \to \infty$. Clear, only the largest order statistics of a sample can be used for testing the distribution tail hypotheses both if $k/n \to 0$ and if $k/n < c$. This problem is reasonable both if only the upper tail of the distribution is of interest and/or if only the largest order statistics of a sample are available.

Let us return to the problem of extremal index estimation. Consider

$$\widetilde{\omega}_L^2(\theta) = \sum_{i=0}^{k-1} \left(\frac{F_\theta(Z_{(L-i)}) - F_\theta(Z_{(L-k)})}{1 - F_\theta(Z_{(L-k)})} - \frac{k - i - 0.5}{k} \right)^2 + \frac{1}{12k}.$$

The following theorem states that under some mild conditions the statistic $\widetilde{\omega}_L^2(\theta)$ with some estimator $\widehat{\theta}_n$ substituted for θ has the same limit distribution as the statistic $\hat{\omega}_k^2$ in Theorem 2 and the classical CMS statistic.

Theorem 3. [15]. *Let the assumptions of Theorem 1 and the condition (5) hold. Assume the extremal index estimator $\widehat{\theta} = \widehat{\theta}_n$ is such that*

$$\sqrt{m_n}(\widehat{\theta}_n - \theta) \xrightarrow{d} \zeta, \quad n \to \infty,$$

where ζ has a non-degenerate cdf H. Assume the sequence of integers (m_n) is such that

$$\frac{k}{m_n} = o(1) \quad and \quad \frac{(\ln L)^2}{m_n} = o(1)$$

as $n \to \infty$. Then

$$\widetilde{\omega}_L^2(\widehat{\theta}_n) \xrightarrow{d} \xi \sim A_1.$$

Remark 1. All extremal index estimators mentioned in Introduction satisfy the assumptions of Theorem 3 with H equal to the normal cdf with zero mean.

Theorem 3 guarantees the correctness of the discrepancy method

$$\widetilde{\omega}_L^2(\widehat{\theta}_n) = \delta, \tag{7}$$

where δ can be selected equal to 0.05, the mode of A_1, and $k/L \to 0$ as $k \to \infty$ and $L \to \infty$. The simulation study provided in [15] shows that $u_{\max} = \max\{u_1, \ldots, u_d\}$ is the best choice for threshold parameter both for the intervals and the K-gaps estimators of the extremal index, where $\{u_1, \ldots, u_d\}$ are solutions of the discrepancy equation (7). The numerical comparison of the proposed method with other methods of threshold selection shows the significant advantage of the developed procedure on a wide class of model processes, see [15] for details. Although the limit distribution of the statistic $\widetilde{\omega}_L^2(\widehat{\theta}_n)$ does not depend on k, the selection of k for samples of moderate sizes remains a problem. The choice $k = \min(\hat{\theta}_0 L, L^\beta)$ with $\beta \in (0, 1)$, where $\hat{\theta}_0$ is some pilot estimate, has proven by simulation study to be the most suitable. But in case of $k/L \to c > 0$ as $k \to \infty$, $L \to \infty$ the distribution of $\widehat{\theta}_n$ affects the limit distribution of the modified CMS statistic $\widetilde{\omega}_L^2(\widehat{\theta}_n)$ in contrast to the case $c = 0$, thus this limit distribution would differ from A_1.

The asymptotic distributions of goodness-of-fit test statistics with parameters of an underlying distribution being estimated were intensively studied in the literature. The starting point for this classical theory was in works [5] and [10], whereas the common method to derive the limit distribution was proposed in [6]. However, this method based on a multivariate central limit theorem and convergence in the Skorokhod space cannot be directly applied to the problem of evaluating the limit distribution of the statistic $\widetilde{\omega}_L^2(\widehat{\theta}_n)$ when the assumption $k/m_n = o(1)$ does not hold ($m_n = O(L)$ for the intervals and K-gaps estimators, thus we can talk about the case $k/L = o(1)$). For this purpose we consider another modification of the CMS statistic, the first analogue of which was introduced in [17],

$$\omega^2_{L,c}(\theta) = L\int_{x(c)}^{\infty} (F^*_L(x) - F_\theta(x))^2 dF_\theta(x), \tag{8}$$

where $F^*_L(x)$ is the empirical distribution function of the sequence $\{Z_i\}_{i\leq L}$ and $x(c) = \inf\{x : F_\theta(x) \geq 1-c\} =: F_\theta^{\leftarrow}(1-c)$, $c \in (0,1)$. In the sequel, we will assume $0 < c < \theta$, therefore $F_\theta(x(c)) = 1-c$. It follows from the results derived in [17] and the assumption (5) that the statistic $\omega^2_{L,c}$ converges in distribution to $\omega^2(c)$, where

$$\omega^2(c) = \int_{1-c}^{1} B^2(t)dt$$

and $B(t)$ is the standard Brownian bridge, i.e. the Gaussian process on the interval $[0,1]$ with mean zero and covariance function $\mathrm{cov}(B_t, B_s) = \min(t,s) - ts$.

Denote $\ell_c(t) = \max(t-(1-c), 0)$. Following the ideas of [6] we introduce the sample process

$$y_{L,c}(t,\theta) = \sqrt{L}\left(\hat{F}_{L,c}(t,\theta) - \ell_c(t)\right), \quad t \in [0,1],$$

where

$$\hat{F}_{L,c}(t,\theta) = \frac{1}{L}\sum_{i=1}^{L} I(1-c < F_\theta(Z_i) \leq t),$$

call it the truncated empirical distribution function of the sequence $\{Z_i\}_{i\leq L}$. Clear, since $\theta > c$ it holds

$$\int_0^1 y^2_{L,c}(t,\theta)dt = \omega^2_{L,c}(\theta).$$

Denote D, the Skorokhod space, i.e. the space of right-continuous functions with left-hand limits on $[0,1]$ and metric $d(x,y)$ (see, e.g., [3], p. 111). The following theorem allows us to find the asymptotic distribution of the statistic $\omega^2_{L,c}(\hat{\theta}_n)$, where $\hat{\theta}_n$ is the intervals estimator (2).

Theorem 4. *Let the sequence $\{Z_i\}$ defined by (4) satisfies the assumptions of Theorem 3.2 [18]. Assume $\theta > 0$. For every $c \in (0,\theta)$ the estimated sample process $\hat{y}_c(t) := y_{L,c}(t,\hat{\theta}_n)$ converges weakly in D as $n \to \infty$ to the Gaussian process $X(t), t \in [1-c, 1]$, with mean zero and covariance function*

$$\begin{aligned} C(t,s) &= \ell_c(\min(t,s)) - (2-2/\theta)\ell_c(t)\ell_c(s) \\ &\quad - \frac{1}{2\theta^2}h_c(t)(2h_c(s)+\tilde{h}_c(s)) - \frac{1}{2\theta^2}h_c(s)(2h_c(t)+\tilde{h}_c(t)) + \frac{v(\theta)}{\theta}h_c(t)h_c(s), \end{aligned} \tag{9}$$

where

$$h_c(t) = (1-t)\log\left(\frac{1-t}{\theta}\right) - c\log\left(\frac{c}{\theta}\right), \quad \tilde{h}_c(t) = (1-t)\log^2\left(\frac{1-t}{\theta}\right) - c\log^2\left(\frac{c}{\theta}\right)$$

and $v(\theta)$ is defined in (3).

We see that the covariance function of the process $X(t)$ depends on θ and c. This fact makes the usage of quantiles of the limit distribution of the statistic $\omega^2_{L,c}(\hat{\theta}_n)$ (or some its appropriate normalization) as δ in the discrepancy method (7) quite inconvenient in practice. However, high efficiency of the discrepancy method apparently means that the values of the mentioned quantiles do not strongly depend on the values of θ and c.

4 Proofs

4.1 Proof of Theorem 2

To prove Theorem 2, we need the following

Lemma 1 *(Lemma 3.4.1, [7]). Let $X, X_1, X_2, ..., X_n$ be i.i.d. random variables with common cdf F, and let $X_{(1)} \leq X_{(2)} \leq ... \leq X_{(n)}$ be the nth order statistics. The joint distribution of $\{X_{(i)}\}_{i=n-k+1}^n$ given $X_{(n-k)} = t$, for some $k = 1, ...,$ $n-1$, equals the joint distribution of the set of order statistics $\{X^*_{(i)}\}_{i=1}^k$ of i.i.d. r.v.s $\{X^*_i\}_{i=1}^k$ with cdf*

$$F_t(x) = P\{X \leq x | X > t\} = \frac{F(x) - F(t)}{1 - F(t)}, \quad x > t.$$

Assume $F_Y(x) = F_0(x)$ for all $x > x_0$ and set $c = 1 - F_0(x_0) - \varepsilon$ for some small $\varepsilon > 0$. Clear, since $k/n < c$ then $P(Y_{(n-k)} > x_0) \to 1$ under the assumptions.

Consider the conditional distribution of $\hat{\omega}^2_k$ given $F_0(Y_{(n-k)}) = t,\ t > 1 - c$. By Lemma 1 and the assumption $k/n < c$, the conditional joint distribution of the set of order statistics $\{F_0(Y_{(i)})\}_{i=n-k+1}^n$ coincides with the joint distribution of the set of order statistics $\{U^*_{(i)}\}_{i=1}^k$ of a sample $\{U^*_i\}_{i=1}^k$ from the uniform distribution on $[t, 1]$. Therefore, it holds

$$\hat{\omega}^2_k \stackrel{d}{=} \frac{1}{(1-t)^2} \left(\sum_{i=1}^k \left(U^*_{(i)} - t - \frac{i - 0.5}{k}(1-t) \right)^2 \right) + \frac{1}{12k}.$$

Next, $V^*_{(i)} = U^*_{(i)} - t,\ 1 \leq i \leq k$, are the order statistics of a sample $\{V^*_i\}_{i=1}^k$ from the uniform distribution on $[0, 1-t]$. Hence, it follows

$$\hat{\omega}^2_k \stackrel{d}{=} \frac{1}{(1-t)^2} \left(\sum_{i=1}^k \left(V^*_{(i)} - \frac{i - 0.5}{k}(1-t) \right)^2 \right) + \frac{1}{12k}.$$

Finally, $W^*_{(i)} = V^*_{(i)}/(1-t),\ 1 \leq i \leq k$, are the order statistics of a sample $\{W^*_i\}_{i=1}^k$ from the uniform distribution on $[0, 1]$. Therefore, we get

$$\hat{\omega}^2_k \stackrel{d}{=} \sum_{i=1}^k \left(W^*_{(i)} - \frac{i - 0.5}{k} \right)^2 + \frac{1}{12k}.$$

It is easy to see, that the latter expression is exactly the CMS statistic and converges in distribution to a random variable ξ with cdf A_1 independently of the value of t.

4.2 Proof of Theorem 4

Assume for convenience that $\hat{\theta}_n = \hat{\theta}_n^2(u_n)$, where $\hat{\theta}_n^2(u_n)$ is defined by (2). For shortening, write $\hat{\theta}$ instead of $\hat{\theta}_n$. We need the following

Lemma 2. *Under the assumptions*

$$\sqrt{L}(\hat{\theta}-\theta) = \frac{\theta}{\sqrt{L}}\sum_{i=1}^{L}\left(2Z_i - \frac{\theta}{2}Z_i^2 - 1\right) + o_P(1),$$

where $o_P(1)$ denotes a sequence of random variables vanishing in probability.

Proof. (of Lemma 2)
Observe that

$$n - \sum_{i=1}^{L} T_i(u_n) \le T_{L+1}(u_n) \stackrel{d}{=} T_1(u_n),$$

where the last relation holds by stationarity and definition of inter-exceedance times. Denote $r_n = n\overline{F}(u_n)$. Theorem 3.2 [18] implies that

$$\sqrt{r_n}\,(L/r_n - 1) \xrightarrow{d} N(0, \theta v(\theta)), \tag{10}$$

thus for all $\varepsilon > 0$

$$P\left(\sqrt{r_n}\Big(1 - \frac{1}{r_n}\sum_{i=1}^{L} Z_i\Big) > \varepsilon\right) = P\left(\sqrt{r_n}\frac{\overline{F}(u_n)}{r_n}\Big(n - \sum_{i=1}^{L} T_i(u_n)\Big) > \varepsilon\right)$$
$$\le P\left(\frac{\overline{F}(u_n)}{\sqrt{r_n}}T_{L+1}(u_n) > \varepsilon\right) = P\left(\frac{1}{\sqrt{r_n}}Z_1 > \varepsilon\right) \to 0, \tag{11}$$

where the last relation holds by Theorem 1. Note that by (10) $L/r_n \xrightarrow{P} 1$, thus

$$\sqrt{L}(\hat{\theta}-\theta) = \sqrt{r_n}(\hat{\theta}-\theta) + o_P(1).$$

The latter relations imply that

$$\sqrt{L}(\hat{\theta}-\theta) = \sqrt{r_n}\left(\frac{2r_n/L}{\frac{1}{r_n}\sum_{i=1}^{L} Z_i^2} - \theta\right) + o_P(1) = \theta\sqrt{r_n}\left(\frac{2r_n/(\theta L) - \frac{1}{r_n}\sum_{i=1}^{L} Z_i^2}{\frac{1}{r_n}\sum_{i=1}^{L} Z_i^2}\right) + o_P(1).$$

Next, it follows from Lemma B.7 [18] that

$$\frac{1}{r_n}\sum_{i=1}^{L} Z_i^2 \xrightarrow{P} \frac{2}{\theta},$$

therefore

$$\sqrt{L}(\hat{\theta}-\theta) = \frac{\theta^2}{2}\sqrt{r_n}\left(\frac{2r_n}{\theta L} - \frac{1}{r_n}\sum_{i=1}^{L} Z_i^2\right) + o_P(1).$$

It immediately follows from (10) and the delta method that

$$\sqrt{r_n}(r_n/L - 1) = \sqrt{r_n}(L/r_n - 1) + o_P(1).$$

Finally, we obtain from (11) and the latter

$$\sqrt{r_n}\left(\frac{2r_n}{\theta L}-\frac{1}{r_n}\sum_{i=1}^{L}Z_i^2\right)=\sqrt{r_n}\left(\frac{4}{\theta}-\frac{2L}{\theta r_n}-\frac{1}{r_n}\sum_{i=1}^{L}Z_i^2\right)+o_P(1)$$
$$=\frac{4}{\theta}\sqrt{r_n}\left(1-\frac{1}{r_n}\sum_{i=1}^{L}Z_i\right)$$
$$+\frac{2}{\theta}\sqrt{r_n}\left(2\frac{1}{r_n}\sum_{i=1}^{L}Z_i-\frac{1}{r_n}\sum_{i=1}^{L}1-\frac{\theta}{2r_n}\sum_{i=1}^{L}Z_i^2\right)+o_P(1)$$
$$=\frac{2}{\sqrt{r_n}\theta}\sum_{i=1}^{L}\left(2Z_i-\frac{\theta}{2}Z_i^2-1\right)+o_P(1).$$

Applying again $L/r_n \xrightarrow{P} 1$, we derive the result.

First of all, we observe that by (11)

$$y_{L,c}(t,\theta)=\sqrt{L}\left(\frac{1}{L}\sum_{i=1}^{L}I(1-c<F_\theta(Z_i)\le t)-\ell_c(t)\right)$$
$$=\frac{1}{\sqrt{L}}\sum_{i=1}^{L}\left(I(1-c<F_\theta(Z_i)\le t)-\ell_c(t)Z_i\right)+\ell_c(t)\sqrt{L}\left(\frac{1}{L}\sum_{i=1}^{L}Z_i-1\right)$$
$$=\frac{1}{\sqrt{L}}\sum_{i=1}^{L}\left(I(1-c<F_\theta(Z_i)\le t)-\ell_c(t)Z_i\right)+o_P(1). \qquad (12)$$

It is also worth noting that we can change all expressions of the form $\frac{1}{\sqrt{L}}\sum_{i=1}^{L}\xi_i$ appearing in our proof with $E\xi_i = 0$ for all i on expressions of the form $\frac{1}{\sqrt{r_n}}\sum_{i=1}^{r_n}\xi_i$ using the same argument as in the proof of Theorem 3.2 [18] based on the formula (10). It means that the randomness of L does not affect the asymptotic of $\hat{y}_c(t)$.

We follow the ideas of the proof of Theorem 2 in [6]. For shortening, write $\hat{\theta}$ instead of $\hat{\theta}_n$. Denote

$$\hat{t}(t)=F_\theta(F_{\hat{\theta}}^{\leftarrow}(t)),\quad t\in[1-c,1].$$

Since $\hat{\theta}$ is a consistent estimator of θ, see [18], we have

$$P(\hat{t}(t)\ge 1-c)\to 1 \qquad (13)$$

for all $t\in(1-c,1]$, and for $t=1-c$ the latter probability tends to 1/2. First, we will show that

$$y_{L,c}(\hat{t}(t),\theta)-y_{L,c}(t,\theta)\xrightarrow{P}0 \qquad (14)$$

uniformly for $t\in[1-c,1]$. We restrict ourselves to the study of the case $t\in(1-c,1]$, the case $t=1-c$ is similar. Denote $x_1(t)=F_{\theta_1}^{\leftarrow}(t)$ and $\tilde{t}(t)=F_{\theta_2}(x_1(t))$ for $\theta_1,\theta_2\ge c$ and $t\in(1-c,1]$. We have

$$\sup_{t\in(1-c,1]}|\tilde{t}-t|\le\sup_{t\in(1-c,1]}|F_{\theta_2}(x_1(t))-F_{\theta_1}(x_1(t))|$$
$$=\sup_{x>F_{\theta_1}^{\leftarrow}(1-c)}\left|(\theta_2-\theta_1)\left.\frac{\partial F_\theta(x)}{\partial\theta}\right|_{\theta=\theta^*}\right|,$$

where θ^* is between θ_1 and θ_2. Since

$$\left|\frac{\partial F_\theta(x)}{\partial \theta}\right| = \left|(\theta^2-1)e^{-\theta x}\right| \le 1$$

for all $\theta \in [0,1]$ and $x > F_{\theta_1}^{\leftarrow}(1-c)$, we derive that $\tilde{t}$ converges uniformly to t as $\theta_1 \to \theta$ and $\theta_2 \to \theta$. Therefore, since $\hat{\theta}$ converges to θ in probability,

$$\sup_{t\in(1-c,1]} |\hat{t}(t) - t| \xrightarrow{P} 0.$$

An appeal to Lemma B.7 [18] gives us that $y_{L,c}(t,\theta) \xrightarrow{d} y(t)$, $t \in (1-c,1]$, in D where $y(t)$ is the Gaussian random process with mean zero and covariance function

$$\text{cov}(y(t), y(s)) = \ell_c(\min(t,s)) - (2-2/\theta)\ell_c(t)\ell_c(s).$$

The rest of the proof of (14) coincides with the corresponding steps in the proof of Lemma 1 [6].

Now let us show that

$$\hat{y}_c(t) = y_{L,c}(t,\theta) - \sqrt{L}(\hat{\theta}-\theta)(g(t,\theta) - g(1-c,\theta)) + o_P(1), \quad t \in [1-c,1], \quad (15)$$

where

$$g(t,\theta) = \frac{1-t}{\theta^2}\log\left(\frac{1-t}{\theta}\right).$$

Note that by definition $\hat{y}_c(1-c) = 0$ a.s., thus it remains to show (15) for $t \in (1-c,1]$. First we find the explicit form of the "estimated" empirical cdf $\hat{F}_{L,c}(\hat{t}(t),\theta)$. We have

$$\begin{aligned}
\hat{F}_{L,c}(\hat{t}(t),\theta) &= \frac{1}{L}\sum_{i=1}^{L} I\left(1-c \le F_\theta(Z_i) < F_\theta(F_{\hat{\theta}}^{\leftarrow}(t))\right) \\
&= \frac{1}{L}\sum_{i=1}^{L} I\left(F_{\hat{\theta}}(F_\theta^{\leftarrow}(1-c)) < F_{\hat{\theta}}(Z_i) \le t\right) \\
&= F_{L,c}(t,\hat{\theta}) - \frac{1}{L}\sum_{i=1}^{L} I\left(1-c \le F_{\hat{\theta}}(Z_i) < F_{\hat{\theta}}(F_\theta^{\leftarrow}(1-c))\right) \\
&= F_{L,c}(t,\hat{\theta}) - \frac{1}{L}\sum_{i=1}^{L} I\left(\hat{t}(1-c) < F_\theta(Z_i) \le 1-c\right).
\end{aligned}$$

Denote

$$\tilde{F}_{L,c}(t,\theta) = \frac{1}{L}\sum_{i=1}^{L} I\left(t < F_\theta(Z_i) \le 1-c\right), \quad t \le 1-c.$$

Therefore we derive for the estimated sample process $\hat{y}_c(t)$

$$\begin{aligned}\hat{y}_c(t) &= \sqrt{L}(F_{L,c}(t,\hat{\theta}) - \ell_c(t)) \\ &= \sqrt{L}\Big(F_{L,c}(\hat{t}(t),\theta) - \ell_c(\hat{t}(t))\Big) + \sqrt{L}\Big(\ell_c(\hat{t}(t)) - \ell_c(t)\Big) + \sqrt{L}\tilde{F}_{L,c}(\hat{t}(1-c),\theta) \\ &= y_{L,c}(\hat{t}(t),\theta) + \sqrt{L}\Big(\ell_c(\hat{t}(t)) - \ell_c(t)\Big) + \sqrt{L}\tilde{F}_{L,c}(\hat{t}(1-c),\theta).\end{aligned}$$

Consider the third summand on the right-hand side. Fix $c_1 \in (c,\theta)$. Note that (14) remains true for all $c_1 < \theta$, thus we derive

$$y_{L,c_1}(\hat{t}(1-c),\theta) - y_{L,c_1}(1-c,\theta) \xrightarrow{P} 0.$$

On the other hand,

$$\begin{aligned} y_{L,c_1}(\hat{t}(1-c),\theta) &- y_{L,c_1}(1-c,\theta) \\ &= \sqrt{L}(F_{L,c_1}(\hat{t}(1-c),\hat{\theta}) - \ell_{c_1}(\hat{t}(1-c))) - \sqrt{L}(F_{L,c_1}(1-c,\hat{\theta}) - \ell_{c_1}(1-c)) \\ &= \sqrt{L}(1-c-\hat{t}(1-c)) - \sqrt{L}\tilde{F}_{L,c}(\hat{t}(1-c),\theta),\end{aligned}$$

therefore we derive

$$\sqrt{L}\tilde{F}_{L,c}(\hat{t}(1-c),\theta) = \sqrt{L}(1-c-\hat{t}(1-c)) + o_P(1).$$

Next, (13) implies that $\sqrt{L}\Big(\ell_c(\hat{t}(t)) - \ell_c(t)\Big) = \sqrt{L}(\hat{t}(t)-t) + o_P(1)$ in case of $t \in (1-c,1]$. We have

$$\begin{aligned}\sqrt{L}\Big(\ell_c(\hat{t}(t)) - \ell_c(t)\Big) &= \sqrt{L}\Big(F_\theta(F_{\hat{\theta}}^{\leftarrow}(t)) - F_{\hat{\theta}}(F_{\hat{\theta}}^{\leftarrow}(t))\Big) + o_P(1) \\ &= \sqrt{L}(\theta-\hat{\theta})\left.\frac{\partial F_\gamma(x)}{\partial\gamma}\right|_{\substack{x=F_{\hat{\theta}}^{\leftarrow}(t)\\ \gamma=\theta^*}} + o_P(1),\end{aligned}$$

where θ^* is between θ and $\hat{\theta}$. Similarly to the corresponding steps in the proof of Lemma 2 [6] we can show that

$$\left.\frac{\partial F_\gamma(x)}{\partial\gamma}\right|_{\substack{x=F_{\hat{\theta}}^{\leftarrow}(t)\\ \gamma=\theta^*}} = \left.\frac{\partial F_\gamma(x)}{\partial\gamma}\right|_{\substack{x=F_{\theta}^{\leftarrow}(t)\\ \gamma=\theta}} + o_P(1),$$

since $\partial F_\gamma(x)/\partial\gamma$ is continuous with respect to (x,γ) for all $x > F_\theta^{\leftarrow}(1-c)$ and $\gamma \in (0,1]$. Clear,

$$\left.\frac{\partial F_\gamma(x)}{\partial\gamma}\right|_{\substack{x=F_{\theta}^{\leftarrow}(t)\\ \gamma=\theta}} = (x-1)e^{-\gamma x}\Big|_{\substack{x=F_{\theta}^{\leftarrow}(t)\\ \gamma=\theta}} = \frac{1-t}{\theta^2}\log\left(\frac{1-t}{\theta}\right) = g(t,\theta).$$

Note that the relation

$$\sqrt{L}(\hat{t}(t)-t) = \sqrt{L}(\theta-\hat{\theta})g(t,\theta) + o_P(1)$$

derived above for $t \in (1-c,1]$ remains true also for $t = 1-c$. Finally, combining the previous relations and using (14), we derive (15).

Define the empirical process

$$z_L(t) = \frac{1}{\sqrt{L}}\sum_{i=1}^{L}\Big(I(1-c< F_\theta(Z_i) \leq t)-\ell_c(t)Z_i-\theta\big(2Z_i-\frac{\theta}{2}Z_i^2-1\big)(g(t,\theta)-g(1-c,\theta))\Big)$$

and notice that $\hat{y}_c(t) = z_L(t) + o_P(1)$ by Lemma 2, (12) and (15). To complete the proof of Theorem 4 we need to prove that

$$(z_L(t_1),\ldots,z_L(t_k)) \xrightarrow{P} (X(t_1),\ldots,X(t_k)), \quad \text{for all } 1-c \leq t_1 < \ldots < t_k \leq 1,$$

where $X(t)$ is the Gaussian process on $[1-c, 1]$ with mean zero and covariance function (9), and justify that the sequence of random elements (z_L) is tight. These parts of the proof are carried out similarly to the proofs of Lemma 3 and Lemma 4 [6], respectively.

5 Conclusion

The paper provides a study of properties of the new threshold selection method for non-parametric estimation of the extremal index of stationary sequences proposed in [15]. We consider a specific normalization of the discrepancy statistic based on some modifications of the Cramér–von Mises–Smirnov statistic ω^2 that is calculated by only k largest order statistics of a sample. We show that the asymptotic distribution of the truncated Cramér–von Mises–Smirnov statistic (8) as $k \to \infty, k/L \to c, L \to \infty$ depends both on c and the limit distribution of the extremal index estimator being substituted in the statistic. We also develop the goodness-of-fit test for distribution tails based on the ω^2 statistic modification, which limit distribution coincides with the limit distribution of the classical Cramér–von Mises–Smirnov statistic under null hypothesis.

References

1. Beirlant, J., Goegebeur, Y., Teugels, J., Segers, J.: Statistics of Extremes: Theory and Applications. Wiley, Chichester (2004)
2. Berghaus, B., Bücher, A.: Weak convergence of a pseudo maximum likelihood estimator for the extremal index. Ann. Stat. **46**(5), 2307–2335 (2018)
3. Billingsley, P.: Convergence of Probability Measures. Wiley, New York (1968)
4. D'Agostino, R.B., Stephens, M.A.: Goodness of Fit Techniques. Marcel Dekker, New York (1986)
5. Darling, D.A.: The Cramér-Smirnov test in the parametric case. Ann. Math. Stat. **26**, 1–20 (1955)
6. Durbin, J.: Weak convergence of the sample distribution function when parameters are estimated. Ann. Stat. **1**(2), 279–290 (1973)
7. de Haan, L., Ferreira, A.: Extreme Value Theory: An Introduction. Springer, Heidelberg (2006). https://doi.org/10.1007/0-387-34471-3
8. Ferro, C.A.T., Segers, J.: Inference for clusters of extreme values. J. R. Stat. Soc. B. **65**, 545–556 (2003)

9. Fukutome, S., Liniger, M.A., Süveges, M.: Automatic threshold and run parameter selection: a climatology for extreme hourly precipitation in Switzerland. Theor. Appl. Climatol. **120**, 403–416 (2015)
10. Kac, M., Kiefer, J., Wolfowitz, J.: On tests of normality and other tests of goodness of fit based on distance methods. Ann. Math. Stat. **26**, 189–211 (1955)
11. Leadbetter, M.R., Lindgren, G., Rootzén, H.: Extremes and Related Properties of Random Sequence and Processes. Springer, New York (1983). https://doi.org/10.1007/978-1-4612-5449-2
12. Markovich, N.M.: Experimental analysis of nonparametric probability density estimates and of methods for smoothing them. Autom. Rem. Contr. **50**, 941–948 (1989)
13. Markovich, N.M.: Nonparametric Analysis of Univariate Heavy-Tailed data: Research and Practice. Wiley, Hoboken (2007)
14. Markovich, N.M.: Nonparametric estimation of extremal index using discrepancy method. In: Proceedings of the X International Conference "System Identification And Control Problems" SICPRO-2015, pp. 160–168. V.A. Trapeznikov Institute of Control Sciences (2015)
15. Markovich, N.M., Rodionov, I.V.: Threshold selection for extremal index estimation. Scand. J. Stat. (2020). Under review. arxiv:2009.02318
16. Northrop, P.J.: An efficient semiparametric maxima estimator of the extremal index. Extremes **18**(4), 585–603 (2015)
17. Pettitt, A.N., Stephens, M.A.: Modified Cramer-von Mises statistics for censored data. Biometrika **63**(2), 291–298 (1976)
18. Robert, C.Y.: Asymptotic distributions for the intervals estimators of the extremal index and the cluster-size probabilities. J. Stat. Plan. Infer. **139**, 3288–3309 (2009)
19. Robert, C.Y., Segers, J., Ferro, C.A.T.: A sliding blocks estimator for the extremal index. Electron. J. Stat. **3**, 993–1020 (2009)
20. Rodionov, I.V.: On discrimination between classes of distribution tails. Probl. Inform. Transm. **54**(2), 124–138 (2018)
21. Süveges, M., Davison, A.C.: Model misspecification in peaks over threshold analysis. Ann. Appl. Stat. **4**(1), 203–221 (2010)
22. Vapnik, V.N., Markovich, N.M., Stefanyuk, A.R.: Rate of convergence in L_2 of the projection estimator of the distribution density. Autom. Rem. Contr. **53**, 677–686 (1992)
23. Voinov, V., Nikulin, M., Balakrishnan, N.: Chi-squared Goodness-of-Fit Tests with Applications. Academic Press, Boston (2013)
24. Weissman, I., Novak, S.Y.: On blocks and runs estimators of the extremal index. J. Stat. Plan. Infer. **66**, 281–288 (1998)

Mean-Square Approximation of Iterated Stochastic Integrals from Strong Exponential Milstein and Wagner-Platen Methods for Non-commutative Semilinear SPDEs Based on Multiple Fourier-Legendre Series

Dmitriy F. Kuznetsov[1(✉)] and Mikhail D. Kuznetsov[2]

[1] Peter the Great Saint-Petersburg Polytechnic University,
Saint-Petersburg 195251, Russian Federation
kuznetsov_df@spbstu.ru

[2] Saint-Petersburg Electrotechnical University,
Saint-Petersburg 197376, Russian Federation

Abstract. This work is devoted to the mean-square approximation of iterated stochastic integrals with respect to the infinite-dimensional Q-Wiener process. These integrals are part of the high-order strong numerical methods (with respect to the temporal discretization) for semilinear stochastic partial differential equations with nonlinear multiplicative trace class noise, which are based on the Taylor formula in Banach spaces and exponential formula for the mild solution of semilinear stochastic partial differential equations. For the approximation of the mentioned stochastic integrals we use the multiple Fourier–Legendre series converging in the sense of norm in Hilbert space. In this article, we propose the optimization of the mean-square approximation procedures for iterated stochastic integrals of multiplicities 1 to 3 with respect to the infnite-dimensional Q-Wiener process.

Keywords: Semilinear stochastic partial differential equation · Infinite-dimensional Q-Wiener process · Nonlinear multiplicative trace class noise · Iterated stochastic integral · Generalized multiple Fourier series · Multiple Fourier–Legendre series · Exponential Milstein scheme · Exponential Wagner–Platen scheme · Legendre polynomial · Mean-square approximation · Expansion

1 Introduction

This paper continues the author's research [1,2] on methods of the mean-square approximation of iterated stochastic integrals (ISIs) with respect to the infinite dimensional Q-Wiener process.

A. N. Shiryaev et al. (Eds.): ICSM-5 2020, PROMS 371, pp. 17–32, 2021.
https://doi.org/10.1007/978-3-030-83266-7_2

It is well-known that one of the effective approaches to the construction of high-order strong numerical methods (with respect to the temporal discretization) for semilinear stochastic partial differential equations (SPDEs) is based on the Taylor formula in Banach spaces and the exponential formula for the mild solution of semilinear SPDE [3,4]. For example, in [3,4] the exponential Milstein and Wagner–Platen methods for semilinear SPDEs under the commutativity conditions were constructed. These methods have strong orders of convergence $1.0-\varepsilon$ and $1.5-\varepsilon$ correspondingly with respect to the temporal variable if the special conditions [3,4] are fullfilled (here ε is an arbitrary small posilive real number). Note that in [5] the convergence of the exponential Milstein scheme for semilinear SPDEs with strong order 1.0 has been proved under additional smoothness assumptions.

An important feature of the mentioned numerical methods is the presence in them the so-called ISIs with respect to the infinite-dimensional Q-Wiener process [6]. The problem of numerical modeling of these ISIs with multiplicities 1 to 3 was solved in [3,4] for the case when special commutativity conditions for SPDE are fulfilled. If the mentioned commutativity conditions are not satisfied, which often corresponds to SPDEs in numerous applications, the numerical modeling of ISIs with respect to the infinite-dimensional Q-Wiener process becomes much more difficult.

Two methods of the mean-square approximation of ISIs from the exponential Milstein scheme for semilinear SPDEs without the commutativity conditions have been considered in [7]. Note that the mean-square approximation error of these ISIs consists of two components [7]. The first component is related with the finite-dimentional approximation of the infinite-dimentional Q-Wiener process while the second one is connected with the approximation of Itô ISIs with respect to the scalar standard Brownian motions. In the author's publication [1] the problem of the mean-square approximation of ISIs with respect to the infinite-dimensional Q-Wiener process in the sense of second component of approximation error (see above) has been investigated for arbitrary multiplicity k ($k \in \mathbb{N}$) of stochastic integrals and without the assumptions of commutativity for SPDE.

In this article, we extend the method [7] for an estimation of the first component of approximation error for ISIs of multiplicities 1 to 3 with respect to the infinite-dimensional Q-Wiener process. In addition, we combine the obtained results with results from [1] and propose the optimization of the mean-square approximation procedures for the mentioned stochastic integrals.

2 Exponential Milstein and Wagner–Platen Numerical Schemes for Non-commutative Semilinear SPDEs

Let U, H be separable $\mathbb{R}$-Hilbert spaces and $L_{HS}(U,H)$ be a space of Hilbert–Schmidt operators from U to H. Let $(\Omega, \mathbf{F}, \mathsf{P})$ be a probability space with a normal filtration $\{\mathbf{F}_t, t\in[0,S]\}$ [6], let $\mathbf{W}_t$ be an U-valued Q-Wiener process with respect to $\{\mathbf{F}_t, t\in[0,S]\}$, which has a covariance trace class operator

$Q \in L(U)$. Here and further $L(U)$ denotes all bounded linear operators on U. Let $U_0 = Q^{1/2}(U)$ be an $\mathbb{R}$-Hilbert space with a scalar product $\langle u, w\rangle_{U_0} = \langle Q^{-1/2}u, Q^{-1/2}w\rangle_U$ for all $u, w \in U_0$ [3,4].

Consider the semilinear parabolic SPDE with multiplicative trace class noise

$$dX_t = (AX_t + F(X_t))\,dt + B(X_t)d\mathbf{W}_t, \quad X_0 = \xi, \quad t \in [0, S], \tag{1}$$

where nonlinear operators F, B ($F : H \to H$, $B : H \to L_{HS}(U_0, H)$), linear operator $A : D(A) \subset H \to H$ as well as the initial value ξ are assumed to satisfy the conditions of existence and uniqueness of the mild solution of (1) (see [4], Assumptions A1–A4).

It is well-known [8] that Assumptions A1–A4 [4] guarantee the existence and uniqueness (up to modifications) of the mild solution $X_t : [0, S] \times \Omega \to H$ of (1)

$$X_t = e^{At}\xi + \int_0^t e^{A(t-\tau)}F(X_\tau)d\tau + \int_0^t e^{A(t-\tau)}B(X_\tau)d\mathbf{W}_\tau \tag{2}$$

with probability 1 (further w. p. 1) for all $t \in [0, S]$, where e^{At} is the semigroup generated by the operator A.

Consider eigenvalues λ_i and eigenfunctions $e_i(x)$ of the covariance operator Q, where $i = (i_1, \ldots, i_d) \in J$, $J = \{i : \ i \in \mathbb{N}^d \text{ and } \lambda_i > 0\}$, and $x = (x_1, \ldots, x_d)$. Note that $e_i(x)$, $i \in J$ form an orthonormal basis of U [6]. The series representation of the Q-Wiener process $\mathbf{W}_t$ has the form [6]

$$\mathbf{W}_t = \sum_{i \in J} e_i \sqrt{\lambda_i}\mathbf{w}_t^{(i)},$$

where $t \in [0, S]$, $\mathbf{w}_t^{(i)}$ $(i \in J)$ are independent standard Wiener processes.

Consider the finite-dimensional approximation of $\mathbf{W}_t$ [6]

$$\mathbf{W}_t^M = \sum_{i \in J_M} e_i \sqrt{\lambda_i}\mathbf{w}_t^{(i)}, \quad t \in [0, S], \tag{3}$$

where $J_M = \{i : \ 1 \le i_1, \ldots, i_d \le M \text{ and } \lambda_i > 0\}$. Obviously, without the loss of generality we can suppose that $J_M = \{1, 2, \ldots, M\}$.

Let $\Delta > 0$, $\tau_p = p\Delta$ ($p = 0, 1, \ldots, N$), and $N\Delta = S$. Consider the exponential Milstein numerical scheme [3]

$$Y_{p+1} = e^{A\Delta}\Biggl(Y_p + \Delta F(Y_p) + \int_{\tau_p}^{\tau_{p+1}} B(Y_p)d\mathbf{W}_s$$
$$+ \int_{\tau_p}^{\tau_{p+1}} B'(Y_p)\int_{\tau_p}^{s} B(Y_p)d\mathbf{W}_\tau d\mathbf{W}_s\Biggr) \tag{4}$$

and Wagner–Platen numerical scheme [4]

$$Y_{p+1} = e^{\frac{A\Delta}{2}}\Biggl(e^{\frac{A\Delta}{2}}Y_p + \Delta F(Y_p) + \int_{\tau_p}^{\tau_{p+1}} B(Y_p)d\mathbf{W}_s$$

$$+\int_{\tau_p}^{\tau_{p+1}} B'(Y_p)\int_{\tau_p}^{s} B(Y_p)d\mathbf{W}_\tau d\mathbf{W}_s$$

$$+\frac{\Delta^2}{2}F'(Y_p)\left(AY_p+F(Y_p)\right)+\int_{\tau_p}^{\tau_{p+1}} F'(Y_p)\int_{\tau_p}^{s} B(Y_p)d\mathbf{W}_\tau ds$$

$$+\frac{\Delta^2}{4}\sum_{i\in J}\lambda_i F''(Y_p)\left(B(Y_p)e_i, B(Y_p)e_i\right)$$

$$+A\left(\int_{\tau_p}^{\tau_{p+1}}\int_{\tau_p}^{s} B(Y_p)d\mathbf{W}_\tau ds-\frac{\Delta}{2}\int_{\tau_p}^{\tau_{p+1}} B(Y_p)d\mathbf{W}_s\right)$$

$$+\Delta\int_{\tau_p}^{\tau_{p+1}} B'(Y_p)\left(AY_p+F(Y_p)\right)d\mathbf{W}_s$$

$$-\int_{\tau_p}^{\tau_{p+1}}\int_{\tau_p}^{s} B'(Y_p)\left(AY_p+F(Y_p)\right)d\mathbf{W}_\tau ds$$

$$+\frac{1}{2}\int_{\tau_p}^{\tau_{p+1}} B''(Y_p)\left(\int_{\tau_p}^{s} B(Y_p)d\mathbf{W}_\tau, \int_{\tau_p}^{s} B(Y_p)d\mathbf{W}_\tau\right)d\mathbf{W}_s$$

$$+\int_{\tau_p}^{\tau_{p+1}} B'(Y_p)\int_{\tau_p}^{s} B'(Y_p)\int_{\tau_p}^{\tau} B(Y_p)d\mathbf{W}_\theta d\mathbf{W}_\tau d\mathbf{W}_s\Bigg) \qquad (5)$$

for SPDE (1), where Y_p is an approximation of X_{τ_p} (mild solution (2) at the time moment τ_p), $p=0,1,\ldots,N$, and B', B'', F', F'' are Fréchet derivatives. Note that in addition to the temporal discretization, the implementation of numerical schemes (4) and (5) also requires a discretization of the infinite-dimensional Hilbert space H and a finite-dimensional approximation of the Q-Wiener process.

Let us focus on the approximation related to the Q-Wiener process. Consider the following Itô ISIs

$$I_{(1)T,t}^{(r_1)}=\int_t^T d\mathbf{w}_{t_1}^{(r_1)},\ I_{(10)T,t}^{(r_10)}=\int_t^T\int_t^{t_2} d\mathbf{w}_{t_1}^{(r_1)}dt_2,\ I_{(01)T,t}^{(0r_2)}=\int_t^T\int_t^{t_2} dt_1 d\mathbf{w}_{t_2}^{(r_2)}, \qquad (6)$$

$$I_{(11)T,t}^{(r_1r_2)}=\int_t^T\int_t^{t_2} d\mathbf{w}_{t_1}^{(r_1)}d\mathbf{w}_{t_2}^{(r_2)},\ I_{(111)T,t}^{(r_1r_2r_3)}=\int_t^T\int_t^{t_3}\int_t^{t_2} d\mathbf{w}_{t_1}^{(r_1)}d\mathbf{w}_{t_2}^{(r_2)}d\mathbf{w}_{t_3}^{(r_3)}, \qquad (7)$$

where $r_1,r_2,r_3\in J_M$, $0\le t<T\le S$, and J_M is defined as in (3).

Let us replace the infinite-dimensional Q-Wiener process in the ISIs from (4), (5) with its finite-dimensional approximation (3). Moreover, replace Y_p with Z, τ_p with t, and τ_{p+1} with T in these integrals. Then we have w. p. 1

$$J_1[B(Z)]_{T,t}^M=\int_t^T B(Z)d\mathbf{W}_s^M=\sum_{r_1\in J_M} B(Z)e_{r_1}\sqrt{\lambda_{r_1}}I_{(1)T,t}, \qquad (8)$$

$$J_2[B(Z)]_{T,t}^M=A\left(\int_t^T\int_t^{s} B(Z)d\mathbf{W}_\tau^M ds-\frac{T-t}{2}\int_t^T B(Z)d\mathbf{W}_s^M\right)$$

$$
= \sum_{r_1 \in J_M} AB(Z)e_{r_1}\sqrt{\lambda_{r_1}}\left(\frac{T-t}{2}I_{(1)T,t}^{(r_1)} - I_{(01)T,t}^{(0r_1)}\right), \tag{9}
$$

$$
J_3[B(Z),F(Z)]_{T,t}^M = (T-t)\int_t^T B'(Z)\left(AZ+F(Z)\right)d\mathbf{W}_s^M
$$

$$
-\int_t^T \int_t^s B'(Z)\left(AZ+F(Z)\right)d\mathbf{W}_\tau^M ds
$$

$$
= \sum_{r_1 \in J_M} B'(Z)\left(AZ+F(Z)\right)e_{r_1}\sqrt{\lambda_{r_1}}I_{(01)T,t}^{(0r_1)}, \tag{10}
$$

$$
J_4[B(Z),F(Z)]_{T,t}^M = \int_t^T F'(Z)\int_t^s B(Z)d\mathbf{W}_\tau^M ds
$$

$$
= \sum_{r_1 \in J_M} F'(Z)B(Z)e_{r_1}\sqrt{\lambda_{r_1}}\left((T-t)I_{(1)T,t}^{(r_1)} - I_{(01)T,t}^{(0r_1)}\right), \tag{11}
$$

$$
I_1[B(Z)]_{T,t}^M = \int_t^T B'(Z)\int_t^s B(Z)d\mathbf{W}_\tau^M d\mathbf{W}_s^M
$$

$$
= \sum_{r_1,r_2 \in J_M} B'(Z)(B(Z)e_{r_1})e_{r_2}\sqrt{\lambda_{r_1}\lambda_{r_2}}I_{(11)T,t}^{(r_1 r_2)}, \tag{12}
$$

$$
I_2[B(Z)]_{T,t}^M = \int_t^T B'(Z)\int_t^s B'(Z)\int_t^\tau B(Z)d\mathbf{W}_\theta^M d\mathbf{W}_\tau^M d\mathbf{W}_s^M
$$

$$
= \sum_{r_1,r_2,r_3 \in J_M} B'(Z)\left(B'(Z)\left(B(Z)e_{r_1}\right)e_{r_2}\right)e_{r_3}\sqrt{\lambda_{r_1}\lambda_{r_2}\lambda_{r_3}}I_{(111)T,t}^{(r_1 r_2 r_3)}, \tag{13}
$$

$$
I_3[B(Z)]_{T,t}^M = \int_t^T B''(Z)\left(\int_t^s B(Z)d\mathbf{W}_\tau^M, \int_t^s B(Z)d\mathbf{W}_\tau^M\right)d\mathbf{W}_s^M
$$

$$
= \sum_{r_1,r_2,r_3 \in J_M} B''(Z)\left(B(Z)e_{r_1}, B(Z)e_{r_2}\right)e_{r_3}\sqrt{\lambda_{r_1}\lambda_{r_2}\lambda_{r_3}}
$$

$$
\times \int_t^T \int_t^s d\mathbf{w}_\tau^{(r_1)} \int_t^s d\mathbf{w}_\tau^{(r_2)} d\mathbf{w}_s^{(r_3)}. \tag{14}
$$

Note that in (9)–(11) we used the Itô formula. Moreover, using the Itô formula we obtain

$$
\int_t^s d\mathbf{w}_\tau^{(r_1)} \int_t^s d\mathbf{w}_\tau^{(r_2)} = I_{(11)s,t}^{(r_1 r_2)} + I_{(11)s,t}^{(r_2 r_1)} + \mathbf{1}_{\{r_1=r_2\}}(s-t) \quad \text{w. p. 1}, \tag{15}
$$

where $\mathbf{1}_A$ is the indicator of the set A. From (15) and (14) we obtain w. p. 1

$$
I_3[B(Z)]_{T,t}^M = \sum_{r_1,r_2,r_3 \in J_M} B''(Z)\left(B(Z)e_{r_1}, B(Z)e_{r_2}\right)e_{r_3}\sqrt{\lambda_{r_1}\lambda_{r_2}\lambda_{r_3}}
$$

$$\times \left(I^{(r_1 r_2 r_3)}_{(111)T,t} + I^{(r_2 r_1 r_3)}_{(111)T,t} + \mathbf{1}_{\{r_1=r_2\}} I^{(0 r_3)}_{(01)T,t} \right). \tag{16}$$

Thus, for the implementation of numerical schemes (4) and (5) we need to approximate the following Itô ISIs $I^{(r_1)}_{(1)T,t}$, $I^{(0r_1)}_{(01)T,t}$, $I^{(r_1 r_2)}_{(11)T,t}$, $I^{(r_1 r_2 r_3)}_{(111)T,t}$, where $r_1, r_2, r_3 \in J_M$, $0 \le t < T \le S$.

3 Approximation of Itô ISIs

Consider an efficient method [9–12] of the mean-square approximation of Itô ISIs of the form

$$J[\psi^{(k)}]^{(i_1 \ldots i_k)}_{T,t} = \int_t^T \psi_k(t_k) \ldots \int_t^{t_2} \psi_1(t_1) d\mathbf{w}^{(i_1)}_{t_1} \ldots d\mathbf{w}^{(i_k)}_{t_k}, \tag{17}$$

where $0 \le t < T \le S$, $\psi_l(\tau)$ $(l = 1, \ldots, k)$ are continuous non-random functions on $[t, T]$, $\mathbf{w}^{(i)}_\tau$ $(i = 1, \ldots, m)$ are independent standard Wiener processes, $\mathbf{w}^{(0)}_\tau = \tau$, $i_1, \ldots, i_k = 0, 1, \ldots, m$.

Suppose that $\{\phi_j(x)\}_{j=0}^{\infty}$ is a complete orthonormal system of functions in the space $L_2([t, T])$ and define the following function on the hypercube $[t, T]^k$

$$K(t_1, \ldots, t_k) = \psi_1(t_1) \ldots \psi_k(t_k) \mathbf{1}_{\{t_1 < \ldots < t_k\}}, \quad t_1, \ldots, t_k \in [t, T] \quad \text{for } k \ge 2 \tag{18}$$

and $K(t_1) \equiv \psi_1(t_1)$ for $t_1 \in [t, T]$, where $\mathbf{1}_A$ is the indicator of the set A.

The function $K(t_1, \ldots, t_k)$ is piecewise continuous on the hypercube $[t, T]^k$. At this situation it is well known that the generalized multiple Fourier series of $K(t_1, \ldots, t_k) \in L_2([t, T]^k)$ converges to $K(t_1, \ldots, t_k)$ on the hypercube $[t, T]^k$ in the mean-square sense, i.e.

$$\lim_{p_1, \ldots, p_k \to \infty} \left\| K(t_1, \ldots, t_k) - \sum_{j_1=0}^{p_1} \ldots \sum_{j_k=0}^{p_k} C_{j_k \ldots j_1} \prod_{l=1}^{k} \phi_{j_l}(t_l) \right\| = 0, \tag{19}$$

where $\| \cdot \|$ is the $L_2([t, T]^k)$-norm and the Fourier coefficient is defined by

$$C_{j_k \ldots j_1} = \int_{[t,T]^k} K(t_1, \ldots, t_k) \prod_{l=1}^{k} \phi_{j_l}(t_l) dt_1 \ldots dt_k. \tag{20}$$

Consider the discretization $\{\tau_j\}_{j=0}^{N}$ of $[t, T]$ such that

$$t = \tau_0 < \ldots < \tau_N = T, \quad \Delta_N = \max_{0 \le j \le N-1} \Delta\tau_j \to 0 \ \text{ if } \ N \to \infty, \tag{21}$$

where $\Delta\tau_j = \tau_{j+1} - \tau_j$.

Theorem 1 [9–12]. *Suppose that $\psi_l(\tau)$ $(l = 1, \ldots, k)$ are continuous non-random functions on the interval $[t, T]$ and $\{\phi_j(x)\}_{j=0}^{\infty}$ is a complete orthonormal system of continuous functions in $L_2([t, T])$. Then*

$$J[\psi^{(k)}]^{(i_1 \ldots i_k)}_{T,t} = \underset{p_1, \ldots, p_k \to \infty}{\text{l.i.m.}} J[\psi^{(k)}]^{(i_1 \ldots i_k) p_1 \ldots p_k}_{T,t}, \tag{22}$$

where

$$J[\psi^{(k)}]_{T,t}^{(i_1\ldots i_k)p_1\ldots p_k}=\sum_{j_1=0}^{p_1}\ldots\sum_{j_k=0}^{p_k}C_{j_k\ldots j_1}\Biggl(\prod_{l=1}^{k}\zeta_{j_l}^{(i_l)}$$

$$-\underset{N\to\infty}{\text{l.i.m.}}\sum_{(l_1,\ldots,l_k)\in\mathrm{G}_k}\phi_{j_1}(\tau_{l_1})\Delta\mathbf{w}_{\tau_{l_1}}^{(i_1)}\ldots\phi_{j_k}(\tau_{l_k})\Delta\mathbf{w}_{\tau_{l_k}}^{(i_k)}\Biggr)\qquad(23)$$

and

$$E_k^{(i_1\ldots i_k)p_1,\ldots,p_k}\le k!\left(I_k-\sum_{j_1=0}^{p_1}\ldots\sum_{j_k=0}^{p_k}C_{j_k\ldots j_1}^2\right),\qquad(24)$$

where $E^{(i_1\ldots i_k)p_1,\ldots,p_k}=\mathsf{M}\left(J[\psi^{(k)}]_{T,t}^{(i_1\ldots i_k)}-J[\psi^{(k)}]_{T,t}^{(i_1\ldots i_k)p_1,\ldots,p_k}\right)^2$, l.i.m. *is a limit in the mean-square sense,* $i_1,\ldots,i_k=1,\ldots,m$ *for* $T-t\in(0,+\infty)$ *and* $i_1,\ldots,i_k=0,1,\ldots,m$ *for* $T-t\in(0,1)$, $I_k^{1/2}$ *is the* $L_2([t,T]^k)$*-norm of* $K(t_1,\ldots,t_k)$,

$$\mathrm{G}_k=\mathrm{H}_k\backslash\mathrm{L}_k,\quad \mathrm{H}_k=\{(l_1,\ldots,l_k):\ l_1,\ldots,l_k=0,\ 1,\ldots,N-1\},$$

$$\mathrm{L}_k=\{(l_1,\ldots,l_k):\ l_1,\ldots,l_k=0,\ 1,\ldots,N-1;\ l_g\ne l_r\ (g\ne r);\ g,r=1,\ldots,k\},$$

$$\zeta_j^{(i)}=\int_t^T\phi_j(s)d\mathbf{w}_s^{(i)}\qquad(25)$$

are independent standard Gaussian random variables for various i *or* j *(in the case when* $i\ne 0$), $C_{j_k\ldots j_1}$ *is the Fourier coefficient* (20), $\Delta\mathbf{w}_{\tau_j}^{(i)}=\mathbf{w}_{\tau_{j+1}}^{(i)}-\mathbf{w}_{\tau_j}^{(i)}$ $(i=0,\ 1,\ldots,m)$, $\{\tau_j\}_{j=0}^{N}$ *is the discretization* (21).

Note that in [9,11,12] some versions and generalizations of Theorem 1 were considered.

Obtain transformed particular cases of Theorem 1 for $k=1,\ldots,4$ [9–12]

$$J[\psi^{(1)}]_{T,t}^{(i_1)}=\underset{p_1\to\infty}{\text{l.i.m.}}\sum_{j_1=0}^{p_1}C_{j_1}\zeta_{j_1}^{(i_1)},\qquad(26)$$

$$J[\psi^{(2)}]_{T,t}^{(i_1i_2)}=\underset{p_1,p_2\to\infty}{\text{l.i.m.}}\sum_{j_1=0}^{p_1}\sum_{j_2=0}^{p_2}C_{j_2j_1}\left(\zeta_{j_1}^{(i_1)}\zeta_{j_2}^{(i_2)}-\mathbf{1}_{\{i_1=i_2\ne0\}}\mathbf{1}_{\{j_1=j_2\}}\right),\qquad(27)$$

$$J[\psi^{(3)}]_{T,t}^{(i_1i_2i_3)}=\underset{p_1,p_2,p_3\to\infty}{\text{l.i.m.}}\sum_{j_1=0}^{p_1}\sum_{j_2=0}^{p_2}\sum_{j_3=0}^{p_3}C_{j_3j_2j_1}\Biggl(\zeta_{j_1}^{(i_1)}\zeta_{j_2}^{(i_2)}\zeta_{j_3}^{(i_3)}$$

$$-\mathbf{1}_{\{i_1=i_2\ne0\}}\mathbf{1}_{\{j_1=j_2\}}\zeta_{j_3}^{(i_3)}-\mathbf{1}_{\{i_2=i_3\ne0\}}\mathbf{1}_{\{j_2=j_3\}}\zeta_{j_1}^{(i_1)}-\mathbf{1}_{\{i_1=i_3\ne0\}}\mathbf{1}_{\{j_1=j_3\}}\zeta_{j_2}^{(i_2)}\Biggr),\qquad(28)$$

$$J[\psi^{(4)}]_{T,t}^{(i_1\ldots i_4)} = \underset{p_1,\ldots,p_4\to\infty}{\text{l.i.m.}} \sum_{j_1=0}^{p_1}\ldots\sum_{j_4=0}^{p_4} C_{j_4\ldots j_1}\Biggl(\prod_{l=1}^{4}\zeta_{j_l}^{(i_l)}$$

$$-\mathbf{1}_{\{i_1=i_2\ne 0\}}\mathbf{1}_{\{j_1=j_2\}}\zeta_{j_3}^{(i_3)}\zeta_{j_4}^{(i_4)} - \mathbf{1}_{\{i_1=i_3\ne 0\}}\mathbf{1}_{\{j_1=j_3\}}\zeta_{j_2}^{(i_2)}\zeta_{j_4}^{(i_4)}$$

$$-\mathbf{1}_{\{i_1=i_4\ne 0\}}\mathbf{1}_{\{j_1=j_4\}}\zeta_{j_2}^{(i_2)}\zeta_{j_3}^{(i_3)} - \mathbf{1}_{\{i_2=i_3\ne 0\}}\mathbf{1}_{\{j_2=j_3\}}\zeta_{j_1}^{(i_1)}\zeta_{j_4}^{(i_4)}$$

$$-\mathbf{1}_{\{i_2=i_4\ne 0\}}\mathbf{1}_{\{j_2=j_4\}}\zeta_{j_1}^{(i_1)}\zeta_{j_3}^{(i_3)} - \mathbf{1}_{\{i_3=i_4\ne 0\}}\mathbf{1}_{\{j_3=j_4\}}\zeta_{j_1}^{(i_1)}\zeta_{j_2}^{(i_2)}$$

$$+\mathbf{1}_{\{i_1=i_2\ne 0\}}\mathbf{1}_{\{j_1=j_2\}}\mathbf{1}_{\{i_3=i_4\ne 0\}}\mathbf{1}_{\{j_3=j_4\}}$$

$$+\mathbf{1}_{\{i_1=i_3\ne 0\}}\mathbf{1}_{\{j_1=j_3\}}\mathbf{1}_{\{i_2=i_4\ne 0\}}\mathbf{1}_{\{j_2=j_4\}}$$

$$+\mathbf{1}_{\{i_1=i_4\ne 0\}}\mathbf{1}_{\{j_1=j_4\}}\mathbf{1}_{\{i_2=i_3\ne 0\}}\mathbf{1}_{\{j_2=j_3\}}\Biggr), \tag{29}$$

where $\mathbf{1}_A$ is the indicator of the set A.

Let us consider the generalization of the formulas (26)–(29) for the case of arbitrary k ($k \in \mathbb{N}$).

Theorem 2 [11,12]. *In the conditions of Theorem 1 the following mean-square converging expansion is valid*

$$J[\psi^{(k)}]_{T,t}^{(i_1\ldots i_k)} = \underset{p_1,\ldots,p_k\to\infty}{\text{l.i.m.}} \sum_{j_1=0}^{p_1}\ldots\sum_{j_k=0}^{p_k} C_{j_k\ldots j_1}\Biggl(\prod_{l=1}^{k}\zeta_{j_l}^{(i_l)} + \sum_{r=1}^{[k/2]}(-1)^r$$

$$\times \sum_{\substack{(\{\{g_1,g_2\},\ldots,\{g_{2r-1},g_{2r}\}\},\{q_1,\ldots,q_{k-2r}\})\\ \{g_1,g_2,\ldots,g_{2r-1},g_{2r},q_1,\ldots,q_{k-2r}\}=\{1,2,\ldots,k\}}} \prod_{s=1}^{r}\mathbf{1}_{\{i_{g_{2s-1}}=\ i_{g_{2s}}\ne 0\}}$$

$$\times\mathbf{1}_{\{j_{g_{2s-1}}=\ j_{g_{2s}}\}}\prod_{l=1}^{k-2r}\zeta_{j_{q_l}}^{(i_{q_l})}\Biggr), \tag{30}$$

where $[\cdot]$ *is an integer part of a real number, the sum in the second line of the formula (30) means the sum with respect to all possible permutations of the set* $(\{\{g_1,g_2\},\ldots,\{g_{2r-1},g_{2r}\}\},\{q_1,\ldots,q_{k-2r}\})$. *At that* $\{g_1,g_2,\ \ldots,g_{2r-1},g_{2r},q_1,$ $\ldots,q_{k-2r}\}$ $=\{1,2,\ldots,k\}$, *braces mean an disordered set, and parentheses mean an ordered set; other notations are the same as in Theorem 1.*

Using Theorem 1 and complete orthonormal system of Legendre polynomials in the space $L_2([t,T])$, we obtain the following approximations of Itô ISIs (6), (7) [9–12]

$$I_{(1)T,t}^{(i_1)} = \sqrt{T-t}\,\zeta_0^{(i_1)}, \tag{31}$$

$$I_{(01)T,t}^{(0i_1)} = \frac{(T-t)^{3/2}}{2}\left(\zeta_0^{(i_1)} + \frac{1}{\sqrt{3}}\zeta_1^{(i_1)}\right), \tag{32}$$

$$I_{(10)T,t}^{(i_10)} = \frac{(T-t)^{3/2}}{2}\left(\zeta_0^{(i_1)} - \frac{1}{\sqrt{3}}\zeta_1^{(i_1)}\right), \tag{33}$$

$$I_{(11)T,t}^{(i_1 i_2)q} = \frac{T-t}{2}\left(\zeta_0^{(i_1)}\zeta_0^{(i_2)} + \sum_{i=1}^{q}\frac{1}{\sqrt{4i^2-1}}\left(\zeta_{i-1}^{(i_1)}\zeta_i^{(i_2)} - \zeta_i^{(i_1)}\zeta_{i-1}^{(i_2)}\right) - \mathbf{1}_{\{i_1=i_2\}}\right), \tag{34}$$

$$I_{(111)T,t}^{(i_1 i_2 i_3)p} = \sum_{j_1,j_2,j_3=0}^{p} C_{j_3 j_2 j_1}\left(\zeta_{j_1}^{(i_1)}\zeta_{j_2}^{(i_2)}\zeta_{j_3}^{(i_3)} - \mathbf{1}_{\{i_1=i_2\}}\mathbf{1}_{\{j_1=j_2\}}\zeta_{j_3}^{(i_3)}\right.$$

$$\left. -\mathbf{1}_{\{i_2=i_3\}}\mathbf{1}_{\{j_2=j_3\}}\zeta_{j_1}^{(i_1)} - \mathbf{1}_{\{i_1=i_3\}}\mathbf{1}_{\{j_1=j_3\}}\zeta_{j_2}^{(i_2)}\right), \tag{35}$$

$$C_{j_3 j_2 j_1} = \frac{1}{8}(T-t)^{3/2}\sqrt{(2j_1+1)(2j_2+1)(2j_3+1)}\,\bar{C}_{j_3 j_2 j_1}, \tag{36}$$

$$\bar{C}_{j_3 j_2 j_1} = \int_{-1}^{1} P_{j_3}(z)\int_{-1}^{z} P_{j_2}(y)\int_{-1}^{y} P_{j_1}(x)dxdydz,$$

where the Gaussian random variable $\zeta_j^{(i)}$ (if $i \neq 0$) is defined by (25) and $P_j(x)$ $(j = 0, 1, 2, \ldots)$ is the Legendre polynomial.

The estimate (24) is rather rough due to the multiplier factor $k!$ Therefore, consider the exact value $E^{(i_1\ldots i_k)p_1,\ldots,p_k} \stackrel{\mathsf{def}}{=} E^{(i_1\ldots i_k)p}$ for $p_1 = \ldots = p_k = p$.

Theorem 3 [10,11]. *Suppose that the conditions of Theorem 1 are satisfied. Then*

$$E_k^{(i_1\ldots i_k)p} = I_k - \sum_{j_1,\ldots,j_k=0}^{p} C_{j_k\ldots j_1}$$

$$\times \mathsf{M}\left(J[\psi^{(k)}]_{T,t}^{(i_1\ldots i_k)}\sum_{(j_1,\ldots,j_k)}\int_t^T \phi_{j_k}(t_k)\ldots\int_t^{t_2}\phi_{j_1}(t_1)d\mathbf{w}_{t_1}^{(i_1)}\ldots d\mathbf{w}_{t_k}^{(i_k)}\right), \tag{37}$$

where $i_1,\ldots,i_k = 1,\ldots,m$; expression $\sum\limits_{(j_1,\ldots,j_k)}$ means the sum with respect to all possible permutations $(j_1,\ldots,j_k)$. At the same time if j_r swapped with j_q in the permutation $(j_1,\ldots,j_k)$, then i_r swapped with i_q in the permutation $(i_1,\ldots,i_k)$; another notations are the same as in Theorem 1.

Note that

$$\mathsf{M}\left(J[\psi^{(k)}]_{T,t}^{(i_1\ldots i_k)}\int_t^T \phi_{j_k}(t_k)\ldots\int_t^{t_2}\phi_{j_1}(t_1)d\mathbf{w}_{t_1}^{(i_1)}\ldots d\mathbf{w}_{t_k}^{(i_k)}\right) = C_{j_k\ldots j_1}. \tag{38}$$

Then we can obtain the following particular cases of Theorem 3 [9–11]

$$E_k^{(i_1\ldots i_k)p} = I_k - \sum_{j_1=0}^{p_1}\ldots\sum_{j_k=0}^{p_k} C_{j_k\ldots j_1}^2 \quad (i_1,\ldots,i_k \text{ are pairwise different}), \tag{39}$$

$$E_k^{(i_1\ldots i_k)p} = I_k - \sum_{j_1,\ldots,j_k=0}^{p} C_{j_k\ldots j_1} \left(\sum_{(j_1,\ldots,j_k)} C_{j_k\ldots j_1} \right) \quad (i_1 = \ldots = i_k),$$

$$E_3^{(i_1 i_2 i_3)p} = I_3 - \sum_{j_1,j_2,j_3=0}^{p} C^2_{j_3 j_2 j_1} - \sum_{j_1,j_2,j_3=0}^{p} C_{j_3 j_1 j_2} C_{j_3 j_2 j_1} \quad (i_1 = i_2 \neq i_3), \quad (40)$$

$$E_3^{(i_1 i_2 i_3)p} = I_3 - \sum_{j_1,j_2,j_3=0}^{p} C^2_{j_3 j_2 j_1} - \sum_{j_1,j_2,j_3=0}^{p} C_{j_2 j_3 j_1} C_{j_3 j_2 j_1} \quad (i_1 \neq i_2 = i_3), \quad (41)$$

$$E_3^{(i_1 i_2 i_3)p} = I_3 - \sum_{j_1,j_2,j_3=0}^{p} C^2_{j_3 j_2 j_1} - \sum_{j_1,j_2,j_3=0}^{p} C_{j_3 j_2 j_1} C_{j_1 j_2 j_3} \quad (i_1 = i_3 \neq i_2). \quad (42)$$

Obviously, the above conditions do not contain multiplier factors $k!$ in contrast to the estimate (24). However, the number of the mentioned conditions is quite large, which is inconvenient for practical calculations. In this paper we propose the hypothesis that all the formulas (40)–(42) can be replaced with the equality (39) for $k = 3$ without noticeable loss of the mean-square accuracy of approximation for Itô ISIs. Section 5 is devoted to the detailed confirmation of the mentioned hypothesis for the case of multiple Fourier–Legendre series.

It should be noted that unlike the method based on Theorem 1, existing approaches to the mean-square approximation of ISIs (see, for example, [13,14]) do not allow to choose different numbers p for approximations of different ISIs. Moreover, the noted approaches [13,14] exclude the possibility for obtaining of approximate and exact expressions for the mean-square approximation error similar to (24), (37).

4 Approximation of ISIs with Respect to the Q-Wiener Process

Consider the following ISI with respect to the Q-Wiener process

$$I[\Phi^{(k)}(Z), \psi^{(k)}]_{T,t} = \int_t^T \psi_k(t_k)\Phi_k(Z) \ldots \int_t^{t_2} \psi_1(t_1)\Phi_1(Z) d\mathbf{W}_{t_1} \ldots d\mathbf{W}_{t_k}, \quad (43)$$

where $Z : \Omega \to H$ is an $\mathbf{F}_t/\mathcal{B}(H)$-measurable mapping, every non-random function $\psi_l(\tau)$ $(l = 1, \ldots, k)$ is continuous on $[t, T]$, and $\Phi_k(v)(\ \ldots (\Phi_1(v)) \ldots\ )$ is a k-linear Hilbert–Schmidt operator mapping from $U_0 \times \ldots \times U_0$ (k times) to H for all $v \in H$.

Let $I[\Phi^{(k)}(Z), \psi^{(k)}]^M_{T,t}$ be an approximation of the ISI (43)

$$I[\Phi^{(k)}(Z), \psi^{(k)}]^M_{T,t} = \int_t^T \psi_k(t_k)\Phi_k(Z) \ldots \int_t^{t_2} \psi_1(t_1)\Phi_1(Z) d\mathbf{W}^M_{t_1} \ldots d\mathbf{W}^M_{t_k}$$
$$= \sum_{r_1,\ldots,r_k \in J_M} \Phi_k(Z)(\ldots(\Phi_1(Z)e_{r_1})\ldots)\, e_{r_k} \sqrt{\lambda_{r_1}\ldots\lambda_{r_k}}\, J[\psi^{(k)}]^{(r_1\ldots r_k)}_{T,t}, \quad (44)$$

where $0 \le t < T \le S$ and $J[\psi^{(k)}]_{T,t}^{(r_1\ldots r_k)}$ is defined by (17).

Let $I[\Phi^{(k)}, \psi^{(k)}]_{T,t}^{M,p_1\ldots,p_k}$ be an approximation of the ISI (44)

$$I[\Phi^{(k)}(Z), \psi^{(k)}]_{T,t}^{M,p_1\ldots,p_k} = \sum_{r_1,\ldots,r_k\in J_M} \Phi_k(Z)\left(\ldots(\Phi_1(Z)e_{r_1})\ldots\right)e_{r_k} \times\sqrt{\lambda_{r_1}\ldots\lambda_{r_k}}\,J[\psi^{(k)}]_{T,t}^{(r_1\ldots r_k)p_1,\ldots,p_k}, \tag{45}$$

where $J[\psi^{(k)}]_{T,t}^{(r_1\ldots r_k)p_1,\ldots,p_k}$ is defined by (23) or as the expression before passing to the limit in (30).

Let $L(U,H)$ be the space of linear and bounded operators mapping from U to H. Let $L(U,H)_0 = \{T|_{U_0} : T \in L(U,H)\}$, where $T|_{U_0}$ is the restriction of the operator T to the space U_0. It is known [6] that $L(U,H)_0$ is a dense subset of the space of Hilbert–Schmidt operators $L_{HS}(U_0,H)$.

Theorem 4 [1,12]. *Let the conditions of Theorem 1 be fulfilled as well as the following conditions:*

1. *$Q \in L(U)$ is a non-negative and symmetric trace class operator (λ_i and e_i ($i \in J$) are its eigenvalues and eigenfunctions correspondingly), $\{\mathbf{W}_\tau, \tau\in[0,S]\}$ is an U-valued Q-Wiener process, and $Z : \Omega \to H$ is an $\mathbf{F}_t/\mathcal{B}(H)$-measurable mapping.*
2. *$\Phi_1 \in L(U,H)_0$, $\Phi_2 \in L(H, L(U,H)_0)$, and $\Phi_k(v)(\ldots(\Phi_1(v))\ldots)$ is a k-linear Hilbert–Schmidt operator mapping from $U_0 \times \ldots \times U_0$ (k times) to H for all $v \in H$ such that $\|\Phi_k(Z)(\ldots(\Phi_1(Z)e_{r_1})\ldots)e_{r_k}\|_H^2 \le L_k < \infty$ w. p. 1 for all $r_1,\ldots,r_k \in J_M$, $M \in \mathbb{N}$. Then*

$$\mathsf{M}\left\|I[\Phi^{(k)}(Z), \psi^{(k)}]_{T,t}^{M} - I[\Phi^{(k)}(Z), \psi^{(k)}]_{T,t}^{M,p_1\ldots p_k}\right\|_H^2 \le L_k(k!)^2\,(\mathrm{tr}Q)^k\left(I_k - \sum_{j_1=0}^{p_1}\ldots\sum_{j_k=0}^{p_k} C_{j_k\ldots j_1}^2\right), \tag{46}$$

where $I_k^{1/2}$ is the $L_2([t,T]^k)$-norm of $K(t_1,\ldots,t_k)$ and $\mathrm{tr}Q = \sum\limits_{i\in J}\lambda_i < \infty$.

Note that the right-hand side of the inequality (46) is independent of M and tends to zero if $p_1,\ldots,p_k \to \infty$ due to the Parseval equality.

Let us consider the approximation of ISIs from (4) and (5). According to (8)–(11), (31), and (32) we can write the following relatively simple formulas

$$J_1[B(Z)]_{T,t}^M = (T-t)^{1/2}\sum_{r_1\in J_M} B(Z)e_{r_1}\sqrt{\lambda_{r_1}}\zeta_0^{(r_1)},$$

$$J_2[B(Z)]_{T,t}^M = -\frac{(T-t)^{3/2}}{2\sqrt{3}}\sum_{r_1\in J_M} AB(Z)e_{r_1}\sqrt{\lambda_{r_1}}\zeta_1^{(r_1)}, \tag{47}$$

$$J_3[B(Z),F(Z)]_{T,t}^M$$
$$=\frac{(T-t)^{3/2}}{2}\sum_{r_1\in J_M}B'(Z)\left(AZ+F(Z)\right)e_{r_1}\sqrt{\lambda_{r_1}}\left(\zeta_0^{(r_1)}+\frac{1}{\sqrt{3}}\zeta_1^{(r_1)}\right), \quad (48)$$

$$J_4[B(Z),F(Z)]_{T,t}^M=\frac{(T-t)^{3/2}}{2}\sum_{r_1\in J_M}F'(Z)B(Z)e_{r_1}\sqrt{\lambda_{r_1}}\left(\zeta_0^{(r_1)}-\frac{1}{\sqrt{3}}\zeta_1^{(r_1)}\right), \quad (49)$$

where $\zeta_0^{(r_1)}$, $\zeta_1^{(r_1)}$ $(r_1\in J_M)$ are independent standard Gaussian random variables.

Further, consider ISIs (12), (13), (16) in detail. Let $I_1[B(Z)]_{T,t}^{M,q}$, $I_2[B(Z)]_{T,t}^{M,p}$, $I_3[B(Z)]_{T,t}^{M,p}$ be approximations of ISIs (12), (13), (16), which have the form

$$I_1[B(Z)]_{T,t}^{M,q}=\sum_{r_1,r_2\in J_M}B'(Z)\left(B(Z)e_{r_1}\right)e_{r_2}\sqrt{\lambda_{r_1}\lambda_{r_2}}I_{(11)T,t}^{(r_1r_2)q},$$

$$I_2[B(Z)]_{T,t}^{M,p}=\sum_{r_1,r_2,r_3\in J_M}B'(Z)\left(B'(Z)\left(B(Z)e_{r_1}\right)e_{r_2}\right)e_{r_3}\sqrt{\lambda_{r_1}\lambda_{r_2}\lambda_{r_3}}I_{(111)T,t}^{(r_1r_2r_3)p}, \quad (50)$$

$$I_3[B(Z)]_{T,t}^{M,p}=\sum_{r_1,r_2,r_3\in J_M}B''(Z)\left(B(Z)e_{r_1},B(Z)e_{r_2}\right)e_{r_3}\sqrt{\lambda_{r_1}\lambda_{r_2}\lambda_{r_3}}$$
$$\times\left(I_{(111)T,t}^{(r_1r_2r_3)p}+I_{(111)T,t}^{(r_2r_1r_3)p}+\mathbf{1}_{\{r_1=r_2\}}I_{(01)T,t}^{(0r_3)}\right), \quad (51)$$

where $p,q\ge 1$, approximations $I_{(11)T,t}^{(r_1r_2)q}$, $I_{(111)T,t}^{(r_1r_2r_3)p}$, $I_{(111)T,t}^{(r_2r_1r_3)p}$ are defined by (34), (35), and $I_{(01)T,t}^{(0r_3)}$ has the form (32).

Let $L_{HS}^{(k)}(U_0,H)$, $k\ge 1$ be the space of k-linear Hilbert–Schmidt operators from $U_0\times\ldots\times U_0$ (k times) to H. Furthermore, let $\|\cdot\|_{L_{HS}^{(k)}(U_0,H)}$ be operator norm in this space.

Let $I_1[B(Z)]_{T,t}$, $I_2[B(Z)]_{T,t}$, $I_3[B(Z)]_{T,t}$ be ISIs which are defined by the equalities (12)–(14) in which the finite-dimensional approximation of the Q-Wiener process depending on M should be replaced with the Q-Wiener process.

Theorem 5 [2]. *Let the condition* 1 *of Theorem 4 as well as the conditions of Theorem 1 be fulfilled. Futhermore, let* $B(v)\in L_{HS}(U_0,H)$, $B'(v)(B(v))\in L_{HS}^{(2)}(U_0,H)$, $B'(v)(B'(v)(B(v)))$, $B''(v)(B(v),B(v))\in L_{HS}^{(3)}(U_0,H)$ *for all* $v\in H$ (*we suppose that Frêchet derivatives* B', B'' *exist; see Sect.* 2). *Moreover, let there exists a constant* C *such that*

$$\left\|B'(Z)(B(Z))Q^{-\alpha}\right\|_{L_{HS}^{(2)}(U_0,H)}+\left\|B'(Z)(B'(Z)(B(Z)))Q^{-\alpha}\right\|_{L_{HS}^{(3)}(U_0,H)}$$
$$+\left\|B(Z)Q^{-\alpha}\right\|_{L_{HS}(U_0,H)}+\left\|B''(Z)(B(Z),B(Z))Q^{-\alpha}\right\|_{L_{HS}^{(3)}(U_0,H)}<C$$

w. p. 1 for some $\alpha > 0$. *Then*

$$\mathsf{M}\left\|I_1[B(Z)]_{T,t} - I_1[B(Z)]_{T,t}^{M,q}\right\|_H^2$$

$$\le (T-t)^2\left(C_0(\mathrm{tr}Q)^2\left(\frac{1}{2} - \sum_{j=1}^{q}\frac{1}{4j^2-1}\right) + K_Q\left(\sup_{i\in J\backslash J_M}\lambda_i\right)^{2\alpha}\right), \quad (52)$$

$$\mathsf{M}\left\|I_2[B(Z)]_{T,t} - I_2[B(Z)]_{T,t}^{M,p}\right\|_H^2 + \mathsf{M}\left\|I_3[B(Z)]_{T,t} - I_3[B(Z)]_{T,t}^{M,p}\right\|_H^2$$

$$\le (T-t)^3\left(C_1\,(\mathrm{tr}Q)^3\left(\frac{1}{6} - \sum_{j_1,j_2,j_3=0}^{p}\hat{C}_{j_3j_2j_1}^2\right) + L_Q\left(\sup_{i\in J\backslash J_M}\lambda_i\right)^{2\alpha}\right), (53)$$

where $\hat{C}_{j_3j_2j_1} = C_{j_3j_2j_1}(T-t)^{-3/2}$, $C_{j_3j_2j_1}$ *is defined by* (36), $p, q \in \mathbb{N}$, C_0, C_1, K_Q, $L_Q < \infty$.

Note that the estimate similar to (52) has been derived in [7] (also see [3]) with the difference related to the first term on the right-hand side of (52). In [7] the authors used the Karhunen–Loeve expansion of the Brownian bridge process for the approximation of Itô ISIs with respect to components of the finite-dimensional Wiener process. In this article we apply Theorem 1 and the system of Legendre polynomials to obtain the first term on the right-hand side of (52). If we assume that $\lambda_i \le C' i^{-\gamma}$ ($\gamma > 1, C' < \infty$) for $i \in J$, then the parameter $\alpha > 0$ obviously increases with decreasing γ [7].

Let $J_2[B(Z)]_{T,t}$, $J_3[B(Z), F(Z)]_{T,t}$ $J_4[B(Z), F(Z)]_{T,t}$ be ISIs which are defined by the equalities (9)–(11) in which the finite-dimensional approximation of the Q-Wiener process depending on M should be replaced with the Q-Wiener process.

Suppose that

$$\mathsf{M}\left\|B'(Z)(AZ + F(Z))Q^{-\alpha}\right\|_{L_{HS}(U_0,H)}^2 + \mathsf{M}\left\|AB(Z)Q^{-\alpha}\right\|_{L_{HS}(U_0,H)}^2 < \infty$$

for some $\alpha > 0$. Then by analogy with the proof of Theorem 5 [2] we obtain

$$\mathsf{M}\left\|J_2[B(Z)]_{T,t} - J_2[B(Z)]_{T,t}^M\right\|_H^2 \le C_2(T-t)^3\left(\sup_{i\in J\backslash J_M}\lambda_i\right)^{2\alpha},$$

$$\mathsf{M}\left\|J_l[B(Z), F(Z)]_{T,t} - J_l[B(Z), F(Z)]_{T,t}^M\right\|_H^2 \le C_3(T-t)^3\left(\sup_{i\in J\backslash J_M}\lambda_i\right)^{2\alpha},$$

where $l = 3, 4$ and $C_2, C_3 < \infty$.

5 Optimization of the Mean-Square Approximation Procedures for Itô ISIs

This section is devoted to the optimization of approximation procedures for Itô ISIs (7) that are used for approximation of ISIs with respect to the infinite-dimensional Q-Wiener process. More precisely, we discuss how to minimize the number p from the approximation (35).

From (39)–(42) for Itô ISIs $I_{(11)T,t}^{(i_1 i_2)}$, $I_{(111)T,t}^{(i_1 i_2 i_3)}$ we obtain [9,11,12]

$$E_2^{(i_1 i_2)p} = \frac{(T-t)^2}{2}\left(\frac{1}{2} - \sum_{i=1}^{p} \frac{1}{4i^2-1}\right) \quad (i_1 \neq i_2), \tag{54}$$

$$E_3^{(i_1 i_2 i_3)p_1} = (T-t)^3\left(\frac{1}{6} - \frac{1}{64} \sum_{j_1,j_2,j_3=0}^{p_1} \prod_{g=1}^{3}(2j_g+1)\, \bar{C}_{j_3 j_2 j_1}^2\right) \tag{55}$$

for $i_1 \neq i_2, i_1 \neq i_3, i_2 \neq i_3$,

$$E_3^{(i_1 i_2 i_3)p_2} = (T-t)^3\left(\frac{1}{6} - \frac{1}{64} \sum_{j_1,j_2,j_3=0}^{p_2} \prod_{g=1}^{3}(2j_g+1)\left(\bar{C}_{j_3 j_2 j_1}^2 + \bar{C}_{j_3 j_1 j_2}\bar{C}_{j_3 j_2 j_1}\right)\right) \tag{56}$$

for $i_1 = i_2 \neq i_3$,

$$E_3^{(i_1 i_2 i_3)p_3} = (T-t)^3\left(\frac{1}{6} - \frac{1}{64} \sum_{j_1,j_2,j_3=0}^{p_3} \prod_{g=1}^{3}(2j_g+1)\left(\bar{C}_{j_3 j_2 j_1}^2 + \bar{C}_{j_2 j_3 j_1}\bar{C}_{j_3 j_2 j_1}\right)\right) \tag{57}$$

for $i_1 \neq i_2 = i_3$,

$$E_3^{(i_1 i_2 i_3)p_4} = (T-t)^3\left(\frac{1}{6} - \frac{1}{64} \sum_{j_1,j_2,j_3=0}^{p_4} \prod_{g=1}^{3}(2j_g+1)\left(\bar{C}_{j_3 j_2 j_1}^2 + \bar{C}_{j_3 j_2 j_1}\bar{C}_{j_1 j_2 j_3}\right)\right) \tag{58}$$

for $i_1 = i_3 \neq i_2$.

Let $p_1, \ldots, p_4$ be minimal natural numbers satisfying the conditions

$$E_3^{(i_1 i_2 i_3)p_j} \leq (T-t)^4, \quad j = 1, \ldots, 4, \tag{59}$$

where the values $E_3^{(i_1 i_2 i_3)p_j}$ $(j = 1, \ldots, 4)$ are defined by (55)–(58).

Let us show by numerical experiments (see Table 1) that in most situations $p_1 \geq p_2, p_3, p_4$. This means that we can use the condition (55) instead of the conditions (55)–(58) for approximation of the Itô ISI $I_{(111)T,t}^{(i_1 i_2 i_3)}$. As a result, we will not get a noticeable increase of the mean-square approximation error (see [12], Sect. 5.4 for details).

Let p_5 be minimal natural number satisfying the condition

$$3!E_3^{(i_1 i_2 i_3)p_5} \leq (T-t)^4, \tag{60}$$

Table 1. The numbers p_1, p_2, p_3, p_4

$T-t$	0.0110	0.0080	0.0045	0.0035	0.0027	0.0025
p_1	12	16	28	36	47	50
p_2	6	8	14	18	23	25
p_3	6	8	14	18	23	25
p_4	12	16	28	36	47	51

Table 2. Comparison of the numbers p_1 and p_5

$T-t$	2^{-1}	2^{-2}	2^{-3}	2^{-4}	2^{-5}	2^{-6}
p_1	0	0	1	2	4	8
p_5	1	3	6	12	24	48

where the value $E_3^{(i_1 i_2 i_3)p_5}$ is defined by (55). Recall that the multiplier factor 3! (see (60)) contains in the estimate (24) for the case $k = 3$.

In Table 2, we can see the numerical comparison of the numbers p_1 and p_5 (the number p_1 is defined as in (59)). Obviously, using the formula (55) significantly reduces the computational costs for approximation of the Itô ISI $I_{(111)T,t}^{(i_1 i_2 i_3)}$, and, as a consequence, for approximation of the integrals (13), (16).

References

1. Kuznetsov, D.F.: Application of the method of approximation of iterated stochastic Itô integrals based on generalized multiple Fourier series to the high-order strong numerical methods for non-commutative semilinear stochastic partial differential equations. Differ. Uravnenia Protsesy Upravlenia. **3**, 18–62 (2019). http://diffjournal.spbu.ru/EN/numbers/2019.3/article.1.2.html
2. Kuznetsov, D.F.: Application of multiple Fourier-Legendre series to strong exponential Milstein and Wagner-Platen methods for non-commutative semilinear stochastic partial differential equations. Differ. Uravnenia Protsesy Upravlenia **3**, 129–162 (2020). http://diffjournal.spbu.ru/EN/numbers/2020.3/article.1.6.html
3. Jentzen, A., Röckner, M.: A Milstein scheme for SPDEs. Found. Comput. Math. **15**(2), 313–362 (2015)
4. Becker, S., Jentzen, A., Kloeden, P.E.: An exponential Wagner-Platen type scheme for SPDEs. SIAM J. Numer. Anal. **54**(4), 2389–2426 (2016)
5. Mishura, Y.S., Shevchenko, G.M.: Approximation schemes for stochastic differential equations in a Hilbert space. Theor. Prob. Appl. **51**(3), 442–458 (2007)
6. Prévôt, C., Röckner, M.: A Concise Course on Stochastic Partial Differential Equations. Lecture Notes in Mathematics, vol. 1905. Springer, Berlin (2007)
7. Leonhard, C., Rößler, A.: Iterated stochastic integrals in infinite dimensions: approximation and error estimates. Stochast. Partial Differ. Eqn.: Anal. Comput. **7**(2), 209–239 (2018). https://doi.org/10.1007/s40072-018-0126-9
8. Da Prato, G., Zabczyk, J.: Stochastic Equations in Infinite Dimensions, 2nd edn. Cambridge University Press, Cambridge (2014)

9. Kuznetsov, D.F.: Numerical integration of stochastic differential equations. 2. Polytechn. Univ. Publ. Saint-Petersburg (2006). https://doi.org/10.18720/SPBPU/2/s17-227
10. Kuznetsov, D.F.: Development and application of the fourier method for the numerical solution of Ito stochastic differential equations. Comput. Math. Math. Phys. **58**(7), 1058–1070 (2018). https://doi.org/10.1134/S0965542518070096
11. Kuznetsov, D.F.: Stochastic differential equations: theory and practice of numerical solution. With MATLAB programs, 6th edn. [In Russian]. Differ. Uravnenia Protsesy Upravlenia. **4**, A.1-A.1073 (2018). http://diffjournal.spbu.ru/EN/numbers/2018.4/article.2.1.html
12. Kuznetsov, D.F.: Strong approximation of iterated Ito and Stratonovich stochastic integrals based on generalized multiple Fourier series. Application to numerical solution of Ito SDEs and semilinear SPDEs. arXiv:2003.14184 [math.PR], pp. 1–741 (2020)
13. Milstein, G.N.: Numerical Integration of Stochastic Differential Equations. Kluwer, Springer, Netherlands (1995)
14. Kloeden, P.E., Platen, E.: Numerical Solution of Stochastic Differential Equations, 2nd edn. Springer, Berlin (1995)

A Sequential Test for the Drift of a Brownian Motion with a Possibility to Change a Decision

Mikhail Zhitlukhin(✉)

Steklov Mathematical Institute of the Russian Academy of Sciences,
8 Gubkina St., Moscow, Russia
mikhailzh@mi-ras.ru

Abstract. We construct a Bayesian sequential test of two simple hypotheses about the value of the unobservable drift coefficient of a Brownian motion, with a possibility to change the initial decision at subsequent moments of time for some penalty. Such a testing procedure allows to correct the initial decision if it turns out to be wrong. The test is based on observation of the posterior mean process and makes the initial decision and, possibly, changes it later, when this process crosses certain thresholds. The solution of the problem is obtained by reducing it to joint optimal stopping and optimal switching problems.

Keywords: Brownian motion · Sequential test · Simple hypothesis · Optimal stopping · Optimal switching

1 Introduction

We consider a problem of sequential testing of two simple hypotheses about the value of the unknown drift coefficient of a Brownian motion. In usual sequential testing problems (see e.g. the seminal works [4,8,10] or the recent monographs [1,9]), a testing procedure must be terminated at some stopping time and a decision about the hypotheses must be made. In contrast, in the present paper we propose a new setting, where a testing procedure does not terminate and it is allowed to change the initial decision (for the price of paying some penalty) if, given later observations, it turns out that it is incorrect.

We will work in a Bayesian setting and assume that the drift coefficient has a known prior distribution on a set of two values. A decision rule consists of an initial decision (τ, d), where τ is the moment at which the decision is made and d is a two-valued function showing which hypothesis is accepted initially, and a sequence of stopping times τ_n, at which the decision can be changed later. The goal is to minimize a penalty function which consists of the three parts: a penalty for the waiting time until the initial decision, a penalty for a

The research was supported by the Russian Science Foundation, project 19-11-00290.

A. N. Shiryaev et al. (Eds.): ICSM-5 2020, PROMS 371, pp. 33–42, 2021.
https://doi.org/10.1007/978-3-030-83266-7_3

wrong decision proportional to the time during which the corresponding wrong hypothesis is being accepted, and a penalty for each change of a decision.

This study was motivated by the paper [6], where a sequential multiple changepoint detection problem was considered. That problem consists in tracking of the value of the unobservable drift coefficient of a Brownian motion, which is modeled by a telegraph process (a two-state Markov process) switching between -1 and $+1$ at random times. In the present paper, we deal with a similar tracking procedure and a penalty function, but the difference is that the unobservable drift coefficient does not change. Among other results on multiple changepoint detection, one can mention the paper [3], where a tracking problem for a general two-state Markov process with a Brownian noise was considered, and the paper [2], which studied a tracking problem for a compound Poisson process.

We solve our problem by first representing it as a combination of an optimal stopping problem and an optimal switching problem (an optimal switching problem is an optimal control problem where the control process assumes only two values). The optimal stopping problem allows to find the initial stopping time, while the subsequent moments when the decision is changed are found from the optimal switching problem. Consequently, the value function of the optimal switching problem becomes the payoff function of the optimal stopping problem. Then both of the problems are solved by reducing them to free-boundary problems associated with the generator of the posterior mean process of the drift coefficient. We consider only the symmetric case (i.e. type I and type II errors are of the same importance), in which the solution turns out to be of the following structure. First an observer waits until the posterior mean process exists from some interval $(-A, A)$ and at that moment of time makes the initial decision. Future changes of the decision occur when the posterior mean process crosses some thresholds $-B$ and B. The constants A, B are found as unique solutions of certain equations.

The rest of the paper consists of the three sections: Sect. 2 describes the problem, Sect. 3 states the main theorem which provides the optimal decision rule, Sect. 4 contains its proof.

2 The Model and the Optimality Criterion

Let $(\Omega, \mathcal{F}, \mathrm{P})$ be a complete probability space. Suppose one can observe a process X_t defined on this probability space by the relation

$$X_t = \mu\theta t + B_t, \tag{1}$$

where B_t is a standard Brownian motion, $\mu > 0$ is a known constant, and θ is a ± 1-valued random variable independent of B_t. It is assumed that neither θ nor B_t can be observed directly. The goal is to find out whether $\theta = 1$ or $\theta = -1$ by observing the process X_t sequentially. Note that the case when the drift coefficient of X_t can take on two arbitrary values $\mu_1 \neq \mu_2$ can be reduced to (1) by considering the process $X_t - \frac{1}{2}(\mu_1 + \mu_2)t$.

We will assume that the prior distribution of θ is known and is characterized by the probability $p = \mathrm{P}(\theta = 1)$. Recall that usual settings of sequential testing problems consist in that an observer must choose a stopping time τ of the (completed and right-continuous) filtration $\mathbb{F}^X = (\mathcal{F}_t^X)_{t\geq 0}$ generated by X_t, at which the observation is stopped, and an $\mathcal{F}_\tau^X$-measurable function d with values -1 or $+1$ that shows which of the two hypotheses is accepted at time τ. The choice of (τ, d) depends on a particular optimality criterion which combines penalties for type I and type II errors, and a penalty for observation duration. But, in any case, a test terminates at time τ.

In this paper we will focus on a setting where an observer can change a decision made initially at time τ and the testing procedure does not terminate.

By a *decision rule* we will call a triple $\delta = (\tau_0, d, T)$, where τ_0 is an $\mathbb{F}^X$-stopping time, d is an $\mathcal{F}_{\tau_0}^X$-measurable function which assumes values ± 1, and $T = (\tau_1, \tau_2 \ldots)$ is a sequence of $\mathbb{F}^X$-stopping times such that $\tau_n \leq \tau_{n+1}$ for all $n \geq 0$. At the moment τ_0, the initial decision d is made. Later, if necessary, an observer can change the decision to the opposite one, and the moments of change are represented by the sequence T. Thus, if, for example, $d = 1$, then at τ_0 an observer decides that $\theta = 1$ and at τ_1 switches the opinion to $\theta = -1$; at τ_2 switches back to $\theta = 1$, and so on. It may be the case that $\tau_n = +\infty$ starting from some n; then the decision is changed only a finite number of times (the optimal rule we construct below will have this property with probability 1).

With a given decision rule δ, associate the $\mathbb{F}^X$-adapted process D_t^δ which expresses the current decision at time t,

$$D_t^\delta = \begin{cases} 0, & \text{if } t < \tau_0, \\ d, & \text{if } t \in [\tau_{2n}, \tau_{2n+1}), \\ -d, & \text{if } t \in [\tau_{2n+1}, \tau_{2n+2}), \end{cases}$$

and define the *Bayesian risk function*

$$R(\delta) = \mathrm{E}\Big(c_0\tau_0 + c_1 \int_{\tau_0}^{\infty} \mathrm{I}(D_t^\delta \neq \theta)dt + c_2 \sum_{t>\tau_0} \mathrm{I}(D_{t-}^\delta \neq D_t^\delta) \Big), \tag{2}$$

where $c_i > 0$ are given constants.

The problem that we consider consists in finding a decision rule δ^* which minimizes R, i.e.

$$R(\delta^*) = \inf_\delta R(\delta).$$

Such a decision rule δ^* will be called *optimal.*

One can give the following interpretation to the terms under the expectation in (2). The term $c_0\tau_0$ is a penalty for a delay until making the initial decision. The next term is a penalty for making a wrong decision, which is proportional to the time during which the wrong hypothesis is being accepted. The last term is a penalty for changing a decision, in the amount c_2 for each change. Note that the problem we consider is symmetric (i.e. type I and type II errors are penalized in the same way); in principle, an asymmetric setting can be studied as well.

3 The Main Result

To state the main result about the optimal decision rule, introduce the posterior mean process

$$M_t = \mathrm{E}(\theta \mid \mathcal{F}_t^X).$$

As follows from known results, the process M_t satisfies the stochastic differential equation

$$dM_t = \mu(1 - M_t^2)d\widetilde{B}_t, \qquad M_0 = 2p - 1, \tag{3}$$

where $\widetilde{B}_t$ is a Brownian motion with respect to $\mathbb{F}^X$ (an *innovation process*, see, e.g., Chap. 7 in [5]), which satisfies the equation

$$d\widetilde{B}_t = dX_t - M_t dt.$$

Representation (3) can be obtained either directly from filtering theorems (see Theorem 9.1 in [5]) or from the known equation for the posterior probability process $\pi_t = \mathrm{P}(\theta = 1 \mid \mathcal{F}_t^X)$ (see Chapter VI in [7]) since $M_t = 2\pi_t - 1$. In the explicit form, M_t can be expressed through the observable process X_t as

$$M_t = 1 - \frac{2(1-p)}{pe^{2\mu X_t} + 1 - p}.$$

Introduce the two thresholds $A, B \in (0,1)$, which depend on the parameters μ, c_0, c_1, c_2 of the problem, and will define the switching boundaries for the optimal decision rule. The threshold B is defined as the solution of the equation

$$\ln\frac{1-B}{1+B} + \frac{2B}{1-B^2} = \frac{2\mu^2 c_2}{c_1}, \tag{4}$$

and the threshold A is defined as the solution of the equation

$$\left(\frac{c_1}{2c_0} - 1\right)\ln\frac{1-A}{1+A} + \frac{2}{1+A}\left(\frac{c_1}{2c_0} + \frac{A}{1-A}\right) = \frac{c_1}{c_0(1-B^2)}. \tag{5}$$

The next simple lemma shows that A and B are well-defined. Its proof is rather straightforward and is omitted.

Lemma 1. *Equations* (4), (5) *have unique solutions* $A, B \in (0,1)$. *If* $c_1 = 2c_0$, *then* $A = B$.

The following theorem, being the main result of the paper, provides the optimal decision rule in an explicit form.

Theorem 1. *The optimal decision rule* $\delta^* = (\tau_0^*, d^*, T^*)$ *consists of the stopping time* τ_0^* *and the decision function* d^* *defined by the formulas*

$$\tau_0^* = \inf\{t \ge 0 : |M_t| \ge A\}, \qquad d^* = \operatorname{sgn} M_{\tau_0},$$

and the sequence of stopping times $T^* = (\tau_n^*)_{n=1}^\infty$ *which on the event* $\{d^* = 1\}$ *are defined by the formulas*

$$\tau_{2k+1}^* = \inf\{t \ge \tau_{2k}^* : M_t \le -B\}, \quad \tau_{2k+2}^* = \inf\{t \ge \tau_{2k+1}^* : M_t \ge B\}, \tag{6}$$

and on the event $\{d = -1\}$ *by the formulas*

$$\tau^*_{2k+1} = \inf\{t \geq \tau^*_{2k} : M_t \geq B\}, \quad \tau^*_{2k+2} = \inf\{t \geq \tau^*_{2k+1} : M_t \leq -B\} \tag{7}$$

(where $\inf \emptyset = +\infty$*).*

Example 1. Figure 1 illustrates how the optimal decision rule works. In this example, we take $p = 0.5$, $\mu = 1/3$, $c_0 = 2/3$, $c_1 = 1$, $c_2 = 3/2$. The thresholds A, B can be found numerically, $A \approx 0.37$, $B \approx 0.55$.

The simulated path on the left graph has $\theta = 1$. The rule δ^* first waits until the process M_t exists from the interval $(-A, A)$. Since in this example it exists through the lower boundary (at τ^*_0), the initial decision is $d^* = -1$ (incorrect). Then the rule waits until M_t crosses the threshold B, and changes the decision to $\theta = 1$ at τ^*_1.

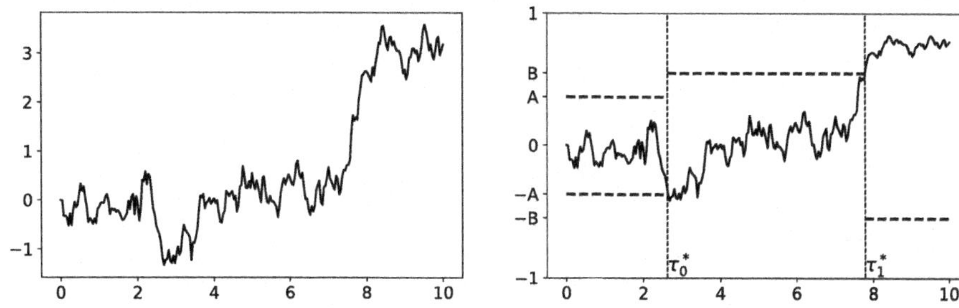

Fig. 1. Left: the process X_t; right: the process M_t. Parameters: $p = 0.5$, $\mu = 1/3$, $c_0 = 2/3$, $c_1 = 1$, $c_2 = 3/2$.

4 Proof of the Main Theorem

Let us denote by P_x and E_x the probability measure and the expectation under the assumption $\mathrm{P}(\theta = 1) = (x + 1)/2$, so the posterior mean process M_t starts from the value $M_0 = x$. It is easy to verify that

$$\mathrm{P}_x(D^\delta_t \neq \theta \mid \mathcal{F}^X_t) = \frac{1 - M_t D^\delta_t}{2},$$

and, by taking intermediate conditioning with respect to $\mathcal{F}^X_t$ in (2), we can see that we need to solve the following problem for $x \in [-1, 1]$

$$V^*(x) = \inf_\delta \mathrm{E}_x \Bigl(c_0 \tau_0 + \frac{c_1}{2} \int_{\tau_0}^\infty (1 - M_t D^\delta_t) dt + c_2 \sum_{t > \tau_0} \mathrm{I}(D^\delta_{t-} \neq D^\delta_t) \Bigr) \tag{8}$$

(by "to solve" we mean to find δ at which the infimum is attained for a given x; in passing we will also find the function $V^*(x)$ in an explicit form).

Observe that there exists the limit $M_\infty := \lim_{t\to\infty} M_t = \theta$ a.s. Hence the solution of problem (8) should be looked for only among decision rules δ such that D_t^δ has a finite number of jumps and $D_\infty^\delta = \theta$ (note that the rule δ^* satisfies these conditions). In view of this, for a stopping time τ_0 denote by $\mathcal{D}(\tau_0)$ the class of all $\mathbb{F}^X$-adapted càdlàg processes D_t such that, with probability 1, they assume values ± 1 after τ_0, have a finite number of jumps, and satisfy the condition $D_\infty = \theta$. Let $U^*(\tau_0)$ be the value of the following optimal switching problem:

$$U^*(\tau_0) = \inf_{D\in\mathcal{D}(\tau_0)} \mathrm{E}_x\Bigl(\frac{c_1}{2}\int_{\tau_0}^{\infty}(1 - M_t D_t)dt + c_2 \sum_{t>\tau_0} \mathrm{I}(D_{t-} \neq D_t)\Bigr). \qquad (9)$$

Consequently, problem (8) can be written in the form

$$V^*(x) = \inf_{\tau_0} \mathrm{E}_x(c_0\tau_0 + U^*(\tau_0)). \qquad (10)$$

Thus, to show that the decision rule δ^* is optimal, it will be enough to show that τ_0^* delivers the infimum in the problem V^*, and D^{δ^*} delivers the infimum in the problem $U^*(\tau_0^*)$. In order to do that, we are going to use a usual approach based on "guessing" a solution and then verifying it using Itô's formula. Since this approach does not show how to actually find the functions V^* and U^*, in the remark after the proof we provide heuristic arguments that can be used for that.

We will first deal with U^*. Let B be the constant from (4). Introduce the "candidate" function $U(x,y)$ $x\in[-1,1]$, $y\in\{-1,1\}$, defined by

$$U(x,1) = \frac{c_1(1-x)}{4\mu^2}\Bigl(\ln\frac{1+x}{1-x} + \frac{2}{1-B^2}\Bigr), \qquad x\in(-B,\ 1], \qquad (11)$$

$$U(x,1) = U(-x,1) + c_2, \qquad x\in[-1,\ -B], \qquad (12)$$

$$U(x,-1) = U(-x,1), \qquad x\in[-1,\ 1] \qquad (13)$$

(see Fig. 2, which depicts the function $U(x,y)$, as well as the function $V(x)$ defined below, with the same parameters as in the example in the previous section).

We are going to show that $U^*(\tau_0) = U(|M_{\tau_0}|,1)$. Let Lf denotes application of the generator of the process M_t to a sufficiently smooth function f, i.e.

$$Lf(x) = \frac{\mu^2}{2}(1-x^2)^2\frac{\partial^2}{\partial x^2}f(x).$$

By U' and ΔU denote, respectively, the derivative with respect to the first argument, and the difference with respect to the second argument of U, i.e.

$$U'(x,y) = \frac{\partial U}{\partial x}(x,y), \qquad \Delta U(x,y) = U(x,y) - U(x,-y).$$

From the above explicit construction (11)–(13), it is not difficult to check that $U(x,y)$ has the following properties:

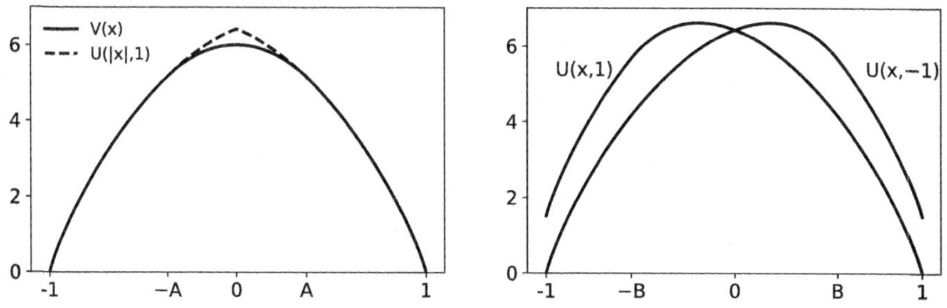

Fig. 2. The functions $V(x)$ and $U(x,y)$. The parameters μ, c_0, c_1, c_2 are the same as in Fig. 1.

(U.1) $U(x,y) \in C^1$ in x for $x \in (-1,1)$, and $U(x,y) \in C^2$ in x except at points $x = -yB$;
(U.2) $(1-x^2)U'(x,y)$ is bounded for $x \in (-1,1)$;
(U.3) $LU(x,y) = -c_1(1-xy)/2$ if $xy > -B$, and $LU(x,y) \geq -c_1(1-xy)/2$ if $xy < -B$;
(U.4) $\Delta U(x,y) = -c_2$ if $xy \geq B$, and $\Delta U(x,y) \geq -c_2$ if $xy < B$.

Consider any process $D \in \mathcal{D}(\tau_0)$ and let $(\tau_n)_{n\geq 1}$ be the sequence of the moments of its jumps after τ_0. Property (U.1) allows to apply Itô's formula to the process $U(M_t, D_t)$, from which for any $s > 0$ we obtain

$$\begin{aligned} U(M_{s\vee\tau_0}, D_{s\vee\tau_0}) &= U(M_{\tau_0}, D_{\tau_0}) \\ &\quad + \sum_{n\,:\,\tau_{n-1}\leq s} \biggl(\int_{\tau_{n-1}}^{s\wedge\tau_n} LU(M_t, D_t)\mathrm{I}(M_t \neq -D_t B)dt \\ &\quad + \mu \int_{\tau_{n-1}}^{s\wedge\tau_n} (1-M_t^2)U'(M_t,D_t)d\widetilde{B}_t + \Delta U(M_{\tau_n}, D_{\tau_n})\mathrm{I}(s \geq \tau_n) \biggr). \qquad (14) \end{aligned}$$

Take the expectation $\mathrm{E}_x(\,\cdot \mid \mathcal{F}^X_{\tau_0})$ of the both sides of (14). By (U.2), the integrand in the stochastic integral is uniformly bounded, so its expectation is zero. Passing to the limit $s \to \infty$ and using the equality $D_\infty = M_\infty$, which implies $U(M_{s\vee\tau_0}, D_{s\vee\tau_0}) \to 0$ as $s \to \infty$, we obtain

$$U(M_{\tau_0}, D_{\tau_0}) \leq \mathrm{E}_x\biggl(\frac{c_1}{2} \int_{\tau_0}^{\infty} (1 - M_t D_t)dt + c_2 \sum_{t>\tau_0} \mathrm{I}(D_t \neq D_{t-}) \biggm| \mathcal{F}^X_{\tau_0} \biggr), \qquad (15)$$

where to get the inequality we used property (U.3) for the first term under the expectation and (U.4) for the second term. Taking the infimum of the both sides of (15) over $D \in \mathcal{D}(\tau_0)$ we find

$$U(M_{\tau_0}, D_{\tau_0}) \leq U^*(\tau_0). \qquad (16)$$

On the other hand, if the process D_t is such that $D_{\tau_0} = \operatorname{sgn} M_{\tau_0}$ (let $\operatorname{sgn} 0 = 1$, if necessary) and its jumps after τ_0 are identified with the sequence $(\tau_n)_{n\geq 1}$

defined as in (6)–(7) but with arbitrary τ_0 in place of τ_0^*, then we would have the equality in (15), as follows from (U.3) and (U.4). Together with (16), this implies that $U^*(\tau_0) = U(M_{\tau_0}, \operatorname{sgn} M_{\tau_0}) = U(|M_{\tau_0}|, 1)$ and the infimum in the definition of $U^*(\tau_0)$ is attained at this process D_t.

Let us now consider the problem V^*. As follows from the above arguments, we can write it in the form

$$V^*(x) = \inf_{\tau_0} \mathrm{E}_x(c_0\tau_0 + U(|M_{\tau_0}|, 1)). \tag{17}$$

It is clear that it is enough to take the infimum only over stopping times with finite expectation.

Let A be the constant defined in (5), and put

$$K = \left(\frac{c_1(1-A)}{4\mu^2} + \frac{c_0 A}{2\mu^2}\right) \ln\frac{1+A}{1-A} + \frac{c_1(1-A)}{2\mu^2(1-B^2)}. \tag{18}$$

Introduce the “candidate” function $V(x)$, $x \in [-1,1]$:

$$V(x) = \frac{c_0 x}{2\mu^2} \ln\frac{1-x}{1+x} + K, \qquad |x| < A, \tag{19}$$

$$V(x) = U(|x|, 1), \qquad |x| \geq A. \tag{20}$$

It is straightforward to check that $V(x)$ has the following properties:

(V.1) $V(x) \in C^1$ in x for $x \in (-1,1)$, and $V(x) \in C^2$ in x except at points $x = \pm A$;
(V.2) $(1-x^2)V'(x)$ is bounded for $x \in (-1,1)$;
(V.3) $LV(x) = -c_0$ if $|x| < A$, and $LV(x) \geq -c_0$ if $|x| > A$;
(V.4) $V(x) = U(|x|,1)$ if $|x| \geq A$, and $V(x) \leq U(|x|,1)$ if $|x| < A$.

Applying Iô’s formula to the process $V(M_t)$ and taking the expectation, for any stopping time τ_0 with $\mathrm{E}\tau_0 < \infty$ we obtain

$$\mathrm{E}_x V(M_{\tau_0}) = V(x) + \mathrm{E}_x \int_0^{\tau_0} LV(M_s)ds$$

(Itô’s formula can be applied in view of (V.1); the expectation of the stochastic integral, which appears in it, is zero in view of (V.2) and the finiteness of $\mathrm{E}\tau_0$).

From (V.3) and (V.4), we find

$$V(x) \leq \mathrm{E}_x(c_0\tau_0 + U(|M_{\tau_0}|, 1)), \tag{21}$$

so, after taking the infimum over τ_0, we get $V(x) \leq V^*(x)$. On the other hand, for the stopping time τ_0^* we have the equality in (21), so $V(x) = V^*(x)$. Consequently, τ_0^* solves the problem V^*.

The proof is complete.

Remark 1. The above proof does not explain how to find the functions $V(x)$ and $U(x,y)$. Here we provide arguments which are based on well-known ideas from the optimal stopping theory and allow to do that. The reader is referred, e.g., to the monograph [7] for details.

Since the process M_t is Markov, we can expect that the optimal process D_t for U^* should depend only on current values of M_t and D_{t-}. Moreover, it is natural to assume that D_t should switch from 1 to -1 when M_t becomes close to -1, and switch from -1 to 1 when M_t becomes close to 1. The symmetry of the problem suggests that there should be a threshold B such that the switching occurs when M_t crosses the levels $\pm B$. This means that the optimal sequence of stopping times T^* is of the form (6)–(7). Consequently, in the set $\{(x,y) : x > -yB\}$, where x corresponds to the value of M_t and y corresponds to the value of D_t, one should continue using the current value of D_t, while in the set $\{(x,y) : x \leq -yB\}$ switch to the opposite one. In what follows, we will call these sets the *continuation set* and the *switching set*, respectively.

Next we need to find B. Introduce the value function $U(x,y)$ (cf. (9); it turns out to be the same function $U(x,y)$ which appears in the proof):

$$U(x,y) = \inf_D \mathrm{E}_x\biggl(\frac{c_1}{2}\int_0^\infty (1 - M_t D_t)dt + c_2 \mathrm{I}(D_0 \neq y) + c_2 \sum_{t>0} \mathrm{I}(D_{t-} \neq D_t)\biggr),$$

where the infimum is taken over all càdlàg processes D_t which are adapted to the filtration generated by M_t, take on values ± 1, and have a finite number of jumps. In the switching set, we have

$$U(x,y) = U(x,-y) + c_2.$$

From the general theory (see Chapter III in [7]), we can expect that the value function $U(x,y)$ in the continuation set solves the ODE

$$LU(x,y) = -\frac{c_1}{2}(1 - xy).$$

Its general solution can be found explicitly:

$$U_{\mathrm{gen}}(x,1) = \frac{c_1(1-x)}{4\mu^2}\ln\frac{1+x}{1-x} + K_1 x + K_2,$$

where K_1 and K_2 are constants. Since we have $U(1,1) = 0$ (if $x = 1$, then $M_t = 1$ for all $t \geq 0$ and the optimal process D is $D_t \equiv 1$), we get $K_2 = -K_1$. To find K_1 and B, we can employ the continuous fit and smooth fit conditions, also known from the general theory, which state that at the boundary of the continuation set, i.e. at the points (x,y) with $x = -yB$, the value function satisfies the equations

$$U(-B,1) = U(-B,-1) + c_2, \qquad U'(-B,1) = U'(-B,-1)$$

(here $x = -B$, $y = 1$; the pair $x = B$, $y = -1$ gives the same equations due to the symmetry of the problem). Solving these equations gives formulas (11)–(13) for $U(x,y)$.

To find the function $V(x)$ we use a similar approach. From the representation as a standard optimal stopping problem (17), we can expect that the optimal stopping time should be the first exit time of the process M_t from some continuation set. Taking into account the original formulation of the problem as a sequential test, it is natural to assume that the initial decision should be made at a moment when the posterior mean becomes close to 1 or -1, i.e. the continuation set for $V(x)$ should be an interval $(-A, A)$. As follows from the general theory, $V(x)$ in the continuation set satisfies the ODE

$$LV(x) = -c_0,$$

which has the general solution

$$V_{\text{gen}}(x) = \frac{c_0 x}{2\mu^2} \ln \frac{1-x}{1+x} + K_3 x + K_4.$$

Due to the symmetry of the problem, we have $V(x) = V(-x)$, so $K_3 = 0$. Then the constants A and K_4 can be found from the continuous fit and smooth fit conditions at $x = A$:

$$V(A) = U(A, 1), \qquad V'(A) = U'(A, 1).$$

These equations give the function $V(x)$ defined in (19)–(20), with $K_4 = K$ from (18).

References

1. Bartroff, J., Lai, T.L., Shih, M.C.: Sequential Experimentation in Clinical Trials: Design and Analysis. Springer Science & Business Media, New York (2012)
2. Bayraktar, E., Ludkovski, M.: Sequential tracking of a hidden Markov chain using point process observations. Stoch. Process. Appl. **119**(6), 1792–1822 (2009)
3. Gapeev, P.V.: Bayesian switching multiple disorder problems. Math. Oper. Res. **41**(3), 1108–1124 (2015)
4. Irle, A., Schmitz, N.: On the optimality of the SPRT for processes with continuous time parameter. Stat. J. Theor. Appl. Stat. **15**(1), 91–104 (1984)
5. Liptser, R.S., Shiryaev, A.N.: Statistics of Random Processes I, II. Springer-Verlag, Berlin (2001)
6. Muravlev, A., Urusov, M., Zhitlukhin, M.: Sequential tracking of an unobservable two-state Markov process under Brownian noise. Seq. Anal. **40**(1), 1–16 (2021)
7. Peskir, G., Shiryaev, A.: Optimal Stopping and Free-Boundary Problems. Birkhäuser Verlag, Basel (2006)
8. Shiryaev, A.N.: Two problems of sequential analysis. Cybernetics **3**(2), 63–69 (1967)
9. Tartakovsky, A., Nikiforov, I., Basseville, M.: Sequential Analysis: Hypothesis Testing and Changepoint Detection. CRC Press, Boca Raton (2014)
10. Wald, A., Wolfowitz, J.: Optimum character of the sequential probability ratio test. Ann. Math. Stat. **19**(3), 326–339 (1948)

Survival Probabilities in Compound Poisson Model with Negative Claims and Investments as Viscosity Solutions of Integro-Differential Equations

Tatiana Belkina(✉)

Central Economics and Mathematics Institute RAS,
Nakhimovsky Prosp. 47, 117418 Moscow, Russia
tbel@cemi.rssi.ru
http://www.cemi.rssi.ru/about/persons/index.php?SECTION_ID=6&ELEMENT_ID=4123

Abstract. This work relates to the problem of the identifying of some solutions to linear integro-differential equations as the probability of survival (non-ruin) in the corresponding collective risk models involving investments. The equations for the probability of non-ruin as a function of the initial reserve are generated by the infinitesimal operators of corresponding dynamic reserve processes. The direct derivation of such equations is usually accompanied by some significant difficulties, such as the need to prove a sufficient smoothness of the survival probability. We propose an approach that does not require a priori proof of the smoothness. It is based on previously proven facts for a certain class of insurance models with investments: firstly, under certain assumptions, the survival probability is at least a viscosity solution to the corresponding integro-differential equation, and secondly, any two viscosity solutions with coinciding boundary conditions are equivalent. We apply this approach, allowing us to justify rigorously the form of the survival probability, to the collective life insurance model with investments.

Keywords: Survival probability · Viscosity solution · Integro-differential equations

1 Introduction

The problem of viscosity solutions of linear integro-differential equations (IDEs) for non-ruin probabilities as a functions of an initial surplus in collective insurance risk models, when the whole surplus is invested into a risky (or risk-free) asset, is considered in [1]. For a rather general model of the resulting surplus process, it is shown that the non-ruin probability always solves corresponding IDE in the viscosity sense. Moreover, for the case when the distributions of claims in the insurance risk process have full support on the half-line, a uniqueness theorem is proved in [1]. In the present paper, we use these results to establish that the solution of some previously formulated and investigated boundary value problem for IDE defines the probability of ruin for the corresponding surplus model.

A. N. Shiryaev et al. (Eds.): ICSM-5 2020, PROMS 371, pp. 43–56, 2021.
https://doi.org/10.1007/978-3-030-83266-7_4

Thus, the uniqueness theorem for a viscosity solution plays the role of a verification argument for the solution of the IDE as the probability of non-ruin for the resulting surplus process in the models with investments. The approach proposed here can be considered as an alternative tool along with traditional verification arguments based on the use of the martingale approach (see, e.g., [2,3] and references therein). It can be used when it is possible to determine a priori the value of the probability of non-ruin at an initial surplus which is equal to zero, and its limiting value when the initial surplus tends to infinity.

The mentioned general model, which is studied in [1], considers an insurance risk in the classical actuarial framework but under the assumptions that the price process of the risky asset is a jump-diffusion process defined by the stochastic exponential of the Lévy process. The classical actuarial framework involves two possible versions of the original model (without investment): the classical Cramér-Lundberg model or the so-called dual risk model (also called compound Poisson model with negative claims [4], or life annuity insurance model [5]). To demonstrate the main idea of this paper, we consider the dual risk model and assume that the insurer's reserve is invested to a risky asset with price modelled by the geometric Brownian motion or it is invested to a risk-free asset. We use this particular case of the model considered in [1], because 1) for the case of an exponential distribution of jumps and risky investments, the existence of a twice continuously differentiable solution to the boundary value problem for the corresponding IDE is proved in [6], where its properties also are studied and the numerical calculations are done; for the risk-free investment, a non-smooth, generally speaking, solution is constructed in [7] and 2) the value of the survival probability at zero surplus level is a priory known (unlike, for example, the Cramér-Lundberg model with investment, where it can be determined only numerically; see, e.g. [8]).

The paper is organized as follows. In Sect. 2 the compound Poisson model with negative claims and investments is described. Then the problem of the identifying of some solutions to linear integro-differential equations as the survival probabilities in this model in two cases: risky and risk-free investments is formulated. In Sect. 3 some preliminary results about survival probabilities as viscosity solutions of IDEs are given. In Sect. 4 a general statement concerning the identifying the survival probability in the considered model (Theorem 3) is proved. In this statement, the uniqueness theorem for a viscosity solution as a verification argument for the survival probability is used. Moreover, the results of Theorem 3 with applying to the case of exponential distribution of premiums size (jumps of the compound Poisson process) are given; here risky investments (Sect. 4.1) as well as risk-free investments (Sect. 4.2) are considered. Section 5 deals with proofs. In Sect. 6 some results of numerical calculations from [7] are presented, and Sect. 7 contains the conclusions.

2 The Model Description and Statement of the Problem

The typical insurance contract for the policyholder in the dual risk model is the life annuity with the subsequent transfer of its property to the benefit of the insurance company. Thus, the surplus of a company in a collective risk model is of the form

$$R_t = u - ct + \sum_{k=1}^{N(t)} Z_k, \quad t \geq 0. \tag{1}$$

Here R_t is the surplus of a company at time $t \geq 0$; u is the initial surplus, $c > 0$ is the life annuity rate (or the pension payments per unit of time), assumed to be deterministic and fixed. $N(t)$ is a homogeneous Poisson process with intensity $\lambda > 0$ that, for any $t > 0$, determines the number of random revenues up to the time t; Z_k $(k = 1, 2, ...)$ are independent identically distributed random variables (r.v.) with a distribution function $F(z)$ $(F(0) = 0,\ \mathbf{E}Z_1 = m < \infty,\ m > 0)$ that determine the revenue sizes (premiums) and are assumed to be independent of $N(t)$. These random revenues arise at the final moments of the life annuity contracts realizations.

We assume also that the insurer's reserve is invested to a risky asset with price S_t modelled by the geometric Brownian motion,

$$dS_t = \mu S_t dt + \sigma S_t dw_t, \quad t \geq 0,$$

where μ is the stock return rate, σ is the volatility, w_t is a standard Brownian motion independent of $N(t)$ and Z_i's.

Then the resulting surplus process X_t is governed by the equation

$$dX_t = \mu X_t dt + \sigma X_t dw_t + dR_t, \quad t \geq 0, \tag{2}$$

with the initial condition $X_0 = u$, where R_t is defined by (1).

Denote by $\varphi(u)$ the survival probability: $\varphi(u) = \mathbf{P}\,(X_t \geq 0,\ t \geq 0)$. Then $\Psi(u) = 1 - \varphi(u)$ is the ruin probability. Then $\tau^u := \inf\{t\colon\ X_t^u \leq 0\}$ is the time of ruin.

Recall at first that the infinitesimal generator $\mathcal{A}$ of the process X_t has the form

$$(\mathcal{A}f)(u) = \frac{1}{2}\sigma^2 u^2 f''(u) + f'(u)(\mu u - c) - \lambda f(u) + \lambda \int_0^\infty f(u+z)\, dF(z), \tag{3}$$

for any function $f(u)$ from a certain subclass of the space $\mathcal{C}^2(\mathbb{R}_+)$ of twice continuously differentiable on $(0, \infty)$ functions (in the case $\sigma > 0$; if $\sigma = 0$ we are dealing with a different class of functions, see [7]).

One of the important questions in this and similar models is the question of whether the survival probability $\varphi(u)$ is a twice continuously differentiable function of the initial capital u on $(0, \infty)$. In the case of a positive answer to this question, we can state that $\varphi(u)$ is a classical solution of the equation

$$(\mathcal{A}f)(u) = 0, \quad u > 0, \tag{4}$$

and the properties of this probability can be investigated as properties of a suitable solution to this equation. In [10], for the case of exponential distribution of Z_k and $\sigma > 0$, such a suitable solution in the set of all solutions of the linear IDE (4) is selected using some results of renewal theory; the regularity (twice continuous differentiability) of $\varphi(u)$ is studied using a method based on integral representations; asymptotic expansions of the survival probability for infinitely large values of the initial capital is obtained.

In contrast to the direct method used in [10], we propose a method based on the assumption of the existence of a classical (or, maybe, viscosity) solution to a boundary value problem for the IDE (4) and verification arguments for the survival probability related to the concept of viscosity. For the case of exponential distribution of the company's random revenues and $\sigma > 0$, the existence theorem for IDE (4) with boundary conditions

$$\lim_{u \to +0} f(u) = 0, \qquad \lim_{u \to +\infty} f(u) = 1, \tag{5}$$

is proved in [6]. The uniqueness of the classical solution is also established, as well as its asymptotic behaviour at zero and at infinity. For the case $\sigma = 0$, a non-smooth (generally speaking) solution is presented in [7].

The problem we are solving here: to prove that if there exists a solution f of the problem (4), (5), then it determines the survival probability of the process (2). For the solving this problem, we use the results of [1] on the survival probability as a viscosity solution to equation (4).

3 Survival Probabilities as Viscosity Solutions of IDEs: Preliminary Results [1]

Let denote by $C_b^2(u)$ the set of bounded continuous functions $f : \mathbb{R} \to \mathbb{R}$ two times continuously differentiable in the classical sense in a neighbourhood of the point $u \in]0, \infty[$ and equal to zero on $]-\infty, 0]$. For $f \in C_b^2(u)$, the value $(\mathcal{A}f)(u)$ is well-defined.

A function $\Phi :]0, \infty[\to [0, 1]$ is called *a viscosity supersolution* of (4) if for every point $u \in]0, \infty[$ and every function $f \in C_b^2(u)$ such that $\Phi(u) = f(u)$ and $\Phi \geq f$ the inequality $(\mathcal{A}f)(u) \leq 0$ holds.

A function $\Phi :]0, \infty[\to [0, 1]$ is called *a viscosity subsolution* of (4) if for every $u \in]0, \infty[$ and every function $f \in C_b^2(u)$ such that $\Phi(u) = f(u)$ and $\Phi \leq f$ the inequality $(\mathcal{A}f)(u) \geq 0$ holds.

A function $\Phi :]0, \infty[\to [0, 1]$ is a *viscosity solution* of (4) if Φ is simultaneously a viscosity super- and subsolution.

From the results of [1], formulated for the more general model of the surplus process, we have that the following propositions are true:

Theorem 1. *The survival probability φ of the process (2) as a function of an initial surplus u is a viscosity solution of IDE (4) with $\mathcal{A}$ defined by (3).*

Theorem 2. *Suppose that the topological support of the measure $dF(z)$ is $\mathbb{R}_+ \setminus \{0\}$. Let Φ and $\tilde{\Phi}$ be two continuous bounded viscosity solutions of (4) with the boundary conditions $\Phi(+0) = \tilde{\Phi}(+0)$ and $\Phi(\infty) = \tilde{\Phi}(\infty)$. Then $\Phi \equiv \tilde{\Phi}$.*

4 Main Results

Theorem 3. *Let the topological support of the measure $dF(z)$ be $\mathbb{R}_+ \setminus \{0\}$ and the survival probability $\varphi(u)$ of the process (2) be continuous on $[0,\infty[$ and not identically zero. Suppose there is a continuous viscosity solution Φ of IDE (4) with the boundary conditions (5). Then $\varphi \equiv \Phi$.*

Proof. First, we note that, as is easy to see, $\varphi(0) = 0$ (see also [6, Lemma 1]). In addition, if $\varphi(u)$ is not identically zero, then

$$\lim_{u\to+\infty} \varphi(u) = 1. \tag{6}$$

Indeed, by the Markov property for any $t, u \geq 0$ we have the identity $\varphi(u) = \varphi(X_{\tau^u \wedge t})$. Using the Fatou lemma and the monotonicity of φ we get, for $t \to \infty$, that $\varphi(u) = \overline{\lim}_t \mathbf{E}\varphi(X_{\tau^u \wedge t}) \leq \mathbf{E}\,\overline{\lim}_t\, \varphi(X_{\tau^u \wedge t}) \leq \mathbf{E}\,\varphi(X_{\tau^u}) I_{\{\tau^u < \infty\}} + {}$ $+ \lim_{u\to+\infty} \varphi(u)\, \mathbf{E}\, I_{\{\tau^u = \infty\}}$. In virtue of definitions, the first term in the right-hand side is zero. Then $\varphi(u) \leq \varphi(u) \lim_{u\to+\infty} \varphi(u)$. Since $\varphi(u)$ is monotone, we conclude from this inequality that if it is not identically zero, then equality (6) is true. In view of Theorem 1 the survival probability φ is the viscosity solution of IDE (4). Therefore, from Theorem 2 on the uniqueness of the viscosity solution with fixed boundary conditions, we have $\varphi \equiv \Phi$.

Remark 1. For the case $\sigma = 0$, the equality (6) is also the consequence of the following relation:

$$\varphi(u) \equiv 1, \;\; u \geq c/\mu, \tag{7}$$

(see Lemma 1 and Remark 2 below).

Next, we consider examples of the application of Theorem 3 in the case of exponential distribution of Z_i.

4.1 The Case of Risky Investments ($\sigma > 0$)

In [6] the following proposition is proved.

Theorem 4. *Let $F(z) = 1 - \exp(-z/m)$, all the parameters in (3): c, λ, m, μ, $\sigma > 0$, and $2\mu > \sigma^2$. Then the following assertions hold:*

(I) *there exists a twice continuously differentiable function f satisfying the equation IDE (4) and conditions (5);*

(II) *this solution may be defined by the formula $f(u) = 1 - \int_u^\infty g(s)\, ds$, where $g(u)$ is the unique solution of the following problem for an ordinary differential equation (ODE):*

$$\frac{1}{2}\sigma^2 u^2 g''(u) + \left(\mu u + \sigma^2 u - c - \frac{1}{2m}\sigma^2 u^2\right) g'(u) + \left(\mu - \lambda - \frac{\mu u - c}{m}\right) g(u) = 0, \quad u > 0, \tag{8}$$

$$\lim_{u \to +0} |g(u)| < \infty, \qquad \lim_{u \to +0} [ug'(u)] = 0, \tag{9}$$

$$\lim_{u \to \infty} [ug(u)] = 0, \qquad \lim_{u \to \infty} [u^2 g'(u)] = 0, \tag{10}$$

with the normalizing condition

$$\int_0^\infty g(s)\, ds = 1. \tag{11}$$

Moreover, in [6], asymptotic representations of the solution f at zero and at infinity are obtained and examples of its numerical calculations by solving the ODE problem (8)–(11) are given.

Theorem 5. *Let the conditions of Theorem 4 be satisfied. Then the function f defined in this theorem is the survival probability for the process (2), i.e., $\varphi \equiv f$.*

4.2 The Case of Risk-Free Investments ($\sigma = 0$)

For this case, our approach can also be applied to a non-smooth (generally speaking) solution constructed in [7] (see also [9,11]).

We assume here that the insurer's reserve is invested to a risk-free asset with price B_t modelled by the equation

$$dB_t = rB_t dt, \quad t \geq 0,$$

where r is the return rate.

Then the resulting surplus process X_t is governed by the equation

$$dX_t = rX_t dt + dR_t, \quad t \geq 0, \tag{12}$$

with the initial condition $X_0 = u$, where R_t is defined by (1).

Recall that, in the case $\sigma = 0$, the infinitesimal generator (3) of the corresponding process X_t takes the form

$$(\mathcal{A}f)(u) = f'(u)(ru - c) - \lambda f(u) + \lambda \int_0^\infty f(u+z)\, dF(z) \tag{13}$$

(here we rename the return rate of the risk-free asset from μ to r). From the results of [7] we have the following

Proposition 1. *Let $F(z) = 1 - \exp(-z/m)$, all the parameters in (13): c, λ, m, $r > 0$. Then the following assertions hold:*

(I) *there exists a continuous function Φ, which is twice continuously differentiable on the interval $(0, c/r)$, satisfying the equation IDE (4) (everywhere, except, perhaps, the point c/r) and the conditions*

$$\lim_{u \to +0} \Phi(0) = 0, \quad \Phi(u) \equiv 1, \;\; u \geq c/r; \tag{14}$$

(II) *on the interval* $(0, c/r)$, *this solution may be defined by the formula* $\Phi(u) = 1 - \int_u^{c/r} g(s)\, ds$, *where*

$$g(u) = \left[\int_0^{c/r} (c/r - u)^{\lambda/r - 1} \exp{(u/m)}\, du \right]^{-1} (c/r - u)^{\lambda/r-1} \exp{(u/m)}; \quad (15)$$

(III) $\Phi(u)$ *is a viscosity solution of IDE (4);*

(IV) *for* $\lambda > 2r$, $\Phi(u)$ *is a twice continuously differentiable on* $(0, \infty)$ *function, i.e., it is a classical solution of IDE (4); in this case* $\lim_{u \uparrow c/r} \Phi''(u) = \lim_{u \uparrow c/r} \Phi'(u) = 0$; *otherwise,* $\Phi(u)$ *satisfies IDE (4) in the classical sense everywhere except for the point* $u = c/r$;

(V) *for* $r < \lambda \leq 2r$, $\Phi(u)$ *is smooth but it is not twice continuously differentiable on* $(0, \infty)$, *since* $\lim_{u \uparrow c/r} \Phi''(u) = -\infty$ *for* $\lambda < 2r$, *and*

$$\lim_{u \uparrow c/r} \Phi''(u) = -m^{-2} \left[\exp\big(c/(rm)\big) - 1 - c/(rm)\right]^{-1} \exp\big(c/(rm)\big) < 0,$$

$\lambda = 2r$;

(VI) *for* $\lambda \leq r$, $\Phi(u)$ *is not smooth, since its derivative is discontinuous at the point* $u = c/r$:

$$\lim_{u \uparrow c/r} \Phi'(u) = m^{-1} \left[\exp\big(c/(rm)\big) - 1\right]^{-1} \exp\big(c/(rm)\big) > 0,$$

$\lambda = r$, *and* $\lim_{u \uparrow c/r} \Phi'(u) = \infty$, *wherein* $\Phi'(u)$ *is integrable at the point* $u = c/r$, $\lambda < r$.

Theorem 6. *Let the conditions of Proposition 1 be satisfied. Then the function* Φ *defined in this proposition is the survival probability for the process (12), i.e.,* $\varphi \equiv \Phi$.

5 Proofs

Let us return to the general case of a process of the form (2) and prove auxiliary statements about non-triviality and continuity of its survival probability (Lemma 1 and Lemma 3 below respectively). We also formulate Lemma 2 about zero value of the survival probability at zero surplus level. Then the statement of Theorem 5 is a consequence of Theorems 3, 4 and Lemmas 1–3. The statement of Theorem 6 is a consequence of Theorem 3, Proposition 1 and the same lemmas.

Lemma 1. *Let*

$$2\mu > \sigma^2. \quad (16)$$

Then the survival probability $\varphi(u)$ *of process (2) is not identically zero. Moreover, if* $\sigma = 0$, *then* $\varphi(u) = 1$, $u \geq c/\mu$.

Proof. 1) The case $\sigma > 0$. Let

$$\mu(x) = \mu x - c, \quad \sigma(x) = \sigma x. \tag{17}$$

Let us consider the process $Y_t = Y_t^u$ given by the equation

$$dY_t = \mu(Y_t)dt + \sigma(Y_t)dw_t, \tag{18}$$

with initial state $Y_0 = u > 0$ and the same standard Brownian motion as in (2). To understand the qualitative behavior of the process (2), we use the corresponding result for the process (18) at first; this result is given in [12, Chapter 4]. Below we use the following functions:

$$\rho(x) = \exp\left(- \int_a^x \frac{2\mu(y)}{\sigma^2(y)} dy \right), \quad x \in [a, \infty), \tag{19}$$

$$s(x) = - \int_x^\infty \rho(y)dy, \quad x \in [a, \infty). \tag{20}$$

It is easy to check that in the case when the functions $\mu(x)$, $\sigma(x)$ are of the form (17) and the relation (16) is valid, we have

$$\int_a^\infty \rho(x)dx < \infty, \quad \int_a^\infty \frac{|s(x)|}{\rho(x)\sigma^2(x)} dx = \infty.$$

Note that the (strong) solution of equation (18) with coefficients defined in (17) and the initial state $Y_0 = u$ can be represented as

$$Y_t^u = \exp(H_t) \left[u - c \int_0^t \exp(-H_s)\, ds \right], \quad t \geq 0, \tag{21}$$

where

$$H_t = \left(\mu - \sigma^2/2\right) t + \sigma w_t.$$

Let us denote $T_a^u := \inf\{t\colon\ Y_t^u \leq a\}$; for the process Y_t^u, the r.v. T_a^u is the moment of its first hitting the level a. Then, for $a < u$, according to [12, Th. 4.2], we conclude that

$$\mathbf{P}\{T_a^u = \infty\} > 0 \tag{22}$$

and $\lim_{t\to\infty} Y_t = \infty$ $\mathbf{P} - a.s.$ on $\{T_a^u = \infty\}$. For the solution of (2) with the same initial state $X_0 = u$ we can write

$$X_t^u = Y_t^u + \exp(H_t) \left[\sum_{i=1}^{N(t)} Z_i \exp\left(-H_{\theta_i}\right) \right], \quad t \geq 0, \tag{23}$$

where θ_i is the moment of the i-th jump of the process $N(t)$. It is clear that

$$X_t^u \geq Y_t^u \quad \mathbf{P} - a.s., \ t \geq 0. \tag{24}$$

Hence, taking into account the relation (22) for $a < u$, we have for the time τ^u of ruin of the process X_t^u that

$$\mathbf{P}\{\tau^u = \infty\} > 0,\ u > 0,$$

i.e., $\varphi(u) > 0, \quad u > 0$.

2) The case $\sigma = 0$. It is clear in this case that for $u \geq c/\mu$ the ruin of the process Y_t^u will never occur and relations (23), (24) are true. Hence, for the process X_t^u we have at least that $\varphi(u) = 1,\ u \geq c/\mu$.

Remark 2. For the survival probability $\varphi(u)$ of process (12) we have clearly from Lemma 1 that $\varphi(u) = 1,\ u \geq c/r$.

Lemma 2. *For $\sigma^2 \geq 0$, the survival probability $\varphi(u)$ of process (2) satisfies the condition*

$$\varphi(0) = 0. \tag{25}$$

For the simple proof of this lemma, see ([6]).

Lemma 3. *Let c, λ, m be positive numbers. Then for arbitrary μ, σ the survival probability $\varphi(u)$ of process (2) is continuous on $[0, \infty[$.*

Proof. Let us prove this statement in the case $\sigma^2 > 0$; otherwise the proof is simpler. Let us show first the continuity of $\varphi(u)$ at zero. In other words, we prove the limit equality

$$\lim_{u \to +0} \varphi(u) = 0. \tag{26}$$

Note that, for $u > 0$ and any fixed $t > 0$,

$$\begin{aligned}
\varphi(u) \leq \mathbf{P}(X_t^u > 0) &= \mathbf{P}\left(u - c\int_0^t \exp(-H_s)\,ds + \sum_{i=1}^{N(t)} Z_i \exp\left(-H_{\theta_i}\right) > 0\right) \\
&\leq \mathbf{P}\left(\sum_{i=1}^{N(t)} Z_i \exp\left(-H_{\theta_i}\right) > 0\right) + \mathbf{P}\left(\int_0^t \exp(-H_s)\,ds < u/c\right) \\
&\leq \mathbf{P}(N(t) \geq 1) + \mathbf{P}\left(t \inf_{s \leq t} \exp(-H_s) < u/c\right).
\end{aligned}$$

Denote $M_s = \exp[(\mu - \sigma^2)s - H_s]$. Clear that

$$M_s = \exp(-\frac{\sigma^2}{2}s - \sigma w_s) \tag{27}$$

is a non-negative martingale with $M_0 = 1$. Hence,

$$\varphi(u) \leq \mathbf{P}(N(t) \geq 1) + \mathbf{P}\left(\inf_{s \leq t} M_s < \frac{u}{ctb(t)}\right), \tag{28}$$

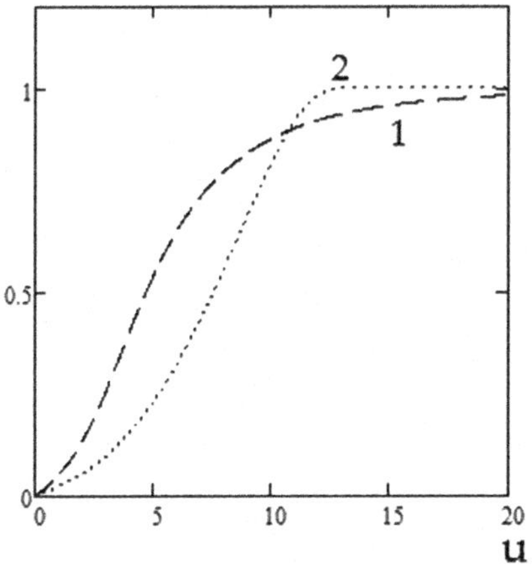

Fig. 1. The case $\lambda > 2r$: r=0.3; $\mu = 0.7$.

where $b(t) = \exp[(\mu - \sigma^2)t]\, I\{\mu > \sigma^2\} + I\{\mu \le \sigma^2\}$ and I is the indicator function of the set. The following inequality is proved in the course of the proof of Lemma 4.2 in [13]. For non-negative supermartingale $M_t, M_0 = 1$, we have

$$\mathbf{P}\left(\inf_{s \le t} M_s < \varepsilon\right) \le 2\mathbf{P}\left(M_t < 2\varepsilon\right), \quad \varepsilon > 0. \tag{29}$$

Setting $\varepsilon = \frac{u}{ctb(t)}$ and applying inequality (29) to the martingale of the form (27), we obtain from (28) that, for any fixed t,

$$\lim_{u \to +0} \varphi(u) \le 1 - \exp\left(-\lambda t\right). \tag{30}$$

Letting $t \to 0$ in (30) and taking into account the non-negativity of φ, we have equality (26).

Let us prove the continuity at any point $u > 0$. Note that the difference between the two processes $X_t(u+\varepsilon) = X_t^{u+\varepsilon}$ and $X_t(u) = X_t^u$ starting at points $u + \varepsilon$ and u respectively, has the form

$$X_t(u + \varepsilon) - X_t(u) = \varepsilon \exp(H_t). \tag{31}$$

For the stopping time $\tau^u \wedge t$, where τ^u is the time of ruin of the process X_t^u, due to the strong Markov property of the process $X_t^{u+\varepsilon}$, we have

$$\begin{aligned}
\varphi(u+\varepsilon) &= \mathbb{E}\varphi(X_{\tau^u \wedge t}(u+\varepsilon)) \\
&= \mathbb{E}[\varphi(X_{\tau^u \wedge t}(u+\varepsilon))I\{\tau^u < \infty\}] + \mathbb{E}[\varphi(X_{\tau^u \wedge t}(u+\varepsilon))I\{\tau^u = \infty\}] \\
&= \mathbb{E}[\varphi(X_t(u+\varepsilon))I\{\tau^u = \infty\}] + \mathbb{E}[\varphi(X_{\tau^u \wedge t}(u+\varepsilon))I\{\tau^u \le t\}] \\
&+\mathbb{E}[\varphi(X_{\tau^u \wedge t}(u+\varepsilon))I\{t < \tau^u < \infty\}] \\
&= \mathbb{E}[\varphi(X_t(u+\varepsilon))I\{\tau^u = \infty\}] + \mathbb{E}[\varphi(X_{\tau^u}(u+\varepsilon))I\{\tau^u \le t\}] \\
&+\mathbb{E}[\varphi(X_t(u+\varepsilon))I\{t < \tau^u < \infty\}].
\end{aligned}$$

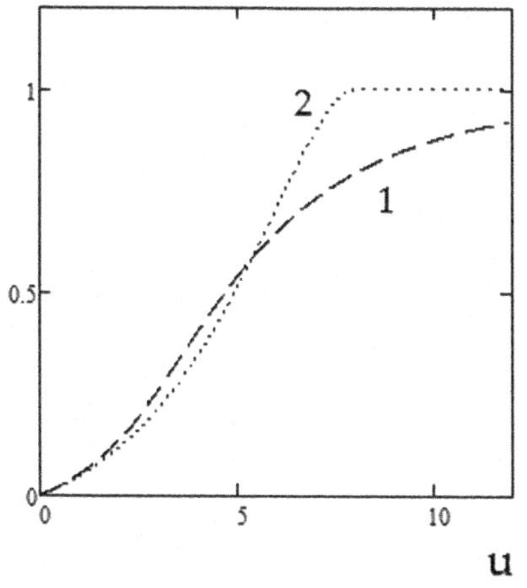

Fig. 2. The case $\lambda = 2r$: r=0.5; $\mu = 0.7$.

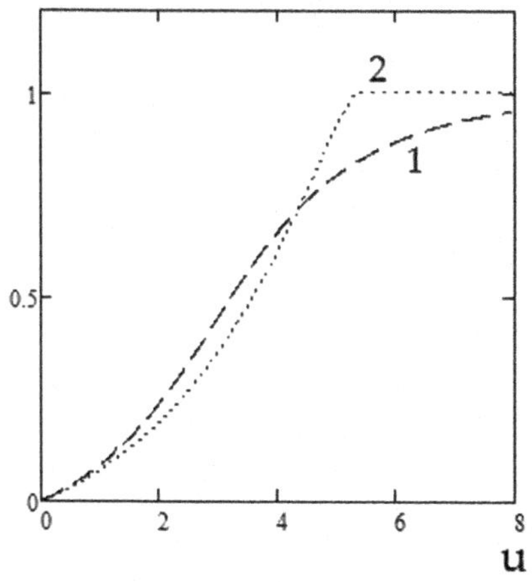

Fig. 3. The case $r < \lambda < 2r$: r=0.75; $\mu = 1$.

For three terms at the end of the last chain of equalities, we have $\mathbb{E}[\varphi(X_t(u+\ \varepsilon))I\{\tau^u = \infty\}] \le \mathbb{P}\{\tau^u = \infty\} = \varphi(u)$,

$$\mathbb{E}[\varphi(X_{\tau^u}(u+\varepsilon))I\{\tau^u \le t\}] = \mathbb{E}[\varphi(\varepsilon \exp(H_{\tau^u})I\{\tau^u \le t\}], \tag{32}$$

$\mathbb{E}[\varphi(X_t(u+\varepsilon))I\{t < \tau^u < \infty\}] \le \mathbb{P}\{t < \tau^u < \infty\}$ (equality (32) is true due to relation (31) and the fact that $X_{\tau^u} = 0$ for the process with positive jumps). Then

$$\varphi(u+\varepsilon) \le \varphi(u) + \mathbb{E}[\varphi(\varepsilon \exp(H_{\tau^u})I\{\tau^u \le t\}] + \mathbb{P}\{t < \tau^u < \infty\}. \tag{33}$$

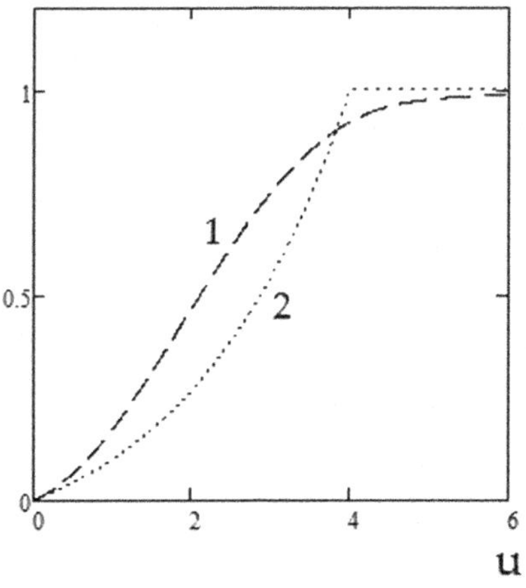

Fig. 4. The case $\lambda = r$: r=1; $\mu = 1.5$.

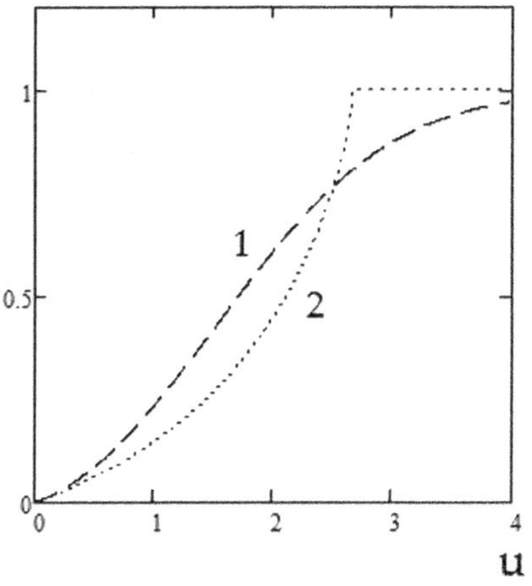

Fig. 5. The case $\lambda < r$: r=1.5; $\mu = 1.75$.

Note that the first term in (33) tends to zero as $\varepsilon \to 0$ due to the proved continuity at zero of $\varphi(u)$, condition (25) and taking into account the dominated convergence theorem. Then, for any t,

$$\lim_{\varepsilon \to +0} (\varphi(u+\varepsilon) - \varphi(u)) \le \mathbb{P}\{t < \tau^u < \infty\}.$$

Letting $t \to \infty$ in the last inequality and taking into account that the survival probability φ is the non-decreasing on the initial state u, we obtain the right-continuity of this function. The left-continuity may be proved analogously.

6 Numerical Results [7]

For the results of numerical calculations (Figs. 1, 2, 3, 4 and 5), the curves with number 1 (2) correspond to the case of risky investments in shares with parameters μ and σ^2 (risk-free ones with return rate r respectively). The figures are presented in order of decreasing smoothness and increasing discontinuity of derivatives for the curves number 2. For all figures, $c = 4$, $m = 2$, $\lambda = 1$, $\sigma^2 = 0.3$ (the parameter values are relative, they are normalized in such a way that $\lambda = 1$).

7 Conclusions

s A new approach to justifying the survival probabilities in dynamic insurance models with investments as the solutions of corresponding IDE problems is proposed. This approach avoids direct proof of the smoothness of the survival probability by using verification arguments based on the uniqueness of the viscosity solution. It can be applied if it has been previously proved, that the survival probability is continuous, not identically equal to zero function, has a known value at zero initial surplus and is a viscosity solution of some IDE problem. The first two facts can be established quite simply, and the last fact can be proved for a whole class of models, as it is done in [1]. In this case, for specific models from this class, it remains only to prove the existence of a solution (classical or in the sense of viscosity) for the corresponding IDE problem. On the other hand, it remains unclear whether this approach can be applied to models in which the corresponding problem for the IDE is not a boundary problem (see, e.g., [8])

Acknowlegements. The author is grateful to Alexander Gushchin for useful discussion on the subject matter.

References

1. Belkina, T.A., Kabanov, Y.M.: Viscosity solutions of integro-differential equations for nonruin probabilities. Theory Probab. Appl. **60**(4), 671–679 (2016)
2. Paulsen, J., Gjessing, H.K.: Ruin theory with stochastic return on investments. Adv. Appl. Probab. **29**(4), 965–985 (1997)
3. Belkina, T.: Risky investment for insurers and sufficiency theorems for the survival probability. Markov Process. Relat. Fields. **20**(3), 505–525 (2014)
4. Asmussen, S., Albrecher, H.: Ruin Probabilities. Advanced Series on Statistical Science and Applied Probability, vol. 14, World Scientific, Singapore (2010)
5. Grandell, J.: Aspects of Risk Theory. Springer, Berlin-New York (1991). https://doi.org/10.1007/978-1-4613-9058-9
6. Belkina, T.A., Konyukhova, N.B., Slavko, B.V.: Solvency of an insurance company in a dual risk model with investment: analysis and numerical study of singular boundary value problems. Comput. Math. Math. Phys. **59**(11), 1904–1927 (2019)

7. Belkina, T.A., Konyukhova, N.B., Slavko, B.V.: Risk-free investments and their comparison with simple risky strategies in pension insurance models: solution of singular problems for integro-differential equations. Comput. Math. Math. Phys. **60**(10), 1621–1641 (2020)
8. Belkina, T.A., Konyukhova, N.B., Kurochkin, S.V.: Dynamical insurance models with investment: constrained singular problems for integrodifferential equations. Comput. Math. Math. Phys. **56**(1), 43–92 (2016)
9. Belkina, T.A., Konyukhova, N.B., Slavko, B.V.: Analytic-numerical investigations of singular problems for survival probability in the dual risk model with simple investment strategies. In: Rykov, V.V., Singpurwalla, N.D., Zubkov, A.M. (eds.) ACMPT 2017. LNCS, vol. 10684, pp. 236–250. Springer, Cham (2017). https://doi.org/10.1007/978-3-319-71504-9_21
10. Kabanov, Yu., Pergamenshchikov, S.: In the insurance business risky investments are dangerous: the case of negative risk sums. Financ. Stochast. **20**(2), 355–379 (2016)
11. Belkina, T.A., Konyukhova, N.B.: Survival probability in the life annuity insurance model as a viscosity solution to an integro-differential equation (in Russian). Vestnik CEMI. 1 (2018). https://doi.org/10.33276/S0000097-9-1
12. Cherny, A.S., Engelbert, H.-J.: Singular Stochastic Differential Equations. Springer, Berlin, Heidelberg (2005). https://doi.org/10.1007/b104187
13. Gushchin, A.A., Valkeila, E.: Approximations and limit theorems for likelihood ratio processes in the binary case. Stat. Decisions **21**, 219–260 (2003)

On the Information Content of Some Stochastic Algorithms

Manuel L. Esquível[1(✉)], Nélio Machado[2], Nadezhda P. Krasii[3], and Pedro P. Mota[1]

[1] FCT Nova and CMA, UNL, 2829-516 Caparica, Portugal
{mle,pjpm}@fct.unl.pt
[2] FCT Nova, UNL, 2829-516 Caparica, Portugal
[3] Don State Technical University, Gagarin Square 1, Rostov-on-Don 344000, Russian Federation

Abstract. We formulate an optimization stochastic algorithm convergence theorem, of Solis and Wets type, and we show several instances of its application to concrete algorithms. In this convergence theorem the algorithm is a sequence of random variables and, in order to describe the increasing flow of information associated to this sequence we define a filtration – or flow of σ-algebras – on the probability space, depending on the sequence of random variables and on the function being optimized. We compare the flow of information of two convergent algorithms by comparing the associated filtrations by means of the Cotter distance of σ-algebras. The main result is that two convergent optimization algorithms have the same information content if both their limit minimization functions generate the full σ-algebra of the probability space.

Keywords: Stochastic algorithms · Global optimization · Convergence of information σ-fields

1 Introduction

There are three main roots we can consider to the present work. The first is a quite general formulation of a stochastic optimization algorithm given in [SW81], studied under a different perspective in [SP99] and [PS00], and then corrected and slightly generalized in [Esq06] and having further developments and extensions in [dC12]. The subject of stochastic optimization – in the perspective adopted in this work – become stabilized with the books [Spa03, Zab03] and the work [Spa04]. The interest on the development of stochastic optimization methods continued, for instance in the work [RS03]. We refer a very effective and general approach to a substantial variety of stochastic optimization problems that takes the denomination of *the cross entropy method* proposed in a unified way in [RK04], further explained in [dBKMR05], with the convergence proved in [CJK07] and further extended in [RK08].

A. N. Shiryaev et al. (Eds.): ICSM-5 2020, PROMS 371, pp. 57–75, 2021.
https://doi.org/10.1007/978-3-030-83266-7_5

The second root originated in [SB98], is detailed in Sect. 4 for the reader's convenience, and may be broadly described as a form of conditioning of the results of any algorithm for global optimization; conditioning in the sense that the algorithm must gather enough information in order to get significant results.

This leads to the third root, namely the formalization of the concept *information*, conveyed by a random variable, as the σ-algebra generated by this random variable. This formalization encompasses many extensions and uses (see [Vid18]) for a recent and thorough account). We may initially refer with introduction of a convergence definition for σ-algebras – the so called *strong convergence* – related to the conditional expectation by Neveu in [Nev65] (or the French version in [Nev64]), with in [Kud74] a very deep study and, with developments, in [Pic98] and [Art01]. Then [Boy71] introducing a different convergence – the *Hausdorff convergence* – with an important observation in [Nev72] and further analysis in [Rog74] and [VZ93]. In the study of convergence of σ-algebras (or fields) there were many noticeable advances – and useful in our perspective – with Cotter in [Cot86] and [Cot87] extended in [ALR03] and detailed in [Bar04] and further extensions in [Kom08].

In the perspective of further developments, we mention [Wan01] and [Yin99], two works that concern the determination of the rates of convergence of stochastic algorithms allowing for the determination of adequate and most effective stopping rules for the algorithm and also [dC11] – and references therein – for a method to obtain confidence intervals for stochastic optimums.

2 Some Random Search Algorithms

We will now develop the following general idea: a convergent stochastic search algorithm for global optimization of a real valued function f defined on a domain $\mathcal{D}$ may be seen simply as a sequence of random variables $\mathbb{Y} = (Y_n)_{n\geq 1}$ such that the sequence $(f(Y_n))_{n\geq 1}$ converges (almost surely or in probability) to a random variable which gives a good estimate of $\min_{x\in\mathcal{D}} f(x)$. This sequence of random variables gives information about f on $\mathcal{D}$. A natural question is how to compare quantitatively the information brought by two different algorithms.

We now describe three algorithms which we will discuss in the following. Important issues for discussion are the convergence of the algorithm and, in case of convergence, the rate of convergence of the algorithm. Let $(\Omega, \mathcal{F}, \mathbb{P})$ be a complete probability space.

2.1 The Pure Random Search Algorithm

For the general problem of minimizing $f : \mathcal{D} \subseteq \mathbb{R}^n \mapsto \mathbb{R}$ over $\mathcal{D}$, a bounded Borel set of $\mathbb{R}^n$, we consider the following natural algorithm.

S.1 Select a point x_1 at random in $\mathcal{D}$. Do $y_1 := x_1$.
S.2 Choose a point x_2 at random in $\mathcal{D}$. Do:

$$y_2 := y_1 \mathbb{1}\{f(y_1) < f(x_2)\} + x_2 \mathbb{1}\{f(y_1) \geq f(x_2)\}.$$

S.3 Repeat S.2.

To this algorithm it corresponds a probabilistic translation given in the following.

Sp.1 Let $X_1, X_2, \ldots, X_n, \ldots$ be independent random variables with common distribution over $\mathcal{D}$ verifying furthermore, with $\mathcal{B}(\mathcal{D})$ the Borel σ-algebra of $\mathcal{D}$:

$$\forall B \in \mathcal{B}(\mathcal{D}) \quad \lambda(B) > 0 \Rightarrow \mathbb{P}[X_1 \in B] > 0. \tag{1}$$

Sp.2 $Y_1 := X_1$
Sp.3 $Y_{n+1} = Y_n \mathbb{I}\{f(Y_n) < f(X_{n+1})\} + X_{n+1}\mathbb{I}\{f(Y_n) \geq f(X_{n+1})\}$

Having no prior information on the minimum set location and for random variables having common distribution on a bounded Borel set, a natural choice for the distribution of the random variables X_j is the uniform distribution. A non uniform distribution will distribute more mass on some particular sub domain. This may entail a loss of efficiency if the minimizer set is not contained in the more charged domain.

Remark 1 (The laws of the random variables of pure random search algorithm). We observe that for $n > 1$ we have:

$$Y_n = \sum_{k=1}^{n} X_k \mathbb{I}_{\bigcap_{j<k}\{f(X_k) \leq f(X_j)\} \cap \bigcap_{j>k}\{f(X_k) < f(X_j)\}},$$

an alternative expression that will allow us to describe the law of Y_n. Let D in the Borel σ-algebra of $\mathcal{D}$ and suppose that the random variables $(X_n)_{n\geq 1}$ are uniformly distributed in $\mathcal{D}$. We have the following disjoint union:

$$\begin{aligned}&\{Y_n \in D\}\\ &= \bigcup_{k=1}^{n} \left(\{X_k \in D\} \cap \bigcap_{1\leq j<k} \{f(X_k) \leq f(X_j)\} \cap \bigcap_{k<j\leq n} \{f(X_k) < f(X_j)\} \right),\end{aligned}$$

which entails, representing by λ the Lebesgue measure over $\mathcal{D}$, that (see the Appendix, page 17, for the complete deduction):

$$\begin{aligned}&\mathbb{P}[Y_n \in D]\\ &= \sum_{k=1}^{n} \left(\frac{1}{\lambda(\mathcal{D})^n} \int_D \lambda(f^{-1}([f(x_k), +\infty[))^{k-1} \lambda(f^{-1}(]f(x_k), +\infty[))^{n-k} d\lambda(x_k) \right),\end{aligned} \tag{2}$$

by Fubini theorem and by the fact that $(X_n)_{n\geq 1}$ is a sequence of independent uniformly distributed random variables on $\mathcal{D}$. Suppose furthermore that for every $x \in \mathcal{D}$ we have $\lambda(f^{-1}(\{f(x)\})) = 0$, we then have:

$$\mathbb{P}[Y_n \in D] = \frac{n}{\lambda(\mathcal{D})^n} \int_D \lambda(f^{-1}([f(x), +\infty[))^{n-1} d\lambda(x),$$

which gives us the density of Y_n with respect to the Lebesgue measure.

2.2 The Random Search Algorithm on (Nearly) Unbounded Domains

In the context of simple random search we may ask what is the natural substitute of the uniform distribution on an unbounded domain? A variant of the algorithm we now present was introduced in [Esq06] having in mind performing global optimization in unbounded domains. For bounded but large domains one may consider an algorithm using, for instance, a Gaussian distribution.

S.1 Select a point x at random in $\mathcal{D} \subset \mathbb{R}$. Do $z := x$.
S.2 Choose a point x at random in $\mathcal{D}$. Choose a point y with distribution $\mathcal{N}(x, \sigma)$ where for instance $\sigma := \text{diam}(\mathcal{D})/10$. Do:

$$z := z\mathbb{1}\{f(z) < f(y)\} + y\mathbb{1}\{f(z) \geq f(y)\}.$$

S.3 Repeat S.2.

The probabilistic recursive translation of this algorithm is the following.

pS.1 Let $X_1, X_2, \ldots, X_n, \ldots$ independent random variables with common uniform distribution over $\mathcal{D}$
pS.2 $Z_1 := X_1$
pS.3 Let $Y_1, Y_2, \ldots, Y_n, \ldots$ be a sequence of independent random variables such that $Y_n \frown \mathcal{N}(X_n, \sigma)$.
pS.4 $Z_{n+1} := Z_n\mathbb{1}\{f(Z_n) < f(Y_{n+1})\} + Y_{n+1}\mathbb{1}\{f(Z_n) \geq f(Y_{n+1})\}$.

2.3 The Zig-Zag Algorithm

The zig-zag algorithm was introduced in [MPB99] (see also for other references and a convergence proof [PM10]) in order to optimize a quadratic function in two sets of multidimensional variables. The main idea of this algorithm may be simply described. In the first step we optimize in one of the sets of variables leaving the variables of the other set unchanged. On the second step, the first set of variables remains unchanged in the optimum value determined in the first step and an optimization is performed in the second set of variables. On the third step, it is now the second set of variables that remains unchanged in the optimum determined in the second step while an optimization is executed in the first set of variables. For the general case, the convergence and – if applicable – the rate of convergence issues were open problems, as far as we know.

One of the possibilities opened by this algorithm is to perform the optimization in sets of strictly smaller linear dimension than the dimension of $\mathcal{D}$. Suppose that $\mathcal{D} \subseteq \mathbb{R}^2$ is bounded.

S.1 Select a point x at random in $\mathcal{D}$. Do $z := x$.
S.2 (Optimization along a lower dimensional subset of the domain)
  S.2.1 Choose a point y at random in $\mathcal{D}$.

S.2.2 Choose, at random, points $\lambda_1, \ldots, \lambda_N \in \mathbb{R}$ such that $\lambda_j z + (1-\lambda_j)y \in \mathcal{D}$ and define x to be such that $f(x) = \min_{1 \leq j \leq N} f(\lambda_j z + (1-\lambda_j)y)$. Do:

$$z := z\mathbb{I}\{f(z) < f(x)\} + x\mathbb{I}\{f(z) \geq f(x)\}.$$

S.3 Repeat S.2

For this algorithm, one probabilistic recursive translation may be the following.

pS.1 Let $Y_1, Y_2, \ldots, Y_n, \ldots$ be a sequence of independent random variables with common uniform distribution over $\mathcal{D}$.
pS.2 $Z_1 := Y_1$
pS.3 For each $n \geq 1$, let $\lambda_1^n, \ldots, \lambda_N^n$ be independent sequences of independent random variables with uniform distribution in $[a, b]$ an interval such that:

$$\forall \lambda \in [a, b] \;\; \forall x, y \in \mathcal{D} \;\; \lambda x + (1-\lambda)y \in \mathcal{D} \,.$$

which is possible as $\mathcal{D}$ is bounded.
pS.4 Define the random variable $X_n^{j_0}$ such that:

$$f(X_n^{j_0}) = \min_{1 \leq j \leq N} f(\lambda_j^n Z_n + (1-\lambda_j^n)Y_{n+1})$$

pS.5 $Z_{n+1} := Z_n \mathbb{I}\{f(Z_n) < f(X_n^{j_0})\} + X_n^{j_0} \mathbb{I}\{f(Z_n) \geq f(X_n^{j_0})\}$.

The main idea of the zig-zag algorithm may, of course, be exploited in several other ways.

3 The Solis and Wets Approach of Random Algorithms

We present this approach following the presentation in [Esq06] – which follows the context formalism of [SW81] – and observe that this approach may be applied to the algorithms described above.

3.1 The Convergence Results

We introduce some definitions which are necessary for the presentation of the convergence results. Let $f : \mathcal{D} \subset \mathbb{R}^n \longrightarrow \mathbb{R}$ be a measurable function defined on a domain $\mathcal{D}$ that can be unbounded. Let $(\Omega, \mathcal{F}, \mathbb{P})$ be a complete probability space. In order to deal with discontinuous functions, such as $\mathbb{I}_{[0,1]\setminus\{1/2\}}$, or unbounded functions, such as $\ln(|x| \, \mathbb{I}_{\mathbb{R}\setminus\{0\}} + (\infty)\mathbb{I}_{\{0\}}$, we need the following notion.

Definition 1 (Essential infimum of f in $\mathcal{D}$).

$$\alpha := \inf\{t \in \mathbb{R} : \lambda(\{x \in \mathcal{D} : f(x) < t\}) > 0\} \tag{3}$$

with λ being the Lebesgue measure on $\mathbb{R}^n$.

The formulation of general hypothesis on the function f in order to obtain the algorithm convergence requires the definition of the sets for $\epsilon > 0$ and $M < 0$.

Definition 2 (Level set of f of height ϵ over α).

$$E_{\alpha+\epsilon,M} := \begin{cases} \{x \in \mathcal{D} : f(x) < \alpha + \epsilon\} & \text{if } \alpha \in \mathbb{R} \\ \{x \in \mathcal{D} : f(x) < M\} & \text{if } \alpha = -\infty \end{cases} \tag{4}$$

The general form of the algorithm may be decomposed into the nuclear part which is a function verifying some condition and the procedure.

Definition 3 (The algorithm).

- *A function* $\psi : \mathcal{D} \times \mathbb{R}^n \mapsto \mathcal{D} \subset \mathbb{R}^n$ *such that the following hypothesis* $[H1]$ *is verified.*

$$[H1] : \begin{cases} \forall t, x \;\; f(\psi(t,x)) \leq f(t) \\ \forall x \in \mathcal{D} \;\; f(\psi(t,x)) \leq f(x) \end{cases} \tag{5}$$

- *A sequence of random variables given by:*

$$\begin{cases} Y_1 = X_1 \\ Y_{n+1} = \psi(Y_n, X_n) \quad \text{for } n \geq 1 \end{cases} \tag{6}$$

where $X_n \frown \mathbb{P}_n$ *satisfying hypothesis in Formula* (1)*, and* $\mathbb{P}_n$ *being a probability measure – the law of* X_n *– that may depend on* $\mathbb{P}_1, \ldots, \mathbb{P}_{n-1}$ *in case of adaptive random search.*

Remark 2 (Examples of stochastic algorithms for global optimization). The pure random search algorithm in Sect. 2.1, the random search on nearly unbounded domains in Sect. 2.2 and the zig-zag algorithm in Sect. 2.3 may be considered as instances of Solis and Wets approach. As presented, the following function obviously describes the algorithms and verifies the hypothesis $H1$ in Formula (5).

$$\psi(t,x) = t\mathbb{I}_{\{f(t)<f(x)\}}(t,x) + x\mathbb{I}_{\{f(t)\geq f(x)\}}(t,x)$$

The following result ensures the convergence of the algorithm under very general hypothesis.

Theorem 1 (A Solis and Wets' type theorem for random search algorithm convergence). *Suppose that f is measurable and bounded from below. Let α be the essential infimum of f in $\mathcal{D}$.*

$H2(\epsilon)$ For pure random search this hypothesis is defined for every $\epsilon > 0$ as:

$$\lim_{k\to+\infty} \prod_{1\leq j\leq k} \mathbb{P}[X_j \in E^c_{\alpha+\epsilon,M}] = \lim_{k\to+\infty} \prod_{1\leq j\leq k} \mathbb{P}_j[E^c_{\alpha+\epsilon,M}] = 0\,.$$

$H'2(\epsilon)$ For adaptive search this hypothesis is defined for every $\epsilon > 0$ as:

$$\lim_{k\to+\infty} \inf_{1\leq j\leq k} \mathbb{P}[X_j \in E^c_{\alpha+\epsilon,M}] = \lim_{k\to+\infty} \inf_{1\leq j\leq k} \mathbb{P}_j[E^c_{\alpha+\epsilon,M}] = 0 \,. \tag{7}$$

If for some $\epsilon > 0$ hypothesis $H2(\epsilon)$ (in case of pure random search) or $H'2(\epsilon)$ (in case of adaptive search) are verified, then:

$$\lim_{n\to+\infty} \mathbb{P}[Y_n \in E_{\alpha+\epsilon,M}] = 1 \,. \tag{8}$$

If for every $\epsilon > 0$ hypothesis $H2(\epsilon)$ (in case of pure random search) or $H'2(\epsilon)$ (in case of adaptive search) are verified, then the sequence $(f(Y_n))_{n\geq 1}$ converges almost surely to a random variable $Min_{f,\mathbb{Y}}$ such that $\mathbb{P}[Min_{f,\mathbb{Y}} \leq \alpha] = 1$.

Proof. A first fundamental observation is that if $Y_n \in E_{\alpha+\epsilon,M}$ or $X_n \in E_{\alpha+\epsilon,M}$, then by hypothesis $H1$ we have that $Y_{n+1} \in E_{\alpha+\epsilon,M}$ and so as $(f(Y_n))_{n\geq 1}$ is decreasing, $Y_{n+k} \in E_{\alpha+\epsilon,M}$ for every $k \geq 1$. As a consequence, for $k > 1$:

$$\{Y_k \in E^c_{\alpha+\epsilon,M}\} \subseteq \{Y_1,\ldots,Y_{k-1} \in E^c_{\alpha+\epsilon,M}\} \cap \{X_1,\ldots,X_{k-1} \in E^c_{\alpha+\epsilon,M}\}.$$

as if it was otherwise we would contradict our first observation. Now, it is clear that:

$$\begin{aligned}\mathbb{P}[Y_k \in E^c_{\alpha+\epsilon,M}] &\leq \mathbb{P}\left[\bigcap_{1\leq j\leq k-1} \{Y_j \in E^c_{\alpha+\epsilon,M}\} \cap \{X_j \in E^c_{\alpha+\epsilon,M}\}\right] \\ &\leq \mathbb{P}\left[\bigcap_{1\leq j\leq k-1} \{X_j \in E^c_{\alpha+\epsilon,M}\}\right] .\end{aligned} \tag{9}$$

On the pure random search scenario we have that $(X_n)_{n\geq 1}$ is a sequence of iid random variables and so:

$$\mathbb{P}\left[\bigcap_{1\leq j\leq k-1} \{X_j \in E^c_{\alpha+\epsilon,M}\}\right] = \prod_{1\leq j\leq k-1} \mathbb{P}\left[X_j \in E^c_{\alpha+\epsilon,M}\right] = \mathbb{P}\left[X_1 \in E^c_{\alpha+\epsilon,M}\right]^{k-1}, \tag{10}$$

implying that

$$1 \geq \mathbb{P}[Y_k \in E_{\alpha+\epsilon,M}] = 1 - \mathbb{P}[Y_k \in E^c_{\alpha+\epsilon,M}] \geq 1 - \mathbb{P}\left[X_1 \in E^c_{\alpha+\epsilon,M}\right]^{k-1} .$$

Now by hypothesis in Formula (1) we have that $\mathbb{P}\left[X_1 \in E^c_{\alpha+\epsilon,M}\right] < 1$ and so conclusion in Formula (8) of the theorem now follows. On the alternative scenario of adaptive random search we still have the same conclusion but now based on the estimate:

$$\mathbb{P}\left[\bigcap_{1\leq j\leq k-1} \{X_j \in E^c_{\alpha+\epsilon,M}\}\right] \leq \inf_{1\leq j\leq k-1} \mathbb{P}\left[X_j \in E^c_{\alpha+\epsilon,M}\right],$$

instead of estimate in Formula (10) used in the pure random search case. For the second conclusion of the proof, observe that the sequence $(f(Y_n))_{n\geq 1}$ being almost surely non increasing, as a consequence of hypothesis $H1$, and bounded below is almost surely convergent to a random variable that we denote by $\mathrm{Min}_{\mathbb{Y}}$. Now, observing that for all $\epsilon > 0$:

$$\lim_{k\to+\infty} \mathbb{P}[Y_k \in E_{\alpha+\epsilon,M}] = \lim_{k\to+\infty} \mathbb{P}[f(Y_k) < \alpha+\epsilon] = 1,$$

in either pure random search or adaptive search, the conclusion follows by a standard argument (see Corollary 1. and Lemma 1. in [Esq06, p. 844]).

Remark 3. Having in mind a characterization of the speed of convergence of the algorithm it may be useful to observe that the following condition $H''2(\epsilon)$ also entails the conclusion of the theorem, although being more stringent than $H'2(\epsilon)$.

$$\lim_{k\to+\infty} \max_{1\leq j\leq k} \mathbb{P}[X_j \in E^c_{\alpha+\epsilon,M}] = \lim_{k\to+\infty} \max_{1\leq j\leq k} \mathbb{P}_j[E^c_{\alpha+\epsilon,M}] = 0. \tag{11}$$

In order to improve Theorem 1 some additional hypothesis are needed. For instance, if the minimizer is not unique then the sequence $(Y_n)_{n\geq 1}$ may not converge. First, let us observe that if the minimizer of f is unique and f is continuous, then the essential minimum of f coincides with the minimum of f.

Proposition 1. *Let f be continuous and admiting an unique minimizer $z \in \mathfrak{D}$ that is such that $f(z) = \min_{x\in\mathfrak{D}} f(x)$. Then $\alpha = \min_{x\in\mathfrak{D}} f(x) =: m$.*

Proof. Let $\epsilon > 0$ be given. There exists $x_\epsilon \in \mathfrak{D}$ such that $m = f(z) < f(x_\epsilon) < m+\epsilon$. By the continuity we have an open neighborhood V of x_ϵ such that for all $x \in V$ we still have $m < f(x) < m+\epsilon$. As a consequence:

$$\lambda(\{x : f(x) < m+\epsilon\}) \geq \lambda(V) > 0,$$

and so $\alpha \leq m+\epsilon$ and, as ϵ is arbitrary, we have $\alpha \leq m$. Consider again a given $\epsilon > 0$. There exists $\alpha < t_\epsilon < \alpha+\epsilon$. As a consequence:

$$\lambda\left(\{x \in \mathfrak{D} : f(x) < t_\epsilon\}\right) > 0,$$

and $m = \min_{x\in\mathfrak{D}} f(x) < \alpha+\epsilon$. As ϵ is arbitrary we have $m \leq \alpha$ and finally the conclusion stated.

Theorem 2. *Suppose the same notations and the same set of hypothesis of Theorem 1, namely that for every $\epsilon > 0$ hypothesis $H2(\epsilon)$ (in case of pure random search) or $H'2(\epsilon)$ (in case of adaptive search) are verified. Suppose, furthermore, that f is continuous and that admits an unique minimizer $z \in \mathfrak{D}$. Then we have almost surely that $\lim_{n\to+\infty} f(Y_n) = f(z)$. If, furthermore, $\mathfrak{D}$ is compact then $\lim_{n\to+\infty} Y_n = z$.*

Proof. Let us first show that the sequence $(f(Y_n))_{n\geq 1}$ converges in probability to $f(z)$. Consider $\epsilon > 0$. As $f(z)$ is now the essential minimum of f in $\mathfrak{D}$ we have that:

$$| f(Y_n) - f(z) |\geq \epsilon \Leftrightarrow \begin{cases} f(Y_n) \leq f(z) - \epsilon & \text{impossible} \\ f(Y_n) \geq f(z) + \epsilon, & \end{cases}$$

the possible case meaning that $Y_n \in E^c_{f(z)+\epsilon}$. Now, by a similar argument as the one used in the proof of Theorem 1, we have that $X_1, \ldots, X_{n-1} \in E^c_{f(z)+\epsilon}$ and so, under each one of the alternative hypothesis, we have:

$$\mathbb{P}\left[| f(Y_n) - f(z) |\geq \epsilon\right] \leq \begin{cases} \mathbb{P}[\{X \in E^c_{f(z)+\epsilon}\}]^{n-1} & \text{under } H2(\epsilon) \\ \inf_{1\leq j\leq n-1} \mathbb{P}[X_j \in E^c_{f(z)+\epsilon}] & \text{under } H'2(\epsilon), \end{cases} \tag{12}$$

thus ensuring that $\lim_{n\to+\infty} \mathbb{P}[| f(Y_n) - f(z) |\geq \epsilon] = 0$. If $H2\epsilon$ (or $H'2\epsilon$) are verified for all $\epsilon > 0$ the convergence in probability follows immediately. Finally, by a standard argument, the convergence almost surely of the sequence $(f(Y_n))_{n\geq 1}$ follows because this sequence is non increasing and convergent in probability. Let us suppose now that $\mathfrak{D}$ is compact and that the sequence $(Y_n)_{n\geq 1}$ does not converge to z almost surely. Then for every ω on a set of positive probability $\Omega' \subset \Omega$:

$$\exists \epsilon > 0 \;\; \forall n \in \mathbb{N} \;\; \exists N_n > n \quad | Y_{N_n}(\omega) - z |> \epsilon. \tag{13}$$

Now for all $\omega \in \Omega'$ the sequence $(Y_n)(\omega)_{n\geq 1}$ is a sequence of points in a compact set $\mathfrak{D}$ and by Bolzano-Weierstrass theorem there is a convergent subsequence $(Y_{n_k})(\omega)_{k\geq 1}$ of $(Y_n)(\omega)_{n\geq 1}$. This subsequence must converge to z because if the limit were y then, by the continuity of f we would have the sequence $(f(Y_{n_k}))(\omega)_{k\geq 1}$ converging to $f(y) = f(z)$. Now as z is an unique minimizer of f in $\mathfrak{D}$ we certainly have $y = z$. Finally observe that the subsequence $(Y_{n_k})(\omega)_{k\geq 1}$ also verifies the condition expressed in Formula (13) for k large enough, which yields the desired contradiction.

3.2 A Preliminary Observation on the Rate of Convergence

Results on the rate of convergence may be used to determine a stopping criterium for the algorithm. As a proxy for the speed of convergence of the algorithms in the context of the proof Theorems 1 and 2, namely for instance Formula (12), we may consider the quantity $\mathbb{P}[X_j \in E^c_{\alpha+\epsilon,M}]$ for various choices of distributions. In case of pure random search we have obviously:

$$\mathbb{P}\left[X_j \in E^c_{\alpha+\epsilon,M}\right] = \frac{\lambda(E^c_{\alpha+\epsilon,M})}{\lambda(\mathfrak{D})} .$$

In case of random search on a nearly unbounded domain we have, (with the notations of Sect. 2.2), that:

$$\mathbb{P}[Y_j \in E^c_{\alpha+\epsilon,M}] = \mathbb{E}\left[\mathbb{E}\left[\mathbb{I}_{\{Y_j \in E^c_{\alpha+\epsilon,M}\}}] \mid X_j\right]\right] .$$

Now as we have that:

$$\mathbb{P}\left[Y_j \in E^c_{\alpha+\epsilon,M} \mid X_j = x\right] = \int_{E^c_{\alpha+\epsilon,M}} \frac{e^{-\frac{(x-u)^2}{2\sigma^2}}}{\sqrt{2\pi\sigma}} du,$$

it follows that,

$$\mathbb{E}\left[\mathbb{1}_{\{Y_j \in E^c_{\alpha+\epsilon,M}\}} \mid X_j\right] = \int_{E^c_{\alpha+\epsilon,M}} \frac{e^{-\frac{(X_j-u)^2}{2\sigma^2}}}{\sqrt{2\pi\sigma}} du,$$

which, in turn, implies that,

$$\mathbb{P}\left[Y_j \in E^c_{\alpha+\epsilon,M}\right] = \mathbb{E}\left[\int_{E^c_{\alpha+\epsilon,M}} \frac{e^{-\frac{(X_j-u)^2}{2\sigma^2}}}{\sqrt{2\pi\sigma}} du\right] = \int_{\mathcal{D}} \int_{E^c_{\alpha+\epsilon,M}} \frac{e^{-\frac{(x-u)^2}{2\sigma^2}}}{\sqrt{2\pi\sigma}} du \frac{dx}{\lambda(\mathcal{D})}$$

where the integral on the right doesn't seem easily estimable, in general. Suppose for simplification that $\mathcal{D} = [-A, +A]$ and that $E^c_{\alpha+\epsilon,M} \subseteq [-a, +a]$ where $0 < a \ll 1 \ll A$. Then, by Fubini theorem,

$$\begin{aligned} \int_{\mathcal{D}} \int_{E^c_{\alpha+\epsilon,M}} \frac{e^{-\frac{(x-u)^2}{2\sigma^2}}}{\sqrt{2\pi\sigma}} du \frac{dx}{\lambda(\mathcal{D})} &\approx \int_{-\infty}^{+\infty} \int_{E^c_{\alpha+\epsilon,M}} \frac{e^{-\frac{(x-u)^2}{2\sigma^2}}}{\sqrt{2\pi\sigma}} du \frac{dx}{\lambda(\mathcal{D})} \\ &= \frac{\lambda(E^c_{\alpha+\epsilon,M})}{\lambda(\mathcal{D})}, \end{aligned}$$

allowing the conclusion that also $\mathbb{P}[Y_j \in E^c_{\alpha+\epsilon,M}] \approx \lambda(E^c_{\alpha+\epsilon,M})/\lambda(\mathcal{D})$ thus showing that the two algorithms, in the special situation assumed for simplification, are comparable in a first approximation.

4 On the Information Content of a Stochastic Algorithm

It is natural to conceive that in order for an algorithm to achieve global stochastic optimization of a function over a domain the algorithm has to collect complete – in some sense – information on the function over the domain. In [SB98] there are some very striking precise results on this idea. Let us detail Stephens and Baritompa's result. Consider a random algorithm described by a sequence of random variable $X_1^f, \ldots, X_n^f, \ldots$ for some function f on a domain $\mathcal{D}$. The closure $\overline{\mathbf{X}^f}$ of the set $\{X_1^f, \ldots, X_n^f, \ldots\}$ is a random set in $\mathcal{D}$.

Theorem 3 (*Global optimization requires global information*). *For any $r \in]0, 1[$, the following are equivalent:*

1. *The probability that the algorithm locate the global minimizers for f as points of $\overline{\mathbf{X}^f}$ is greater or equal than r, for any f in a sufficiently rich class of functions.*

2. *The probability that $x \in \overline{\mathbf{X}^f}$ is greater or equal than r, for any $x \in \mathcal{D}$ and f in a sufficiently rich class of functions.*

That is, roughly speaking, an algorithm works on any rich class of functions if and only if we have $\mathbb{P}[\overline{\mathbf{X}^f} = \mathcal{D}] = 1$. In the case of deterministic search the result is as expected, namely that the algorithm *sees* – in an intuitive yet precise sense – the global optimum for a class of functions in a domain if and only if the closure of the set of finite testing sequences, for any function, is dense in the domain. The extension of this result to the stochastic case gives the necessary and sufficient condition, in Theorem 3, that the lower bound of the probability of a stochastic algorithm *seing* the global optimum is the same as the lower bound of the probability of an arbitrary point of the domain belonging to the closure of the (random) set of finite testing sequences.

Having in mind the study of the limitations of an effective global optimization stochastic algorithm we address the problem of studying the information content of an algorithm. We recall that – as in Theorem 1 – a random algorithm may be identified with a sequence of random variables. The flow of information gained through a sequential observation of the sequence of random variables is usually described by the natural filtration associated with the sequence. In order to compare, in the information sense, two sequences of random variables we need to compare the associated natural filtrations.

In Theorem 5 below, by resorting to a natural defined notion of the information content of a stochastic algorithm, we obtain the result that two convergent algorithms have the same information content if the information generated by their respective minimizing functions is the whole available information in the probability space. So, the connection between the function and the stochastic set-up to generate stochastic algorithms for its global optimization - namely, probability space, probability laws of the algorithm – deserves to be further investigated.

In the following Sect. 4.1 we briefly recall results from [Cot86, Cot87, ALR03, Kud74, Bar04] on the set of complete sub σ-algebras of $\mathcal{F}$ as a topological metric space.

4.1 The Cotter Metric Space of the Complete σ-algebras

Recall that all random variables are defined on a complete probability space $(\Omega, \mathcal{F}, \mathbb{P})$. We now consider $\mathfrak{F}^\star$, the set of all σ-algebras $\mathcal{G} \subseteq \mathcal{F}$ which are complete with respect to $\mathbb{P}$.

Remark 4. We may define an equivalence relation $\mathcal{R}$ on $\mathfrak{F}^\star$ by considering an equivalence relation $\sim$ for sets in $\mathcal{F}$ defined for all $G, H \in \mathcal{F}$ by:

$$G \sim H \Leftrightarrow \mathbb{P}[G \setminus H \cup H \setminus G] = 0. \qquad (14)$$

As so, the quotient class $\mathfrak{F} := \mathfrak{F}^\star / \mathcal{R}$ is the class of all sub-σ-algebras of $\mathcal{F}$ with elements identified up to sets of probability zero.

Strong convergence in $L^1(\Omega, \mathcal{F}, \mathbb{P})$ – and also in $L^p(\Omega, \mathcal{F}, \mathbb{P})$ for $p \in [1, +\infty[$ – of a sequence $(\mathcal{G}_n)_{n\geq 1}$ of σ-algebras to $\mathcal{G}_\infty$ was introduced by Neveu in 1970 (see [Nev64, pp. 117–118]) with the condition that:

$$\forall X \in L^1(\Omega, \mathcal{F}, \mathbb{P}) \lim_{n\to+\infty} \|\mathbb{E}[X \mid \mathcal{G}_n] - \mathbb{E}[X \mid \mathcal{G}_\infty]\|_{L^1(\Omega,\mathcal{F},\mathbb{P})} = 0, \tag{15}$$

noticing that for the sequence $(\mathcal{G}_n)_{n\geq 1}$ to converge it suffices that for all $F \in \mathcal{F}$ the sequence $(\mathbb{E}[\mathbb{1}_F \mid \mathcal{G}_n])_{n\geq 1}$ converges in probability. In 1985, Cotter showed that this notion of convergence defines a topology which is metrizable (see [Cot87]). The Cotter distance d_c is defined on $\mathfrak{F} \times \mathfrak{F}$ by:

$$\begin{aligned} d_c(\mathcal{H}, \mathcal{G}) &= \sum_{i=1}^{+\infty} \frac{1}{2^i} \min\left(\mathbb{E}\left[|\mathbb{E}[X_i \mid \mathcal{H}] - \mathbb{E}[X_i \mid \mathcal{G}]|\right], 1\right) \\ &= \sum_{i=1}^{+\infty} \frac{1}{2^i} \min\left(\|\mathbb{E}[X_i \mid \mathcal{H}] - \mathbb{E}[X_i \mid \mathcal{G}]\|_1, 1\right). \end{aligned} \tag{16}$$

with $\mathcal{H}, \mathcal{G} \in \mathfrak{F}$, $\|X\|_1$ the $L^1(\Omega, \mathcal{F}, \mathbb{P})$ norm of X, with $(X_i)_{i\in\mathbb{N}}$ a dense denumerable set in $L^1(\Omega, \mathcal{F}, \mathbb{P})$. We have that $(\mathfrak{F}, d_c)$ is a complete metric space.

We will need a consequence of the definition of the Cotter distance (see Corollary III.35, in [Bar04, p. 36]) that we quote for the reader's convenience.

Proposition 2. *Consider $\mathcal{G}_1 \subset \mathcal{G}_2 \subset \mathcal{G}_3$ in $\mathfrak{F}$, Then we have that:*

$$d_c(\mathcal{G}_2, \mathcal{G}_3) \leq 2 d_c(\mathcal{G}_1, \mathcal{G}_3).$$

We will also need a remarkable result of Cotter (see Corollary 2.2 and Corollary 2.4 in [Cot87, p. 42]) that we formulate next.

Theorem 4. *Let $\mathcal{L}_{\mathbb{P}}$ be the metric space of the real valued random variables defined on the probability space $(\Omega, \mathcal{F}, \mathbb{P})$ with the metric of the convergence in probability. Let $\sigma : \mathcal{L}_{\mathbb{P}} \mapsto \mathfrak{F}$ that to each random variable X associates $\sigma(X) = \{X^{-1}(B) : B \in \mathcal{B}(\mathbb{R})\}$ the sigma-algebra generated by X. Then, considering the metric space $(\mathfrak{F}, d_c)$ with d_c the Cotter distance defined in Formulas* (16), *we have that σ is continuous at $X \in \mathcal{L}_{\mathbb{P}}$ if and only if $\sigma(X) = \mathcal{F}$.*

This result on the continuity of the map σ between metric spaces $\mathcal{L}_{\mathbb{P}}$ and $(\mathfrak{F}, d_c)$ will be applied to convergent sequences.

4.2 The Information Content of a Random Algorithm

Let $\mathbb{Y} = (Y_n)_{n\geq 1}$ be a stochastic algorithm for the minimization of f on a domain $\mathcal{D}$. According to Theorem 1 we may define a convergent algorithm for the minimization problem of f on the domain $\mathcal{D}$.

Definition 4. *Let α be the essential infimum of f on $\mathcal{D}$ defined in Formula* (3)*. Following Theorem 1, the algorithm $\mathbb{Y}$ **converges** on $\mathcal{D}$ if the sequence $(f(Y_n))_{n\geq 1}$ converges almost surely to a random variable $Min_{f,\mathbb{Y}}$ such that:*

$$\mathbb{P}[Min_{f,\mathbb{Y}} \leq \alpha] = 1.$$

Now given a random algorithm $\mathbb{Y} = (Y_n)_{n\geq 1}$ we define the flow of information associated to this algorithm.

Definition 5. *The* ***flow of information*** *associated to the algorithm* $\mathbb{Y} = (Y_n)_{n\geq 1}$ *for the global minimization of the function* f *is given by the natural filtration of* $(f(Y_n))_{n\geq 1}$*, which is the increasing sequence of* σ*-algebras defined by:*

$$\mathcal{F}_n^{\mathbb{Y}} := \sigma\left(f(Y_1), \ldots, f(Y_n)\right).$$

The terminal σ-algebra associated to this algorithm, $\mathcal{F}_\infty^{\mathbb{Y}}$, is naturally defined as (in the two usual notations):

$$\mathcal{F}_\infty^{\mathbb{Y}} := \sigma\left(\bigcup_{n=1}^{+\infty} \mathcal{F}_n^{\mathbb{Y}}\right) = \bigvee_{n=1}^{+\infty} \mathcal{F}_n^{\mathbb{Y}}.$$

As an immediate result we have that the filtration converges in the Cotter distance to the terminal σ-algebra.

Proposition 3. *For every stochastic algorithm* $\mathbb{Y} = (Y_n)_{n\geq 1}$*,*

$$\lim_{n\to+\infty} d_c\left(\mathcal{F}_n^{\mathbb{Y}}, \mathcal{F}_\infty^{\mathbb{Y}}\right) = 0.$$

Proof. Let's first observe that by Proposition 2.2 of Cotter (see again [Cot86]) any increasing sequence of σ-algebras converges in the Cotter distance. In fact, by a standard argument we have that:

$$\bigcap_{n=1}^{+\infty} \bigvee_{m=n}^{+\infty} \mathcal{F}_m^{\mathbb{Y}} = \mathcal{F}_\infty^{\mathbb{Y}} = \bigvee_{n=1}^{+\infty} \bigcap_{m=n}^{+\infty} \mathcal{F}_m^{\mathbb{Y}}$$

and by the result quoted this suffices to ensure that the filtration associated to the algorithm converges. Now, it is a well known fact (see [Bil95, p. 470]) that, by the definitions above, we have that almost surely:

$$\forall Z \in L^1(\Omega, \mathcal{F}, \mathbb{P}) \quad \lim_{n\to+\infty} \mathbb{E}\left[Z \mid \mathcal{F}_n^{\mathbb{Y}}\right] = \mathbb{E}\left[Z \mid \mathcal{F}_\infty^{\mathbb{Y}}\right] \tag{17}$$

as the sequence $(\mathbb{E}\left[Z \mid \mathcal{F}_n^{\mathbb{Y}}\right])_{n\geq 1}$ is uniformly integrable, (17) implies that

$$\forall Z \in L^1(\Omega, \mathcal{F}, \mathbb{P}) \quad \lim_{n\to+\infty} \left\|\mathbb{E}\left[Z \mid \mathcal{F}_n^{\mathbb{Y}}\right] - \mathbb{E}\left[Z \mid \mathcal{F}_\infty^{\mathbb{Y}}\right]\right\|_{L^1(\Omega,\mathcal{F},\mathbb{P})} = 0,$$

and this is just definition given by Formula (15).

We now compare the information content of two stochastic algorithms by comparing their information induced filtrations.

Definition 6. *Two algorithms* $\mathbb{Y}^1$ *and* $\mathbb{Y}^2$ *are* ***informationally asymptotically equivalent*** *(IAE) if and only if:*

$$\lim_{n\to+\infty} d_c\left(\mathcal{F}_n^{\mathbb{Y}^1}, \mathcal{F}_n^{\mathbb{Y}^2}\right) = 0.$$

As an easy first observation we have that two algorithms are informationally asymptotically equivalent if and only if the Cotter distance of their terminal σ-algebras is zero, that is:

Proposition 4. *Let $\mathbb{Y}^1$ and $\mathbb{Y}^2$ be two algorithms, Then:*

$$\mathbb{Y}^1 IAE\, \mathbb{Y}^2 \Leftrightarrow d_c\left(\mathcal{F}_\infty^{\mathbb{Y}^1}, \mathcal{F}_\infty^{\mathbb{Y}^2}\right) = 0. \tag{18}$$

Proof. If the two algorithms are informationally asymptotically equivalent then the condition about the terminal σ-algebras is verified as an immediate consequence of Proposition 3. In fact,

$$d_c\left(\mathcal{F}_\infty^{\mathbb{Y}^1}, \mathcal{F}_\infty^{\mathbb{Y}^2}\right) \leq d_c\left(\mathcal{F}_\infty^{\mathbb{Y}^1}, \mathcal{F}_n^{\mathbb{Y}^1}\right) + d_c\left(\mathcal{F}_n^{\mathbb{Y}^1}, \mathcal{F}_n^{\mathbb{Y}^2}\right) + d_c\left(\mathcal{F}_n^{\mathbb{Y}^2}, \mathcal{F}_\infty^{\mathbb{Y}^2}\right).$$

Now, for the converse suppose that $d_c\left(\mathcal{F}_\infty^{\mathbb{Y}^1}, \mathcal{F}_\infty^{\mathbb{Y}^2}\right) = 0$ and that the algorithms are not IAE. Then, for some $\epsilon > 0$ there exists an increasing integer sequence $(n_k^\epsilon)_{k\in\mathbb{N}}$ such that

$$\forall k \in \mathbb{N},\ \ d_c\left(\mathcal{F}_{n_k^\epsilon}^{\mathbb{Y}^1}, \mathcal{F}_{n_k^\epsilon}^{\mathbb{Y}^2}\right) \geq \epsilon.$$

We then have that for all $k \geq 1$,

$$\begin{aligned}
\epsilon &\leq d_c\left(\mathcal{F}_{n_k^\epsilon}^{\mathbb{Y}^1}, \mathcal{F}_{n_k^\epsilon}^{\mathbb{Y}^2}\right) \leq d_c\left(\mathcal{F}_{n_k^\epsilon}^{\mathbb{Y}^1}, \mathcal{F}_\infty^{\mathbb{Y}^1}\right) + d_c\left(\mathcal{F}_\infty^{\mathbb{Y}^1}, \mathcal{F}_\infty^{\mathbb{Y}^2}\right) + d_c\left(\mathcal{F}_\infty^{\mathbb{Y}^1}, \mathcal{F}_{n_k^\epsilon}^{\mathbb{Y}^2}\right) \\
&= d_c\left(\mathcal{F}_{n_k^\epsilon}^{\mathbb{Y}^1}, \mathcal{F}_\infty^{\mathbb{Y}^1}\right) + d_c\left(\mathcal{F}_\infty^{\mathbb{Y}^1}, \mathcal{F}_{n_k^\epsilon}^{\mathbb{Y}^2}\right) \\
&\leq \limsup_{n\to+\infty}\left(d_c\left(\mathcal{F}_{n_k^\epsilon}^{\mathbb{Y}^1}, \mathcal{F}_\infty^{\mathbb{Y}^1}\right) + d_c\left(\mathcal{F}_\infty^{\mathbb{Y}^1}, \mathcal{F}_{n_k^\epsilon}^{\mathbb{Y}^2}\right)\right) = 0,
\end{aligned}$$

again, by Proposition 3, which is a contradiction.

Our purpose now is to illustrate the intuitive idea that a convergent algorithm for minimizing a function must recover all available information about the function. For the first result we require that the algorithm exhausts all the available information in the probability space. We will suppose that the two algorithms $\mathbb{Y}^1$ and $\mathbb{Y}^2$ both converge. We will show next that, if we suppose,

$$\sigma\left(\mathrm{Min}_{f,\mathbb{Y}^1}\right) = \mathcal{F} = \sigma\left(\mathrm{Min}_{f,\mathbb{Y}^2}\right), \tag{19}$$

then, these algorithms, $\mathbb{Y}^1$ and $\mathbb{Y}^2$, are informationally asymptotic equivalent.

Theorem 5. *With the notations of Definition 4, let $\mathbb{Y}^1$ and $\mathbb{Y}^2$ be two algorithms that converge. We have that:*

$$\sigma\left(Min_{f,\mathbb{Y}^1}\right) = \mathcal{F} = \sigma\left(Min_{f,\mathbb{Y}^2}\right) \Rightarrow \mathbb{Y}^1 IAE\, \mathbb{Y}^2.$$

Proof. The proof is a consequence of the continuity of the operator that maps each random variable to the σ-algebra it generates formulated in Cotter's Theorem 4. We have that the sequences,

$$\left(\sigma\left(f(Y_n^1)\right)\right)_{n\geq 1}, \left(\sigma\left(f(Y_n^2)\right)\right)_{n\geq 1},$$

both converge in the Cotter distance to $\mathcal{F}$ by reason of the hypothesis. Now, by definition, as we have that for all $n \geq 1$,

$$\sigma\left(f(Y_n^1)\right) \subset \mathcal{F}_n^{\mathbb{Y}^1} \subset \mathcal{F} \ , \ \sigma\left(f(Y_n^2)\right) \subset \mathcal{F}_n^{\mathbb{Y}^2} \subset \mathcal{F},$$

by Proposition 2, we have:

$$d_c\left(\mathcal{F}_n^{\mathbb{Y}^1}, \mathcal{F}\right) \leq 2d_c\left(\sigma\left(f(Y_n^1)\right), \mathcal{F}\right) \ , \ d_c\left(\mathcal{F}_n^{\mathbb{Y}^2}, \mathcal{F}\right) \leq 2d_c\left(\sigma\left(f(Y_n^2)\right), \mathcal{F}\right)$$

and so we also have that the sequences,

$$\left(\mathcal{F}_n^{\mathbb{Y}^1}\right)_{n\geq 1}, \left(\mathcal{F}_n^{\mathbb{Y}^2}\right)_{n\geq 1},$$

converge in the Cotter distance to $\mathcal{F}$. Finally, as we have:

$$d_c\left(\mathcal{F}_n^{\mathbb{Y}^1}, \mathcal{F}_n^{\mathbb{Y}^2}\right) \leq d_c\left(\mathcal{F}_n^{\mathbb{Y}^1}, \mathcal{F}\right) + d_c\left(\mathcal{F}, \mathcal{F}_n^{\mathbb{Y}^2}\right),$$

we have the condition of Formula (19) appearing in Theorem 5.

Remark 5. If condition in Formula (19), essential in Theorem 5, is not verified – then by Cotter's theorem quoted in Theorem 4 – the map σ is not continuous at $\sigma\left(\mathrm{Min}_{f,\mathbb{Y}^1}\right)$ and $\sigma\left(\mathrm{Min}_{f,\mathbb{Y}^2}\right)$ and so – it is in general not true that the sequences $\left(\sigma\left(f(Y_n^1)\right)\right)_{n\geq 1}$ and $\left(\sigma\left(f(Y_n^2)\right)\right)_{n\geq 1}$ converge. As a consequence, despite the fact that, by Proposition 3, the sequences $\left(\mathcal{F}_n^{\mathbb{Y}^1}\right)_{n\geq 1}$ and $\left(\mathcal{F}_n^{\mathbb{Y}^2}\right)_{n\geq 1}$ both converge – to $\mathcal{F}_\infty^{\mathbb{Y}^1}$ and $\mathcal{F}_\infty^{\mathbb{Y}^2}$, respectively – we can not ensure that the condition given by Formula (18) in Proposition 4 is verified and so, we can not conclude that the two algorithms are IAE.

If moreover the algorithms are informationally asymptotic equivalent, and their associated limit minimum functions take a denumerable set of values, then their associated limit minimum functions will coincide almost surely thus saying, essentially, that two IAE convergent algorithms carry the same information content with respect to the minimization function.

Theorem 6. *With the notations of Definition 4, let $\mathbb{Y}^1$ and $\mathbb{Y}^2$ be two algorithms that converge. Let us suppose that the set $Min_{f,\mathbb{Y}^1}(\Omega) \cup Min_{f,\mathbb{Y}^2}(\Omega)$ is denumerable. We then have that:*

$$\mathbb{Y}^1 IAE\ \mathbb{Y}^2 \Rightarrow Min_{f,\mathbb{Y}^1} = Min_{f,\mathbb{Y}^2} \ a.\ s. \tag{20}$$

Proof. The announced result is a consequence of Proposition 4. In fact, if $\mathbb{Y}^1$ and $\mathbb{Y}^2$ are IAE then this means that:

$$\mathcal{F}_\infty^{\mathbb{Y}^1} \sim \mathcal{F}_\infty^{\mathbb{Y}^2},$$

and so by (14), for every B in the Borel σ-algebra of the reals $\mathcal{B}(\mathbb{R})$,

$$\mathbb{P}\left[\mathrm{Min}_{f,\mathbb{Y}^1}^{-1}(B) \setminus \mathrm{Min}_{f,\mathbb{Y}^2}^{-1}(B)\right] = 0 = \mathbb{P}\left[\mathrm{Min}_{f,\mathbb{Y}^2}^{-1}(B) \setminus \mathrm{Min}_{f,\mathbb{Y}^1}^{-1}(B)\right] \tag{21}$$

Now, consider $B = \{x\} \in \mathcal{B}(\mathbb{R})$. Formulas (21) imply that:

$$\mathbb{P}\left[\{\omega \in \Omega \mid \mathrm{Min}_{f,\mathbb{Y}^1}(\omega) \neq x\} \cup \{\omega \in \Omega \mid \mathrm{Min}_{f,\mathbb{Y}^2}(\omega) = x\}\right] = 1,$$

and also

$$\mathbb{P}\left[\{\omega \in \Omega \mid \mathrm{Min}_{f,\mathbb{Y}^2}(\omega) \neq x\} \cup \{\omega \in \Omega \mid \mathrm{Min}_{f,\mathbb{Y}^1}(\omega) = x\}\right] = 1.$$

Now, by considering the intersection

$$\left(\{\mathrm{Min}_{f,\mathbb{Y}^1} \neq x\} \cup \{\mathrm{Min}_{f,\mathbb{Y}^2} = x\}\right) \cap \left(\{\mathrm{Min}_{f,\mathbb{Y}^2} \neq x\} \cup \{\mathrm{Min}_{f,\mathbb{Y}^1} = x\}\right),$$

which is is a set of probability one, we get by expanding that for every $x \in \mathbb{R}$:

$$\mathbb{P}\left[\{\mathrm{Min}_{f,\mathbb{Y}^1} \neq x \wedge \mathrm{Min}_{f,\mathbb{Y}^2} \neq x\} \cup \{\mathrm{Min}_{f,\mathbb{Y}^1} = x \wedge \mathrm{Min}_{f,\mathbb{Y}^2} = x\}\right] = 1.$$

And so by considering the denumerable set $\mathrm{Im} = \mathrm{Min}_{f,\mathbb{Y}^1}(\Omega) \cup \mathrm{Min}_{f,\mathbb{Y}^2}(\Omega)$, as

$$\begin{aligned}
&\{\mathrm{Min}_{f,\mathbb{Y}^1} \neq \mathrm{Min}_{f,\mathbb{Y}^2}\} \\
&\subseteq \bigcup_{x \in \mathrm{Im}} \{\mathrm{Min}_{f,\mathbb{Y}^1} = x \wedge \mathrm{Min}_{f,\mathbb{Y}^2} \neq x\} \cup \{\mathrm{Min}_{f,\mathbb{Y}^1} \neq x \wedge \mathrm{Min}_{f,\mathbb{Y}^2} = x\} \\
&= \bigcup_{x \in \mathrm{Im}} \left(\{\mathrm{Min}_{f,\mathbb{Y}^1} \neq x \wedge \mathrm{Min}_{f,\mathbb{Y}^2} \neq x\} \cup \{\mathrm{Min}_{f,\mathbb{Y}^1} = x \wedge \mathrm{Min}_{f,\mathbb{Y}^2} = x\}\right)^c
\end{aligned}$$

we will have that $\mathbb{P}\left[\mathrm{Min}_{f,\mathbb{Y}^1} \neq \mathrm{Min}_{f,\mathbb{Y}^2}\right] = 0$, as wanted.

The particular case of an unique minimizer of a continuous function on a compact domain deserves special mention as a case where two algorithms having IAE lead to the same minimizing function almost surely.

Proposition 5. *With the notations of definition 4, let $\mathbb{Y}^1$ and $\mathbb{Y}^2$ be two algorithms that converge. Suppose, furthermore, that f is continuous, that f admits an unique minimizer z and $\mathcal{D}$ is compact. Then we have that:*

$$\mathbb{Y}^1 \, IAE \, \mathbb{Y}^2 \Rightarrow \begin{cases} \lim_{n \to +\infty} f(Y_n^1) = f(z) = \lim_{n \to +\infty} f(Y_n^2) \ a.\ s. \\ \lim_{n \to +\infty} Y_n^1 = z = \lim_{n \to +\infty} Y_n^2 \ a.\ s. \end{cases}.$$

Proof. As we have $\lim_{n \to +\infty} f(Y_n^1) = f(z) = \lim_{n \to +\infty} f(Y_n^2)$ and $\lim_{n \to +\infty} Y_n^1 = z = \lim_{n \to +\infty} Y_n^2$, by Theorem 2 and Proposition 1, we also have that $\mathrm{Min}_{f,\mathbb{Y}^1} = f(z) = \mathrm{Min}_{f,\mathbb{Y}^2}$ almost surely and so, by Theorem 6, we have the announced result.

Remark 6. Let us observe, with respect to Proposition 5, that under the hypothesis stated in that proposition, that is, if we have almost surely,

$$\mathrm{Min}_{f,\mathbb{Y}^1} = \mathrm{Min}_{f,\mathbb{Y}^2} = f(z),$$

then, by modifying $\text{Min}_{f,\mathbb{Y}^1}$ and $\text{Min}_{f,\mathbb{Y}^2}$ on sets of probability zero we would have that:

$$\sigma\left(\text{Min}_{f,\mathbb{Y}^1}\right) = \{\emptyset, \Omega\} = \sigma\left(\text{Min}_{f,\mathbb{Y}^2}\right).$$

By Remark 4, in general, under the hypothesis of Proposition 5, the two σ-algebras $\sigma\left(\text{Min}_{f,\mathbb{Y}^1}\right)$ and $\sigma\left(\text{Min}_{f,\mathbb{Y}^1}\right)$ are equal to $\{\emptyset, \Omega\}$ in $\mathfrak{F} := \mathfrak{F}^\star/\sim$ – the class of all sub-σ-algebras of $\mathcal{F}$ identified up to sets of probability zero – and so the condition in Formula (19) – which is essential in Theorem 5 – may be verified only for deterministic algorithms (as in this case all random variables are constant).

Acknowledgements. This work was partially supported, for the first and fourth authors, through the project UID/MAT/00297/2020 of the Centro de Matemática e Aplicações, financed by the Fundação para a Ciência e a Tecnologia (Portuguese Foundation for Science and Technology) and, for the third author, through RFBR (Grant n. 19-01-00451).

A Appendix

Deduction of Formula (2). Let λ_x denote the Lebesgue measure over $\mathcal{D}$ applied to the set defined by the variable x.

$$\begin{aligned}
&\mathbb{P}[Y_n \in D] \\
&= \sum_{k=1}^{n} \mathbb{P}\left[\{X_k \in D\} \cap \bigcap_{1\leq j<k} \{f(X_k) \leq f(X_j\} \cap \bigcap_{k<j\leq n} \{f(X_k) < f(X_j)\}\right] \\
&= \sum_{k=1}^{n} \left(\frac{1}{\lambda(\mathcal{D})^n} \int_{\mathcal{D}^n} \mathbb{I}_{\{x_k \in D\}} \prod_{1\leq j<k} \mathbb{I}_{\{f(x_k)\leq f(x_j\}} \right.\\
&\left.\times \prod_{k<j\leq n} \mathbb{I}_{\{f(x_k)<f(x_j)\}} d\lambda(x_1)\dots d\lambda(x_n)\right) \\
&= \sum_{k=1}^{n} \left(\frac{1}{\lambda(\mathcal{D})} \int_{\mathcal{D}} d\lambda(x_k) \frac{1}{\lambda(\mathcal{D})^{k-1}} \prod_{1\leq j<k} \int_{\mathcal{D}^{k-1}} \mathbb{I}_{\{f(x_k)\leq f(x_j\}} d\lambda(x_j) \right.\\
&\left.\times \frac{1}{\lambda(\mathcal{D})^{n-k}} \prod_{k<j\leq n} \int_{\mathcal{D}^{n-k}} \mathbb{I}_{\{f(x_k)<f(x_j)\}} d\lambda(x_j)\right) \\
&= \sum_{k=1}^{n} \left(\frac{1}{\lambda(\mathcal{D})} \int_{\mathcal{D}} \mathbb{I}_{\{x_k \in D\}} \frac{\lambda_x(\{f(x_k) \leq f(x\})^{k-1}}{\lambda(\mathcal{D})^{k-1}} \right.\\
&\left.\times \frac{\lambda_x(\{f(x_k) < f(x\})^{n-k}}{\lambda(\mathcal{D})^{n-k}} d\lambda(x_k)\right) \\
&= \sum_{k=1}^{n} \left(\frac{1}{\lambda(\mathcal{D})^n} \int_{D} \lambda(f^{-1}([f(x_k), +\infty[)^{k-1} \lambda(f^{-1}(]f(x_k), +\infty[)^{n-k} d\lambda(x_k)\right).
\end{aligned}$$

References

[ALR03] Appel, M.J., LaBarre, R., Radulović, D.: On accelerated random search. SIAM J. Optim. **14**(3), 708–731 (2003). (Electronic)

[Art01] Artstein, Z.: Compact convergence of σ-fields and relaxed conditional expectation. Probab. Theory Relat. Fields **120**(3), 369–394 (2001)

[Bar04] Barty, K.: Contributions à la discrétisation des contraintes de mesurabilité pour les problèmes d'optimisation stochastique. Ph.D. thesis, École Nationale des Ponts et Chaussées, Paris, France, June 2004

[Bil95] Billingsley, P.: Probability and Measure. Wiley Series in Probability and Mathematical Statistics, 3rd edn. Wiley, New York (1995)

[Boy71] Boylan, E.S.: Equiconvergence of martingales. Ann. Math. Stat. **42**, 552–559 (1971)

[CJK07] Costa, A., Jones, O.D., Kroese, D.: Convergence properties of the cross-entropy method for discrete optimization. Oper. Res. Lett. **35**(5), 573–580 (2007)

[Cot86] Cotter, K.D.: Similarity of information and behavior with a pointwise convergence topology. J. Math. Econ. **15**(1), 25–38 (1986)

[Cot87] Cotter, K.D.: Convergence of information, random variables and noise. J. Math. Econ. **16**(1), 39–51 (1987)

[dBKMR05] de Boer, P.-T., Kroese, D.P., Mannor, S., Rubinstein, R.Y.: A tutorial on the cross-entropy method. Ann. Oper. Res. **134**, 19–67 (2005)

[dC11] de Carvalho, M.: Confidence intervals for the minimum of a function using extreme value statistics. Int. J. Math. Modell. Numer. Optim. **2**(9), 288–296 (2011)

[dC12] de Carvalho, M.: A generalization of the Solis and Wets method. J. Stat. Plann. Inference **142**(3), 633–644 (2012)

[Esq06] Esquível, M.L.: A conditional Gaussian martingale algorithm for global optimization. In: Gavrilova, M., et al. (eds.) ICCSA 2006, Part III. LNCS, vol. 3982, pp. 841–851. Springer, Heidelberg (2006). https://doi.org/10.1007/11751595_89

[Kom08] Komisarski, A.: Distances between σ-fields on a probability space. J. Theoret. Probab. **21**(4), 812–823 (2008)

[Kud74] Kudō, H.: A note on the strong convergence of σ-algebras. Ann. Probab. **2**(1), 76–83 (1974)

[MPB99] Mexia, J.T., Pereira, D., Baeta, J.: L_2 environmental indexes. Listy Biom. **36**(2), 137–143 (1999)

[Nev64] Neveu, J.: Bases mathématiques du calcul des probabilités. Masson et Cie, Éditeurs, Paris (1964)

[Nev65] Neveu, J.: Mathematical Foundations of the Calculus of Probability. Translated by Amiel Feinstein. Holden-Day Inc., San Francisco (1965)

[Nev72] Neveu, J.: Note on the tightness of the metric on the set of complete sub σ-algebras of a probability space. Ann. Math. Stat. **43**, 1369–1371 (1972)

[Pic98] Piccinini, L.: Convergence of nonmonotone sequence of sub-σ-fields and convergence of associated subspaces $L^p(\mathcal{B}_n)(p \in [1, +\infty])$. J. Math. Anal. Appl. **225**(1), 73–90 (1998)

[PM10] Pereira, D.G., Mexia, J.T.: Comparing double minimization and zigzag algorithms in joint regression analysis: the complete case. J. Stat. Comput. Simul. **80**(1–2), 133–141 (2010)

[PS00] Peng, J., Shi, D.: Improvement of pure random search in global optimization. J. Shanghai Univ. **4**(2), 92–95 (2000)

[RK04] Rubinstein, R.Y., Kroese, D.P.: The Cross-Entropy Method. Information Science and Statistics, Springer, New York (2004). https://doi.org/10.1007/978-1-4757-4321-0. A Unified Approach to Combinatorial Optimization, Monte-Carlo Simulation, and Machine Learning

[RK08] Rubinstein, R.Y., Kroese, D.P.: Simulation and the Monte Carlo Method. Wiley Series in Probability and Statistics, 2nd edn. Wiley-Interscience, Hoboken (2008)

[Rog74] Rogge, L.: Uniform inequalities for conditional expectations. Ann. Probab. **2**, 486–489 (1974)

[RS03] Raphael, B., Smith, I.F.C.: A direct stochastic algorithm for global search. Appl. Math. Comput. **146**(2–3), 729–758 (2003)

[SB98] Stephens, C.P., Baritompa, W.: Global optimization requires global information. J. Optim. Theory Appl. **96**(3), 575–588 (1998)

[SP99] Shi, D., Peng, J.: A new theoretical framework for analyzing stochastic global optimization algorithms. J. Shanghai Univ. **3**(3), 175–180 (1999)

[Spa03] Spall, J.C.: Introduction to Stochastic Search and Optimization. Wiley-Interscience Series in Discrete Mathematics and Optimization. Wiley-Interscience, Hoboken (2003). Estimation, Simulation, and Control

[Spa04] Spall, J.C.: Stochastic optimization. In: Handbook of Computational Statistics, pp. 169–197. Springer, Berlin (2004)

[SW81] Solis, F.J., Wets, R.J.-B.: Minimization by random search techniques. Math. Oper. Res. **6**(1), 19–30 (1981)

[Vid18] Vidmar, M.: A couple of remarks on the convergence of σ-fields on probability spaces. Statist. Probab. Lett. **134**, 86–92 (2018)

[VZ93] Van Zandt, T.: The Hausdorff metric of σ-fields and the value of information. Ann. Probab. **21**(1), 161–167 (1993)

[Wan01] Wang, X.: Convergence rate of conditional expectations. Sci. Math. Jpn. **53**(1), 83–87 (2001)

[Yin99] Yin, G.: Rates of convergence for a class of global stochastic optimization algorithms. SIAM J. Optim. **10**(1), 99–120 (1999). (Electronic)

[Zab03] Zabinsky, Z.B.: Stochastic Adaptive Search for Global Optimization. Nonconvex Optimization and its Applications, vol. 72. Kluwer Academic Publishers, Boston (2003)

On Modification of the Law of Large Numbers and Linear Regression of Fuzzy Random Variables

Vladimir L. Khatskevich(✉)

Air Force Academy named after N.E. Zhukovsky and Y.U. Gagarin, Voronezh, Russian Federation

Abstract. Extreme properties of the average characteristics of fuzzy random variables are given. A new form of the law of large numbers for fuzzy random variables is established. An optimal linear regression of fuzzy random variables is constructed, in which the coefficients are similar to the Fourier coefficients. It is shown that under certain conditions, the optimal regression has the maximum cosine of the angle with the predicted fuzzy random variable.

Keywords: Fuzzy random variables · Means · Covariance · Variance · Law of large numbers · Linear regression

1 Introduction

Fuzzy random variables originated as a branch of fuzzy mathematics in [1–3]. They are widely used in financial mathematics, forecasting, decision theory, and others. In particular, the mathematical model of a random experiment with fuzzy outcomes is interpreted as a fuzzy random variable. The current state of the theory of fuzzy random variables is reflected in [4–7] and others.

In this paper, a new definition of the quasi-scalar product between fuzzy random variables is introduced, and its relationship with the covariance of fuzzy random variables proposed in [8] is revealed. Extreme properties of expectations and fuzzy expectations of fuzzy random variables are discussed.

The main results of this work are devoted to the law of large numbers (LLN) for fuzzy random variables and linear regression of fuzzy random variables. The difference between our result and those known from the LLN consists in using a special metric associated with the quasi-scalar product introduced by the author. The specificity of our result on linear regression is to derive a formula for optimal linear regression coefficients similar to the Fourier coefficients for an orthonormal system in Hilbert space. This is provided by introduction the definition quasi-scalar product.

In addition, it is shown that under certain conditions, the optimal solution has the maximum cosine of the angle with the predicted fuzzy random variable in the class of linear estimates.

A. N. Shiryaev et al. (Eds.): ICSM-5 2020, PROMS 371, pp. 76–91, 2021.
https://doi.org/10.1007/978-3-030-83266-7_6

Below, by the fuzzy set A given on the universal space U, we mean the set of ordered pairs $(u, \mu_A(u))$, where the membership function $\mu_A : U \to [0,1]$, determines the degree to which $\forall u \in U$ belongs to the set A.

We rely on the following definition of a fuzzy number (cf. [9] chap. 2–4). A fuzzy number is called a fuzzy set whose universal set is the set of real numbers R, and which additionally satisfies the following conditions:

1) the support (supp) of a fuzzy number is a closed and bounded (compact) set of real numbers;
2) the membership function of a fuzzy number is convex;
3) the membership function of a fuzzy number is normal, i.e. the supremum of the membership function is equal to one;
4) the membership function of a fuzzy number is semi-continuous from above.

We will use the interval representation of fuzzy numbers. Namely, we will assign a set of α-intervals to each fuzzy number.

The set of fuzzy numbers satisfying the conditions 1)–4) is denoted by J.

As known, the set α-level of a fuzzy number $\tilde{z} \in J$ with the membership function $\mu_{\tilde{z}}(x)$ is defined by the relation

$$Z_\alpha = \{x | \mu_{\tilde{z}}(x) \geq \alpha\} \ \ (\alpha \in (0,1]), \ \ Z_0 = supp(\tilde{z}).$$

According to the above assumptions 1)–4) all α-levels of a fuzzy number are closed and bounded intervals at the real axis. Let's denote the left (lower) border of the interval $z^-(\alpha)$, and the right (upper) - $z^+(\alpha)$, i.e. $Z_\alpha = [z^-(\alpha), z^+(\alpha)]$. Sometimes $z^-(\alpha)$ and $z^+(\alpha)$ they are called the left and right indices of a fuzzy number, respectively.

On a set of fuzzy numbers, you can enter the definition of distances between them in different ways. The interval approach sometimes uses the Hausdorff distance, which for fuzzy numbers $\tilde{z}$, $\tilde{u} \in J$ with α-level sets $Z_\alpha = [z^-(\alpha), z^+(\alpha)]$ and $U_\alpha = [u^-(\alpha), u^+(\alpha)]$ in accordance with [5] is defined by the formula $\rho_H(\tilde{z}, \tilde{u}) = \sup\limits_{0<\alpha\leq 1} d_H(Z_\alpha, U_\alpha)$, where

$$d_H(Z_\alpha, U_\alpha) = \max\left[\sup_{z\in Z_\alpha} \inf_{u\in u_\alpha} |z-u|, \sup_{u\in u_\alpha} \inf_{z\in z_\alpha} |z-u|\right]$$

- Hausdorff metric. Some other distances are also considered (see, for example., [7,8,10]).

Denote by J_0 - the set of fuzzy numbers in the interval representation of which the left and right indices are quadratically summable.

We use the distance $\rho(\tilde{z}, \tilde{u})$ between the fuzzy numbers $\tilde{z}$ and $\tilde{u}$ from J_0 with α-level sets $Z_\alpha = [z^-(\alpha), z^+(\alpha)]$ and $U_\alpha = [u^-(\alpha), u^+(\alpha)]$, which is defined by the formula

$$\rho(\tilde{z}, \tilde{u}) = \left(\int_0^1 (z^-(\alpha) - u^-(\alpha))^2 + (z^+(\alpha) - u^+(\alpha))^2 d\alpha\right)^{\frac{1}{2}}. \qquad (1)$$

Here and below, integration is understood by Lebesgue.

This kind of distance was previously used, for example, in [11].

Let the fuzzy number $\tilde{z}$ correspond to α - levels $Z_\alpha = [z^-(\alpha), Z^+(\alpha)]$, with $\alpha \in (0, 1]$. Suppose, as is customary in interval analysis,

$$midZ_\alpha = \frac{1}{2}(z^+(\alpha) + z^-(\alpha)), \;\; radZ_\alpha = \frac{1}{2}(z^+(\alpha) - z^-(\alpha)).$$

For fuzzy numbers $\tilde{z}$ and $\tilde{u}$ from J_0, with sets of α - levels $[z^-(\alpha), z^+(\alpha)]$ and $[u^-(\alpha), u^+(\alpha)]$, we define quasi-scalar product

$$\langle \tilde{z}, \tilde{u} \rangle = \int_0^1 (midZ_\alpha, midU_\alpha + radU_\alpha radZ_\alpha) d\alpha$$

$$= 0.5 \int_0^1 (z^+(\alpha)u^+(\alpha) + z^-(\alpha)u^-(\alpha)) d\alpha. \tag{2}$$

The quasinorm $\tilde{z}$ is $\langle \tilde{z}, \tilde{z} \rangle^{1/2}$.

According to (1), (2) the distance between the fuzzy numbers $\tilde{z}$ and $\tilde{u}$ from J_0 with sets of α - levels $[z^-(\alpha), z^+(\alpha)]$ and $[u^-(\alpha), u^+(\alpha)]$ matches the quasinorm of a fuzzy number whose left index is $z^-(\alpha) - u^-(\alpha)$, and the right index $z^+(\alpha) - u^+(\alpha)$.

Under the sum of the fuzzy numbers $\tilde{z}$ and $\tilde{u}$ we will understand a fuzzy number with α - levels $[z^-(\alpha)+u^-(\alpha), z^+(\alpha)+u^+(\alpha)]$. The product of a fuzzy number $\tilde{z}$ by a positive number c is a fuzzy number with α - levels $[cz^-(\alpha), cz^+(\alpha)]$. In the case of $c < 0$ - a fuzzy number with α - levels $[cz^+(\alpha), cz^-(\alpha)]$.

The following properties of the introduced quasi-scalar product are valid:

1) $\langle \tilde{z}, \tilde{u} \rangle = \langle \tilde{u}, \tilde{z} \rangle$ $(\forall \tilde{z}, \tilde{u} \in J_0)$;
2) $\langle c_1\tilde{z}, c_2\tilde{u} \rangle = c_1c_2 \langle \tilde{z}, \tilde{u} \rangle$ $(\forall \tilde{z}, \tilde{u} \in J_0)$, provided that the product of the numbers $c_1c_2 > 0$;
3) $\langle \tilde{z}_1 + \tilde{z}_2, \tilde{u} \rangle = \langle \tilde{z}_1, \tilde{u} \rangle + \langle \tilde{z}_2, \tilde{u} \rangle$ $(\forall \tilde{z}_1, \tilde{z}_2, \tilde{u} \in J_0)$;
4) $\langle \tilde{z}, \tilde{z} \rangle \geq 0$ $(\forall \tilde{z} \in J_0)$, and the condition $\langle \tilde{z}, \tilde{z} \rangle = 0$ is equivalent to vanishing the left and right indexes $\tilde{z}$;
5) Cauchy-Bunyakovsky Inequality $| \langle \tilde{z}, \tilde{u} \rangle | \leq \langle \tilde{z}, \tilde{z} \rangle^{1/2} \langle \tilde{u}, \tilde{u} \rangle^{1/2}$ $(\forall \tilde{z}, \tilde{u} \in J_0)$.

The quasi-scalar product of the form $\langle \tilde{z}, \tilde{u} \rangle_1 = \int_0^1 (mid\ Z_\alpha\, mid\ U_\alpha) d\alpha$ is considered in [6]. It is easy to see that in this case, turning the $\langle \tilde{z}, \tilde{z} \rangle_1^{1/2}$ to zero does not guarantee that the left and right indexes of $\tilde{z}$ are equal to zero.

Other definitions of the scalar product of fuzzy numbers are also found in the literature (see, for example, [10]).

Note the following relationship between the quasi-scalar product (2) and the distance (1).

Set the fuzzy number $\tilde{z}$ with indexes $z^-(\alpha)$ and $z^+(\alpha)$ match the vector function $\bar{z}(\alpha) = (z^-(\alpha), z^+(\alpha))^T$. Scalar product of $\langle \bar{z}, \bar{u} \rangle$ vector functions $\bar{z}$ and $\bar{u}$ define by equality (2). Then $\rho(\tilde{z}, \tilde{u}) = ||\bar{z} - \bar{u}||$.

We introduce the concept of the cosine of the angle between fuzzy numbers. For fuzzy numbers $\tilde{z}$, $\tilde{u} \in J_0$, we put

$$cos(\tilde{z}, \tilde{u}) = \frac{\langle \tilde{z}, \tilde{u} \rangle}{\langle \tilde{z}, \tilde{z} \rangle^{1/2} \langle \tilde{u}, \tilde{u} \rangle^{1/2}}.$$

Note the properties of the cosine.

1. $|cos(\tilde{z}, \tilde{u})| \leq 1 \ (forall \tilde{z}, \tilde{u} \in J_0)$.
   This follows from the Cauchy-Bunyakovsky inequality.
2. $cos(\tilde{z}, \tilde{u}) = 0$, if and only if $\tilde{z}$ and $\tilde{u}$ quasi-orthogonal.
3. $cos(\tilde{z}, \tilde{u}) = 1$, if and only if $\tilde{z}$ and $\tilde{u}$ are collinear, i.e. there is a number $\lambda > 0$ such that $\tilde{z} = \lambda \tilde{u}$.

Indeed, $cos(\tilde{z}, \tilde{u})$ matches $cos(\bar{z}, \bar{u})$, where $\bar{z}, \bar{u}$ are vector functions corresponding to $\tilde{z}$ and $\tilde{u}$, respectively. Then the condition $cos(\tilde{z}, \tilde{u}) = cos(\bar{z}, \bar{u}) = 1$ means that $\bar{z} = \lambda \bar{u}$, where $\lambda > 0$. This is equivalent to $\tilde{z} = \lambda \tilde{u}$, i.e. the fuzzy numbers $\tilde{z}$ and $\tilde{u}$ collinear.

2 Fuzzy Random Variables and Their Averages

Let (Ω, Σ, P) be a probability space, where Ω is a set of elementary events, Σ is a σ-algebra consisting of subsets of the set Ω, and P is a probability measure.

A measurable map $\tilde{X} : \Omega \to J_0$ is called a fuzzy random variable if, for any $\omega \in \Omega$, the set $\tilde{X}(\omega)$ is a fuzzy number from J_0.

Consider the intervals of α - levels of a fuzzy random variable $\tilde{X}$ for a fixed ω. Namely, $X_\alpha(\omega) = \{t \in R : \mu_{\tilde{X}(\omega)} \geq \alpha\}$, where $\mu_{\tilde{X}(\omega)}$ - membership function of a fuzzy number $\tilde{X}(\omega)$, and $\alpha \in (0, 1]$. The interval $X_\alpha(\omega)$ represent as $X_\alpha(\omega) = [X^-(\omega, \alpha), X^+(\omega, \alpha)]$, where the boundaries are $X^-(\omega, \alpha)$ and $X^+(\omega, \alpha)$ - random variables. They are called, respectively, the left and right index of the fuzzy random variable $\tilde{X}$.

Below, we will consider the class $\mathfrak{X}$ of fuzzy random variables $\tilde{X}$, for which indexes $X^-(\omega, \alpha)$ and $X^+(\omega, \alpha)$ are functions that are quadratically summable by $\Omega \times [0, 1]$.

Put

$$x^-(\alpha) = \int_\Omega X^-(\omega, \alpha) dP, \quad x^+(\alpha) = \int_\Omega X^+(\omega, \alpha) dP. \tag{3}$$

A fuzzy number $\tilde{x}$ with indexes defined by formula (3) is called the fuzzy expectation of a fuzzy random variable $\tilde{X}$.

Let $X_\alpha(\omega) = [X^-(\omega, \alpha), X^+(\omega, \alpha)]$ - interval α - level of the fuzzy random variable $\tilde{X}$. Put $mid\, X_\alpha(\omega) = \frac{1}{2}(X^+(\omega, \alpha) + X^-(\omega, \alpha))$ and $rad\, X_\alpha(\omega) = \frac{1}{2}(X^+(\omega, \alpha) - X^-(\omega, \alpha))$.

Expectation $E(\tilde{X})$ a fuzzy random variable $\tilde{X}$ is a number defined by the expression

$$E(\tilde{X}) = \int_0^1 \int_\Omega mid\ X_\alpha(\omega) dP d\alpha = 0.5 \int_0^1 \int_\Omega (X^-(\omega,\alpha) + X^+(\omega,\alpha)) dP d\alpha. \quad (4)$$

Note the equality

$$E(\tilde{X}) = \int_0^1 mid\ X_\alpha d\alpha = 0.5 \int_0^1 (X^-(\alpha) + X^+(\alpha)) d\alpha,$$

where $X^-(\alpha)$ and $X^+(\alpha)$ are determined by formulas (3).

We define a quasi-scalar product for fuzzy random variables $\tilde{X}$ and $\tilde{Y}$ with α - level sets $X_\alpha(\omega) = [X^-(\omega,\alpha), X^+(\omega,\alpha)]$ and $Y_\alpha(\omega) = [Y^-(\omega,\alpha), Y^+(\omega,\alpha)]$ formula

$$\left\langle \tilde{X}, \tilde{Y} \right\rangle = \int_0^1 \int_\Omega (mid\ X_\alpha(\omega)\, mid\ Y_\alpha(\omega) + rad\ X_\alpha(\omega)\, rad\ Y_\alpha(\omega)) dP d\alpha$$

$$= 0.5 \int_0^1 \int_\Omega (X^+(\omega,\alpha) Y^+(\omega,\alpha) + X^-(\omega,\alpha) Y^-(\omega,\alpha)) dP d\alpha. \quad (5)$$

In this case, the quasinorm of the fuzzy random variable $\tilde{X}$ will be denoted $||\tilde{X}|| = \left\langle \tilde{X}, \tilde{X} \right\rangle^{1/2}$.

Note that the same properties 1)–5) hold for the quasi-scalar product (5) as for the quasi-scalar product of fuzzy numbers.

Some other definitions of the scalar product of fuzzy random variables may be found in [6,10], and others.

Fuzzy random variables $\tilde{X}$ and $\tilde{Y}$ with α-level intervals $[X(\omega,\alpha)^-, X(\omega,\alpha)^+]$ and $[Y(\omega,\alpha)^-, Y(\omega,\alpha)^+]$ are called independent if the random variables $X(\omega,\alpha)^-$ and $Y(\omega,\alpha)^-$, as well as $X(\omega,\alpha)^+$ and $Y(\omega,\alpha)^+$ are pairwise independent for all $\alpha \in (0,1]$.

It is easy to check.

Statement 1. *for independent fuzzy random variables* $\tilde{X}$ *and* $\tilde{Y}$ *their quasiscalar product* $\left\langle \tilde{X}, \tilde{Y} \right\rangle = \langle \tilde{x}, \tilde{y} \rangle$, *where* $\tilde{x}, \tilde{y}$*-fuzzy expectations* $\tilde{x}$ *and* $\tilde{y}$, *respectively.*

Define the distance between fuzzy random variables $\tilde{X}$ and $\tilde{Y}$ of class $\mathfrak{X}$ expression

$$d(\tilde{X}, \tilde{Y}) = (\int_0^1 \int_\Omega ([X^-(\omega,\alpha) - Y^-(\omega,\alpha)]^2$$

$$+ \left[X^+(\omega,\alpha) - Y^+(\omega,\alpha)\right]^2)dPd\alpha)^{1/2}. \quad (6)$$

Definition (6) corresponds to the definition of the distance between fuzzy numbers (1). Other definitions of the distance between fuzzy-random variables are used, for example, in the works [3,7,10].

It turns out that the expectation $E(\tilde{X})$ and the fuzzy expectation $\tilde{x}$ of a fuzzy random variable $\tilde{X}$ have certain extreme properties with respect to distances (1) and (6), respectively.

Denote by $\hat{y}$ a singleton corresponding to the number $y \in R$, i.e. a fuzzy number characterized by the membership function $\mu_{\hat{y}}(x)$ equal to 1 for $x = y$ and zero in other cases. By definition, all left and right indexes of $\hat{y}$ are equal to y.

The following statements take place.

Statement 2. *For a given fuzzy random variable $\tilde{X}$ with indexes $X^-(\omega,\alpha)$, $X^+(\omega,\alpha)$ its expectation is $E(\tilde{X})$ is the only solution to the extreme problem*

$$d(\tilde{X},\hat{y}) \to \min \quad (\forall y \in R),$$

where the distance is $d(\tilde{X},\hat{y})$ is defined by the formula (6).

Statement 3. *Expectation $E(\tilde{X})$ is the only solution to the following extreme problem*

$$\rho(\tilde{x},\hat{y}) \to \min \ (\forall y \in R),$$

where is the distance ρ defined by the formula (2).

These statements are verified by applying an extreme sign for scalar differentiable functions $f(y) = d^2(\tilde{X},y)$ and $g = \rho^2(\tilde{x},y)$, respectively, taking into account the expectation definition $E(\tilde{X})$ and fuzzy expectation $\tilde{x}$ of a fuzzy random variable $\tilde{X}$.

The following theorem is true.

Theorem 1. *The fuzzy expectation $\tilde{x}$ of a fuzzy random variable $\tilde{X}$ is the solution to the following extreme problem*

$$d(\tilde{X},\tilde{y}) \to \min \ (\forall \tilde{y} \in J_0).$$

The proof is just to check equality

$$d^2(\tilde{X},\tilde{y}) = d^2(\tilde{X},\tilde{x}) + \rho^2(\tilde{x},\tilde{y}) \ \ (\forall \tilde{y} \in J_0).$$

We emphasize that the average fuzzy random variables in various aspects are widely discussed in the literature. However, their extreme properties were not previously observed.

3 The Law of Large Numbers

According to [8] we define the covariance between fuzzy random variables $\tilde{X}$ and $\tilde{Y}$ with intervals of α-level $[X^-(\omega,\alpha)X^+(\omega,\alpha)]$ and $[Y^-(\omega,\alpha),Y^+(\omega,\alpha)]$ by formula

$$cov[\tilde{X},\tilde{Y}] = 0.5\int_0^1\int_\Omega((X^-(\omega,\alpha)-x^-(\alpha))(Y^-(\omega,\alpha)-y^-(\alpha))$$

$$+(X^+(\omega,\alpha)-x^+(\alpha))(Y^+(\omega,\alpha)-y^+(\alpha)))dPd\alpha. \quad (7)$$

where $x^-(\alpha)$ and $x^+(\alpha)$ defined by formulas (3) and similarly $y^-(\alpha)$ and $y^+(\alpha)$.

This definition is convenient for us because it is closely related to the quasi-scalar product (5) and distance (6) that we have introduced. Various definitions of covariance of fuzzy random variables are found in the literature. In particular, in [6], the covariance $cov[\tilde{X},\tilde{Y}]$ fuzzy random variables $\tilde{X}$, $\tilde{Y}$ is similar to (7) expression

$$cov[\tilde{X},\tilde{Y}] = 0.25\int_0^1\int_\Omega(X^-(\omega,\alpha)+X^+(\omega,\alpha)-x^-(\alpha)$$

$$-x^+(\alpha))(Y^-(\omega,\alpha)+Y^+(\omega,\alpha)-y^-(\alpha)-y^+(\alpha))dPd\alpha.$$

However, it is easy to see that this formula actually includes covariances of random variables $mid\ X_\alpha(\omega)$ and $mid\ Y_\alpha(\omega)$, but not for $rad\ X_\alpha(\omega)$ and $rad\ Y_\alpha(\omega)$.

In this sense, the formula (7) used below more adequately reflects the structure of fuzzy random variables.

The definition of covariance (7) has a lot of properties, that are such a modification of the case of real random variables (see, [8]).

$$1)\ cov[\tilde{X}+\tilde{Z},\tilde{Y}] = cov[\tilde{X},\tilde{Y}] + cov[\tilde{Z},\tilde{Y}];$$

$$2)\ cov[c_1\tilde{X},c_2\tilde{Y}] = c_1c_2 cov[\tilde{X},\tilde{Y}],$$

for any real $c_1,c_2 \in R$ such that $c_1c_2 > 0$.

This definition of sum fuzzy random variables and product of fuzzy random variable with real number understands as the respective definition for fuzzy numbers above.

A specific property of the covariance of fuzzy random variables defined by formula (7) with the quasi-scalar product (5) introduced by us (and not noted in [8]) is the following

$$3.\ cov[\tilde{X},\tilde{Y}] = \left\langle\tilde{X},\tilde{Y}\right\rangle - \langle\tilde{x},\tilde{y}\rangle.$$

This property (for other definitions of covariance) was considered, for example, in [6,10].

As usual, fuzzy random variables $\tilde{X}_1, \tilde{X}_2$ are called uncorrelated if $cov[\tilde{X}_1, \tilde{X}_2] = 0$.

Remark 1. If the fuzzy random variables $\tilde{X}_1, \tilde{X}_2$ are independent, they are uncorrelated.

This follows from property 3 of the covariance given statement 1.

Remark 2. If the fuzzy random variables $\tilde{X}_1, \tilde{X}_2$ are uncorrelated, then $\left\langle \tilde{X}_1, \tilde{X}_2 \right\rangle = \langle \tilde{x}_1, \tilde{x}_2 \rangle$, where $\tilde{x}_1, \tilde{x}_2$ - fuzzy expectations $\tilde{x}_1$ and $\tilde{x}_2$. Conversely, if the previous equality is satisfied, the fuzzy random variables $\tilde{X}_1, \tilde{X}_2$ are uncorrelated.

This follows from property 3 of the covariance.

We define the variance of the fuzzy random variable $\tilde{X}$ the equation $D(\tilde{X}) = cov[\tilde{X}, \tilde{X}]$ and note its properties (cf. [8]):

1. $D(c\tilde{X}) = c^2 D(\tilde{X})$ for any real number c.
2. $D(\tilde{X} + \tilde{Y}) = D(\tilde{X}) + D(\tilde{Y}) + 2cov[\tilde{X}, \tilde{Y}]$ for $\forall \tilde{X}, \tilde{Y} \in \mathfrak{X}$.
3. $D(\tilde{z}) = 0$ for any fuzzy number $\tilde{z} \in J_0$.

   Important for us is the following special property of the dispersion
4. $$D(\tilde{X}) = \frac{1}{2} d^2(\tilde{X}, \tilde{x}) \ \ (\forall \tilde{X} \in \mathfrak{X}),$$

where $\tilde{x}$ is the fuzzy expectation of a fuzzy random variable $\tilde{X}$, and $d^2(\tilde{X}, \tilde{x})$ is the distance defined by the formula (6).

It follows from the equality $D(\tilde{X}) = cov[\tilde{X}, \tilde{X}]$ and the definitions (6), (7).

Consider for fuzzy random variables looks Chebyshev's inequality (see, e.g., [12], Chap. 6, Sect. 32 to "normal" random variables).

Lemma 1 (Chebyshev's Inequality). *For a fuzzy random variable $\tilde{X}$ with a fuzzy expectation $\tilde{x}$ and a given $\varepsilon > 0$, the inequality occurs*

$$P(d(\tilde{X}, \tilde{x}) \geq \varepsilon) \leq \frac{2}{\varepsilon^2} D(\tilde{X}). \tag{8}$$

Indeed, by the probability properties

$$P(d(\tilde{X}, \tilde{x}) \geq \varepsilon) = \int\limits_{d(\tilde{X}, \tilde{x}) \geq \varepsilon} dP.$$

Since in the integration domain $\frac{d^2(\tilde{X}, \tilde{x})}{\varepsilon^2} \geq 1$, then

$$\int\limits_{d(\tilde{X}, \tilde{x}) \geq \varepsilon} dP \leq \frac{1}{\varepsilon^2} \int\limits_{\Omega} d^2(\tilde{X}, \tilde{x}) dP = \frac{1}{\varepsilon^2} d^2(\tilde{X}, \tilde{x}).$$

Whence, taking into account the property 4 of the variance, follows (8).

Inequality (8) is similar to the corresponding inequality from [8], but it uses a different definition of distance.

Let's look how the law of large numbers turns out in the case of fuzzy random variables. There are a significant number of publications on this subject (see, for example, [7,13–15]). The main difference is in determining the distance between fuzzy numbers (respectively, between fuzzy random variables).

Theorem 2 (Law of large numbers). *Let $\tilde{X}_1, \tilde{X}_2, ..., \tilde{X}_n$ be a collection of pairwise uncorrelated fuzzy random variables with fuzzy expectations $\tilde{x}_i$. Let their variances be uniformly bounded, i.e. there is a constant $c > 0$ such that $D(X_i) \leq c$ $(i = 1, ..., n)$. Then the relation is valid*

$$P(d(\frac{1}{n}\sum_{i=1}^{n}\tilde{X}_i, \frac{1}{n}\sum_{i=1}^{n}\tilde{x}_i) \geq \varepsilon) \leq \frac{2c}{n\varepsilon^2}. \tag{9}$$

Indeed, putting the Chebyshev's inequality $\tilde{X} = \frac{1}{n}\sum\limits_{i=1}^{n}\tilde{X}_i$, get

$$P(d(\frac{1}{n}\sum_{i=1}^{n}\tilde{X}_i, \frac{1}{n}\sum_{i=1}^{n}\tilde{x}_i) \leq \frac{2}{\varepsilon^2}D(\frac{1}{n}\sum_{i=1}^{n}\tilde{X}_i).$$

Further, under the properties 1, 2 of the variance we have

$$D(\frac{1}{n}\sum_{i=1}^{n}\tilde{X}_i) = \frac{1}{n^2}D(\sum_{i=1}^{n}\tilde{X}_i) = \frac{1}{n^2}\sum_{i=1}^{n}D(\tilde{X}_i) \leq \frac{c}{n}.$$

Hence the result.

Inequality (9) implies

Corollary 1. *In the conditions of Theorem 2 the relation is valid*

$$P(d(\frac{1}{n}\sum_{i=1}^{n}\tilde{X}_i, \frac{1}{n}\sum_{i=1}^{n}\tilde{x}_i) < \varepsilon) \geq 1 - \frac{2c}{n\varepsilon^2}. \tag{10}$$

The law of large numbers means that the probability on the left in (10) tends to 1 for $n \to \infty$.

Let's consider an important special case of the law of large numbers. It is said that fuzzy random variables $\tilde{X}$ and $\tilde{Y}$ with intervals of α - levels $[X^-(\omega,\alpha), X^+(\omega,\alpha)]$ and $[Y^-(\omega,\alpha), Y^+(\omega,\alpha)]$ are equally distributed if $X^-(\omega,\alpha)$ and $Y^-(\omega,\alpha)$, and $X^+(\omega,\alpha)$ and $Y^+(\omega,\alpha)$, are equally distributed for all $\alpha \in [0,1]$.

It is said that $\tilde{X}_1, \tilde{X}_2, ..., \tilde{X}_n$ is a fuzzy random sample if $\tilde{X}_i$ are independent and equally distributed. Theorem 2 implies

Corollary 2. *Let $\tilde{X}_1, \tilde{X}_2, ..., \tilde{X}_n$ be a fuzzy random sample and $\tilde{x}$ be a fuzzy expectation for each of the fuzzy random variables $\tilde{X}_i$. Then*

$$P(d(\frac{1}{n}\sum_{i=1}^{n}\tilde{X}_i,\tilde{x}) < \varepsilon) \geq 1 - \frac{2c}{n\varepsilon^2},$$

where c is the variance of the fuzzy random variable $\tilde{X}_i$.

Moreover, under conditions of Corollary 2 and properties 1)–4) of the variance the convergence on metric (6) of $\frac{1}{n}\sum_{i=1}^{n}\tilde{X}_i$ to $\tilde{x}$ is valid, when $n \to \infty$.

4 Linear Regression

Let's consider the optimal linear approximation of a (predicted) fuzzy random variable $\tilde{Y}$ using a system of (predictive) fuzzy 1random variables $\tilde{X}_1, \tilde{X}_2, ..., \tilde{X}_n$. In a number of works [8,11,15–17] and other tasks of this kind were considered. In this case, the specifics of the problem are determined by the choice of the distance to be minimized. We investigate the question of approximating a fuzzy random variable $\tilde{Y}$ - the linear combinations $\sum_{i=1}^{n}\beta_i\tilde{X}_i$ with real coefficients β_i $(i = 1, ..., n)$ by the criterion of minimizing the distance (6).

Consider first the extreme challenge with nonnegative coefficients $\beta_i \geq 0$ $(i = 1, ..., n)$

$$d(\tilde{Y},\sum_{i=1}^{n}\beta_i\tilde{X}_i) \to \min \quad (\forall\beta_i \geq 0). \tag{11}$$

Takes place

Lemma 2. *Let the fuzzy random variables $\tilde{X}_i$ be quasi-orthogonal for $i \neq j$, and their quasinorms $\varkappa_j := \left\langle \tilde{X}_j, \tilde{X}_j \right\rangle^{1/2} \neq 0$ $(j = 1, ..., n)$. Let the condition $b_i = \left\langle \tilde{Y}, \tilde{X}_i \right\rangle \geq 0$ $(i = 1, ..., n)$. then problem (11) has a non-negative solution, and the only one. It has the form $\beta_i^* = \frac{b_i}{\varkappa_i^2}$, $(i = 1, ..., n)$.*

Indeed, due to the assumption that the coefficients β_i are non-negative, the left index $\sum_{i=1}^{n}\beta_i\tilde{X}_i$ is equal to $\sum_{i=1}^{n}\beta_i X_i^-(\omega,\alpha)$, and the right one is $\sum_{i=1}^{n}\beta_i X_i^+(\omega,\alpha)$. We will omit the arguments ω, α in the proof below. Put

$$F(\beta_1, ..., \beta_n) = d^2(\tilde{Y},\sum_{i=1}^{n}\beta_i\tilde{X}_i)$$

$$= \int_0^1\int_\Omega((Y^+ - \sum_{i=1}^{n}\beta_i X_i^+)^2 + (Y^- - \sum_{i=1}^{n}\beta_i X_i^-)^2)dPd\alpha. \tag{12}$$

This is a quadratic form in $\beta_1, ..., \beta_n$.

Differentiate with respect to (12) for β_j and equate the derivative to zero

$$\frac{\partial F}{\partial \beta_j} = -2\int_0^1\int_\Omega ((Y^+ - \sum_{i=1}^n \beta_i X_i^+)X_j^+ + (Y^- - \sum_{i=1}^n \beta_i X_i^-)X_j^-)dPd\alpha = 0.$$

Hence, for every $j = 1, 2, ..., n$ we have

$$\sum_{i=1}^n \beta_i \int_0^1\int_\Omega (X_i^+ X_j^+ + X_i^- X_j^-)dPd\alpha = \int_0^1\int_\Omega (Y^+ X_j^+ + Y^- X_j^-)dPd\alpha,$$

i.e.

$$\sum_{i=1}^n \beta_i \left\langle \tilde{X}_i, \tilde{X}_j \right\rangle = \left\langle \tilde{Y}, \tilde{X}_j \right\rangle \ (j = 1, ..., n), \tag{13}$$

We introduce the following notation. Vector B with coefficients $b_i = \left\langle \tilde{Y}, \tilde{X}_i \right\rangle$, matrix A with coefficients $A_{ij} = \left\langle \tilde{X}_i, \tilde{X}_j \right\rangle$, vector β with coefficients β_i. In vector form, system (13) has the form $A\beta = B$. Matrix A due to the quasi-orthogonality of the system $\{\tilde{X}_i\}$ has a diagonal form, with positive numbers on the main diagonal $\varkappa_i^2$ $(i = 1, ..., n)$. then the solution is $\beta^* = A^{-1}B$, i.e. $\beta_i^* = \frac{b_i}{\varkappa_i^2}$ $(i = 1...n)$.

The nonnegativity of the obtained coefficients β_i^* is provided by the condition $b_i = \left\langle \tilde{Y}, \tilde{X}_i \right\rangle \geq 0$ $(i = 1, ..., n)$.

To verify that $\beta^* = A^{-1}B$ is the minimum point, consider the second derivative

$$\frac{\partial^2 F}{\partial \beta_j \partial \beta_s} = 2\int_0^1\int_\Omega 4(X_s^+ X_j^+ + X_s^- X_j^-)dPd\alpha = \left\langle \tilde{X}_s, \tilde{X}_j \right\rangle \text{ when } s \neq j.$$

$$\frac{\partial^2 F}{\partial \beta_j^2} = 2\int_0^1\int_\Omega ((X_j^+)^2 + (X_j^-)^2)dPd\alpha = 4\left\langle \tilde{X}_j, \tilde{X}_j \right\rangle \text{ when } s = j.$$

A sufficient sign of a minimum is the positive definiteness of the Hesse matrix $\{\frac{\partial^2 F}{\partial \beta_j \partial \beta_s}\}$. And this is provided by the quasi-orthogonality of the system $\{\tilde{X}_j\}$.

Remark 3. Condition $\left\langle \tilde{Y}, \tilde{X}_i \right\rangle > 0$ means that there is an acute angle between the fuzzy random variables $\tilde{Y}$ and $\tilde{X}_i$. In other words, the fuzzy-random variables $\tilde{Y}$ and $\tilde{X}_i$ increase (in this sense) or decrease at the same time.

Remark 4. In the conditions of Lemma 2, we can reject the requirement of pairwise quasi-orthogonality of fuzzy random variables $\tilde{X}_1, ..., \tilde{X}_n$. It is sufficient to require positive invertibility of their Gram matrix A with coefficients $a_{ij} = \left\langle \tilde{X}_i, \tilde{X}_j \right\rangle$. In the sense that the inverse matrix A^{-1} exists and converts vectors with non-negative coordinates back to vectors with non-negative coordinates.

Remark 5. If, under Lemma 2, we reject the requirement of pairwise quasi-orthogonality of fuzzy random variables $\tilde{X}_1, ..., \tilde{X}_n$, but additionally assume their pairwise uncorrelability, then it is sufficient to require positive invertibility of the Gram matrix from their fuzzy expectations $\langle \tilde{x}_i, \tilde{x}_j \rangle$.

We emphasize that the coefficients β_i^* are analogous to the Fourier coefficients when decomposing in an orthogonal system in a Hilbert space (see, for example, [17], Chap. II, Sect. 11 for random variables). This is due to the relationship of the metric to be minimized in problem (11) with quasi-scalar product (5).

The proximity of fuzzy random variables (as well as any space with a scalar product) can be characterized by the cosine of the angle between them.

Define the cosine between the fuzzy random variables $\tilde{Y}, \tilde{Z}$ by the equality

$$\cos(\tilde{Y}, \tilde{Z}) = \frac{\left\langle \tilde{Y}, \tilde{Z} \right\rangle}{||\tilde{Y}||||\tilde{Z}||}. \tag{14}$$

According to the definition (14) and the properties of the cosine of the angle between the fuzzy numbers $|cos(\tilde{Y}, \tilde{Z})| \leq 1$. In this case, $cos(\tilde{Y}, \tilde{Z}) = 0$, if and only, if $\tilde{Y}$ and $\tilde{Z}$ are quasi-orthogonal. And $cos(\tilde{Y}, \tilde{Z}) = 1$, if and only, if $\tilde{Z} = \lambda\tilde{Y}$ are collinear ($\lambda > 0$).

Denote, as in Lemma 2, $\beta_i^* = \frac{1}{\varkappa_i^2} \left\langle \tilde{Y}, \tilde{X}_i \right\rangle$ and consider

$$\tilde{Z}_n^* = \sum_{i=1}^{n} \beta_i^* \tilde{X}_i \tag{15}$$

- an optimal estimate of the predicted fuzzy random variable $\tilde{Y}$ from Lemma 2.

Theorem 3. *Let the conditions of Lemma 2. Then the optimal estimate (15) has the maximum cosine with the predicted fuzzy random variable $\tilde{Y}$ in the class of linear estimates of the form $Z_n = \sum\limits_{i=1}^{n} \beta_i \tilde{X}_i$ $(\beta_i \geq 0)$.*

Indeed, we will show that

$$|cos(\tilde{Y}, \tilde{Z}_n)| \leq cos(\tilde{Y}, \tilde{Z}_n^*).$$

Due to the properties of the quasi-scalar product and the non-negativity of the coefficients $\beta_i^* \geq 0$ we have

$$\left\langle \tilde{Y}, \tilde{Z}_n^* \right\rangle = \left\langle \tilde{Y}, \sum_{i=1}^{n} \beta_i^* \tilde{X}_i \right\rangle = \sum_{i=1}^{n} \beta_i^* \left\langle \tilde{Y}, \tilde{X}_i \right\rangle = \sum_{i=1}^{n} (\beta_i^*)^2 \varkappa_i^2.$$

In this case, due to the pairwise quasi-orthogonality of the system $||Z_n^*||^2 = \sum\limits_{i=1}^{n} (\beta_i^*)^2 \varkappa_i^2$. Then

$$cos(\tilde{Y}, \tilde{Z}_n^*) = \frac{\sum\limits_{i=1}^{n} \varkappa_i^2 (\beta_i^*)^2}{||\tilde{Y}||(\sum\limits_{i=1}^{n} \varkappa_i^2 (\beta_i^*)^2)^{1/2}} = \frac{1}{||\tilde{Y}||} (\sum_{i=1}^{n} \varkappa_i^2 (\beta_i^*)^2)^{1/2}.$$

Consider

$$\left\langle \tilde{Y}, \tilde{X}_i \right\rangle = \sum_{i=1}^{n} \beta_i \left\langle \tilde{Y}, \tilde{X}_i \right\rangle = \sum_{i=1}^{n} \beta_i \beta_i^* \varkappa_i^2$$

and $||\tilde{Z}_n||^2 = \sum_{i=1}^{n} \beta_i^2 \varkappa_i^2$.

Then

$$cos(\tilde{Y}, \tilde{Z}_n) = \frac{\sum_{i=1}^{n} \varkappa_i^2 \beta_i \beta_i^*}{||\tilde{Y}||(\sum_{i=1}^{n} \beta_i^2 \varkappa_i^2)^{1/2}}.$$

By the Cauchy-Schwarz inequality

$$|cos(\tilde{Y}, \tilde{Z}_n)| \leq \frac{(\sum_{i=1}^{n} \varkappa_i^2 \beta_i^2)^{1/2} (\sum_{i=1}^{n} \varkappa_i^2 (\beta_i^*)^2)^{1/2}}{||\tilde{Y}||(\sum_{i=1}^{n} \varkappa_i^2 \beta_i^2)^{1/2}}$$

$$= \frac{1}{||\tilde{Y}||} (\sum_{i=1}^{n} \varkappa_i^2 (\beta_i^*)^2)^{1/2} = cos(\tilde{Y}, \tilde{Z}_n^*),$$

which was required to be proved.

Let's consider the optimal regression problem in a situation where all linear approximation coefficients are not assumed to be nonnegative and the condition $\left\langle \tilde{Y}, \tilde{X}_i \right\rangle \geq 0$ $(i = 1, ..., n)$ is met.

Note that the explicit form of the formula for the distance $d(\tilde{Y}, \sum_{i=1}^{n} \beta_i \tilde{X}_i)$ in the case of coefficients β_i of an arbitrary sign is inconvenient for research, since in this case the product of the interval α - the level of a fuzzy number $\tilde{z}$ by a clear number β is given by the cumbersome expression

$$\beta[z^-, z^+] = [\min\{\beta z^-, \beta z^+\}, \max\{\beta z^-, \beta z^+\}].$$

However, in the general situation, the following statement is true. Let's say $c_* = \max_{j=1,...,n} \{\frac{||\tilde{Y}||}{||\tilde{X}_j||}\}$.

Theorem 4. *Let the fuzzy-random variables* $\tilde{X}_i$ *be quasi-orthogonal for* $i \neq j$, *and all their quasinorms* $\varkappa_i \neq 0$ $(i = 1, ..., n)$. *then the problem is*

$$d(\tilde{Y}, \sum_{i=1}^{n} \beta_i \tilde{X}_i) \to \min \; (\beta_i \in [-c_*, \infty)) \tag{16}$$

has a solution, and the only one. It has the form $\beta_i^* = \frac{b_i}{\varkappa_i^2}$ $(i = 1, ..., n)$.

Note though problem (16) does not assume that the coefficients β_i are positive, the formula for the coefficients β_i has the same form as in Lemma 2. At the same time the coefficients b_i in the condition of Theorem 4 may have different signs.

In the proof of Theorem 4, the following special property of the distance (6) between fuzzy random variables will be used.

Lemma 3. *For any fuzzy random variables* $\tilde{X}$, $\tilde{Y}$ *and* $\tilde{W}$ *in* $\mathfrak{X}$ *the next equality holds*

$$d(\tilde{X} + \tilde{W}, \tilde{Y} + \tilde{W}) = d(\tilde{X}, \tilde{Y}).$$

In fact, this is true because subject to the rules of interval addition on the left are

$$(\tilde{X} + \tilde{W})^- = X^- + W^-,\ (\tilde{Y} + \tilde{W})^- = Y^- + W^-,$$

and similarly for the right indexes.

After substituting the corresponding expressions in (6), we obtain the required equality.

Proof of Theorem 4. Let the condition $\left\langle \tilde{Y}, \tilde{X}_i \right\rangle \geq 0$ not be satisfied for at least one j. Consider the fuzzy random variable $\tilde{Z} = \tilde{Y} + c_* \sum_{i=1}^{n} \tilde{X}_i$. According to the definition $c_* > 0$, $\left\langle \tilde{Z}, \tilde{X}_j \right\rangle \geq 0$ $(j = 1, ..., n)$. Consider for $\tilde{Z}$ task (11). Let $\gamma_i \geq 0$ be the optimal coefficients of a linear combination $\sum_{i=1}^{n} \gamma_i \tilde{X}_i$ for $\tilde{Z}$, obtained by solving problem (11). The vector γ with coordinates γ_i is defined by the formula $\gamma = A^{-1} f$, for $f_i = \left\langle \tilde{Z}, \tilde{X}_j \right\rangle$.

We show that the coefficients $\gamma_i - c_*$ are optimal for linear approximation of a fuzzy random variable $\tilde{Y}$ by the system $\{\tilde{X}_i\}$.

Consider the distance $d(\tilde{Y}, \sum_{i=1}^{n} (\gamma_i - c_*) X_i)$. By Lemma 3 and taking into account the definition of $\tilde{Z}$, we have

$$d(\tilde{Y}, \sum_{i=1}^{n} (\gamma_i - c_*) X_i) = d(\tilde{Y} + \sum_{i=1}^{n} c_* \tilde{X}_i, \sum_{i=1}^{n} \gamma_i \tilde{X}_i) = d(\tilde{Z}, \sum_{i=1}^{n} \gamma_i \tilde{X}_i). \qquad (17)$$

Since $\{\gamma_i\}$ - solution of problem (11) for $\tilde{Z}$, then in accordance with Lemma 2 for any set of numbers $\xi_i \geq 0$ $(i = 1, ..., n)$ can record

$$d(\tilde{Z}, \sum_{i=1}^{n} \gamma_i \tilde{X}_i) \leq d(\tilde{Z}, \sum_{i=1}^{n} \xi_i \tilde{X}_i) = d(\tilde{Y} + \sum_{i=1}^{n} c \tilde{X}_i, \sum_{i=1}^{n} (\xi_i - c_*) X_i + \sum_{i=1}^{n} c_* \tilde{X}_i)$$

Using Lemma 3 again, we get

$$d(\tilde{Z}, \sum_{i=1}^{n} \gamma_i \tilde{X}_i) \leq d(\tilde{Y}, \sum_{i=1}^{n} (\xi_i - c_*) \tilde{X}_i).$$

Then (17) implies the inequality

$$d(\tilde{Y}, \sum_{i=1}^{n} (\gamma_i - c_*) X_i) \leq d(\tilde{Y}, \sum_{i=1}^{n} (\xi_i - c_*) X_i).$$

Since here $(\xi_i - c_*)$ - arbitrary coefficients from a closed interval $[-c_*, \infty)$, then $(\gamma_i - c_*)$ - optimal coefficients.

Note that by definition $\tilde{Z}$ and according to Lemma 2

$$\gamma_j = \frac{1}{\varkappa_j^2} \left\langle \tilde{Z}, \tilde{X}_j \right\rangle = \frac{1}{\varkappa_j^2} \left\langle (\tilde{Y} + \sum_{i=1}^{n} c_* \tilde{X}_i), \tilde{X}_j \right\rangle.$$

Then, taking into account quasiorthogonality system $\{\tilde{X}_i\}$ will receive

$$\gamma_j = \frac{1}{\varkappa_j^2} \left(b_j + c_* \left\langle \tilde{X}_j, \tilde{X}_j \right\rangle \right) = \frac{b_j}{\varkappa_j^2} + c_*.$$

Hence, the optimal coefficients of $\tilde{\beta}_j$ for $\tilde{X}_j$ in the linear approximation $\tilde{Y}$, having the form $(\gamma_j - c_*)$, defined by the equality $\frac{b_j}{\varkappa_j^2}$ $(j = 1, ..., n)$. this is what the statement implies.

Remark 6. Similarly to Remark 4, under the conditions of theorem 4, one can reject the quasi-orthogonality of the system $\{\tilde{X}_i\}$ and instead assume positive invertibility of their Gram matrix. In addition, it is required that the sum of elements of all columns of the Gram matrix be positive.

References

1. Kwakernaak, H.: Fuzzy random variables - I. Definitions and theorems. Inf. Sci. **15**, 1–29 (1978)
2. Nahmias, S.: Fuzzy variables in a random environment. In: Advanced in Fuzzy Sets Theory. NHPC, Amsterdam (1979)
3. Puri, M.L., Ralesku, D.A.: Fuzzy random variables. J. Math. Anal. Appl. **114**, 409–422 (1986)
4. Nguyen, H.T., Wu, B.: Fundamentals of Statistics with Fuzzy Data, 204 p. Springer, Berlin (2006). https://doi.org/10.1007/11353492
5. Viertl, R.: Statistical Methods for Fuzzy Data, p. 268. Wiley, Chichester (2011)
6. Shvedov, A.S.: Estimating the means and the covariances of fuzzy random variables. Appl. Econ. **42**, 121–138 (2016)
7. Colubi, A.: Statistical inference about the means of fuzzy random variables: applica analysis of fuzzy-and real-valued data. Fuzzy Sets Syst. **160**, 344–356 (2009)

8. Feng, Y., Hu, L., Shu, H.: The variance and covariance of fuzzy random variables. Fuzzy Syst. **120**, 487–497 (2001)
9. Piegat, A.: Fuzzy Modeling and Control, 728 p. Springer, Heidelberg (2001). https://doi.org/10.1007/978-3-7908-1824-6
10. Nather, W.: Regression with fuzzy data. Comput. Stat. Data Anal. **51**(1), 235–252 (2006)
11. Bargiela, A., Pedrycz, W., Nakashima, T.: Multiple Regression with fuzzy data. Fuzzy Sets Syst. **158**, 2169–2188 (2007)
12. Gnedenko, B.V.: Theory of Probability, 520 p. CRC Press (1998)
13. Li, S., Ogura, Y., Kreinovich, V.: Limit Theorems and Applications of Set Valued and Fuzzy Valued Random Variables, p. 399. Kluwer Academic Publishers, Dordrecht (2002)
14. Colubi, A., Coppi, R., D'Urso, P., Gil, M.A.: Statistics with fuzzy random variables. Metron - Int. J. Stat. **65**, 277–303 (2007)
15. Akbari, M.G.H., Rezaei, A.H., Waghei, Y.: Statistical inference about the variance of fuzzy random variables. Sankhy: Indian J. Stat. **71–B**(Part 2), 206–221 (2009)
16. Gonzalez-Rodriguez, G., Blanco, A., Colubi, A., Lubiano, M.A.: Estimation of a simple linear regression model for fuzzy random variables. Fuzzy Sets Syst. **160**, 357–370 (2009)
17. Shiryaev, A.N.: Probability, 403 p. Springer, Heidelberg (1996)

Stochastic Approach to the Vanishing Viscosity Method

Yana Belopolskaya(✉)

SPbGASU, St. Petersburg 190005, Russia

Abstract. We derive stochastic counterparts for solutions of the forward Cauchy problem for two classes of nonlinear parabolic equations. We refer to the first class parabolic systems such that coefficients of the higher order terms are the same for each equation in the system and to the second class parabolic systems with different higher order term coefficients. With a simple substitution we reduce a system of the first class to a system which may be interpreted as a system of backward Kolmogorov equations and construct a probabilistic representation of its solution. A different approach based on interpretation of a system under consideration as a system of forward Kolmogorov equations is developed to deal with stochastic counterparts of the second class systems. These approaches allow to reduce the investigation of the vanishing viscosity limiting procedure to a stochastic level which makes its justification to be much easier.

Keywords: Systems of parabolic equations · Stochastic differential equations · Vanishing viscosity

1 Probabilistic Counterparts of Nonlinear Parabolic Systems

In this paper we consider the Cauchy problem for two types of nonlinear second order parabolic equations and construct probabilistic representations of solutions to the Cauchy problem for these systems. Namely, we consider the Cauchy problem for systems of the form

$$\frac{\partial u_m}{\partial t} + \sum_{i=1}^{d} \frac{\partial f_m^i(x,u)}{\partial x_i} = \frac{1}{2}\sum_{i,j=1}^{d} G^{ij}(x,u)\frac{\partial^2 u_m}{\partial x_i \partial x_j} + \sum_{q=1}^{d_1} c_{mq}(x,u)u_q, \qquad (1)$$

$$u_m(0,x) = u_{0m}(x),\ m = 1,\dots,d_1,$$

called reaction-diffusion systems and for systems of the form

$$\frac{\partial v_m}{\partial t} + \sum_{i=1}^{d} \frac{\partial f_m^i(y,v)}{\partial y_i} = \frac{1}{2}\sum_{q=1}^{d_1}\sum_{i,j=1}^{d} \nabla^2_{y_i,y_j}[G^{ij}_{mq}(y,v)v_q] + \sum_{q=1}^{d_1} c_{mq}(y,v)v_q, \qquad (2)$$

A. N. Shiryaev et al. (Eds.): ICSM-5 2020, PROMS 371, pp. 92–106, 2021.
https://doi.org/10.1007/978-3-030-83266-7_7

$$v_m(0,y) = u_{0m}(y),$$

called systems with cross-diffusion.

Our aim is to construct probabilistic representations of the Cauchy problem solutions for these systems and justify the existence of their vanishing viscosity limits.

A probabilistic approach based on the theory of stochastic differential equations (SDEs) to scalar nonlinear parabolic equations was started in papers by McKean [1] and Freidlin [2]. The approach suggested by McKean was developed to construct a solution of a nonlinear parabolic equation called the Vlasov equation which arises in plasma physics. The theory of the McKean-Vlasov type equations now is a well developed theory with many applications (see [3]). Some results concerning systems of McKean-Vlasov equations with coefficients which are functionals of equation solutions were obtained in [4–6]. On the other hand Freidlin's approach allows to deal with nonlinear scalar parabolic equations with coefficients depending on an unknown function pointwise. In general both approaches allow to construct stochastic equations for stochastic processes to be used in probabilistic representations of the original nonlinear Cauchy problem solution.

To extend the theory to the case of systems of nonlinear parabolic equations one meets some additional problems. In particular, Freidlin's approach was extended to systems of nonlinear parabolic equations by Dalecky and the author in [7,8]. In this paper we construct stochastic representations for solutions of the Cauchy problem for systems of nonlinear parabolic equations in the framework of both approaches and apply these representations to justify the vanishing viscosity method which allow to construct the Cauchy problem solutions for systems of hyperbolic equations.

To construct a probabilistic representation of a solution to a system of the form (1) we reduce it to a system of backward Kolmogorov equations. To this end we set

$$div[f_m(x,u)] = \sum_{i=1}^{d}\sum_{q=1}^{d_1} \frac{\partial f_m^i(x,u)}{\partial u_q}\frac{\partial u_q}{\partial x_i} + \sum_{i=1}^{d}\frac{\partial f_m^i(x,u)}{\partial x_i}$$

$$= \sum_{i=1}^{d}\sum_{q=1}^{d_1} B_{mq}^i(x,u)\frac{\partial u_q}{\partial x_i} + \kappa_m(x,u)$$

and rewrite (1) in a suitable form

$$\frac{\partial u_m}{\partial t} + \sum_{i=1}^{d}\sum_{q=1}^{d_1} B_{mq}^i(x,u)\frac{\partial u_q}{\partial x_i} = \frac{1}{2}\sum_{i,j=1}^{d} G^{ij}(x,u)\frac{\partial^2 u_m}{\partial x_i \partial x_j} + \sum_{q=1}^{d_1} c_{mq}(x,u)u_q \quad (3)$$

$$-\kappa_m(x,u), \quad u_m(0,x) = u_{0m}(x), \quad m = 1,\dots,d_1.$$

Then we introduce functions $g_m(T-t,x) = u_m(t,x)$ and reduce (3) to the backward Cauchy problem

$$\frac{\partial g_m}{\partial t} + \frac{1}{2}\sum_{i,j,k=1}^{d} A^{ik}(y,g)\frac{\partial^2 g_m}{\partial x_i \partial x_j}A^{kj}(y,g) - \sum_{q=1}^{d_1}\sum_{i=1}^{d} B^i_{mq}(x,g)\frac{\partial g_q}{\partial x_i} \tag{4}$$

$$+\sum_q c_{mq}(x,g)g_q - \kappa_m(x,g) = 0, \quad g_m(T,x) = u_{0m}(x),$$

with respect to g assuming that $G^{ij} = \sum_{k=1}^d A^{ik}A^{kj}$.

Below we will use notations $\nabla_{x_i} = \frac{\partial}{\partial x_i}, \nabla^2_{x_i,x_j} = \frac{\partial^2}{\partial x_i \partial x_j}$ and $TrA\nabla^2 gA = \sum_{i,j,k=1}^d A^{ik}\frac{\partial^2 g}{\partial x_i \partial x_j}A^{kj}$.

We say that condition **C 1** holds if the functions $A(x,u) \in R^d \otimes R^d, c(x,u) \in R^{d_1} \otimes R^{d_1}, C(x,u) \in R^d \otimes R^{d_1} \otimes R^{d_1}$ satisfy the estimates

$$|A(x,u) - A(x_1,u_1)| \le L\|x-x_1\|^2 + L_1|u-u_1|^2, \quad x \in R^d, u \in R^{d_1}$$

$$|A(x,u)|^2 \le K[1 + \|x\|^2 + |u|^2],$$

$$|c(x,u) - c(x^1,u^1)|^2 \le L\|x-x^1\|^2 + L_1\|u-u^1\|^2,$$

$$|[C(x,u) - C(x^1,u^1)]y|^2 \le L\|x-x^1\|^2 + L_1\|u-u^1\|^2\|y\|^2,$$

$$|C(x,u)y\|^2 \le \rho\|u\|^2\|y\|^2, \quad y \in R^d, \; c(x,u)h\cdot h \le [\rho_0 + \rho\|u\|^2]\|h\|^2, h \in R^{d_1},$$

where $L_1 > 0$ depends on $\max(|u|,|u^1|)$, $u_0(x) \in R^{d_1}$, $\kappa(x,u) \in R^{d_1}$ are bounded and differentiable in their arguments.

Here $h\cdot u = \sum_{j=1}^{d_1} h_j u_j$, $|A| = \sum_{k=1}^d \|Ae_k\|^2$, where $\{e_k\}_{k=1}^d$ is an orthonormal basis in R^d, and

$$sup_x\|u_0(x)\|^2 \le K_0, \quad sup_x\|\nabla u_0(x)\|^2 \le K_0^1.$$

We say that condition $\mathbf{C2_k}$ holds if coefficients and initial data are k times differentiable and satisfy **C1** along with their derivatives.

To simplify the problem we assume first that $f(x,u) = f(u)$ and hence $\kappa(x,u) \equiv 0$.

Denote by $(\Omega, \mathcal{F}, P)$ a probability space and let $w(t) \in R^d$ denote the Wiener process defined on this probability space.

Consider a stochastic system

$$d\xi(s) = A(\xi(s), g(T-s,\xi(s)))dw(s), \quad \xi(t) = x \in R^d, s \ge t, \tag{5}$$

$$d\eta(s) = c^*(\xi(s), g(T-s,\xi(s)))\eta(s)ds - C^*(g(T-s,\xi(s)))(\eta(s),dw(s)), \tag{6}$$

$$\eta(t) = h \in R^{d_1},$$

$$h\cdot g(T-t,x) = E[\eta_{t,h}(T)\cdot u_0(\xi_{t,x}(T))]. \tag{7}$$

Here $c^*h\cdot g = h\cdot cg$, $B = CA$ and $[C^i]^*h\cdot g = h\cdot C^i g\cdot h, i = 1,\ldots d$.

Theorem 1. *Assume that* $\mathbf{C2_1}$ *hold. Then there exists an interval* $[T_1, T] \subset [0, T]$ *with* $\tau = T - T_1$, $T_1 \geq 0$, *satisfying*

$$\tau < \frac{1}{2\rho_0} \ln\left(1 + \frac{2\rho_0}{3\rho K_0}\right) \tag{8}$$

such that there exists a solution to (5)–(7) *for all* $t \in [T_1, T]$ *and functions* $g_m(t, x)$ *are classical solutions of the Cauchy problem* (4).

The proof of the theorem one can find in [4].

Consider the Cauchy problem with a small positive parameter ϵ

$$\frac{\partial g_m^\epsilon}{\partial t} + \frac{\epsilon^2}{2} \sum_{i,j,k=1}^{d} A^{ik}(x, g^\epsilon) \nabla^2_{x_i,x_j} g_m^\epsilon A^{kj}(x, g^\epsilon) - \sum_{q=1}^{d_1} \sum_{i=1}^{d} B^i_{mq}(g^\epsilon) \nabla_{x_i} g_q^\epsilon \tag{9}$$

$$+ \sum_q c_{mq}(x, g^\epsilon) g_q^\epsilon = 0, \quad g_m(T, x) = u_{0m}(x).$$

Our aim is to prove that solutions $g_m^\epsilon(t, x)$ of (9) converge to solutions of the hyperbolic system

$$\frac{\partial g_m}{\partial t} - \sum_{q=1}^{d_1} \sum_{i=1}^{d} B^i_{mq}(g) \nabla_{x_i} g_q + \sum_q c_{mq}(x, g) g_q = 0, \quad g_m^\epsilon(T, x) = u_{0m}(x). \tag{10}$$

To this end we need some additional conditions.

We say that condition **C 3** holds if $d_1 \times d_1$-matrices $B^i, i = 1, \ldots, d$, have a simple spectrum $\sigma^i = \{\lambda_1^i, \ldots, \lambda_{d_1}^i\}$ corresponding to eigenvectors $h_1, \ldots, h_{d_1}$, $B^i(g) h_m = \lambda_m^i(g) h_m$ and $\lambda_m(g)$ satisfy $\mathbf{C2_1}$.

First we consider a stochastic system associated with (9) in the case $c \equiv 0$,

$$d\xi^\epsilon(s) = \epsilon A(\xi^\epsilon(s), g^\epsilon(T - s, \xi^\epsilon(s))) dw(s), \quad \xi^\epsilon(t) = x \in R^d, s \geq t, \tag{11}$$

$$d\eta^\epsilon(s) = -\epsilon^{-1} A^{-1}(\xi^\epsilon(s), g^\epsilon(T - s, \xi^\epsilon(s))) B^*(g^\epsilon(T - s, \xi^\epsilon(s)))(\eta(s), dw(s)), \tag{12}$$

$$\eta(t) = h \in R^{d_1},$$

$$h \cdot g^\epsilon(T - t, x) = E[\eta^\epsilon_{t,h}(T) \cdot u_0(\xi^\epsilon_{t,x}(T))]. \tag{13}$$

Theorem 2. *Assume that* $\mathbf{C2_1}$ *and* **C 3** *hold. Then functions* g_m^ϵ *satisfying* (11)–(13) *converge in sup norm uniformly on compacts to functions* $g_m(T - t, x)$ *defined by a system*

$$dx_m(s) = -\lambda_m(g(T - s, x_m(s))) ds, \quad x_m(t) = x, \tag{14}$$

$$g_m(T - t, x) = u_{0m}(x_m(T)). \tag{15}$$

Proof. Let

$$Z^\epsilon(\tau) = A^{-1}(\xi^\epsilon(\tau), g^\epsilon(T-\tau, \xi^\epsilon(\tau)))B^*(g^\epsilon(T-\tau, \xi^\epsilon(\tau)))$$

A solution to the linear SDE (12) has the form

$$\eta^\epsilon(t) = \exp\left\{-\int_s^t \epsilon^{-1} Z^\epsilon(\tau)\cdot dw(\tau) - \frac{1}{2}\int_s^t \epsilon^{-2}[Z^\epsilon(\tau)]^2 d\tau\right\} h.$$

Let eigenvectors $h_m, m = 1, \dots, d_1$, of the matrices B^i corresponding to eigenvalues λ^i_m serve as initial data in (12). Then we get

$$\eta^\epsilon_m(s) = \exp\left\{-\int_t^s \epsilon^{-1} A^{-1}(\xi^\epsilon(\tau), g^\epsilon(T-\tau, \xi^\epsilon(\tau)))\lambda_m(g_\epsilon(T-\tau, \xi^\epsilon(\tau)))\cdot\right.$$

$$\left.\cdot dw(\tau) - \frac{1}{2}\int_t^s \epsilon^{-2}\|A^{-1}(\xi^\epsilon(\tau), g^\epsilon(T-\tau, \xi^\epsilon(\tau)))\lambda_m(g^\epsilon(T-\tau, \xi^\epsilon(\tau)))\|^2 d\tau\right\} h_m.$$

Note that under the theorem assumptions we can verify that

$$\kappa^\epsilon_m(\tau) = \epsilon^{-1} A^{-1}(\xi^\epsilon(\tau), g^\epsilon(T-\tau, \xi^\epsilon(\tau)))\lambda_m(g^\epsilon(T-\tau, \xi^\epsilon(\tau)))$$

satisfies Novikov's condition. Hence, setting $dQ^\epsilon_m = \eta^\epsilon_m(T)dP$ we deduce that $\tilde{w}^m(t) = -\int_0^t \kappa^\epsilon_m(s)ds + w(t)$ is a Brownian motion with respect to Q^ϵ_m. In addition the process $\xi^\epsilon(t)$ satisfying (10) by the Girsanov theorem solves an SDE

$$d\tilde{\xi}^\epsilon_m(s) = -\lambda_m(g_\epsilon(T-s, \tilde{\xi}^\epsilon_m(s)))ds + \epsilon A(\tilde{\xi}^\epsilon_m(s), g^\epsilon(T-s, \tilde{\xi}^\epsilon_m(s)))d\tilde{w}^m(s), \quad (16)$$

$$\tilde{\xi}^\epsilon_m(t) = x.$$

Thus, Q^ϵ_m- law of $\tilde{\xi}^\epsilon_m(s)$ is the same as the P-law of $\xi^\epsilon(s)$ that yields

$$\sum_m h_m g^\epsilon_m(t,x) = E^P[\eta^\epsilon_{s,h}(t)\cdot u_0(\xi^\epsilon_{t,x}(T))] = \sum_m h_m E^{Q^\epsilon_m}[u_{0m}(\tilde{\xi}^\epsilon_{t,x}(T))]. \quad (17)$$

Since under theorem assumptions $g^\epsilon_m(t,x)$ satisfying (13) and (17) are proved to be bounded and Lipschitz continuous functions on $[T_1, T]$ we obtain

$$E\|\tilde{\xi}^\epsilon_m(T) - x_m(T)\|^2 \le 2TE^{Q^m_\epsilon}\int_t^T \|\lambda_m(g^\epsilon(T-s, \tilde{\xi}^\epsilon_m(s))) - \lambda_m(g(T-s, x_m(s)))\|^2 ds$$

$$+2\epsilon^2\int_t^T E^{Q^m_\epsilon}\|A(\tilde{\xi}^\epsilon_m(s), g^\epsilon(T-s, \tilde{\xi}^\epsilon_m(s)))\|^2 ds$$

$$\le 2T\int_t^T L_\lambda E^{Q^m_\epsilon}\|\tilde{\xi}^\epsilon_m(\tau) - x_m(\tau)\|^2 L_g d\tau + 2T\int_t^T L_\lambda\|g^\epsilon(T-s, x_m(s))$$

$$-g(T-s, x_m(s))\|^2 ds$$

By the Gronwall lemma we get

$$E^{Q^m_\epsilon}\|\tilde{\xi}^\epsilon_m(T) - x_m(T)\|^2 \le e^{CT}2T\int_t^T L_\lambda\|g^\epsilon(T-s, x_m(s)) - g(T-s, x_m(s))\|^2 ds$$

$$+2\epsilon^2 \int_t^T E^{Q^m_\epsilon} \|A(\tilde{\xi}^\epsilon_m(s), g^\epsilon(T-s, \tilde{\xi}^\epsilon_m(s)))\|^2 ds.$$

where $C = 2TL_\lambda L_g$. Moreover,

$$\|g^\epsilon_m(T-t,x) - g_m(T-t,x)\|^2 \le E^{Q^m_\epsilon} \|u_{0m}(\tilde{\xi}^\epsilon_m(T)) - u_{0m}(x_m(T))\|^2$$
$$\le L_0 \|\tilde{\xi}^\epsilon_m(T) - x_m(T)\|^2$$
$$\le L_0 e^{CT} 2T \int_t^T L_\lambda \|g_m(T-s, \tilde{\xi}_m(s)) - g_m(T-s, \tilde{\xi}_m(s))\|^2 ds$$
$$+ 2\epsilon^2 \int_t^T E^{Q^m_\epsilon} \|A(\tilde{\xi}^\epsilon_m(s), g_\epsilon(T-s, \tilde{\xi}^\epsilon_m(s)))\|^2 ds,$$

Applying Gronwall's lemma once again we get

$$\|g^\epsilon_m(T-t,x) - g_m(T-t,x)\|^2 \le e^{C_1 T} \epsilon^2 \int_t^T E^{Q^m_\epsilon} \|A(\tilde{\xi}^\epsilon_m(s), g^\epsilon(T-s, \tilde{\xi}^\epsilon_m(s)))\|^2 ds,$$

where $C_1 = 2TL_0 e^{CT} L_\lambda$. Since the integral in the right hand side of the last inequality is bounded for $t \in [T_1, T]$ we obtain

$$\lim_{\epsilon \to 0} \sup_{(t,x)\in[T_1,T]\times K} \|g^\epsilon_m(T-t,x) - g_m(T-t,x)\|^2 = 0$$

for any compact $K \subset R^d$. Thus, functions $g^\epsilon_m(t,x)$ given by (13) converge in the sup-norm as $\epsilon \to 0$ to functions $g_m(t,x)$ satisfying (15) and the processes $\tilde{\xi}^\epsilon_m(s)$ satisfying (12) converge to $x_m(s)$ satisfying (14) with probability 1.

To apply a similar approach to the case when $c \ne 0$ we proceed as follows.

Consider a stochastic system of the form

$$d\xi^\epsilon(s) = \epsilon A(\xi^\epsilon(t), g^\epsilon(T-s, \xi^\epsilon(s)))dw(s), \quad \xi^\epsilon(t) = x \in R^d, s \ge t, \tag{18}$$

$$d\eta^\epsilon(s) = -\epsilon^{-1} A^{-1}(\xi^\epsilon(s), g^\epsilon(T-s, \xi^\epsilon(s))) B^*(\xi^\epsilon(s), g^\epsilon(T-s, \xi^\epsilon(s)))(\eta(s), dw(s)), \tag{19}$$

$$\eta(t) = h \in R^{d_1},$$

$$h \cdot g^\epsilon(T-t,x) = E\left[\eta_{t,h}(T) \cdot u_0(\xi^\epsilon_{t,x}(T))\right. \tag{20}$$

$$\left. + \int_t^T \eta_{t,h}(\tau) \cdot c(\xi^\epsilon(s), g^\epsilon(T-s, \xi^\epsilon(s))) g^\epsilon(T-s, \xi^\epsilon(s)) ds \right],$$

Theorem 3. *Assume that* $\mathbf{C2_1}$ *and* **C 3** *hold and* $f = f(g)$. *Then there exists an interval* $[T_1, T] \subset [0, T]$ *such that for a fixed* ϵ *there exists a solution of* (18)–(20) *for all* $t \in [T_1, T]$. *In addition, functions* g^ϵ_m *satisfying* (18)–(20) *uniformly on compacts converge to functions* $g_m(T-t,x)$ *satisfying a system*

$$dx_m(s) = -\lambda_m(g(T-s, x_m(s)))ds, \quad x_m(t) = x, \tag{21}$$

$$g_m(T-t,x) = u_{0m}(x_m(T)) + \int_t^T c(\xi(s), g(T-s, x_m(s)) g(T-s, x_m(s)) ds. \tag{22}$$

Proof. By Theorem 1 we know that for any fixed ϵ there exists a solution of the system (18)–(20). defined on the interval $[T_1, T]$ and $g^\epsilon(T-t,x)$ is a differentiable bounded function. Besides we may consider and alternative stochastic system associated with (9) which includes SDE (18), SDE

$$d\eta^\epsilon(\tau) = -Z^\epsilon(\tau)(\eta^\epsilon_m(\tau), dw(\tau)) \tag{23}$$

with initial data $\eta^\epsilon_m(s) = h$ and a closing relation

$$h \cdot g^\epsilon(T-t,x) = E\left[\eta_{t,h}(T) \cdot u_0(\xi^\epsilon_{t,x}(T))\right. \tag{24}$$

$$\left. + \int_t^T \eta^\epsilon_{t,h}(s) \cdot c(\xi^\epsilon_{t,x}(s), g^\epsilon(T-s, \xi^\epsilon_{t,x}(s)))g^\epsilon(T-s, \xi^\epsilon_{t,x}(s))ds\right].$$

To verify that a classical solution $g^\epsilon(T-t,x)$ of (9) admits a representation (24) we consider a process $\gamma_m(s) = \eta^\epsilon_m(s) \cdot g^\epsilon_m(T-s, \xi^\epsilon(s))$ and compute its stochastic differential applying Ito's formula

$$d\gamma(s) = d\eta^\epsilon(s) \cdot g^\epsilon(s, \xi^\epsilon(s)) + \eta^\epsilon(s) \cdot dg^\epsilon(T-s, \xi^\epsilon(s)) + d\eta^\epsilon(s) \cdot dg^\epsilon(T-s, \xi^\epsilon(s))$$

$$= -Z^\epsilon(s)))(\eta(s), dw(s)) \cdot g^\epsilon(s, \xi^\epsilon(s))$$

$$+\eta^\epsilon(s) \cdot \left[\frac{\partial g^\epsilon}{\partial s} + \frac{\epsilon^2}{2} TrA\nabla^2 g^\epsilon A^* + B \cdot \nabla g^\epsilon_m\right](T-s, \xi^\epsilon(s))ds].$$

Integrating the last relation in time from t to T and computing expectation we get

$$E[\eta^\epsilon(T) \cdot u_0(\xi^\epsilon_{t,x}(T))] - h \cdot g^\epsilon(t,x) = \int_t^T E\eta^\epsilon(s) \cdot \left[\frac{\partial g^\epsilon(T-s, \xi^\epsilon_{t,x}(s))}{\partial s}\right.$$

$$+\frac{\epsilon^2}{2} TrA(\xi^\epsilon_{t,x}(s), g^\epsilon(T-s, \xi^\epsilon_{t,x}(s)))\nabla^2 g^\epsilon(s, \xi^\epsilon(s))A^*(\xi^\epsilon_{t,x}(s), g^\epsilon(T-s, \xi^\epsilon_{t,x}(s)))$$

$$\left. +B(g^\epsilon(T-s, \xi^\epsilon_{t,x}(s))) \cdot \nabla g^\epsilon(T-s, \xi^\epsilon_{t,x}(s)))\right] ds. \tag{25}$$

Let $h = h_m$ be eigenvectors of the matrix B^i corresponding to eigenvalues λ^i_m. Denote by

$$\Phi^\epsilon_m(t, T, \xi(\cdot)) = \exp\left\{\int_t^T [-\epsilon^{-1} A^{-1}(\xi^\epsilon(s), g^\epsilon(T-s, \xi^\epsilon(s)))\lambda_m(\xi^\epsilon(s),\right.$$

$$g^\epsilon(T-s, \xi^\epsilon(s))) \cdot dw(s) - \int_t^T \epsilon^{-2} \|A^{-1}(\xi^\epsilon(s), g^\epsilon(T-s, \xi^\epsilon(s)))\lambda(\xi^\epsilon(s),$$

$$\left. g^\epsilon(T-s, \xi^\epsilon(s)))\|^2 ds\right\}.$$

Keeping in mind that g^ϵ satisfies (9) we deduce from (25)

$$g^\epsilon_m(T-t,x) = E\left[\Phi^\epsilon_m(t, T, \xi(\cdot))u_{0m}(\xi^\epsilon(T)) - \right.$$

$$\left.\int_t^T \Phi_m^\epsilon(s,T,\xi(\cdot))\sum_{q=1}^{d_1} c_{mq}(\xi^\epsilon(s),g^\epsilon(T-s,\xi^\epsilon(s)))g_q^\epsilon(T-s,\xi^\epsilon(s))ds\right].$$

Next we have to apply the Girsanov formula to obtain

$$g_m^\epsilon(T-t,x) = E^{Q_\epsilon^m}\left[u_{0m}(\tilde{\xi}^\epsilon(T))\right]$$

$$-E^{Q_\epsilon^m}\left[\int_t^T \sum_{q=1}^{d_1} c_{mq}(\tilde{\xi}^\epsilon(s),g^\epsilon(T-s,\tilde{\xi}^\epsilon(s)))g_q^\epsilon(T-s,\tilde{\xi}^\epsilon(s))ds\right],$$

where $\tilde{\xi}_m(t)$ satisfy (14). To prove that g_m^ϵ satisfying (13) converge in the sup-norm as $\epsilon \to 0$ to functions $g_m(t,x)$ satisfying (15) and the processes $\tilde{\xi}_m^\epsilon(s)$ satisfying (12) converge to $x_m(s)$ satisfying (14) with probability 1 we use the same reasons as in Theorem 2.

Corollary. Assume that **C2**$_3$ and **C3** hold. Then the functions g_m^ϵ given by (13) are unique classical solutions of (9), while the functions g given by (15) are unique classical solution of the system (10).

Remark 1. We can generalise the above results to include the case $f = f(x,g)$. In this case one has to study a stochastic system including (5) and equations

$$d\eta(s) = c^*(\xi(s),g(T-s,\xi(s)))\eta(s)ds - C^*(\xi(s),g(T-s,\xi(s)))(\eta(s),dw(s)),$$

$$h\cdot g(T-t,x) = E\left[\eta_{t,h}(T)\cdot u_0(\xi_{t,x}(T)) - \int_t^T [\eta_{t,h}(\tau)\cdot \kappa(\tau,\xi_{t,x}(\tau))d\tau\right],$$

$$\eta(s) = h \in R^{d_1}.$$

Extending **C 3** to this case one can prove that both Theorems 2 and 3 are valid.

2 Stochastic System Associated with Nonlinear Parabolic Systems with Cross-Diffusions

In this section we consider systems of nonlinear parabolic equations which admit a natural interpretation as systems of forward Kolmogorov equations. Actually, we consider a simplified version of a system of the form (2), namely, a system

$$\frac{\partial \tilde{v}_m}{\partial t} + \sum_{i=1}^{d} \nabla_{y_i} f_m^i(y,\tilde{v}) = \frac{1}{2}\sum_{i,j=1}^{d} \nabla^2_{y_i,y_j}[G_m^{ij}(y,\tilde{v})\tilde{v}_m] + c_m(y,\tilde{v})\tilde{v}_m, \tag{26}$$

$$\tilde{v}_m(0,y) = u_{0m}(y)$$

assuming that $G_m^{ij} = \sum_{k=1}^{d} A^{ik}A^{kj}$. For simplicity we assume that u_{0m} and all coefficients in (26) are smooth and bounded.

We say that a function $\tilde{v}_m(t,y)$ is a weak solution of (26) if an integral identity

$$\frac{\partial}{\partial t}\int_{R^d} h(y)\tilde{v}_m(t,y)dy = \int_{R^d} \tilde{v}_m(t,y)\left[\frac{1}{2}TrA_m(y,\tilde{v})\nabla^2 h(y)A^*_m(y,\tilde{v})\right.$$

$$\left.+f_m(y,\tilde{v})\cdot\nabla h(y) + c_m(y,\tilde{v})h(y)\right] dy$$

holds for any $h \in C_0^\infty(R^d)$.

To have a possibility to treat (26) as a system of forward Kolmogorov equations for distributions of some stochastic processes we will consider its mollification to make coefficients to be functionals of the required measures. To this end we choose a mollifier that is a function $\rho \in C_0^\infty(R^d)$ with $\int_{R^d}\rho(y)dy = 1$ and $\rho * \tilde{v}(y) = \int_{R^d}\rho(y-x)\tilde{v}(dx)$ and consider a parabolic system

$$\frac{\partial v_m}{\partial t} = \frac{1}{2}\sum_{i,j=1}^{d}\nabla^2_{y_i,y_j}[G^{ij}_m(y,\rho * v)v_m] - div f_m(y,\rho * v) + c_m(y,\rho * v)v_m, \quad (27)$$

$$v_m(0,dy) = u_{0m}(dy).$$

One can see that if $c \equiv 0$ then (27) is a system of McKean-Vlasov equations. Assume that there exists a unique positive bounded integrable weak solution to (26) and construct its probabilistic representation. To this end we need more notations.

Let $\mathcal{P} = \mathcal{P}(R^d)$ be a family of Borel probability measures on R^d and $\mathcal{P}_2 = \{\mu \in \mathcal{P} : \|\mu\|^2 = \int_{R^d}\|y\|^2\mu(dy) < \infty.\}$

Denote by $\mathcal{C}^d = C([0,T];R^d)$ and let $\xi_m = \xi_m(t,\omega)$ be a canonical process on $\Omega = \mathcal{C}^d$, that is $\xi(t,\omega)$ is the value $\omega(t)$ of $\omega \in \mathcal{C}^d$ at $t \in [0,T]$. We set $\mathcal{F}$ and $\mathcal{F}_t$ the smallest σ-algebras generated by $\{\xi_m(s) : s \le T\}$ and $\{\xi_m(s) : s \le t\}$ respectively. Let $\mathcal{P}_2 = B([0,T];\mathcal{P}_2)$ be the space of $\mathcal{P}_2$ valued Borel measures.

Let $d_1 = 1$ then to construct a measure valued weak solution $\mu(t,dy)$ of (29) given $\mu_0(dy) = u_0(y)dy$ one should construct a probability measure γ on $(\mathcal{C}^d,\mathcal{F})$ [9] such that:

(i) The distribution $\mu^\gamma(t) = \gamma\circ\xi^{-1}(t)$ of the process $\xi(t)$ under γ, that is $\mu(t,dy) = \gamma\{\xi(t)\in dy\}$ belongs to $\mathcal{P}_2$;
(ii) $\mu^\gamma(0,dy) = \mu_0(dy)$;
(iii) for every $h \in C_0^\infty(R^d)$, $h(\xi(t)) - \int_0^t \mathcal{A}^u h(\xi(s))ds$ is a martingale with respect to $(\gamma,\mathcal{F}_t)$, where

$$\mathcal{A}^u h = \frac{1}{2}TrA(y,u)\nabla^2 h(y)A^*(y,u) + f_m(y,u)\cdot\nabla h(y).$$

To extend this theory to the case of parabolic systems we consider a stochastic system of the form

$$d\xi_m(s) = f_m(\xi_m(s),u(s,\xi_m(s)))ds + A_m(\xi_m(s),u(s,\xi_m(s)))dw_m(s), \quad (28)$$

$$\xi_m(0) = \xi_{0m}, \quad \xi_m(0) \sim u_{0m}(y)dy = \gamma_m\{\xi_m(0)\in dy\},$$

$$u_m(t,y) = E\left[\rho(y - \xi_m(t)) \exp\left\{\int_0^t c_m(\xi_m(s), u(s, \xi_m(s)))ds\right\}\right] \tag{29}$$

$$= \int_{\mathcal{C}^d} \rho(y - \xi_m(t,\omega)) \exp\left\{\int_0^t c_m(\xi_m(s,\omega), u(s, \xi_m(s,\omega)))ds\right\} \gamma_m(d\omega),$$

$$\mathcal{L}(\xi_m) = \gamma_m.$$

Here $w_m(t) \in R^d$ are independent Wiener processes, ξ_{0m} are independent random variables with distribution μ_{0m} which do not depend on w_m.

Stochastic system (28), (29) is a closed system with respect to a couple $(\xi_m(t), u_m(t,y))$ and under certain conditions we can prove the existence and uniqueness of its solution (see [10,11]). In this paper for simplicity we assume that coefficients $f_m, A_m c_m$ and initial functions u_{0m} are bounded smooth functions. To verify that there exists a connection between $u_m(t,y)$ defined by (29) and a solution of (27) we construct measures $\mu_m(t,dy)$ satisfying (27) such that $\mu_m(t,dy) = u_m(t,y)dy$.

To obtain the required measures we note that for any $h \in C_b(R^d)$ an expression

$$E\left[\int_{R^d} h(\xi_m(t)) \exp\left\{\int_0^t c_m(\xi_m(s), \rho * \mu(s, \xi_m(s)))ds\right\}\right],$$

where $\xi_m(t)$ is a solution of a stochastic equation (28) is a bounded linear functional over the space $C_b(R^d)$ of continuous bounded functions and hence, by the Riesz theorem [12] there exists a probability measure $\mu_m(t,dy)$ such that

$$\int_{R^d} h(y)\mu_m(t,dy) = E\left[\int_{R^d} h(\xi_m(t)) \exp\left\{\int_0^t c_m(\xi_m(s), \rho * \mu(s, \xi_m(s)))ds\right\}\right]. \tag{30}$$

Next we consider a system

$$d\xi_m(s) = f_m(\xi_m(s), \rho * \mu^\gamma(s, \xi_m(s)))ds + A_m(\xi_m(s), \rho * \mu^\gamma(s, \xi_m(s)))dw_m(s), \tag{31}$$

$$\xi_m(0) = \xi_{0m},$$

$$\int_{R^d} h(y)\mu_m^\gamma(t,dy) = E\left[\int_{R^d} h(\xi_m(t)) \exp\left\{\int_0^t c_m(\xi_m(s), \rho * \mu^\gamma(s, \xi_m(s)))ds\right\}\right] \tag{32}$$

$$= \int_{\mathcal{C}_d} h(\xi_m(t,\omega)) \exp\left\{\int_0^t c_m(\xi_m(s,\omega), \rho * \mu^\gamma(s, \xi_m(s,\omega)))ds\right\} \gamma_m(d\omega)$$

$$\mathcal{L}(\xi_m) = \gamma_m$$

and verify that if there exists a solution $(\xi_m(t), u_m(t,y))$ of (28), (29) then u_m is connected to a solution v_m of (27) by the relation $u_m(t,y) = \rho * v_m(t,y) = \int_{R^d} \rho(y-x)v_m(t,dx)$. To this end we establish a correspondence between (28), (29) and (31), (32).

Theorem 4. *The existence of a solution to the McKean type system (28), (29) is equivalent to the existence of a solution to the system (31), (32). More precisely, given a solution (ξ_m, μ_m^γ) of (31), (32) we can verify that the pair (ξ_m, u_m^γ), where $u_m^\gamma = \rho * \mu_m^\gamma$ solves (28), (29) and vice versa.*

Proof. Let $(\xi_m(t), u_m^\gamma(t))$ be a solution to (28), (29). Denote by

$$F(h)(z) = \frac{1}{(2\pi)^{\frac{d}{2}}} \int_{R^d} h(y) e^{-z\cdot y} dy$$

the Fourier transform of a function $h \in S(R^d)$. Since $\rho \in L^1(R^d)$, we may apply the Fourier transform to the function $u_m^\gamma(t, y)$ satisfying (29) and obtain

$$F(u_m^\gamma)(t, z) \tag{33}$$

$$= F(\rho)(z) \int_{C^d} e^{-iz\cdot\xi_m(t,\omega)} \exp\left\{ \int_0^t c_m(u^{\gamma_m}(s, \xi_m(s,\omega)))ds \right\} \gamma_m(d\omega).$$

We deduce from the Lebesgue dominated convergence theorem that

$$g_m^\gamma(t, z) = \int_{C^d} e^{-iz\cdot\xi_m(t,\omega)} \exp\left\{ \int_0^t c_m(u^{\gamma_m}(s, \xi_m(s,\omega)))ds \right\} \gamma_m(d\omega) \tag{34}$$

is continuous and bounded for $t \in [0, T]$ since c_m is bounded.

Let $\{\beta_k\}_{k=1,\dots,d}$ be a sequence of complex numbers and $\{y_k\}_{k=1,\dots,d} \in (R^d)^d$. Note that for all $z \in R^d$ we have the equality

$$\sum_{k=1}^d \sum_{j=1}^d \beta_k \bar{\beta}_j e^{-iz\cdot(y_k - y_j)} = \left(\sum_{k=1}^d \beta_k e^{-iz\cdot y_k} \right) \overline{\left(\sum_{j=1}^d \beta_j e^{-iz\cdot y_j} \right)}$$

$$= |\sum_{k=1}^d \beta_k e^{-iz\cdot y_k}|^2,$$

and thus, g_m^γ is non-negative definite. Then the Bochner theorem states that there exists a finite non-negative Borel measure $\nu_m^\gamma(t)$ on R^d, such that for all $z \in R^d$

$$g_m^\gamma(t, z) = \frac{1}{(\sqrt{2\pi})^d} \int_{R^d} e^{-iz\cdot y} \nu_m^\gamma(t, dy). \tag{35}$$

Let us verify that $\mu_m^\gamma(t, dy) \equiv \nu_m^\gamma(t, dy)$ satisfy (32). As far as $\mu_m^\gamma(t, dy)$ is a finite non-negative Borel measure we can treat it as an element of the Schwartz space $S'(R^d)$. Hence the equality $F^{-1}(g_m^\gamma(t)) = \mu_m^\gamma(t)$ holds and for any $\phi \in \mathcal{S}(R^d)$ the estimate

$$|\int_{R^d} \phi(x) \mu_m^\gamma(t, dx)| \le \|\phi\|_\infty \mu_m^\gamma(t, R^d) < \infty$$

is valid. From (33) and (35) we deduce $F(u_m^\gamma)(t, \cdot) = F(\rho) F(\mu_m^\gamma(t))$ that yields $u_m^\gamma(t, \cdot) = \rho * \mu_m^\gamma(t, \cdot)$. Denote by $\langle \phi, \mu \rangle = \int_{R^d} \phi(y) \mu(t, dy)$, $\phi \in S(R^d)$. Applying the Fubini theorem and the equality $u_m^\gamma(t,\cdot) = \rho * \mu_m^\gamma(t,\cdot)$, we obtain for all $\phi \in \mathcal{S}(R^d)$

$$\langle \phi, \mu_m^\gamma(t)\rangle = \langle \phi, F^{-1}(g_m^\gamma)(t)\rangle = \langle F^{-1}(\phi), g_m^\gamma(t)\rangle$$

$$= \int_{R^d} F^{-1}(\phi)(z)\left(\int_{C^d} e^{-iz\cdot\xi_m(t,\omega)} e^{\int_0^t c_m(u^\gamma(s,\xi_m(s,\omega)))ds}\gamma_m(d\omega)\right) dz$$

$$= \int_{C^d}\left(\int_{R^d} F^{-1}(\phi)(z)e^{-iz\cdot\xi_m(t,\omega)}dz\right) e^{\int_0^t c_m(u^\gamma(s,\xi_m(s,\omega)))ds}\gamma_m(d\omega)$$

$$= \int_{C^d}\left(\int_{R^d} F^{-1}(\phi)(z)e^{-iz\cdot\xi_m(t,\omega)}dz\right) e^{\int_0^t c_m(\rho*\mu^\gamma(s,\xi_m(s,\omega)))ds}\gamma_m(d\omega)$$

$$= \int_{C^d} \phi(\xi_m(t,\omega))e^{\int_0^t c_m(\rho*\mu_m^\gamma(s,\xi_m(s,\omega)))ds}\gamma_m(d\omega).$$

Thus, $(\xi_m(t), u_m^\gamma(t))$ is a solution of (28), (29), if $(\xi_m(t),\ \mu_m^\gamma(t))$ is a solution of (31), (32).

To prove an inverse statement assume that (ξ_m, μ_m^γ) satisfy (31), (32). Set $u_m^\gamma(t,y) = \rho * \mu_q^\gamma(t,y)$ and note that $(\xi_m(t), u_m^\gamma(t))$ satisfy (28), (29). Since $\mu^\gamma(t)$ is a finite measure to verify (32) it is enough to put $\phi = \rho(x-\cdot)$ in it. Thus we have proved the theorem.

Next we state a link between the above stochastic systems and (27).

Theorem 5. *Measures $\mu_m^\gamma(t,dy)$ satisfying (32) satisfy the Cauchy problem (27) in a weak sense.*

Proof. Denote by $\Phi_m(t,\rho*\mu^\gamma(\xi_m)) = e^{\int_0^t c_m(\rho*\mu^\gamma(s,\xi_m(s,\omega)))ds}$. Let us apply the Ito formula to the process $\zeta_m(t) = h(\xi_m(t))\Phi_m(t,\rho*\mu^\gamma(\xi_m))$ where $\xi_m(t)$ satisfy (31) and $h \in C_0^\infty(R^d)$. Then we get

$$E[h(\xi_m(t))\Phi_m(t,\rho*\mu^\gamma(\xi_m))] = Eh(\xi_{0m})$$

$$+\int_0^t E[h(\xi_m(s))c_m(\rho*\mu^\gamma(s,\xi_m(s,\omega)))\Phi_m(s,\rho*\mu^\gamma(\xi_m))]ds$$

$$+\int_0^t E\left[\sum_{i,j=1}^d G^{ij}(\rho*\mu^\gamma(s,\xi_m(s)))\frac{\partial^2 h(\xi_m(s))}{\partial y_i \partial y_j}\Phi_m(s,\rho*\mu^\gamma(\xi_m))\right] ds$$

$$+\int_0^t E\left[\sum_{i=1}^d f_m^i(\rho*\mu^\gamma(s,\xi_m(s)))\frac{\partial h(\xi_m(s))}{\partial y_i}\Phi_m(s,\rho*\mu^\gamma(\xi_m))\right] ds.$$

Thus, by definition of the measure μ_m^γ in (32) we have

$$\int_{R^d} h(y)\mu_m^\gamma(t,dy) = \int_{R^d} h(y)\mu_{0m}^\gamma(dy)$$

$$+\int_0^t\int_{R^d}[h(y)c_m(\rho*\mu^\gamma(s,y))\mu^\gamma(s,dy)ds$$

$$+\frac{1}{2}\int_0^t\int_{R^d}\sum_{i,j=1}^d G^{ij}(\rho*\mu^\gamma(s,y))\frac{\partial^2 h(y)}{\partial y_i\partial y_j}\mu^\gamma(s,dy)ds$$

$$+\int_0^t\int_{R^d}\sum_{i=1}^d f_m^i(\rho*\mu^\gamma(s,y))\frac{\partial h(y)}{\partial y_i}\mu^\gamma(s,dy)ds.$$

At the end of the section we consider the Cauchy problem

$$\frac{\partial v_m^\epsilon}{\partial t}=\frac{\epsilon^2}{2}\sum_{i,j=1}^d\frac{\partial^2}{\partial y_i\partial y_j}G^{ij}(y,\rho*v_m^\epsilon)v_m^\epsilon-div f_m(x,\rho*v_m^\epsilon)+c_m(y,\rho*v_m^\epsilon)v_m^\epsilon, \quad (36)$$

$$v_m^\epsilon(0,y)=u_{0m}(y)$$

with a small positive parameter ϵ and study limiting behaviour of its solution as $\epsilon\to 0$.

Theorem 6. *Assume that there exists a unique solution $(\xi_{m,\epsilon}(t),u_{m,\epsilon}(t,y))$ to the system (28), (29) with $A_{m,\epsilon}=\epsilon A_m$ such that $\xi_{m,\epsilon}(t)\in R^d$ is a Markov process and $u_{m,\epsilon}(t)\in L^1(R^d)\cap Lip(R^d)$. Then the couple $(\xi_{m,\epsilon}(t),u_{m,\epsilon}(t,y))$ converges to a solution of the system*

$$dx_m(s)=f_m(v(s,x_m(s)))ds,\quad x_m(0)=\xi_{0m}, \quad (37)$$

$$v_m(t,y)=\left[\rho(y-x_m(t))\exp\left\{\int_0^t c_m(x_m(s),v(s,x_m(s)))ds\right\}\right] \quad (38)$$

and $u_m(t)=\rho*v_m(t,y)$ where v_m is a weak solution of a system

$$\frac{\partial v_m}{\partial t}+div f_m(v)=\sum_{q=1}^{d_1}c_{mq}(y,v)v_q,\quad v_m(0,y)=u_{0m}(y).$$

Proof. Consider a system

$$d\xi_{m,\epsilon}^\gamma(s)=f_m(\xi_{m,\epsilon}^\gamma(s),u_\epsilon^\gamma(s,\xi_{m,\epsilon}(s)))ds+\epsilon A_m(\xi_{m,\epsilon}^\gamma(s),u_\epsilon^\gamma(s,\xi_{m,\epsilon}^\gamma(s)))dw_m(s), \quad (39)$$

$$\xi_m(0)=\xi_{0m},\quad \mathcal{L}(\xi_m)=\gamma_m,$$

$$u_{m,\epsilon}^\gamma(t,y)=E\left[\rho(y-\xi_{m,\epsilon}^\gamma(t))\exp\left\{\int_0^t c_m(\xi_{m,\epsilon}^\gamma(s),u^\gamma(s,\xi_{m,\epsilon}^\gamma(s)))ds\right\}\right] \quad (40)$$

$$=\int_{C_d}\rho(y-\xi_{m,\epsilon}^\gamma(t,\omega))\exp\left\{\int_0^t c_m(\xi_{m,\epsilon}^\gamma(s,\omega),u(s,\xi_{m,\epsilon}^\gamma(s,\omega)))ds\right\}\gamma_{m,\epsilon}(d\omega),$$

and estimate a difference $\alpha_m(t)=E\|\xi_{m,\epsilon}^\gamma(t)-x_m(t)\|^2$, where $x_m(t)$ satisfies (36). Keeping in mind properties of coefficients f_m,A_m and ρ and assuming Lipschitz continuity of u_m we deduce

$$\alpha_m(t)\le\int_0^t L_f[1+L_u]\alpha_m(s)ds+\epsilon^2\int_0^t\|A_m(\xi_{m,\epsilon}^\gamma(s),u_\epsilon^\gamma(s,\xi_{m,\epsilon}(s)))\|^2ds.$$

Applying the Gronwall lemma we get

$$\alpha_m(t) \leq \epsilon^2 \int_0^t E\|A_m(\xi^\gamma_{m,\epsilon}(s), u^\gamma_\epsilon(s, \xi^\gamma_{m,\epsilon}(s)))e^{L_f[1+L_u][T-T_1]}. \tag{41}$$

Next we derive an estimate for

$$\beta_m(t) = \|u^\gamma_{m,\epsilon}(t,y) - v_m(t,y)\|^2$$

$$= \|E\left[\rho(y - \xi^\gamma_{m,\epsilon}(t)) \exp\left\{\int_0^t c_m(\xi^\gamma_{m,\epsilon}(s), u^\gamma(s, \xi^\gamma_{m,\epsilon}(s)))ds\right\}\right.$$

$$\left. -\rho(y - x_m(t)) \exp\left\{\int_0^t c_m(x_m(s), v(s, x_m(s)))ds\right\}\right] \|^2$$

$$\leq L_\rho E\|\xi^\gamma_{m,\epsilon}(t) - x_m(t)\|^2 + K_\rho K_1 L_c \int_0^t E[(1 + L_{u_\epsilon})\|\xi^\gamma_{m,\epsilon}(s) - x_m(s)\|^2]ds$$

$$+K_\rho L_c e^{K_c T} \int_0^t \|u^\gamma_\epsilon(s,y) - v(s,y)\|^2]ds.$$

Thus,

$$\|u^\gamma_\epsilon(t,y) - v(t,y)\|^2 = \sum_{m=1}^{d_1} \|u^\gamma_{m,\epsilon}(t,y) - v_m(t,y)\|^2$$

$$\leq M \sup_{s\in[T_1,T]} \|\xi^\gamma_{m,\epsilon}(s) - x_m(s)\|^2 + K_\rho L_c e^{K_c T} \int_0^t \|u^\gamma_\epsilon(s,y) - v(s,y)\|^2]ds$$

and by the Gronwall lemma we get

$$\|u^\gamma_\epsilon(t,y) - v(t,y)\|^2 \leq M sup_{s\in[T_1,T]}\alpha_m(s)e^{M_1 T}$$

where $M = d_1[L_\rho + K_\rho K_1 L_c T], M_1 = K_\rho L_c e^{K_c T}$. Setting

$$K = Me^{M_1 T}e^{L_f[1+L_u]T}$$

and keeping in mind the estimate (41) we derive

$$\sup_{t\in[T_1,T]} \|u^\gamma_\epsilon(t,y) - v(t,y)\|^2$$

$$\leq \epsilon^2 d_1 K \int_0^T E\|A_m(\xi^\gamma_{m,\epsilon}(s), u^\gamma_\epsilon(s, \xi^\gamma_{m,\epsilon}(s)))\|^2 ds \to 0$$

as $\epsilon \to 0$ since A_m are bounded.

Next we deduce from Theorem 4 that along with the couple $(\xi^\gamma_{m,\epsilon}(t), u^\gamma_\epsilon(t,y))$ there exists a couple $(\xi^\gamma_{m,\epsilon}(t), \mu^\gamma_\epsilon(t,dy))$ satisfying

$$d\xi^\gamma_{m,\epsilon}(s) = f_m(\xi^\gamma_{m,\epsilon}(s), \rho*\mu^\gamma_\epsilon(s, \xi^\gamma_{m,\epsilon}(s)))ds + \epsilon A_m(\xi^\gamma_{m,\epsilon}(s), \rho*\mu^\gamma_\epsilon(s, \xi^\gamma_{m,\epsilon}(s)))dw_m(s), \tag{42}$$

$$\xi_m(0) = \xi_{0m},$$

$$\int_{R^d} h(y)\mu^\gamma_{m,\epsilon}(t,dy) = \tag{43}$$

$$E\left[\int_{R^d} h(\xi^\gamma_{m,\epsilon}(t)) \exp\left\{\int_0^t c_m(\xi^\gamma_{m,\epsilon}(s), \rho * \mu^\gamma_\epsilon(s, \xi^\gamma_{m,\epsilon}(s)))ds\right\}\right]$$

for any $h \in C_b(R^d)$. Since due to the above estimates we know that $\xi^\gamma_{m,\epsilon}(t)$ with probability 1 converges to $x_m(t)$ satisfying (37) we deduce that letting $\epsilon \to 0$ we obtain that $\int_{R^d} h(y)\mu^\gamma_{m,\epsilon}(t,dy)$ converge to $\int_{R^d} h(y)v_m(t,y)dy$, where $v_m(t,y)dy = \mu_m(t,dy)$ satisfy (38). Thus we prove that $\mu^\gamma_{m,\epsilon}(t,dy)$ converge to $\mu_m(t,dy)$, $m = 1,\ldots,d_1$ in a weak sense.

References

1. McKean, H.P.: A class of Markov processes associated with non-linear parabolic equations. Proc. Natl. Acad. Sci. **562**(6), 1907–1911 (1966)
2. Freidlin, M.: Quasilinear parabolic equations and measures in functional spaces. Funct. Anal. Appl. **1**(3), 237–240 (1967)
3. Carmona, R., Delarue, F.: Probabilistic Theory of Mean Field Games with Applications I-II. Springer, New York (2018). https://doi.org/10.1007/978-3-319-58920-6
4. Talay, D., Tomašević, M.: A new McKean-Vlasov stochastic interpretation of the parabolic-parabolic Keller-Segel model: the one-dimensional case. Bernoulli **26**(2), 1323–1353 (2020)
5. Belopolskaya, Y.: Stochastic interpretation of quasilinear parabolic systems with cross diffusion. Theory Probab. Appl. **61**(2), 208–234 (2017)
6. Belopolskaya, Ya.: Probabilistic interpretations of quasilinear parabolic systems. In: CONM AMS, vol. 734, pp. 39–56 (2019)
7. Belopolskaya, Ya.I., Dalecky, Yu.L.: Investigation of the Cauchy problem with quasilinear systems with finite and infinite number of arguments by means of Markov random processes. Izv. VUZ Math. **38**(12), 6–17 (1978)
8. Belopolskaya, Ya.I., Dalecky, Yu.L.: Stochastic Equations and Differential Geometry. Kluwer Academic Publishers, Amsterdam (1990)
9. Funaki, T.: A certain class of diffusion processes associated with nonlinear parabolic equations Z. Wahrscheinlichkeitstheorie verw. Gebiete **67**, 331–348 (1984)
10. Belopolskaya, Y.: Probabilistic interpretation of the Cauchy problem solution for systems of nonlinear parabolic equations. Lobachevskii J. Math. **41**(4), 597–612 (2020). https://doi.org/10.1134/S1995080220040046
11. Belopolskaya, Y.I., Stepanova, A.O.: Stochastic interpretation of the MHD-Burgers system. J. Math. Sci. **244**(5), 703–717 (2020). https://doi.org/10.1007/s10958-020-04643-1
12. Dunford, N., Schwartz, J.: Linear Operators, Part 1: General Theory, vol. 1. Interscience, New York (1988)

Generalization Bound for Imbalanced Classification

Evgeny Burnaev(✉)

Skoltech, Moscow 121205, Russia
e.burnaev@skoltech.ru
http://adase.group

Abstract. The oversampling approach is often used for binary imbalanced classification. We demonstrate that the approach can be interpreted as the weighted classification and derived a generalization bound for it. The bound can be used for more accurate re-balancing of classes. Results of computational experiments support the theoretical estimate of the optimal weighting.

Keywords: Imbalanced classification · Generalization bound · Resampling amount · Weighted classification

1 Introduction

In this paper, we consider the imbalanced binary classification, i.e. the case of two-class classification when one class (a minor class) has much less representatives in the available dataset than the other class (a major class). Many real-world problems have unavoidable imbalances due to properties of data sources, e.g. network intrusion detection and maintenance [1–3,8], damage detection from satellite images [12,13,17], prediction and localization of failures in technical systems [4,5,7,21,22], etc. In these examples target events (diseases, failures, etc.) are rare and presented only in a small fraction of available data.

Often the main goal of the imbalanced classification is to accurately detect the minor class [11]. However, standard classification approaches (logistic regression, SVM, etc.) are often based on the assumption that all classes as equally represented [10]. As a consequence the resulting classification model is biased towards the major class. E.g., if we predict an event occurring in just 1% of all cases and the classification model always gives a “no-event” prediction, then the model error is equal to 1%. Therefore, the average accuracy of the classifier is high, although it can not be used for the minor class detection.

An efficient way to deal with the problem is to *resample* the dataset in order to decrease the class imbalance, as it was discussed in [6,20]. In practice we can perform *oversampling*, i.e., add synthesized elements to the minor class, or perform *undersampling*, i.e., delete particular elements from the major class; or do the both types of samplings. There also exist other more delicate approaches to resampling.

A. N. Shiryaev et al. (Eds.): ICSM-5 2020, PROMS 371, pp. 107–119, 2021.
https://doi.org/10.1007/978-3-030-83266-7_8

Most of the resampling approaches takes as input the *resampling amount*, which defines how many observations we have to add or delete. In [6,20] they demonstrated that there is no "universal" choice of the resampling amount and the final classification accuracy significantly depends on a particular value we select for a dataset at hand.

The authors of [6,20] proposed to use either the cross-validation procedure [10] or the meta-learning procedure to select the resampling method and the resampling amount. However, these approaches are purely empirical and require to spend significant time for additional computational experiments due to the exhaustive search.

In this work we argue that the resampling approaches can be considered as a specific variant of the weighted classification: so to deal with a possible class imbalance when constructing a classifier we use a weighted error (risk) to stress the most important class. The question is how to select an appropriate weight value to up-weight the minor class. For that we estimate the theoretical generalization ability of the classifier with the weighted loss function and explore how it depends on the weighting scheme. We discuss how these findings can be used in practice when solving the imbalanced classification problem. In Sect. 2 we introduce the main notations and provide a theoretical problem statement. In Sect. 3 we prove the main result of the paper, namely, we obtain the generalization bound for the weighted binary classification and obtain an optimal weighting scheme. We propose the algorithm for the weighted classification based on the derived optimal weighting, and evaluate its empirical performance in Sect. 4. Results of computational experiments demonstrate usefulness of the proposed approach. We discuss conclusions in Sect. 5.

2 Problem Statement

Let us consider the formal binary classification problem statement, discuss how it can be interpreted as the weighted classification task in case we use the standard oversampling technique, and estimate the corresponding excess risk. Thanks to the estimate, we can characterize the influence of the weight (playing a role of the resampling amount) on the generalization ability of the classifier.

We denote by

- $\mathcal{F} \subseteq \mathcal{Y}^{\mathcal{X}}$ a class of binary classifiers with a multi-dimensional input space $x \in \mathcal{X}$ and an output label space $\mathcal{Y}$. Here we consider $\mathcal{Y} = \{-1, +1\}$ for simplicity. E.g.

$$\mathcal{F} = \{f_{a,b} : f_{a,b}(x) = 2\mathbb{I}(\langle a, x\rangle + b \geq 0) - 1\};$$

- $\mathbb{P}$ a distribution on $\mathcal{X} \times \mathcal{Y}$;
- π a prior probability of a positive class, i.e.

$$\mathbb{P} = \pi\mathbb{P}_{x|y=+1} + (1-\pi)\mathbb{P}_{x|y=-1};$$

- $\mathcal{D} = \{(x_i, y_i)\}_{i=1}^N$ a training sample, $x_i \in \mathcal{X}$, $y_i \in \mathcal{Y}$;

- $\mathcal{R}_N(\mathcal{F})$ a Rademacher complexity of $\mathcal{F}$ [15]. Recall that the empirical Rademacher complexity of some family of functions $\mathcal{G}$ from $\mathcal{Z}$ to $[a,b]$ for a fixed sample $S = (z_1, \dots, z_m)$ is equal to

$$\hat{\mathcal{R}}_S(\mathcal{G}) = \mathbb{E}_\sigma \left[\sup_{g \in \mathcal{G}} \frac{1}{m} \sum_{i=1}^{m} \sigma_i g(z_i) \right],$$

where $\sigma = (\sigma_1, \dots, \sigma_m)$ are Rademacher random variables. Then the Rademacher complexity of $\mathcal{G}$ w.r.t. some distribution $\mathbb{P}$ on $\mathcal{Z}$ is defined as

$$\mathcal{R}_m(\mathcal{G}) = \mathbb{E}_{S \sim \mathbb{P}^m} \left[\hat{\mathcal{R}}_S(\mathcal{G}) \right].$$

We consider a zero-one loss function $L(\hat{y}, y) = \mathbb{I}_{\hat{y} \neq y}$. The theoretical risk is equal to $\mathbb{E}_\mathbb{P} L(f(x), y)$, so that the theoretically optimal classifier

$$f^* = \arg\min_{f \in \mathcal{F}} \mathbb{E}_\mathbb{P} L(f(x), y).$$

The empirical risk has the form

$$\mathbb{E}_\mathcal{D} L(f(x), y) = \frac{1}{N} \sum_{j=1}^{N} L(f(x_j), y_j).$$

If we perform the oversampling the empirical risk can be represented as

$$\frac{1}{N} \sum_{j=1}^{N} u_j L(f(x_j), y_j),$$

where $u_j \geq 1$ is equal to the number of times the object x_j from the initial sample $\mathcal{D}$ is selected in the oversampling procedure (we count x_j as well). Thus the binary classification problem in case of the oversampling can be interpreted as a classification problem with a weighted empirical loss: we optimize the weighted empirical risk when training a classifier and measure its accuracy using a non-weighted theoretical risk.

Therefore, we define some (fixed) weighting function

$$u : (\mathcal{X} \times \mathcal{Y}) \to (0, +\infty).$$

The weighted empirical risk is equal to

$$\mathbb{E}_\mathcal{D} u(x, y) L(f(x), y) = \frac{1}{N} \sum_{i=1}^{N} u(x_i, y_i) L(f(x_i), y_i),$$

so that the empirical classifier

$$\hat{f} = \arg\min_{f \in \mathcal{F}} \mathbb{E}_\mathcal{D} u(x, y) L(f(x), y). \tag{1}$$

We would like to derive an upper bound for the excess risk

$$\Delta(\mathcal{F},\mathbb{P}) = \sup_{f\in\mathcal{F}} \left(\mathbb{E}_{\mathbb{P}} L(f(x),y) - \mathbb{E}_{\mathcal{D}} u(x,y) L(f(x),y)\right),$$

which characterizes a generalization ability of the classifier. In particular, high values of the excess risk means that the function class $\mathcal{F}$ is "too complex" for the considered problem.

There exist theoretical results about the classification performance when the classifier is trained with the weighted loss. E.g. in [9] a bayesian framework for imbalanced classification with a weighted risk is proposed, [19] investigated the calibration of asymmetric surrogate losses, [16] considered the case of cost-sensitive learning with noisy labels. The case of weighted risk for the one-dimensional classification based on probability density functions estimates is considered in [14].

However, to the best of our knowledge, there is no studied upper bound for the excess risk with explicitly derived dependence on the class imbalance value π and the weighting scheme $u(\cdot)$ that quantifies their influence on the overall classification performance.

3 Generalization Bound

To derive explicit expressions we use an additional assumption, namely, we consider

$$u(x,y) = (1+g_+(w))\mathbb{I}_{\{y=+1\}} + (1+g_-(w))\,\mathbb{I}_{\{y=-1\}}$$

for some positive weighting functions $g_+(w)$ and $g_-(w)$ of the weight value $w \geq 0$. We can tune w to re-balance the proportion between classes and decrease $\Delta(\mathcal{F},\mathbb{P})$.

Theorem 1. *With probability* $1-\delta$, $\delta > 0$ *for* $\mathcal{D} \sim \mathbb{P}^N$ *the excess risk* $\Delta(\mathcal{F},\mathbb{P})$ *is upper bounded by*

$$\overline{\Delta}(w) = 3\left(g_+(w)\pi + g_-(w)(1-\pi)\right) + \mathcal{R}_N(\mathcal{F}) + \left(2 + g_+(w) + g_-(w)\right)\alpha_N, \quad (2)$$

where $\alpha_N = \sqrt{\frac{\log \delta^{-1}}{2N}}$.

Proof. Let

$$\mathcal{L} = \{(x,y) \to L(f(x),y) : f \in \mathcal{F}\}$$

be a composite loss class. For any $L \in \mathcal{L}$ we get that

$$\begin{aligned}\mathbb{E}_{\mathbb{P}} L - \mathbb{E}_{\mathcal{D}} uL &= \mathbb{E}_{\mathbb{P}} L - \mathbb{E}_{\mathbb{P}} uL + \mathbb{E}_{\mathbb{P}} uL - \mathbb{E}_{\mathcal{D}} uL \\ &\leq \mathbb{E}_{\mathbb{P}} |(1-u)L| + (\mathbb{E}_{\mathbb{P}} uL - \mathbb{E}_{\mathcal{D}} uL). \quad (3)\end{aligned}$$

Since any $L \in \mathcal{L}$ is bounded from above by 1 for the first term in (3) we obtain

$$\begin{aligned}\mathbb{E}_{\mathbb{P}} |(1-u)L| &\leq \mathbb{E}_{\mathbb{P}} g_+(w)\mathbb{I}_{\{y=+1\}} + \mathbb{E}_{\mathbb{P}} g_-(w)\mathbb{I}_{\{y=-1\}} \\ &= g_+(w)\pi + g_-(w)(1-\pi). \quad (4)\end{aligned}$$

Thanks to McDiarmid'd concentration inequality [15], applied to the function class $\mathcal{L}_u = \{uL : L \in \mathcal{L}\}$, with probability $1-\delta$, $\delta > 0$ for $\mathcal{D} \sim \mathbb{P}^N$ we get the upper bound on the excess risk

$$\sup_{L\in\mathcal{L}}(\mathbb{E}_{\mathbb{P}}uL - \mathbb{E}_{\mathcal{D}}uL) \le 2\mathcal{R}_N(\mathcal{L}_u) + \max[(1+g_+(w)),(1+g_-(w))]\alpha_N \le$$
$$\le 2\mathcal{R}_N(\mathcal{L}_u) + (2+g_+(w)+g_-(w))\alpha_N. \tag{5}$$

Let us find a relation between $\mathcal{R}_N(\mathcal{L}_u)$ and $\mathcal{R}_N(\mathcal{L})$. We denote by z_i a pair $z_i = (x_i, y_i)$. By the definition (see [15]) the empirical Rademacher complexity

$$\begin{aligned}
\hat{\mathcal{R}}_{\mathcal{D}}(\mathcal{L}_u) &= \frac{1}{N}\mathbb{E}_\sigma \sup_{L\in\mathcal{L}_u}\sum_{i=1}^N \sigma_i u(z_i)L(z_i)\\
&\le \frac{1}{N}\mathbb{E}_\sigma \sup_{L\in\mathcal{L}_u}\sum_{i=1}^N \sigma_i L(z_i) + \frac{g_+(w)}{N}\mathbb{E}_\sigma \sup_{L\in\mathcal{L}_u}\sum_{i:y_i=+1}\sigma_i L(z_i)\\
&+ \frac{g_-(w)}{N}\mathbb{E}_\sigma \sup_{L\in\mathcal{L}_u}\sum_{i:y_i=-1}\sigma_i L(z_i)\\
&\le \hat{\mathcal{R}}_{\mathcal{D}}(\mathcal{L}) + \frac{g_+(w)}{N}\mathbb{E}_\sigma \sup_{L\in\mathcal{L}_u}\sum_{i:y_i=+1}\sigma_i L(z_i)\\
&+ \frac{g_-(w)}{N}\mathbb{E}_\sigma \sup_{L\in\mathcal{L}_u}\sum_{i:y_i=-1}\sigma_i L(z_i).
\end{aligned}$$

For the zero-one loss

$$\mathbb{E}_\sigma \sup_{L\in\mathcal{L}_u}\sum_{i:y_i=+1}\sigma_i L(z_i) \le \#\{i : y_i = +1\},$$

and

$$\mathbb{E}_\sigma \sup_{L\in\mathcal{L}_u}\sum_{i:y_i=-1}\sigma_i L(z_i) \le \#\{i : y_i = -1\}.$$

The Rademacher complexity

$$\begin{aligned}
\mathcal{R}_N(\mathcal{L}_u) &= \mathbb{E}_{\mathcal{D}\sim\mathbb{P}^N}\hat{\mathcal{R}}_{\mathcal{D}}(\mathcal{L}_u)\\
&\le \mathbb{E}_{\mathcal{D}\sim\mathbb{P}^N}\left[\hat{\mathcal{R}}_{\mathcal{D}}(\mathcal{L}) + \frac{g_+(w)}{N}\#\{i : y_i = +1\} + \frac{g_-(w)}{N}\#\{i : y_i = -1\}\right]\\
&= \mathcal{R}_N(\mathcal{L}) + g_+(w)\pi + g_-(w)(1-\pi).
\end{aligned} \tag{6}$$

Using the fact that $\mathcal{R}_N(\mathcal{L}) = \frac{1}{2}\mathcal{R}_N(\mathcal{F})$, substituting inequalities (4), (5) and (6) into (3), we get that

$$\Delta(\mathcal{F},\mathbb{P}) \le 3\,(g_+(w)\pi + g_-(w)(1-\pi)) + \mathcal{R}_N(\mathcal{F}) + (2+g_+(w)+g_-(w))\,\alpha_N.$$

By collecting the terms with w in $\overline{\Delta}(w)$ (2) we get

$$g_+(w)\,(3\pi+\alpha_N) + g_-(w)\,(3(1-\pi)+\alpha_N),$$

and so minimizing this quantity w.r.t. w we can make the upper bound $\overline{\Delta}(w)$ tighter. For example, in case we set

$$g_+(w) = w \;\; g_-(w) = 1/w,$$

the optimal weight

$$w^{opt} = \sqrt{\frac{3(1-\pi)+\alpha_N}{3\pi+\alpha_N}} \approx \sqrt{\frac{1-\pi}{\pi}}, \tag{7}$$

where $\alpha_N \approx 0$ for $N \gg 1$. For such optimal w^{opt} we get

$$\overline{\Delta}^{opt} = \overline{\Delta}(w^{opt}) = 6\sqrt{\pi(1-\pi)} + \mathcal{R}_N(\mathcal{F}) + \alpha_N\left(2 + \frac{1}{\sqrt{\pi(1-\pi)}}\right).$$

Thus we obtain an estimate on how the weighting influences the classification accuracy: e.g. in the imbalanced case (when $\pi \approx 0$ or $\pi \approx 1$) for $N \gg 1$ by selecting the weight optimally we reduce the generalization gap almost to zero, as $\overline{\Delta}^{opt} \approx 0$; at the same time not optimal weight can lead to overfitting.

As we already discussed, under some mild modeling assumptions the binary classification problem in case of the oversampling can be interpreted as the classification problem with the weighted loss. Therefore not correctly selected resampling amount has the same negative effect as not optimal weight value for the classification with the weighted loss function. If we know the class imbalance, we can use the optimal value w^{opt} either to set the weight in case we use the weighted classification scheme, or as a reference value when selecting the resampling amount in case we use the oversampling approach—this should help to reduce the number of steps of the exhaustive search, used in [6,20].

4 Empirical Results

Let us perform an empirical evaluation of the obtained estimate (7). We expect that for the optimal weight value w^{opt} the classifier achieves better accuracy on the test when being trained by minimizing the weighted empirical loss. We consider the following protocol of experiments:

1. Consider different values of the weight $w \in W_K = \{w_1, \dots, w_K\}$;
2. Train a classifier $f_w(x)$ by minimizing a weighted empirical loss (1) for the particular weight value $w = w_i$;
3. Estimate accuracy on the test set and find the weight $w^* \in W_K$ for which accuracy is the highest;
4. Compare the best obtained weight with the theoretical weight calculated using the formula (7).

We generated artificial datasets as pairs of 2D Gaussian samples with various means and covariance matrices and sample sizes, where each Gaussian sample

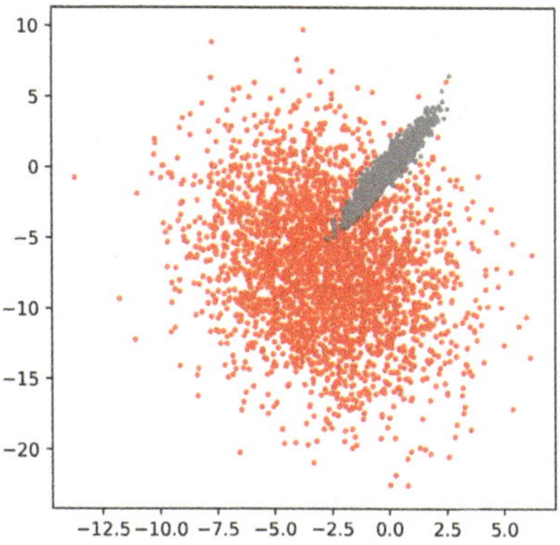

Fig. 1. Example of a toy dataset 1

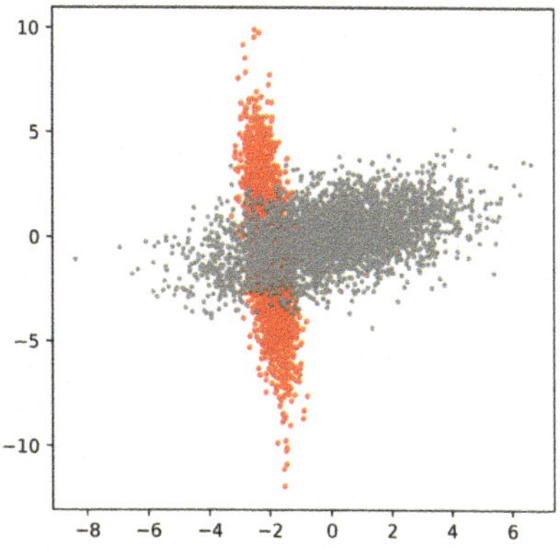

Fig. 2. Example of a toy dataset 2

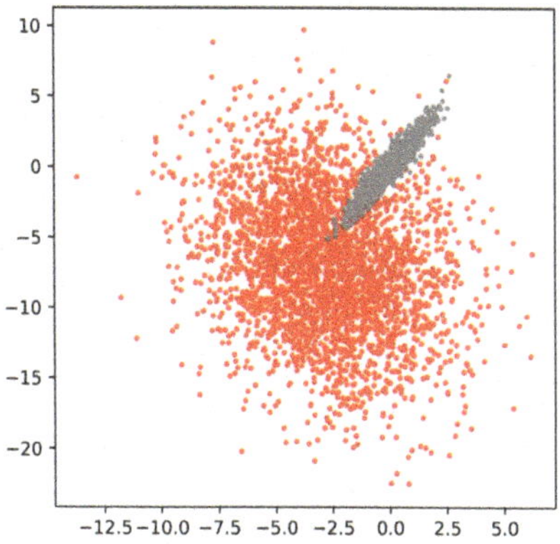

Fig. 3. Example of a toy dataset 3

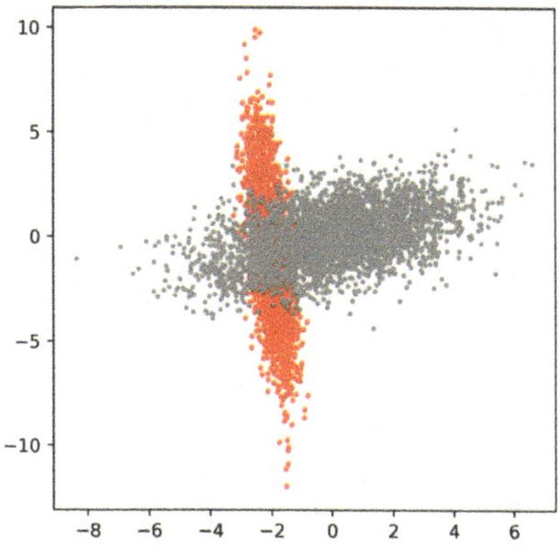

Fig. 4. Example of a toy dataset 4

corresponds to some class. Examples of artificial datasets 1, 2, 3 and 4 are shown in Figs. 1, 2, 3 and 4.

We took real datasets from Penn Machine Learning Benchmark repository [18]: we selected diabetes, german, waveform-40, satimage, splice, spambase, hypothyroid, and mushroom, that have various types of data and features. Due to multiclass data, we took class 0 for waveform-40 and splice, class 1 for satimage and class 2 for diabetes as a positive, whereas other classes were combined into a negative one.

To obtain a specific balance between classes in experiments, we used undersampling of an excess class. In this way we can get learning samples $\mathcal{D}$ corresponding to different values of π. Using this method, we varied the positive class share to test the dependence of the results on π.

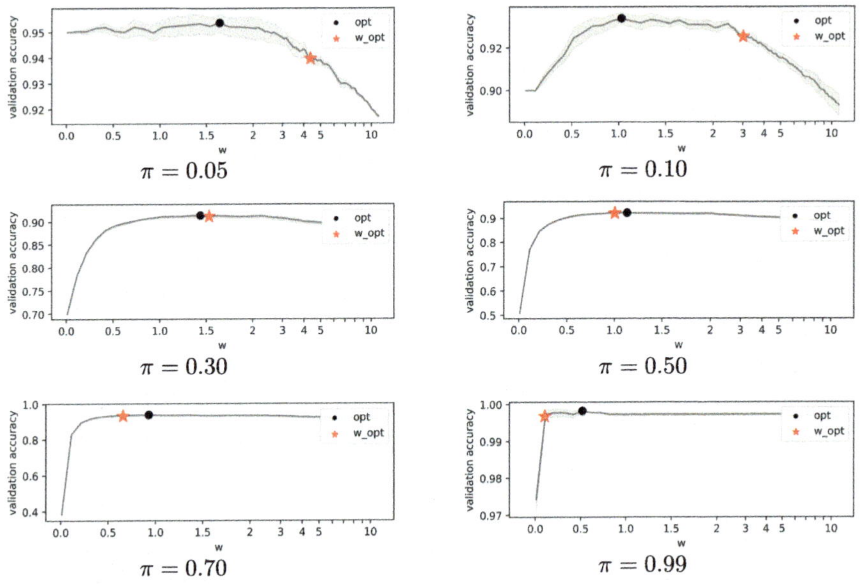

Fig. 5. w^* (black dot) vs. w^{opt} (red star) for the toy dataset 1 and different values of π

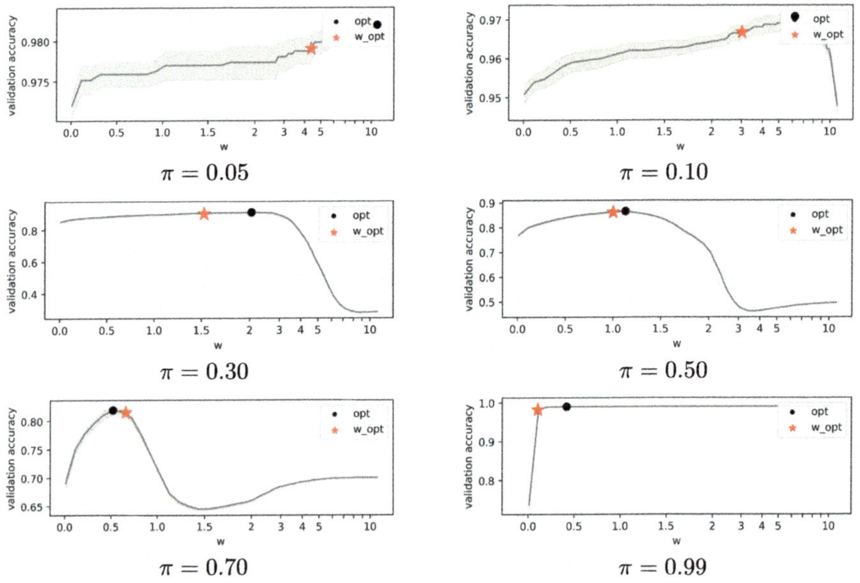

Fig. 6. w^* (black dot) vs. w^{opt} (red star) for the toy dataset 2 and different values of π

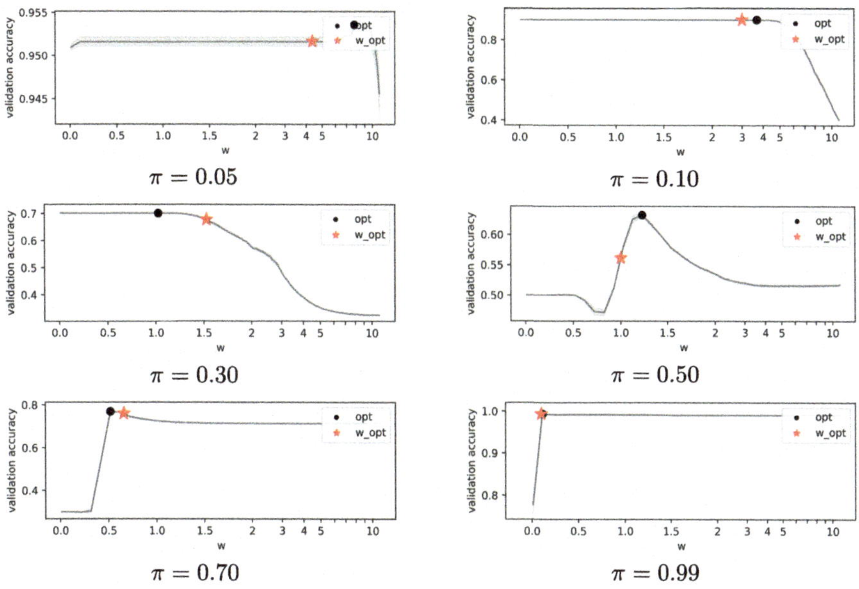

Fig. 7. w^* (black dot) vs. w^{opt} (red star) for the toy dataset 3 and different values of π

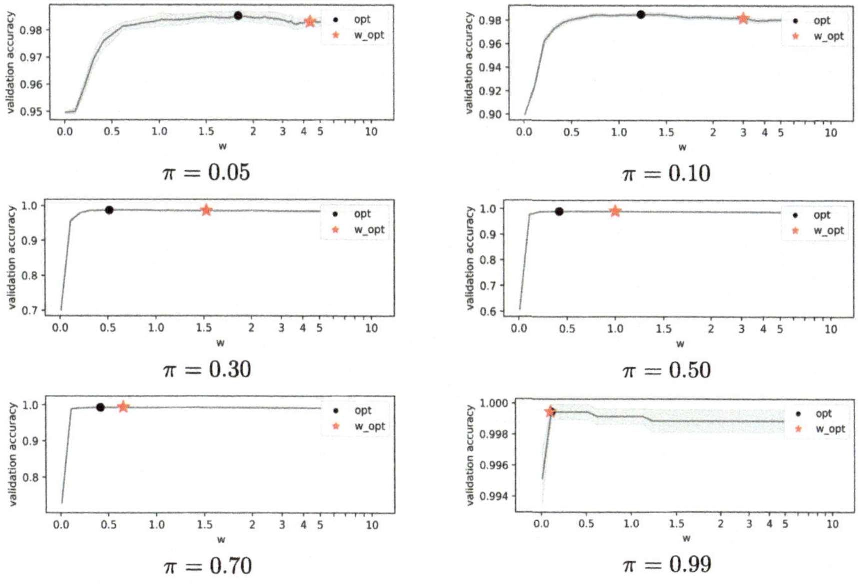

Fig. 8. w^* (black dot) vs. w^{opt} (red star) for the toy dataset 4 and different values of π

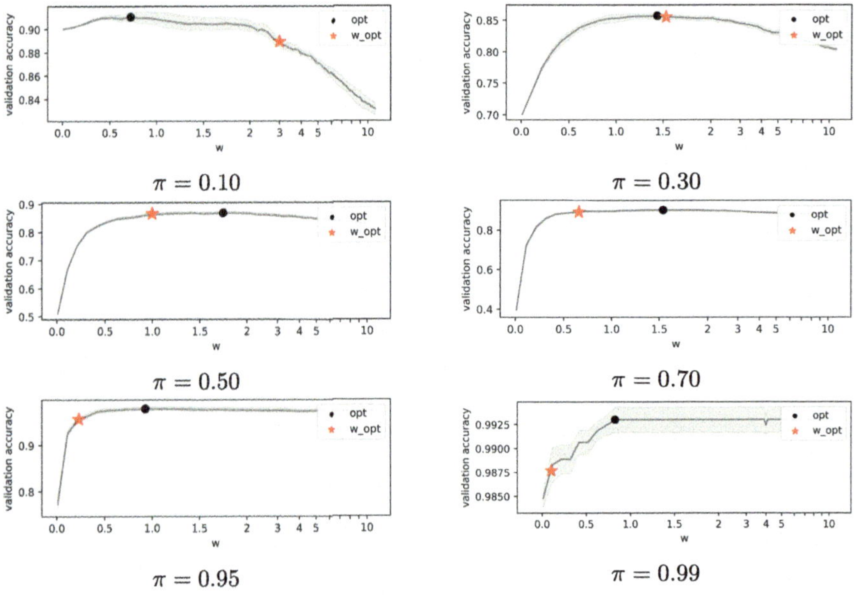

Fig. 9. w^* (black dot) vs. w^{opt} (red star) for the real dataset waveform-40 and different values of π

To measure the performance of the method, we conducted 5-fold cross-validation of a Logistic Regression classifier [10]. We provide examples of typical results on different datasets and for different positive class shares π: in Figs. 5, 6, 7, 8 there are results for toy datasets, and in Figs. 9 and 10 there are results for two real datasets—waveform-40 and hypothyroid). In particular, we show how the average validation accuracy depends on the weight w; we indicate empirically optimal values of w^* by black dots, and indicate theoretically optimal values of w^{opt} by red stars. We can observe that for most of the cases estimates w^* and w^{opt} agree rather well. Moreover, although the estimate w^{opt} is obtained under general conditions from the rather loose bound (2), still the classifier with $w = w^{opt}$ provides often quite good accuracy even if there is a big difference between w^* and w^{opt}.

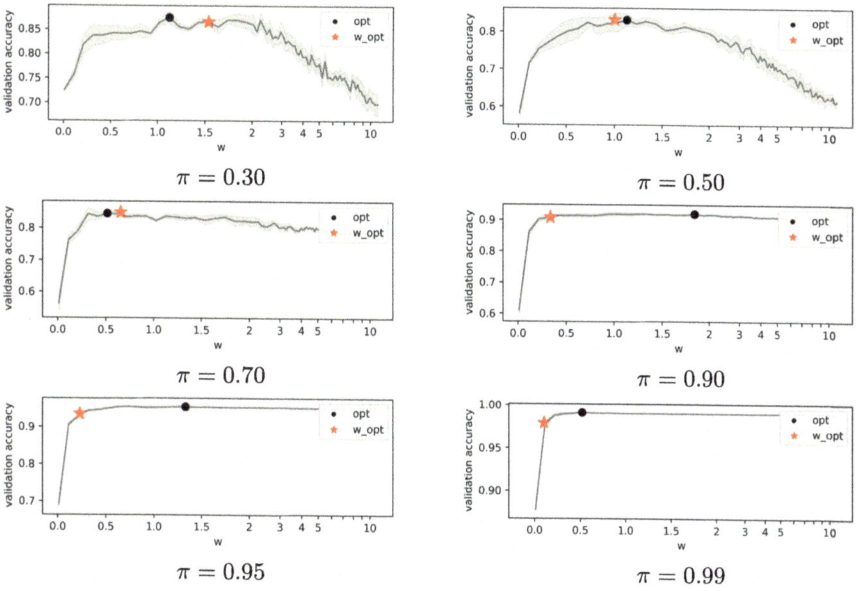

Fig. 10. w^* (black dot) vs. w^{opt} (red star) for the real dataset hypothyroid and different values of π

5 Conclusion

We considered the binary classification problem in the imbalanced setting. We showed that the oversampling approach under somewhat realistic assumptions can be interpreted as the weighted classification. We derived the generation bound for the weighted classification and discussed what connection the bound has with the selection of the resampling amount. We proposed the algorithm based on the derived optimal weighting. Results of the computational experiments demonstrated usefulness of the proposed approach.

Acknowledgments. E. Burnaev would like to thank Ivan Nazarov for fruitfull discussions on the problem statement and the derivation of the generalization bounds, Nikita Klyuchnikov for his help with computational experiments. E. Burnaev was supported by the Russian Fund of Basic Research under Grants 20-01-00203 and 21-51-12005 NNIO_a.

References

1. Artemov, A., Burnaev, E.: Ensembles of detectors for online detection of transient changes. In: Proceedings of the SPIE, vol. 9875, p. 9875–9875 - 5 (2015)
2. Artemov, A., Burnaev, E.: Detecting performance degradation of software-intensive systems in the presence of trends and long-range dependence. In: IEEE 16th International Conference on Data Mining Workshops (ICDMW), pp. 29–36 (2016)
3. Artemov, A., Burnaev, E., Lokot, A.: Nonparametric decomposition of quasi-periodic time series for change-point detection. In: Proceedings of the SPIE, vol. 9875, p. 9875–9875 - 5 (2015)
4. Burnaev, E.: On construction of early warning systems for predictive maintenance in aerospace industry. J. Commun. Technol. Electron. **64**(12), 1473–1484 (2019)
5. Burnaev, E.: Rare failure prediction via event matching for aerospace applications. In: Proceedings of the 3rd International Conference on Circuits, System and Simulation (ICCSS-2019), pp. 214–220 (2019)
6. Burnaev, E., Erofeev, P., Papanov, A.: Influence of resampling on accuracy of imbalanced classification. In: Proceedings of the SPIE, vol. 9875, p. 9875–9875 - 5 (2015)
7. Burnaev, E., Erofeev, P., Smolyakov, D.: Model selection for anomaly detection. In: Proceedings of the SPIE, vol. 9875, p. 9875–9875 - 6 (2015)
8. Burnaev, E., Smolyakov, D.: One-class SVM with privileged information and its application to malware detection. In: IEEE 16th International Conference on Data Mining Workshops (ICDMW), pp. 273–280 (2016)
9. Dupret, G., Koda, M.: Bootstrap re-sampling for unbalanced data in supervised learning. Eur. J. Oper. Res. **134**(1), 141–156 (2001)
10. Hastie, T., Tibshirani, R., Friedman, J.: The Elements of Statistical Learning: Data Mining, Inference and Prediction. Springer, Heidelberg (2009)
11. He, H., Garcia, E.A.: Learning from imbalanced data. IEEE Trans. Knowl. Data Eng. **21**(9), 1263–1284 (2009)
12. Ignatiev, V., Trekin, A., Lobachev, V., Potapov, G., Burnaev, E.: Targeted change detection in remote sensing images. In: Proceedings of the SPIE (2019)
13. Kolos, M., Marin, A., Artemov, A., Burnaev, E.: Procedural synthesis of remote sensing images for robust change detection with neural networks. In: Lu, H., Tang, H., Wang, Z. (eds.) ISNN 2019. LNCS, vol. 11555, pp. 371–387. Springer, Cham (2019). https://doi.org/10.1007/978-3-030-22808-8_37
14. Maiboroda, R., Markovich, N.: Estimation of heavy-tailed probability density function with application to web data. Comput. Stat. **19**, 569–592 (2004)
15. Mohri, M., Rostamizadeh, A., Talwalkar, A.: Foundations of Machine Learning. MIT Press, Cambridge (2018)
16. Natarajan, N., et al.: Cost-sensitive learning with noisy labels. JMLR **18**(1), 5666–5698 (2018)
17. Novikov, G., Trekin, A., Potapov, G., Ignatiev, V., Burnaev, E.: Satellite imagery analysis for operational damage assessment in emergency situations. In: Abramowicz, W., Paschke, A. (eds.) BIS 2018. LNBIP, vol. 320, pp. 347–358. Springer, Cham (2018). https://doi.org/10.1007/978-3-319-93931-5_25
18. Olson, R.S., La Cava, W., Orzechowski, P., Urbanowicz, R.J., Moore, J.H.: PMLB: a large benchmark suite for machine learning evaluation and comparison. BioData Min. **10**(1), 36 (2017)
19. Scott, C.: Calibrated asymmetric surrogate losses. Electron. J. Stat. **6**, 958–992 (2012)

20. Smolyakov, D., Korotin, A., Erofeev, P., Papanov, A., Burnaev, E.: Meta-learning for resampling recommendation systems. In: Proceedings of the SPIE 11041, 11th International Conference on Machine Vision (ICMV 2018), p. 110411S (2019)
21. Smolyakov, D., Sviridenko, N., Burikov, E., Burnaev, E.: Anomaly pattern recognition with privileged information for sensor fault detection. In: Pancioni, L., Schwenker, F., Trentin, E. (eds.) ANNPR 2018. LNCS (LNAI), vol. 11081, pp. 320–332. Springer, Cham (2018). https://doi.org/10.1007/978-3-319-99978-4_25
22. Smolyakov, D., Sviridenko, N., Ishimtsev, V., Burikov, E., Burnaev, E.: Learning ensembles of anomaly detectors on synthetic data. In: Lu, H., Tang, H., Wang, Z. (eds.) ISNN 2019. LNCS, vol. 11555, pp. 292–306. Springer, Cham (2019). https://doi.org/10.1007/978-3-030-22808-8_30

Resonance in Large Finite Particle Systems

Alexandr Lykov, Vadim Malyshev, Margarita Melikian(✉), and Andrey Zamyatin

Faculty of Mechanics and Mathematics, Lomonosov Moscow State University, Moscow 119991, Russia

Abstract. We consider general linear system of Hamiltonian equations. The corresponding linear operator is assumed to be positive definite for the particles could not escape to infinity. However, there are also external driving forces, that could make the solution unbounded. It is assumed that driving force depends only on time, it can be periodic, almost-periodic and random. Moreover, it acts only on one coordinate. Our main problem here is to understand what restrictions on driving force and/or what dissipative force could be added to escape resonance (unbounded trajectories). Various conditions for existence and non-existence of resonance are obtained, for any number of particles.

Keywords: Linear systems · Hamiltonian dynamics · Resonance · Boundedness

1 The Model

There are extreme models in non-equilibrium statistical physics. First one is the ideal (or almost ideal) gas where the particles are free that is do not interact (or almost do not interact) with each other. The second is when the particles interact but each particle moves inside its own potential well, which also moves due to interaction of particles. The simplest such model is the general linear model with quadratic potential interaction energy. However, besides interaction, there can be also external driving and dissipative forces. There are many different qualitative phenomena concerning such systems. One of the most important is resonance phenomena. That is when the particles start to leave their potential wells, the system becomes unstable and the dynamics becomes unbounded. Here we study the models with large number of particles where the resonance can occur even if the external forces act only on one fixed particle.

We consider general linear system of N_0 point particles in $\mathbb{R}^d$ with $N = dN_0$ coordinates $q_j \in \mathbb{R}$, $j = 1,..,N$. Let

$$v_j = \frac{dx_j}{dt}, p_j = m_j v_j,\ j = 1,..,N,$$
$$q = (q_1, ..., q_N)^T, p = (p_1, ..., p_N)^T, \psi(t) = (q_1, ..., q_N, p_1, ..., p_N)^T.$$

A. N. Shiryaev et al. (Eds.): ICSM-5 2020, PROMS 371, pp. 120–130, 2021.
https://doi.org/10.1007/978-3-030-83266-7_9

Here m_j is the mass of the particle having q_j as one of its coordinate. Further on we put $m_j = 1$ and thus $p_j = v_j$. Potential and kinetic energies are:

$$U(\psi(t)) = \frac{1}{2} \sum_{1 \leq j,l \leq N} V_{j,l} q_j q_l = \frac{1}{2}(q, Vq), \; T(\psi(t)) = \sum_{j=1}^{N} \frac{p_j^2}{2} = \frac{1}{2}(p, p),$$

where the matrix $V = (V_{ij})$ is always assumed to be positive definite. Then $q_j(t), v_j(t)$ are bounded for any initial conditions.

If there are also external forces $f_j(t, v_j)$, acting correspondingly on the coordinates j, then we have the following system of equations:

$$\ddot{q}_j = -\sum_l V_{j,l} q_l + f_j(t, v_j), \; j = 1, .., N,$$

or in the first order form:

$$\ddot{q}_j = \dot{v}_j = \sum_l V_{j,l} q_l + f_j(t, v_j).$$

or

$$\dot{\psi} = A_0 \psi + F, \tag{1}$$

where

$$A_0 = \begin{pmatrix} 0 & E \\ -V & 0 \end{pmatrix}, \tag{2}$$

$$F = (0, ..., 0, f_1(t, v_1), ..., f_N(t, v_N))^T \in \mathbb{R}^{2N}.$$

We shall consider the case:

$$f_j(t, v_j) = f(t)\delta_{j,n} - \alpha v_j \delta_{j,k}, \tag{3}$$

where the time dependent external force $f(t)$ acts only on fixed coordinate n, and the dissipative force $-\alpha v_k$ acts only on coordinate k.

For fixed initial conditions and parameters (F, V), we say that resonance takes place if the solutions $x_j(t), v_j(t)$ are not bounded in $t \in [0, \infty)$ at least for one $j = 1, ..., N$.

2 Main Results

As matrix V is positively definite, its eigenvalues are strictly positive and it is convenient to denote them as $a_k = \nu_k^2, k = 1, ..., N$, furthermore, it is convenient to consider all ν_k positive too. Corresponding system of eigenvectors we denote as $\{u_k, k = 1, ..., N\}$ and this system can always be assumed to be orthonormal.

2.1 Periodic Driving Force

Here we consider the case (3) with $f(t) = a \sin \omega t$ and $\alpha = 0$. And for general periodic force one could just (due to linearity of equations) consider its Fourier series.

Denote Ω_N the set of all positive-definite $(N \times N)$-matrices. It is an **open** subset in R^M (the set of all symmetric matrices), where $M = N + \frac{N^2 - N}{2} = \frac{N(N+1)}{2}$. It is open because any sufficiently small perturbation does not change positive definiteness. Denote μ the Lebesgue measure on Ω_N, and let (for fixed ω) $\Omega(\omega)$ be the subset of Ω_N such that $\omega^2 = \nu_l^2$ at least for one $l \in \{1, ..., N\}$. It is an algebraic manifold in $\mathbb{R}^M$ and thus $\mu(\Omega(\omega)) = 0$.

Theorem 1. *1). Assume that $V \notin \Omega(\omega)$ that is, for all $j \in \{1, ..., N\}$, $\nu_j^2 \neq \omega^2$. Then for all $j \in \{1, ..., N\}$ and all $t \geq 0$:*

$$|q_j(t)| \leq 2d_j\beta, \ |p_j(t)| \leq 2d_j\beta\omega,$$

where

$$\beta = \max_r \frac{1}{|\omega^2 - \nu_r^2|}, d_j = |a| \sum_{k=1}^{N} |(u_k, e_n)(u_k, e_j)|. \tag{4}$$

In other words, there will not be resonance for almost all matrices V;

2). Assume $\omega^2 = \nu_l^2$ at least for one $l \in \{1, ..., N\}$. Then $q_j(t)$, $p_j(t)$ are bounded uniformly in $t \geq 0$ if and only if for this j holds:

$$\sum_{k \in I(\omega)} (u_k, e_n)(u_k, e_j)) = 0, \tag{5}$$

where $I(\omega) = \{k \in \{1, ..., N\} : \omega^2 = \nu_k^2\}$. Otherwise resonance occurs. Moreover, for all $j \in \{1, ..., N\}$:

$$\liminf_{t \to +\infty} q_j(t) = -\infty, \ \limsup_{t \to +\infty} q_j(t) = +\infty,$$
$$\liminf_{t \to +\infty} p_j(t) = -\infty, \ \limsup_{t \to +\infty} p_j(t) = +\infty,$$

and

$$\limsup_{t \to +\infty} \frac{T(\psi(t))}{t^2} = \limsup_{t \to +\infty} \frac{U(\psi(t))}{t^2} = \lim_{t \to +\infty} \frac{H(\psi(t))}{t^2} = C,$$

where

$$C = \frac{a^2}{8} \sum_{k \in I(\omega)} (u_k, e_n)^2.$$

2.2 Arbitrary Driving Force and Dissipation

Now we consider the force (3), that is the equation:

$$\ddot{q}_j = -\sum_l V_{j,l}q_l + f(t)\delta_{j,n} - \alpha\dot{q}_k\delta_{j,k},\ j = 1,..,N,\ \alpha > 0.$$

In the vector form it can be written as:

$$\dot{q}_j = p_j,$$
$$\dot{p}_j = -\sum_l V_{j,l}q_l + f(t)\delta_{j,n} - \alpha p_k\delta_{j,k},$$

or

$$\dot{\psi} = A\psi + f(t)g_n, \tag{6}$$

where

$$A = \begin{pmatrix} 0 & E \\ -V & -\alpha D \end{pmatrix} \tag{7}$$

is the $2N \times 2N$-matrix with $N \times N$ blocks, E is the unit $N \times N$-matrix,

$$D = D_k = diag(\delta_{1,k},..,\delta_{N,k}),$$

$$g_n = (\overline{0}, e_n)^T \in \mathbb{R}^{2N},\ e_n = (\delta_{1,n},..,\delta_{N,n}),\ \overline{0} = (0,...,0) \in \mathbb{R}^N, \tag{8}$$

and again we consider zero initial conditions.

Put

$$S = \begin{pmatrix} 0 & 0 \\ 0 & \alpha D \end{pmatrix}.$$

Then

$$A = A_0 - S,$$

where A_0 was defined in (2). It is known that $Re\,\nu \leq 0$ for all eigenvalues of A [see [2]].

Theorem 2. *1). Assume that the function $f(t)$ (defined by (3)) grows in time t on $[0,\infty)$ not faster than the power function. Then if the spectrum of A does not have pure imaginary eigenvalues, then the solution of the system (6) is bounded on $[0,\infty)$.*

2). Let $\Lambda_N \subset \Omega_N \subset \mathbb{R}^M$ be the set of matrices V such that all eigenvalues of the matrix A lie inside left halfplane. Then the Lebesgue measure $\mu(\Omega_N \backslash \Lambda_N) = 0$, that is for almost all matrices V the spectrum of the corresponding matrices $A = A(V)$ lies inside left halfplane.

2.3 Almost-Periodic Force

Suppose the force $f(t)$ has the form:

$$f(t) = \int_R a(\omega)\cos(\omega t)d\omega,$$

where $a(\omega) \in l_1(\mathbb{R})$ is a sufficiently smooth function. Then the function $f(t)$ on $\mathbb{R}$ is almost periodic.

Theorem 3. *Assume $\alpha = 0$, that is there is no dissipative force. Then for any initial data the solutions $\{x_k(t), v_k(t)\}$ are bounded on the time interval $[0, \infty)$.*

3 Proofs

3.1 Proof of Theorem 1

Note first that the eigenvalues of matrix A_0 are $\pm i\nu_1, ..., \pm i\nu_N$. In fact, if $u = u_k$ - eigenvector of V corresponding to the eigenvalue ν_k^2, $k = 1, ..., N$, (i.e. $Vu = \nu_k^2 u$) then vector $x_\pm = \begin{pmatrix} u \\ \lambda_\pm u \end{pmatrix}$, where $\lambda_\pm = \pm i\nu_k$ is the eigenvector of A_0, corresponding to the eigenvalue $\lambda_\pm = \pm i\nu_k$:

$$A_0 x_\pm = \begin{pmatrix} 0 & E \\ -V & 0 \end{pmatrix}\begin{pmatrix} u \\ \lambda_\pm u \end{pmatrix} = \begin{pmatrix} \lambda_\pm u \\ -Vu \end{pmatrix} = \begin{pmatrix} \lambda_\pm u \\ -\nu_k^2 u \end{pmatrix} = \begin{pmatrix} \lambda_\pm u \\ \lambda_\pm^2 u \end{pmatrix} = \lambda_\pm x_\pm.$$

It is well-known that the solution of equation (1) can be written as:

$$\psi(t) = e^{A_0 t}\left(\int_0^t f(s)e^{-A_0 s} g_n ds + \psi(0)\right). \tag{9}$$

It is easy to prove that:

$$e^{A_0 t} = \begin{pmatrix} \cos(\sqrt{V}t) & (\sqrt{V})^{-1}\sin(\sqrt{V}t) \\ -\sqrt{V}\sin(\sqrt{V}t) & \cos(\sqrt{V}t) \end{pmatrix},$$

where trigonometric functions of matrices are defined by the corresponding power series. Then we can find $q(t)$, $p(t)$ explicitely:

$$\begin{aligned} q(t) = \textstyle\int_0^t f(s)(\sqrt{V})^{-1}\sin(\sqrt{V}(t-s))e_n ds + \cos(\sqrt{V}t)q(0) \\ +(\sqrt{V})^{-1}\sin(\sqrt{V}t)p(0), \end{aligned} \tag{10}$$

$$p(t) = \int_0^t f(s)\cos(\sqrt{V}(t-s))e_n ds - \sqrt{V}\sin(\sqrt{V}t)q(0) + \cos(\sqrt{V}t)p(0). \tag{11}$$

Let us expand vectors $e_n, q(0), p(0)$ in the orthonormal basis of eigenvectors of V:

$$e_n = \sum_{k=1}^N (u_k, e_n)u_k,\ q(0) = \sum_{k=1}^N (u_k, q(0))u_k,\ p(0) = \sum_{k=1}^N (u_k, p(0))u_k.$$

Then as

$$(\sqrt{V})^{-1}u_k = \frac{1}{\nu_k}u_k,\ \sin(\sqrt{V}t)u_k = u_k\sin(\nu_k t),$$
$$\cos(\sqrt{V}t)u_k = u_k\cos(\nu_k t),$$

we have:

$$q(t) = \sum_{k=1}^{N}[(u_k, e_n)(\int_0^t f(s)\frac{\sin(\nu_k(t-s))}{\nu_k}ds)$$
$$+ (u_k, q(0))\cos(\nu_k t) + (u_k, p(0))\frac{\sin(\nu_k t)}{\nu_k}]u_k, \tag{12}$$

$$p(t) = \sum_{k=1}^{N}[(u_k, e_n)(\int_0^t f(s)\cos(\nu_k(t-s))ds)$$
$$-(u_k, q(0))\nu_k\sin(\nu_k t) + (u_k, p(0))\cos(\nu_k t)]u_k. \tag{13}$$

Thus we reduced the question of boundedness to the question of boundedness of the functions:

$$\widetilde{q}_k(t) = \int_0^t f(s)\sin(\nu_k(t-s))ds,$$
$$\widetilde{p}_k(t) = \int_0^t f(s)\cos(\nu_k(t-s))ds.$$

For those $j \in \{1, ..., N\}$, where $\omega^2 \neq \nu_j^2$:

$$\widetilde{q}_j(t) = \int_0^t \sin(\omega s)\sin(\sqrt{a_j}(t-s))ds = \frac{\sqrt{a_j}}{\omega^2 - a_j}(\sin(\omega t) - \sin(\sqrt{a_j}t)),$$
$$\widetilde{p}_j(t) = \int_0^t \sin(\omega s)\cos(\sqrt{a_j}(t-s))ds = \frac{\omega}{\omega^2 - a_j}(\cos(\sqrt{a_j}t) - \cos(\omega t)).$$

It follows:

$$|q_j(t)| = |(q, e_j)| = |a\sum_{k=1}^{N}(u_k, e_n)\frac{1}{\nu_k}\widetilde{q}_k(t)(u_k, e_j)| \leq 2d_j\beta,$$
$$|p_j(t)| = |(p, e_j)| = |a\sum_{k=1}^{N}(u_k, e_n)\widetilde{p}_k(t)(u_k, e_j)| \leq 2d_j\beta\omega.$$

Now consider $j \in \{1, ..., N\}$, where $\nu_j^2 = \nu_l^2 = \omega^2$:

$$\widetilde{q}_j(t) = \int_0^t \sin(\omega s)\sin(\sqrt{a_j}(t-s))ds = \tfrac{\sin(\omega t)}{2\omega} - \tfrac{t\cos(\omega t)}{2},$$
$$\widetilde{p}_j(t) = \int_0^t \sin(\omega s)\cos(\sqrt{a_j}(t-s))ds = \tfrac{t\sin(\omega t)}{2},$$
$$q_j(t) = (q, e_j) = a\sum_{k=1}^{N}(u_k, e_n)\tfrac{1}{\nu_k}\widetilde{q}_k(t)(u_k, e_j)$$
$$= (a\textstyle\sum_{k\in I(\omega)}(u_k, e_n)(u_k, e_j))(-\tfrac{t\cos(\omega t)}{2\omega}) + \underline{O}(1) = B_j(-\tfrac{t\cos(\omega t)}{2\omega}) + \underline{O}(1),\ t \to +\infty,$$

where $I(\omega) = \{j \in \{1, ..., N\} : \omega^2 = \nu_j^2\}$.

Similarly for $p_j(t)$:

$$p_j(t) = (p, e_j) = a \sum_{k=1}^{N} (u_k, e_n)\widetilde{p}_k(t)(u_k, e_j)$$

$$= (a \sum_{k \in I(\omega)} (u_k, e_n)(u_k, e_j))\frac{t\sin(\omega t)}{2} + \underline{O}(1) = B_j \frac{t\sin(\omega t)}{2} + \underline{O}(1),\ t \to +\infty.$$

The theorem follows.

3.2 Proof of Theorem 2

In [1–3] the following subspaces of

$$L = \{\psi = (q, p),\ q, p \in \mathbb{R}^N\}$$

were defined (with $H = U + T$):

$$L_- = \{\psi \in L : H(e^{At}\psi) \longrightarrow 0,\ t \longrightarrow +\infty\},$$
$$L_0 = \{\psi \in L : \tfrac{d}{dt} H(e^{At}\psi) = 0\ \forall t\},$$

and was proved that:

1). L_-, L_0 are linear orthogonal subspaces;
2). $L = L_- \oplus L_0$;
3). both L_-, L_0 are invariant with respect to dynamics;
4). $L_0 = \{0\}$ iff A does not have pure imaginary eigenvalues;
5). A does not have pure imaginary eigenvalues iff the vectors $e_n, Ve_n, ..., V^{N-1}e_n$ are linear independent.

Note that resonance is possible for pure imaginary eignvalues as in the integrals, introduced below, secular terms like $t\cos(\omega t)$, $t\sin(\omega t)$ can appear.

The first statement of the theorem follows from Theorem 4.1. (in [4],p.88), where the solution of the system:

$$\dot{\psi} = B\psi + F(t), \tag{14}$$

where B is some linear operator and $F(t)$ is vector function, is considered.

We cite this theorem almost literally.

Theorem 4. *(see [4], Theorem 4.1, p.88)*

In order for there to correspond to any bounded-on-the-real-line continuous vector function $F(t)$ one and only one bounded-on-the-real-line solution of (14) it is necessary and sufficient that the spectrum $\sigma(B)$ not intersect the imaginary axis.

The solution is given by formula:

$$x(t) = \int_{\mathbb{R}} G_B(t-s)F(s)ds,$$

where $G_B(t)$ is principal Green function for equation.

In our case $F(t) = f(t)g_n, t \geq 0$, and

$$G_B(t) = e^{Bt}P_-,$$

where P_- is the spectral projection corresponding to the spectrum of B in the left (negative) halfplane.

3.3 Proof of Theorem 3

Using formula (12) we want to prove boundedness in $t \in [0, \infty)$ of the function

$$I(t) = \int_0^t f(s) \sin(\nu_k(t-s))ds.$$

We have:

$$\begin{aligned} \int_0^t \sin(\nu_k(t-s))f(s)ds &= \int_R a(\omega)d\omega \int_0^t \sin(\nu_k(t-s)) \cos(\omega s)ds \\ &= \nu_k \int_R a(\omega) \tfrac{\cos \omega t - \cos \nu_k t}{\nu_k^2 - \omega^2} d\omega \\ &= 2 \int_R a(\omega) \tfrac{\sin((\omega+\nu_k)t) \sin((\omega-\nu_k)t)}{(\omega+\nu_k)(\omega-\nu_k)} d\omega. \end{aligned}$$

We see that unboundedness in time can only arise when we integrate in a small neighborhood of ν_k. Denoting $\omega = \nu_k + x$ we get as $\epsilon \to 0$:

$$2 \int_{-\epsilon}^{\epsilon} a(\nu_k + x) \frac{\sin((x + 2\nu_k)t) \sin(xt)}{(x + 2\nu_k)x} dx \sim \frac{a(\nu_k) \sin(2\nu_k t)}{\nu_k} \int_{-\epsilon}^{\epsilon} \frac{\sin xt}{x} dx.$$

At the same time we have that the integral:

$$\int_{-\epsilon}^{\epsilon} \frac{\sin xt}{x} dx = \int_{-t\epsilon}^{t\epsilon} \frac{\sin x}{x} dx$$

is bounded uniformly in t. Indeed, on arbitrary period $(N, N+2\pi)$ put $x = N+y$, then

$$\frac{1}{x} = \frac{1}{N} \frac{1}{1 + \frac{y}{N}} = \frac{1}{N} - \frac{y}{N^2} + \ldots$$

The first term gives 0 in the integrals for such periods, and the rest will give a convergent sum.

4 Conclusion

Note first that the solution boundedness problem with the external force $f(t)$ was in fact reduced to the problem of boundedness of the integral:

$$I(t) = \int_0^t f(s) \sin \omega s ds. \tag{15}$$

Now we want to formulate simple and more difficult problems concerning general situation with resonances.

Let us summarize now how to get rid of resonances without self-isolation from external influence. If the external force is periodic, there are following possibilities:

1). one should choose his own oscillation frequency sufficiently far from the external frequency;
2). use external "smooth" almost-periodic force like in Theorem 3;
3). if all previous is impossible one should be simultaneously be under influence of some external dissipative force;
4). what will be if the external force is neither periodic nor almost-periodic. In particular, what will be if $f(s)$ is a random stationary process.

If it is unbounded then the solution also will be unbounded. If $f(s)$ is bounded, consider the following cases.
5). If $f(s)$ is stationary with fast correlation function decay, then the solution will be unbounded. In fact, let $\tau = \frac{2\pi}{\omega}$ be the period in (15). Consider the sequence of random variables

$$\xi_k = \int_{k\tau}^{(k+1)\tau} f(s) \sin \omega s ds$$

and their sums

$$S_N = \xi_1 + ... + \xi_N.$$

If ξ_k are independent or have sufficient decay of correlations, then just by central limit theorem there cannot be boundedness. Interesting question is to formulate general conditions when, keeping the randomness and wihout dissipation forces, one can have bounded solutions.
6). Complete different situation will be for infinite collection of particles. Namely, in many cases there will not be resonance (unbounded graph) due to phenomenon that energy escaped to infinity.

We consider countable number of point particles (with unit masses) on the real axis $x_k \in \mathbb{R}, k \in \mathbb{Z}$. Intuitively, we would like that each particle $x_k(t)$ were close to $ak \in \mathbb{R}$ for some $a > 0$. That is why we introduce the formal Hamiltonian:

$$H(q,p) = \frac{1}{2} \sum_{k \in \mathbb{Z}} p_k^2 + \frac{1}{2} \sum_{k,j \in \mathbb{Z}} b(k-j) q_k q_j,$$

where $q_k = x_k - ak$, and $p_k(t) = \dot{q}_k(t)$ are momenta of the particles k. The real function $v(k)$ is assumed to satisfy the following conditions:

1. symmetry: $b(k) = b(-k), b(0) > 0$;
2. boundedness of the support, that is there exists $r \in \mathbb{N}$ such that $b(k) = 0$ for any $k, |k| > r$;
3. for any $\lambda \in \mathbb{R}$:

$$\omega^2(\lambda) = \sum_{k \in \mathbb{Z}} b(k) e^{ik\lambda} > 0.$$

It follows that the linear operator V in $l_2(\mathbb{Z})$ with elements $V_{jk} = b(k-j)$ (in the standard orthonormal basis $e_n \in l_2(\mathbb{Z})$, $e_n(j) = \delta_{j,n}$) is a positive definite self-adjoint operator.

The trajectories of the system are defined by the following system of equations:

$$\ddot{q}_j = -\sum_k b(k-j)q_k + f(t)\delta_{j,n},\ j \in \mathbb{Z},$$

where $f(t)$ is some external force which acts only on the particle n, $\delta_{j,n}$ is the Kronecker symbol. We will always assume zero initial conditions:

$$q_k(0) = 0,\ p_k(0) = 0,\ k \in \mathbb{Z}.$$

We can rewrite it in the Hamiltonian form:

$$\dot{q}_j = p_j, \dot{p}_j = -\sum_k b(k-j)q_k + f(t)\delta_{j,n}. \tag{16}$$

In $l_2(\mathbb{Z} \times \mathbb{Z})$ define the (state) vector $\psi(t) = \begin{pmatrix} q(t) \\ p(t) \end{pmatrix}$ and the linear operator A_0 which was defined in (2). Then the system can be rewritten as follows:

$$\dot{\psi} = A_0\psi + f(t)g_n, \tag{17}$$

where g_n is defined in (8).

Here we assume that $f(t)$ is a real-valued stationary random process (in the wider sense) with zero mean and covariance function $B(s)$, so that:

$$Ef(t) = 0, Ef(t)f(s) = B(t-s).$$

Also assume that there exist random measure $Z(dx)$ and (spectral) measure $\mu(dx)$ such that for any Borel set $D \subset \mathbb{R}$:

$$EZ(D) = 0,\ E|Z|^2(D) = \mu(D),\ EZ(D_1)Z^*(D_2) = 0$$

for nonintersecting D_1 and D_2, and moreover:

$$B(s) = \int_{\mathbb{R}} e^{isx}\mu(dx),\ f(s) = \int_{\mathbb{R}} e^{isx} Z(dx). \tag{18}$$

We assume also that the support of the random measure is "separated" from the spectrum of A_0. Then the following assertion holds.

Theorem 5. *Solution $\psi(t)$ of the system (16) can be presented as the sum of two centered random processes:*

$$\psi(t) = \zeta(t) + \eta(t),$$
$$\eta(t) = -e^{A_0 t} \int_{\mathbb{R}} e^{itx} R_{A_0}(ix) Z(dx) g,$$
$$\zeta(t) = e^{A_0 t} \int_{\mathbb{R}} R_{A_0}(ix) Z(dx) g = -e^{A_0 t}\eta(0),$$

where $R_A(z) = (A_0 - zI)^{-1}$ is the resolvent of the operator A_0 (I is the unit operator in $l_2(\mathbb{Z} \times \mathbb{Z})$. Moreover, components of $\eta(t)$ are stationary (in wider sense) random processes, and each component of $\zeta(t) \to 0$ a.s. as $t \to +\infty$.

Proof of this theorem and the development of this theme will be given elsewhere.

References

1. Lykov, A.A., Malyshev, V.A.: Harmonic chain with weak dissipation. Markov Process. Relat. Fields **18**, 1–10 (2012)
2. Lykov, A.A., Malyshev, V.A.: Convergence to Gibbs equilibrium - unveiling the mystery. Markov Process. Relat. Fields **19**, 643–666 (2013)
3. Lykov, A.A., Malyshev, V.A., Muzychka, S.A.: Linear Hamiltonian systems under microscopic random influence. Theory Probab. Appl. **57**(4), 684–688 (2013)
4. Daleckii, J.L., Krein, M.G.: Stability of Solutions of Differential Equations in Banach Space. Amer. Mathem. Society, Providence, Rhode Island (1974)

Cycles in Spaces of Finitely Additive Measures of General Markov Chains

Alexander I. Zhdanok[1,2](✉)

[1] Institute for Information Transmission Problems of the RAS, Moscow, Russia
[2] Tuvinian Institute for Exploration of Natural Resources of the Siberian Branch RAS, Kyzyl, Russia

Abstract. General Markov chains in an arbitrary phase space are considered in the framework of the operator treatment. Markov operators continue from the space of countably additive measures to the space of finitely additive measures. Cycles of measures generated by the corresponding operator are constructed, and algebraic operations on them are introduced. One of the main results obtained is that any cycle of finitely additive measures can be uniquely decomposed into the coordinate-wise sum of a cycle of countably additive measures and a cycle of purely finitely additive measures. A theorem is proved (under certain conditions) that if a finitely additive cycle of a Markov chain is unique, then it is countably additive.

Keywords: General Markov chains · Markov operators · Finitely additive measures · Cycles of measures · Decomposition of cycles

1 Introduction

The considered general Markov chains (MC) are random processes with an arbitrary phase space, with discrete time, and homogeneous in time. MCs are given by the usual transition probability, countably additive in the second argument, which generates two Markov operators T and A in the space of measurable functions and in the space of countably additive measures, respectively. Thus, we use the operator treatment in the theory of general MCs, proposed in 1937 by N. Kryloff and N. Bogolyuboff, and developed in detail in the article [1]. Later, in a number of works by different authors, an extension of the Markov operator A to the space of finitely additive measures was carried out, which turned A into an operator topologically conjugate to the operator T, and opened up new possibilities in the development of the operator treatment. Within the framework of such a scheme, we carry out here the study of cycles of measures of general MC. In this case, we use a number of information on the general theory of finitely additive measures from the sources [2] and [3].

In the ergodic theory of MC, one usually distinguishes in the space of its states ergodic classes and their cyclic subclasses, if such exist (see, for example, [4]). However, in the general phase space, the study of such sets has its natural

A. N. Shiryaev et al. (Eds.): ICSM-5 2020, PROMS 371, pp. 131–143, 2021.
https://doi.org/10.1007/978-3-030-83266-7_10

limitations. Therefore, in some cases it is more convenient to use not cycles of sets, but cycles of measures generated by the Markov operator A.

In this paper we propose a corresponding construction for cyclic finitely additive measures of MC on an arbitrary measurable space. We study cycles of countably additive and purely finitely additive measures, and their relationship. A number of theorems on the properties of cycles are proved. In particular, an analogue of the Alexandroff-Yosida-Hewitt expansion for cycles of finitely additive measures is constructed.

In the proof of the theorems presented here, we also use some results of papers [5] and [6].

2 Finitely Additive Measures and Markov Operators

Let X be an arbitrary infinite set and Σ the sigma-algebra of its subsets containing all one-point subsets from X. Let $B(X, \Sigma)$ denote the Banach space of bounded Σ-measurable functions $f : X \to R$ with sup-norm.

We also consider Banach spaces of bounded measures $\mu : \Sigma \to R$, with the norm equal to the total variation of the measure μ (but you can also use the equivalent sup-norm):

$ba(X, \Sigma)$ is the space of finitely additive measures,
$ca(X, \Sigma)$ is the space of countably additive measures.

If $\mu \geq 0$, then $||\mu|| = \mu(X)$.

Definition 1 ([2]). *A finitely additive nonnegative measure μ is called purely finitely additive (pure charge, pure mean) if any countably additive measure λ satisfying the condition $0 \leq \lambda \leq \mu$ is identically zero. An alternating measure μ is called purely finitely additive if both components of its Jordan decomposition are purely finitely additive.*

Any finitely additive measure μ can be uniquely expanded into the sum $\mu = \mu_1 + \mu_2$, where μ_1 is countably additive and μ_2 is a purely finitely additive measure (the Alexandroff-Yosida-Hewitt decomposition, see [2] and [3]).

Purely finitely additive measures also form a Banach space $pfa(X, \Sigma)$ with the same norm, $ba(X, \Sigma) = ca(X, \Sigma) \oplus pfa(X, \Sigma)$.

Examples 1. Here are two examples of purely finitely additive measures.

Let $X = [0, 1] \subset R$ $(R = (-\infty; +\infty))$ and $\Sigma = B$ (Borel sigma algebra). There is (proved) a finitely additive measure $\mu\colon B \to R$, $\mu \in S_{ba}$, such that for any $\varepsilon > 0$ the following holds:

$$\mu((0, \varepsilon)) = 1, \;\; \mu([\varepsilon, 1]) = 0, \;\; \mu(\{0\}) = 0.$$

We can say that the measure μ fixes the unit mass arbitrarily close to zero (on the right), but not at zero. According to [2], such a measure is purely finitely additive, but it is not the only one. It is known that the cardinality of a family of such measures located "near zero (on the right)" is not less than $2^{2^{\aleph_0}} = 2^c$ (hypercontinuum). And the same family of purely finitely additive measures exists "near each point $x_0 \in [0,1]$ (to the right, or to the left, or both there, and there)".

Examples 2. Let $X = R = (-\infty; +\infty)$ and $\Sigma = B$. There is (proved) a finitely additive measure μ: $B \to R$, $\mu \in S_{ba}$, such that for any $x \in R$ the following holds:

$$\mu((x,\infty)) = 1, \quad \mu((-\infty, x)) = 0, \quad \mu(\{x\}) = 0.$$

We can say that the measure μ fixes the unit mass arbitrarily far, "near infinity". This measure is also purely finitely additive. And there are also a lot of such measures.

We denote the sets of measures:

$S_{ba} = \{\mu \in ba(X,\Sigma) : \mu \geq 0, ||\mu|| = 1\}$, $S_{ca} = \{\mu \in ca(X,\Sigma) : \mu \geq 0, ||\mu|| = 1\}$,

$S_{pfa} = \{\mu \in pfa(X,\Sigma) : \mu \geq 0, ||\mu|| = 1\}$.

All measures from these sets will be called probabilistic.

Markov chains (MC) on a measurable space (X,Σ) are given by their transition function (probability) $p(x,E), x \in X, E \in \Sigma$, under the usual conditions:

1. $0 \leq p(x,E) \leq 1, p(x,X) = 1, \forall x \in X, \forall E \in \Sigma$;
2. $p(\cdot,E) \in B(X,\Sigma), \forall E \in \Sigma$;
3. $p(x,\cdot) \in ca(X,\Sigma), \forall x \in X$.

We emphasize that our transition function is a countably additive measure in the second argument, i.e. we consider classical MCs.

The transition function generates two Markov linear bounded positive integral operators:

$T : B(X,\Sigma) \to B(X,\Sigma), (Tf)(x) = Tf(x) = \int_X f(y)p(x,dy)$,

$\forall f \in B(X,\Sigma), \forall x \in X$;

$A : ca(X,\Sigma) \to ca(X,\Sigma), (A\mu)(E) = A\mu(E) = \int_X p(x,E)\mu(dx)$,

$\forall \mu \in ca(X,\Sigma), \forall E \in \Sigma$.

Let the initial measure be $\mu_0 \in S_{ca}$. Then the iterative sequence of countably additive probability measures $\mu_n = A\mu_{n-1} \in S_{ca}, n \in N$, is usually identified with the Markov chain.

Topologically conjugate to the space $B(X,\Sigma)$ is the (isomorphic) space of finitely additive measures: $B^*(X,\Sigma) = ba(X,\Sigma)$ (see, for example, [3]). Moreover, the operator $T^* : ba(X,\Sigma) \to ba(X,\Sigma)$ is topologically conjugate to the operator T:

$$T^*\mu(E) = \int_X p(x,E)\mu(dx), \forall \mu \in ba(X,\Sigma), \forall E \in \Sigma.$$

The operator T^* is the only bounded continuation of the operator A to the entire space $ba(X, \Sigma)$ while preserving its analytic form. The operator T^* has its own invariant subspace $ca(X, \Sigma)$, i.e. $T^*[ca(X, \Sigma)] \subset ca(X, \Sigma)$, on which it matches the original operator A. The construction of the Markov operators T and T^* is now functionally closed. We shall continue to denote the operator T^* as A.

In such a setting, it is natural to admit to consideration also the Markov sequences of probabilistic finitely additive measures $\mu_0 \in S_{ba}, \mu_n = A\mu_{n-1} \in S_{ba}, n \in N$, keeping the countable additivity of the transition function $p(x, \cdot)$ with respect to the second argument.

3 Cycles of Measures and Their Properties

Definition 2. *If $A\mu = \mu$ holds for some positive finitely additive measure μ, then we call such a measure invariant for the operator A (and for the Markov chain).*

We denote the sets of all probability invariant measures for the operator A:

$\Delta_{ba} = \{\mu \in S_{ba} : \mu = A\mu\}$,
$\Delta_{ca} = \{\mu \in S_{ca} : \mu = A\mu\}$, $\Delta_{pfa} = \{\mu \in S_{pfa} : \mu = A\mu\}$.

A classical countably additive Markov chain may or may not have invariant countably additive probability measures, i.e. possibly $\Delta_{ca} = \emptyset$ (for example, for a symmetric walk on Z).

In [7, Theorem 2.2] Šidak proved that any countably additive MC on an arbitrary measurable space (X, Σ) extended to the space of finitely additive measures has at least one invariant finitely additive measure, i.e. always $\Delta_{ba} \neq \emptyset$. Šidak in [7, Theorem 2.5] also established in the general case that if a finitely additive measure μ is invariant $A\mu = \mu$, and $\mu = \mu_1 + \mu_2$ is its decomposition into are countably additive and purely finitely additive components, then each of them is also invariant: $A\mu_1 = \mu_1$, $A\mu_2 = \mu_2$. Therefore, it suffices to study invariant measures from Δ_{ca} and from Δ_{pfa}, separately.

Definition 3. *A finite numbered set of pairwise different positive finitely additive measures $K = \{\mu_1, \mu_2, ..., \mu_m\}$ will be called a cycle measures of an operator A of a given Markov chain (or a cycle of measures MC) if*

$$A\mu_1 = \mu_2, A\mu_2 = \mu_3, ..., A\mu_{m-1} = \mu_m, A\mu_m = \mu_1.$$

Such cycles will be called finitely additive. The number $m \geq 1$ will be called the *cycle period*, and the measures $\mu_1, \mu_2, ..., \mu_m$ – *cyclic measures*. Unnormalized cycles will also be used below.

If $K = \{\mu_1, \mu_2, ..., \mu_m\}$ is a MC cycle, then, obviously,

$$A^m\mu_1 = \mu_1, A^m\mu_2 = \mu_2, ..., A^m\mu_m = \mu_m,$$

i.e. all cyclic measures μ_i are invariant for the operator A^m and $A^m(K) = K$.

The following well-known statement is obvious. Let $K = \{\mu_1, \mu_2, ..., \mu_m\}$ be a cycle of finitely additive measures. Then the measure

$$\mu = \frac{1}{m}\sum_{k=1}^{m}\mu_k = \frac{1}{m}\sum_{k=1}^{m}A^{k-1}\mu_1$$

is invariant for the operator A, i.e. $A\mu = \mu$ (here A^0 is the identity operator).

Definition 4. *The measure constructed above will be called the mean cycle measure K.*

Definition 5. *We call each method of choosing a measure μ_1 in K an operation renumbering a cycle K.*

Definition 6. *We say that two cycles of the same period K^1 and K^2 are identical if there is a renumbering of cycles K^1 or K^2 such that all their cyclic measures with the same numbers match. In this case, we will write $K^1 = K^2$. Instead of the words "identical cycles", we will still say the words "equal cycles".*

Obviously, for the cycles to be equal, it is sufficient that their first measures coincide.

Hereinafter, it is convenient to call cyclic measures $\mu_i, i = 1, ..., m$, *cycle coordinates* K.

Definition 7. *By the operation of multiplying a cycle of measures $K = \{\mu_1, \mu_2, ..., \mu_m\}$ by a number $\gamma > 0$ we mean the construction of a cycle of measures $\gamma K = \{\gamma\mu_1, \gamma\mu_2, ..., \gamma\mu_m\}$.*

Since the operator A is isometric in the cone of positive measures, all cyclic measures of one cycle $K = \{\mu_1, \mu_2, ..., \mu_m\}$ have the same norm $\|\mu_1\| = \|\mu_2\| =, ..., = \|\mu_m\| = \|\mu\|$, which is naturally called the norm $\|K\|$ of the cycle K itself.

To give the cycle a probabilistic meaning, it is sufficient to multiply it coordinatewise by the normalizing factor $\gamma = \frac{1}{\|\mu\|}$: $\hat{K} = \gamma \cdot K = \{\gamma\mu_1, \gamma\mu_2, ..., \gamma\mu_m\}$. We obtain a probability cycle with the norm $\|\hat{K}\| = 1, \hat{K} \subset S_{ba}$.

Definition 8. *Let there be given two cycles of measures of the same MC $K^1 = \{\mu_1^1, \mu_2^1, ..., \mu_m^1\}$ and $K^2 = \{\mu_1^2, \mu_2^2, ..., \mu_m^2\}$ of the same period m. We call the sum of cycles K^1 and K^2 the following set of measures $K = K^1 + K^2 = \{\mu_1^1 + \mu_1^2, ..., \mu_m^1 + \mu_m^2\}$ derived from K^1 and K^2 coordinatewise addition.*

The measure spaces are semi-ordered by the natural order relation. In them one can introduce the notion of infimum $inf\{\mu_1, \mu_2\} = \mu_1 \wedge \mu_2$ and supremum $sup\{\mu_1, \mu_2\} = \mu_1 \vee \mu_2$, which are also contained in these spaces. Thus, the measure spaces $ba(X, \Sigma), ca(X, \Sigma)$ and $pfa(X, \Sigma)$ are lattices (K-lineals).

The exact formulas for constructing the ordinal infimum and supremum of two finitely additive measures are given, for example, in [2].

Definition 9. *Two positive measures $\mu_1, \mu_2 \in ba(X, \Sigma)$ are called disjoint if $\mu_1 \wedge \mu_2 = 0$.*

Definition 10. *Two positive measures $\mu_1, \mu_2 \in ba(X, \Sigma)$ are called singular if there are two sets $D_1, D_2 \subset X$, $D_1, D_2 \in \Sigma$, such that $\mu_1(D_1) = \mu_1(X)$, $\mu_2(D_2) = \mu_2(X)$ and $D_1 \cap D_2 = \emptyset$.*

Countably additive measures μ_1, μ_2 are disjunct if and only if they are singular (see [2]).

If the measures μ_1 and μ_2 are singular, then they are also disjoint (see [2]).

Definition 11. *A cycle $K = \{\mu_1, \mu_2, ..., \mu_m\}$ is called a cycle of disjoint measures if all its cyclic measures are pairwise disjoint, i.e. $\mu_i \wedge \mu_j = 0$ for all $i \neq j$.*

Definition 12. *Two cycles of measures K^1, K^2 are called disjoint, if each measure from the cycle K^1 is disjoint with each measure from the cycle K^2.*

If the cycle of disjoint measures $K = \{\mu_1, \mu_2, ..., \mu_m\}$ is countably additive, then all its cyclic measures are pairwise singular and have pairwise disjoint supports (sets of full measure) $D_1, D_2, ..., D_m \in \Sigma$, that is, $\mu_i(D_i)$, $i = 1, ..., m$, and $D_i \cap D_j = \emptyset$ for $i \neq j$.

If we do not require pairwise disjointness (singularity) of the measures of a countably additive cycle, then new, somewhat unexpected objects may appear in the state space of a MC. Let's give a suitable simple example.

Examples 3. Let the MC be finite, having exactly three states $X = \{x_1, x_2, x_3\}$ with transition probabilities:

$$p(x_1, x_1) = 1, p(x_2, x_3) = 1, p(x_3, x_2) = 1.$$

This means that the MC has in the state space X one stationary state $\{x_1\}$ (we can say that this is a cycle of period $m = 1$) and one cycle $\{x_2, x_3\}$ of the period $m = 2$. Within the framework of the operator approach, it is more convenient for us to translate what has been said into the language of measures as follows.

Let $x \in X$ and $E \subset X (E \in \Sigma = 2^X)$. Then $p(x_1, E) = \delta_{x_1}(E)$, $p(x_2, E) = \delta_{x_3}(E)$, $p(x_3, E) = \delta_{x_2}(E)$, where $\delta_{x_i}(\cdot)$, $i = 1, 2, 3$, are the Dirac measures at the points x_1, x_2, x_3. For an operator A such a MC we have: $A\delta_{x_1} = \delta_{x_1}$, $A\delta_{x_2} = \delta_{x_3}$, $A\delta_{x_3} = \delta_{x_2}$, i.e. the family of measures $K = \{\delta_{x_2}, \delta_{x_3}\}$ is a cycle according to Definition 3, and the cyclic measures δ_{x_2} and δ_{x_3} are singular.

Consider one more family of measures $\tilde{K} = \{\frac{1}{2}\eta_1, \frac{1}{2}\eta_2\}$, where $\eta_1 = \delta_{x_1} + \delta_{x_2}$, $\eta_2 = \delta_{x_1} + \delta_{x_3}$. Then $A\eta_1 = A(\delta_{x_1} + \delta_{x_2}) = A\delta_{x_1} + A\delta_{x_2} = \delta_{x_1} + \delta_{x_3} = \eta_2$ and similarly $A\eta_2 = \eta_1$. Since the measures η_1 and η_2 are different, then by Definition 3, the family of measures $\tilde{K}$ is also a MC cycle different from K. Moreover, the measures η_1 and η_2 are not disjoint: $\eta_1 \wedge \eta_2 = \delta_{x_1} \neq 0$. These measures are not singular: their supports $\{x_1, x_2\}$ and $\{x_1, x_3\}$ intersect, i.e. $\{x_1, x_2\} \cap \{x_1, x_3\} = \{x_1\} \neq \emptyset$.

Remark 1. Such cycles with intersecting cyclic sets of states, as in Example 3, are usually not considered in the classical theory of MC.

However, we believe that the study of intersecting cycles of sets is very useful in general theory. Research of such cycles is more productive for us in terms of measure cycles. In this case, instead of intersecting sets of measures, one should consider cycles of measures that are not disjoint. Our Theorems 1, 2, 3, and 5 (proved in Sect. 4) do not require pairwise disjointness (or singularity) of cyclic measures in measure cycles.

4 Main Results

Theorem 1. *Any finitely additive cycle of measures for an arbitrary MC is a linearly independent set in the linear space* $ba(X, \Sigma)$.

Proof. We prove by induction.

Consider first two arbitrary different measures $\mu_1, \mu_2 \in S_{ba}$ (not necessarily cyclic), for which $\|\mu_1\| = \|\mu_2\| = 1$.

They are obviously linearly independent.

In particular the cycle $K = \{\mu_1, \mu_2\}$ consisting of two different measures from S_{ba}, is linearly independent.

Now let the cycle consist of three pairwise different measures: $K = \{\mu_1, \mu_2, \mu_3\} \subset S_{ba}$. As we found out above, any two measures of them are linearly independent.

Suppose that one of these three measures is linearly dependent on the other two, let it be the measure μ_3 (the number is not important here). Then there exist numbers $\alpha_1, \alpha_2, 0 \leq \alpha_1, \alpha_2 \leq 1, \alpha_1 + \alpha_2 = 1$, such that the measure μ_3 is uniquely representable as a linear combination $\mu_3 = \alpha_1\mu_1 + \alpha_2\mu_2$.

Let $\alpha_1 = 0$. Then $\alpha_2 = 1$ and $\mu_3 = \mu_2$ which contradicts the pairwise difference of the three measures. Similarly for $\alpha_1 = 1$. Therefore, we can assume that $0 < \alpha_1, \alpha_2 < 1$.

By cycle conditions

$$\mu_1 = A\mu_3 = A(\alpha_1\mu_1 + \alpha_2\mu_2) = \alpha_1 A\mu_1 + \alpha_2 A\mu_2 = \alpha_1\mu_2 + \alpha_2\mu_3.$$

Since $\alpha_2 \neq 0$ from this we get $\mu_3 = \frac{1}{\alpha_2}\mu_1 - \frac{\alpha_1}{\alpha_2}\mu_2$. Since the decomposition of μ_3 is unique, we have $\alpha_1 = \frac{1}{\alpha_2}, \alpha_2 = -\frac{\alpha_1}{\alpha_2} < 0$. Since α_1 and α_2 are positive, we obtain a contradiction in the second equality. Therefore, all three measures μ_1, μ_2 and μ_3 are linearly independent.

We turn to the general case.

Let be a cycle of measures $K = \{\mu_1, \mu_2, ..., \mu_m\}$ with an arbitrary period $m \geq 3$. We assume that the sets of any $m - 1$ pieces of measures μ_i from K are linearly independent. Assume that the measure μ_m (the number is not important) depends linearly on the measures $\mu_1, \mu_2, ..., \mu_{m-1}$. Then the measure μ_m is uniquely represented as

$$\mu_m = \sum_{i=1}^{m-1} \alpha_i\mu_i,$$

where $0 \leq \alpha_i \leq 1$ for $i = 1, 2, ..., m$, $\sum_{i=1}^{m-1} \alpha_i = 1$.

Assume that for some $t \in \{1, 2, ..., m-1\}$, $\alpha_t = 0$ is executed. Then the measure μ_m is linearly expressed in terms of $m-2$ pieces of measures μ_i, all of them together with μ_m will be $m-1$ piece. This contradicts the assumption that the sets of any $m-1$ pieces of measures μ_i from K are linearly independent. Therefore, all $\alpha_i > 0$, $i = 1, 2, ..., m-1$.

Now let $t \in \{1, 2, ..., m-1\}$ be $\alpha_t = 1$. Then all other $\alpha_i = 0$ $(i \neq t)$ and $\mu_m = \alpha_t \cdot \mu_t = \mu_t$, which contradicts the condition of pairwise difference of all measures from the cycle.

So, for all coefficients in the linear decomposition of the measure μ_m we have $0 < \alpha_i < 1$, $i = 1, 2, ..., m-1$.

We apply the operator A to this decomposition of the measure μ_m and obtain:

$$\mu_1 = A\mu_m = \sum_{i=1}^{m-1} \alpha_i A\mu_i = \sum_{i=1}^{m-1} \alpha_i \mu_{i+1} = \alpha_1\mu_2 + \alpha_2\mu_3 + ... + \alpha_{m-1}\mu_m.$$

Therefore, we have $(\alpha_{m-1} \neq 0)$:

$$\mu_m = \frac{1}{\alpha_{m-1}}\mu_1 - \frac{1}{\alpha_{m-1}} \sum_{i=1}^{m-2} \alpha_i \mu_{i+1}.$$

Since the representation for the measure μ_m is unique, here and above we obtain the following relations for the coefficients of the measure μ_2:

$$0 < \alpha_2 = -\frac{\alpha_1}{\alpha_{m-1}} < 0.$$

It follows from the contradiction obtained that the measure μ_m is linearly independent of the other measures of the cycle. Consequently, any other measure $\mu_i \in K$ is linearly independent of the other measures of the cycle K. The theorem is proved.

Theorem 2. *Let $K = \{\mu_1, \mu_2, ..., \mu_m\}$ be a finitely additive cycle of measures for an arbitrary MC. If at least one cyclic measure μ_i is countably additive, then all other cyclic measures in K and their mean measures are also countably additive. Such cycles will be called countably additive.*

Proof. Since $\mu_{i+1} = A\mu_i, i = 1, 2, ..., m-1$ and $\mu_1 = A\mu_m$ then the statement of the theorem follows from the fact that the operator A has the space $ca(X, \Sigma)$ as its invariant subspace in $ba(X, \Sigma)$, that is, transforms countably additive measures into countably additive ones. The countable additivity of the mean measure follows from the fact that $ca(X, \Sigma)$ is a linear space, i.e. the sum of countably additive measures is also countably additive and a countably additive measure multiplied by a number is also countably additive.

The theorem is proved.

Proposition 1. *There exist classical Markov chains with purely finitely additive cycles of measures with period $m \geq 2$.*

Examples 4. An example of a classical MC is constructed, for which the existence of a purely finitely additive cycle of measures is proved.

For simplicity, we take a deterministic MC generated by a point transformation.

Let $X = (0, 1) \cup (1, 2)$, $\Sigma = B_X$ (Borel σ-algebra on X). Denote $D_1 = (0, 1)$, $D_2 = (1, 2)$. Then $D_1 \cup D_2 = X$, $D_1 \cap D_2 = \emptyset$.

Let's define the transition function of the Markov chain according to the rules:

$p(x, \{1 + x^2\}) = 1$, if $x \in (0, 1)$;

$p(y, \{(y - 1)^2\}) = 1$, if $y \in (1, 2)$.

Then $p(x, D_2) = 1$, if $x \in D_1$; $p(x, D_1) = 1$, if $x \in D_2$.

Therefore, the sets of states D_1 and D_2 are cyclic and form a singular cycle $S = \{D_1, D_2\}$ with period $m = 2$.

Note that for any trajectory of the Markov chain beginning at the point $x_0 \in (0, 1)$, its subsequence with even numbers tends to one from the right:

$$1 + x_0^2,\ 1 + x_0^{16}, 1 + x_0^{64}, ... \to 1,$$

and the subsequence with odd numbers tends to zero from the right:

$$x_0,\ x_0^4, x_0^{32}, ... \to 0$$

(and vice versa, for $x_0 \in (1, 2)$).

By Šidak's theorem (see [7, Theorem 2.2]) for a given MC there exists an invariant finitely additive measure $\mu = A\mu \in S_{ba}$. It can be shown that for her $\mu(D_1) = \mu(D_2) = \frac{1}{2} > 0$.

We construct two new measures μ_1 and μ_2 as the restriction of the measure μ to the sets D_1 and D_2: $\mu_1(E) = \mu(E \cap D_1)$, $\mu_2(E) = \mu(E \cap D_2)$ for all $E \subset X$, $E \in \Sigma$, and $\mu = \mu_1 + \mu_2$. The measures μ_1 and μ_2 are singular and have supports D_1 and D_2. It can be proved that $A\mu_1 = \mu_2$ and $A\mu_2 = \mu_1$. This means that the measures μ_1 and μ_2 form a disjoint cycle of finitely additive measures $K = \{\mu_1, \mu_2\}$.

Let $0 < \varepsilon < 1$ and $D_1^\varepsilon = (0, \varepsilon)$, $D_2^\varepsilon = (1, 1 + \varepsilon)$. We can get that for any ε, $\mu_1(D_1^\varepsilon) = 1/2$, $\mu_2(D_2^\varepsilon) = 1/2$. This means that the measures μ_1 and μ_2 and their mean measure are purely finitely additive. The constructed MC has no invariant countably additive measures.

It can be shown that the singular sets D_1^ε and D_2^ε for any ε form a cycle of states $S^\varepsilon = \{D_1^\varepsilon, D_2^\varepsilon\}$ and are also supports of measures μ_1 and μ_2.

It can be proved that the family of all pairwise disjoint invariant finitely additive measures of a given MC has cardinality at least a continuum, i.e. $2^{\aleph_0}$.

Let us modify the considered MC - add the points 0 and 1 to $X = (0, 1) \cup (1, 2)$ and get $X = [0, 2)$. Let us determine the possible transitions from these points using the same formulas as the original MC. We get:

$p(0, \{1\}) = 1$, $p(1, \{0\}) = 1$.

This means that the family of state sets $S_0 = \{\{0\}, \{1\}\}$ for the new MC is a new singular cycle of dimension $m = 2$.

It corresponds to a new singular cycle of countably additive measures $K_0 = \{\delta_0, \delta_1\}$, where δ_0 and δ_1 are Dirac measures at the points 0 and 1, respectively. Their mean measure $\mu = \frac{1}{2}\delta_0 + \frac{1}{2}\delta_1$ is countably additive and is the only invariant measure of the modified Markov chain in the class of countably additive measures.

Note that the whole infinite family of disjoint purely finitely additive cycles of measures K considered above on $X = (0, 1) \cup (1, 2)$ remains the same for the new MC.

Theorem 3. *Let $K = \{\mu_1, \mu_2, ..., \mu_m\}$be a finitely additive cycle of measures for an arbitrary MC. If at least one cyclic measure μ_i is purely finitely additive, then all other cyclic measures in K and their mean measure are also purely finitely additive. Such cycles will be called purely finitely additive.*

Proof. Let the cyclic measure μ_1 be purely finitely additive (the number is not important here) and $\mu_2 = A\mu_1$.

Suppose that the measure μ_2 is not purely finitely additive. We decompose the measure μ_2 into two components $\mu_2 = \lambda_{ca} + \lambda_{pfa}$ where λ_{ca} is a countably additive measure, and λ_{pfa} is purely finitely additive. By assumption $\mu_2 \neq \lambda_{pfa}$ whence $\lambda_{ca} \neq 0$, $\lambda_{ca} \geq 0, \lambda_{ca}(X) = \gamma > 0$.

We apply the operator A to the measure μ_2

$$A\mu_2 = A\lambda_{ca} + A\lambda_{pfa} = \mu_3.$$

The operator A takes countably additive measures to the same ones and is isometric in the cone of positive measures. It follows from this that the measure $A\lambda_{ca}$ is countably additive, positive, and $A\lambda_{ca}(X) = \gamma > 0$. This means that the measure μ_3 also has a positive countably additive component $A\lambda_{ca}$.

Continuing this procedure further at the last step we get the decomposition

$$\mu_1 = A\mu_m = A^{m-2}\lambda_{ca} + A^{m-2}\lambda_{pfa},$$

where the measure $A^{m-2}\lambda_{ca}$ is countably additive, positive, and $A^{m-2}\lambda_{ca}(X) = \gamma > 0$.

Thus, the initial measure μ_1 has a nonzero countably additive component and, thus, is not purely finitely additive, which contradicts the conditions of the theorem. Therefore, the measure μ_2 is also purely finitely additive.

Repeating this procedure sequentially for all the other cyclic measures μ_3, $\mu_4, ..., \mu_m$ we get that they are all purely finitely additive.

It remains to prove that the mean cyclic measure is also purely finitely additive. But this follows from the fact that the space of purely finitely additive measures $pfa(X, \Sigma)$ is also linear, which is proved nontrivially in [2, Theorem 1.17]. The theorem is proved.

Now let us present an extended cyclic analogue of the Alexandroff-Yosida-Hewitt decomposition given in Sect. 2.

Theorem 4. *Let $K = \{\mu_1, \mu_2, ..., \mu_m\}$ be a finitely additive cycle of measures of pairwise disjoint measures with period m of an arbitrary MC and $\mu_i = \mu_i^{ca} + \mu_i^{pfa}$ a decomposition of cyclic measures into a countably additive component μ_i^{ca} and a purely finitely additive component μ_i^{pfa}, $i = 1, 2, ..., m$. Then these components are also cyclic, form the cycles K^{ca} and K^{pfa}, the cycle K is the coordinate sum of these cycles $K = K^{ca} + K^{pfa}$, and the mean measure of the cycle K is uniquely representable as the sum of its countably additive and purely finitely additive components, which coincide with the mean measures of the cycles K^{ca} and K^{pfa}, respectively. Moreover, the cycles K^{ca} and K^{pfa} consist of pairwise disjoint measures and are disjoint with each other, i.e. every measure from K^{ca} is disjoint with every measure from K^{pfa}.*

Proof. We denote tuples of countably additive and purely finitely additive components of cyclic measures of a cycle K by the symbols $K^{ca} = \{\mu_1^{ca}, \mu_2^{ca}, ..., \mu_m^{ca}\}$ and $K^{pfa} = \{\mu_1^{pfa}, \mu_2^{pfa}, ..., \mu_m^{pfa}\}$. The coordinate-wise sum of these two tuples gives the original cycle $K = K^{ca} + K^{pfa}$. Now we need to show that the measures μ_i^{ca} and μ_i^{pfa} are cyclic, that is, the tuples K^{ca} and K^{pfa} are cycles.

Let us prove the theorem step by step.

Assume that some of the measures μ_i^{ca} is zero. Then $\mu_i = \mu_i^{pfa}$, and according to Theorem 3 all other measures $\mu_j = \mu_j^{pfa}$, i.e., the cycle $K = K^{pfa}$, and the theorem is proved. Similarly, for $\mu_i^{pfa} = 0$, the cycle K is countably additive by Theorem 2, $K = K^{ca}$, and the present theorem is proved. The main case remains when all $\mu_i^{ca} \neq 0$ and all $\mu_i^{pfa} \neq 0$, which is what we assume below.

Take two arbitrary measures μ_i^{ca} and μ_j^{ca} $(i \neq j)$ from K^{ca}.

Then

$$0 \leq \mu_i^{ca} \wedge \mu_j^{ca} \leq (\mu_i^{ca} + \mu_i^{pfa}) \wedge (\mu_j^{ca} + \mu_j^{pfa}) = \mu_i \wedge \mu_j.$$

By the conditions of the theorem, all measures from K are pairwise disjoint. Therefore, $\mu_i \wedge \mu_j = 0$ and $\mu_i^{ca} \wedge \mu_j^{ca} = 0$, i.e., all measures from K^{ca} are pairwise disjoint. And since, as we now assume, all measures from the tuple K^{ca} are nonzero, then they are all pairwise distinct.

Similarly, we obtain that all measures from the tuple K^{pfa} are also pairwise disjoint and distinct.

We emphasize that the tuples of measures K^{ca} and K^{pfa} have dimensions m, which coincides with the period m of the original cycle K.

By the conditions of the theorem, the cycle K has an (arbitrary) period $m \in N$. Consequently, each cyclic measure μ_i of the cycle K is an invariant measure of the operator A^m, that is, $\mu_i = A^m \mu_i$, $i = 1, 2, ..., m$. Take the first cyclic measure with its Alexandroff-Yosida-Hewitt decomposition [2] $\mu_1 = \mu_1^{ca} + \mu_1^{pfa}$. By Šidak's Theorem ([7, Theorem 2.5]) both components of the measure μ_1 are also invariant measures for the operator A^m, that is, $\mu_1^{ca} = A^m \mu_1^{ca}$, $\mu_1^{pfa} = A^m \mu_1^{pfa}$.

Each of these components generates its own cycle

$$\hat{K}^{ca} = \{\mu_1^{ca}, A\mu_1^{ca}, ..., A^{m-1}\mu_1^{ca}\},$$

$$\hat{K}^{pfa} = \{\mu_1^{pfa}, A\mu_1^{pfa}, ..., A^{m-1}\mu_1^{pfa}\}.$$

Obviously, the coordinate-wise sum of these two cycles gives the whole cycle $K = \hat{K}^{ca} + \hat{K}^{pfa}$.

Since the measure μ_1^{ca} is countably additive, then, according to Theorem 2, all other cyclic measures of the cycle $\hat{K}^{ca}$ are countably additive. Since the measure μ_1^{pfa} is purely finitely additive, then, according to Theorem 3, all other cyclic measures of the cycle $\hat{K}^{pfa}$ are purely finitely additive.

By the uniqueness of the decomposition of any measure into countably additive and purely finitely additive components (see [2]), we obtain the following equalities (here the symbol A^0 means the identical operator):

$$\mu_1^{ca} = A^0\mu_1^{ca}, \ \ \mu_2^{ca} = A\mu_1^{ca}, \ ..., \ \ \mu_m^{ca} = A^{m-1}\mu_1^{ca},$$

where on the left are the measures of the tuple K^{ca}, and on the right are the cyclic measures of the cycle $\hat{K}^{ca}$.

Similar equalities are also true for purely finitely additive components.

From this we get that $K^{ca} = \hat{K}^{ca}$, $K^{pfa} = \hat{K}^{pfa}$, i.e. tuples K^{ca} and K^{pfa} are cycles, and $K = K^{ca} + K^{pfa}$. Note that this decomposition of the cycle K is unique. The main statement of the theorem is proved.

Now the corresponding equalities for the mean measures of cycles are obvious.

In [2] (Theorem 1.16) it was proved that any countably additive measure is disjoint with any purely finitely additive measure. Therefore, the cycles of the measures K^{ca} and K^{pfa} are disjoint. Above we showed that all measures from K^{ca} and K^{pfa} are also pairwise disjoint. The theorem is proved.

Corollary 1. *A finitely additive cycle of measures K is countably additive if and only if its mean measure is countably additive.*

Corollary 2. *A finitely additive cycle of measures K is purely finitely additive if and only if its mean measure is purely finitely additive.*

Under the conditions of Theorem 4 just proved, the requirement of pairwise disjointness of cyclic measures in the cycle K is essential. If we remove it, then the theorem becomes incorrect.

Theorem 5. *Let an arbitrary MC have one finitely additive cycle of measures K of any period and its mean measure μ is the only invariant finitely additive measure for the operator A. Then the cycle K and its mean measure μ are countably additive.*

Proof. Consider a cycle of finitely additive measures $K = \{\mu_1, \mu_2, ..., \mu_m\}$ and its mean measure $\mu = \frac{1}{m}\sum_{i=1}^{m}\mu_i$. In Sect. 3 shows that the mean measure μ of the cycle K is invariant for the operator A, i.e. $\mu \in \Delta_{ba}$. By the condition of the theorem, this measure is unique in Δ_{ba}, i.e. $\Delta_{ba} = \{\mu\}$.

In ([5], Theorem 8.3), it is proved that if a MC has in S_{ba} a unique invariant measure μ, i.e. $\Delta_{ba} = \{\mu\}$, then this measure is countably additive. Therefore, by Theorem 2 and Corollary 1, the cycle K is countably additive.

The theorem is proved.

Therefore, it follows (under the above conditions) that there are no "single" purely finitely additive cycles.

Acknowledgments. This work was supported by the Russian Foundation for Basic Research (project No. 20-01-00575-a).

References

1. Yosida, K., Kakutani, S.: Operator-theoretical treatment of Markoff's processes and mean ergodic theorem. Ann. Math. (2) **42**(1), 188–228 (1941)
2. Yosida, K., Hewitt, E.: Finitely additive measures. Trans. Am. Math. Soc. **72**(1), 46–66 (1952)
3. Dunford, N., Schwartz, J.: Linear Operatiors, Part I: General Theory. Interscience Publisher, Geneva (1958)
4. Revuz, D.: Markov Chains. North-Holland Mathematical Library, Oxford (1984)
5. Zhdanok, A.I.: Finitely additive measures in the ergodic theory of Markov chains I. Sib. Adv. Math. **13**(1), 87–125 (2003). Zhdanok, A.I.: Konechno-additivnyye mery v ergodicheskoy teorii tsepey Markova I. Matematicheskiye trudy **4**(2), 53–95 (2001). (in Russian)
6. Zhdanok, A.I.: Finitely additive measures in the ergodic theory of Markov chains II. Sib. Adv. Math. **13**(2), 108–125 (2003). Zhdanok, A.I.: Konechno-additivnyye mery v ergodicheskoy teorii tsepey Markova II. Matematicheskiye trudy **5**(1), 45–65 (2002). (in Russian)
7. Šidak, Z.: Integral representations for transition probabilities of Markov chains with a general state space. Czechoslov. Math. J. **12**(4), 492–522 (1962)

Branching Walks with a Finite Set of Branching Sources and Pseudo-sources

Elena Yarovaya(✉), Daria Balashova, and Ivan Khristolyubov

Department of Probability Theory, Faculty of Mathematics and Mechanics, Lomonosov Moscow State University, Moscow, Russia
yarovaya@mech.math.msu.su

Abstract. Branching random walks play a key role in modeling the evolutionary processes with birth and death of particles depending on the structure of a medium. The branching random walk on a multidimensional lattice with a finite number of branching sources of three types is investigated. It is assumed that the intensities of branching in the sources can be arbitrary. The principal attention is paid to the analysis of spectral characteristics of the operator describing evolution of the mean numbers of particles both at an arbitrary point and on the entire lattice. The obtained results provide an explicit conditions for the exponential growth of the numbers of particles without any assumptions on jumps variance of the underlying random walk.

Keywords: Branching random walks · Equations in Banach spaces · Non-homogeneous environments · Positive eigenvalues · Population dynamics

1 Introduction: Model of BRW/r/k/m

We present results for continuous-time *branching random walks* (BRWs) on the lattice $\mathbf{Z}^d$, $d \in \mathbf{N}$, with a finite number of lattice sites in which the generation of particles can occur, which are called *branching sources*. By a BRW we mean a stochastic process that combines branching (birth or death) of particles at certain points on $\mathbf{Z}^d$ with their random walk on $\mathbf{Z}^d$. The goal of the paper is to study the distributions of the particle population $\mu_t(y)$ at every point $y \in \mathbf{Z}^d$ and $\mu_t = \sum_{y\in\mathbf{Z}^d} \mu_t(y)$ over the lattice $\mathbf{Z}^d$ for a BRW with branching sources of different type without any assumptions on the variance of jumps of the underlying random walk.

Suppose that there is a single particle at the moment $t = 0$ on the lattice situated at the point $x \in \mathbf{Z}^d$. Each particle moves on the lattice $\mathbf{Z}^d$ until it reaches a source where its behavior changes. There are three types of branching sources, depending on whether branching takes place or not and on whether random walk symmetry is violated or not. At sources of the first type, particles die or are born, and random walk symmetry is maintained, see, e.g., [1,2,11]. At sources of the second type, walk symmetry is violated through an increase

A. N. Shiryaev et al. (Eds.): ICSM-5 2020, PROMS 371, pp. 144–163, 2021.
https://doi.org/10.1007/978-3-030-83266-7_11

in the degree of branching or walk dominance, see, e.g., [9]. Sources of the third type should be called "pseudo-sources," because at these sources only the walk symmetry is violated, with no particle births or deaths ever occurring. BRWs with r sources of the first type, k of the second type, and m of the third type are denoted BRW/r/k/m and introduced in [12]. Particles exist on $\mathbf{Z}^d$ independently of each other and of their antecedent history.

We define random walk by its generator

$$A = \mathscr{A} + \sum_{j=1}^{k+m} \zeta_j \Delta_{u_j} \mathscr{A} \tag{1}$$

where $\mathscr{A} = (a(x,y))_{x,y\in\mathbf{Z}^d}$ satisfies the *regularity property* $\sum_{y\in\mathbf{Z}^d} a(x,y) = 0$ for all x, where $a(x,y) \geq 0$ for $x \neq y$, $-\infty < a(x,x) < 0$. From this it follows that A itself satisfies this regularity property [12,13]. Additionally, we assume that the intensities $a(x,y)$ are *symmetric* and *spatially homogeneous*, that is, $a(x-y) := a(x,y) = a(y,x) = a(0,y-x)$. Thus we can denote $a(y,x)$, $a(0,y-x)$, that is, $a(x-y) := a(x,y) = a(y,x) = a(0,y-x)$. The matrix $\mathscr{A}$ under consideration is irreducible, so for any $z \in \mathbf{Z}^d$ there is such a set of vectors $z_1, \dots, z_k \in \mathbf{Z}^d$ that $z = \sum_{i=1}^k z_i$ and $a(z_i) \neq 0$ for $i = 1, \dots, k$. It is fairly clear that the irreducibility property is inherited by the perturbed matrix A. This, however, does not hold true for the properties of spatial homogeneity and, most importantly, symmetry. We will, however, make use of the structure of A and the symmetry of the underlying matrix $\mathscr{A}$ in order to overcome this complication.

According to the axiomatics outlined in [3, Ch. III, §2], the probabilities $p(h,x,y)$ of a particle at $x \notin \{v_1, v_2, \dots, v_{k+r}\}$ to jump to a point y over a short period of time h can be presented as $p(h,x,y) = a(x,y)h + o(h)$ for $y \neq x$ and $p(h,x,x) = 1 + a(x,x)h + o(h)$ for $y = x$. From these equalities, see, for instance, [3, Ch. III], we obtain the *Kolmogorov backward equations*:

$$\frac{\partial p(t,x,y)}{\partial t} = \sum_{x'} a(x,x') p(t,x',y), \qquad p(0,x,y) = \delta(x-y), \tag{2}$$

where $\delta(\cdot)$ is the discrete Kronecker δ-function on $\mathbf{Z}^d$.

Infinitesimal generating functions $f(u,v_i) = \sum_{n=0}^{\infty} b_n(v_i) u^n$, $0 \leq u \leq 1$, govern branching process at each of the sources $v_1, v_2, \dots, v_{k+r}$. We denote *source intensities* $\beta_i := \beta_i^{(1)} = f^{(1)}(1,v_i) = (-b_1(v_i)) \left(\sum_{n\neq 1} n b_n(v_i)/(-b_1(v_i)) - 1 \right)$ where the sum is the average number of descendants a particle has at the source v_i.

If the particle is not in the branching source, then its random walk occurs in accordance with the above rules. Consider a combination of branching and walking processes observed when a particle is in one of the branching sources $v_1, v_2, \dots, v_{k+r}$. In this case, the following possible transitions, which can occur in a short period of time h, are as follows: the particle will either move to a point $y \neq v_i$ with the probability of $p(h,v_i,y) = a(v_i,y)h + o(h)$, or will remain at the source and produce $n \neq 1$ descendants with the probability of

$p_*(h, v_i, n) = b_n(v_i)h + o(h)$ (we assume that the particle itself is included in these n descendants and we say that the particle dies if $n = 0$), or no changes will occur to the particle at all, which has the probability of $1 - \sum_{y \neq v_i} a(v_i, y)h - \sum_{n \neq 1} b_n(v_i)h + o(h)$. Thus, the time spent by the particle in the source v_i is exponentially distributed with the parameter $-(a(v_i, v_i) + b_1(v_i))$. The evolution of each new particle obeys the same law and does not depend on the evolution of other particles.

Let us introduce the moments of the random variables $\mu_t(y)$ and μ_t as $m_n(t, x, y) = \mathsf{E}_x \mu_t^n(y)$ and $m_n(t, x) = \mathsf{E}_x \mu_t^n$, respectively, where n is the order of the moment and E_x is the mean on condition $\mu_0(\cdot) = \delta_x(\cdot)$.

In BRW$/r/k/m$ more general multi-point perturbations of the self-adjoint operator $\mathscr{A}$ generated of the symmetric random walk are used than in BRW$/r/0/0$ or in BRW$/0/k/0$, see, e.g., [13]. This follows from the statement, see [12], that the mean number of particles $m_1(t) = m_1(t, \cdot, y)$ at a point $y \in \mathbf{Z}^d$ in BRW$/r/k/m$ is governed by:

$$\frac{dm_1(t)}{dt} = \mathscr{Y} m_1(t), \quad m_1(0) = \delta_y,$$

where

$$\mathscr{Y} = \mathscr{A} + \left(\sum_{s=1}^{r} \beta_s \Delta_{z_s} \right) + \left(\sum_{i=1}^{k} \zeta_i \Delta_{x_i} \mathscr{A} + \sum_{i=1}^{k} \eta_i \Delta_{x_i} \right) + \left(\sum_{j=1}^{m} \chi_j \Delta_{y_j} \mathscr{A} \right). \tag{3}$$

Here, $\mathscr{A} : l^p(\mathbf{Z}^d) \to l^p(\mathbf{Z}^d)$, $p \in [1, \infty]$, is a symmetric operator, $\Delta_x = \delta_x \delta_x^T$, and $\delta_x = \delta_x(\cdot)$ denotes a column-vector on the lattice taking the unit value at the point x and vanishing at other points, β_s, ζ_i, η_i, and χ_j are some constants. The same equation is also valid for the mean number of particles (the mean population size) over the lattice $m_1(t) = m_1(t, \cdot)$ with the initial condition $m_1(0) = 1$ in $l^\infty(\mathbf{Z}^d)$. Operator (3) can be written as

$$\mathscr{Y} = \mathscr{A} + \sum_{i=1}^{k+m} \zeta_i \Delta_{u_i} \mathscr{A} + \sum_{j=1}^{k+r} \beta_j \Delta_{v_j}. \tag{4}$$

In each of the sets $U = \{u_i\}_{i=1}^{k+m}$, and $V = \{v_j\}_{j=1}^{k+r}$, the points are pairwise distinct, but U and V may have a nonempty intersection. The points from $V \setminus U$ correspond to r sources of the first type; those from $U \cap V$ to k sources of the second type; and those from $U \setminus V$ to m sources of the third type.

Denote the largest positive eigenvalue of the operator Y by λ_0. In contrast to [7] we consider BRW$/r/k/m$ instead of $BRW/r/0/0$ and assume that in (4) the parameters β_j are real ($\beta_j \in \mathbb{R}$) instead of being positive ($\beta_j > 0$). Under this assumption, we conclude that if λ_0 exitsts then it is simple, strictly positive and guarantees an exponential growth of the first moments m_1 of particle both at an arbitrary point y and on the entire lattice.

Theorem 1. *Let for BRW$/r/k/m$ under consideration the operator $\mathscr{Y}$ have an isolated eigenvalue $\lambda_0 > 0$, and let the remaining part of its spectrum be located* *on the halfline* $\{\lambda \in \mathbb{R} : \lambda \leqslant \lambda_0 - \epsilon\}$, *where* $\epsilon > 0$. *If* $\beta_i^{(r)} = O(r!r^{r-1})$ *for all* $i = 1, \dots, N$ *and* $r \in \mathbb{N}$, *then the following statements hold in the sense of convergence in distribution*

$$\lim_{t\to\infty} \mu_t(y)e^{-\lambda_0 t} = \psi(y)\xi, \quad \lim_{t\to\infty} \mu_t e^{-\lambda_0 t} = \xi, \tag{5}$$

where $\psi(y)$ *is the eigenfunction corresponding to the eigenvalue* λ_0 *and* ξ *is a nondegenerate random variable.*

One approach to analysing Eqs. (2) and evolutionary equations for mean numbers of particles $m_1(t, x, y)$ and $m_1(t, x)$ is to treat them as differential equations in Banach spaces. To apply this approach to our case, we introduce the operators

$$(\mathscr{A}u)(x) = \sum_{x'} a(x - x')u(x'), \qquad (\Delta_{x_i}u)(x) = \delta(x - x_i)u(x), \quad i = 1, \dots, N.$$

on functions set $u(x)$, $x \in \mathbf{Z}^d$. We can represent the operator (4) in a more convenient form:

$$\mathscr{Y} = \mathscr{Y}_{\beta_1,\dots,\beta_{k+r}} = A + \sum_{i=1}^{k+r} \beta_i \Delta_{v_i} \tag{6}$$

where $\beta_i \in \mathbb{R}$, $i = 1, \dots, \beta_{k+r}$. All operators in (6) can be considered as linear continuous operators in any of the spaces $l^p(\mathbf{Z}^d)$, $p \in [1, \infty]$. Note that the operator $\mathscr{A}$ is self-adjoint in $l^2(\mathbf{Z}^d)$ [12–14].

Now, treating for each $t \geq 0$ and each $y \in \mathbf{Z}^d$ the $p(t, \cdot, y)$ and $m_1(t, \cdot, y)$ as elements of $l^p(\mathbf{Z}^d)$ for some p, we can write (see, for example, [12]) the following differential equations in $l^p(\mathbf{Z}^d)$:

$$\frac{dp(t, x, y)}{dt} = (Ap(t, \cdot, y))(x), \qquad p(0, x, y) = \delta(x - y),$$
$$\frac{dm_1(t, x, y)}{dt} = (\mathscr{Y} m_1(t, \cdot, y))(x), \qquad m_1(0, x, y) = \delta(x - y),$$

and for $m_1(t, x)$ the following differential equation in $l^\infty(\mathbf{Z}^d)$:

$$\frac{dm_1(t, x)}{dt} = (\mathscr{Y} m_1(t, \cdot))(x), \qquad m_1(0, x) \equiv 1.$$

Point out that for large t the asymptotic behaviour of the transition probabilities $p(t, x, y)$, as well as of the mean particle numbers $m_1(t, x, y)$ and $m_1(t, x)$ is tightly connected with operators $\mathscr{A}$ and $\mathscr{Y}$ spectral properties.

The properties of $p(t, x, y)$ can be expressed in terms of the Green's function which can be defined [11, § 2.2] as the Laplace transform of the transition probability $p(t, x, y)$ or through the resolvent form:

$$G_\lambda(x, y) := \int_0^\infty e^{-\lambda t} p(t, x, y) dt = \frac{1}{(2\pi)^d} \int_{[-\pi,\pi]^d} \frac{e^{i(\theta, y-x)}}{\lambda - \phi(\theta)} d\theta, \qquad \lambda \geq 0.$$

where $x, y \in \mathbf{Z}^d$, $\lambda \geq 0$, and $\phi(\theta)$ is the transition intensity $a(z)$ Fourier transform:

$$\phi(\theta) := \sum_{z \in \mathbf{Z}^d} a(z) e^{i(\theta, z)} = \sum_{x \in \mathbf{Z}^d} a(x) \cos(x, \theta), \qquad \theta \in [-\pi, \pi]^d. \tag{7}$$

The meaning of the function $G_0(x, y)$ is as follows: it represents the mean amount of time spent by a particle at at $y \in \mathbf{Z}^d$ as $t \to \infty$ provided that at the initial moment $t = 0$ the particle was at $x \in \mathbf{Z}^d$. The asymptotic behaviour of the mean numbers of particles $m_1(t, x, y)$ and $m_1(t, x)$ as $t \to \infty$ can be described in terms of the function $G_\lambda(x, y)$, see, e.g., [11]. Lastly, BRW asymptotic behaviour depends strongly on whether $G_0 := G_0(0, 0)$ is finite, it was shown in [10].

The approach presented in this section is based on representing the BRW evolution equations as differential equations in Banach spaces. It can also be applied to a wide range of problems, including the description of the evolution of higher-order moments of particle numbers (see, e.g., [11,12]).

2 Key Equations and Auxiliary Results

We start off with a crucial remark. Since the operator $\mathscr{Y}$ is in general not self-adjoint, the vast analytical apparatus, developed in [13] and relying heavily on the self-adjointness of the operators involved, is not applicable here directly. Due to the structure of $\mathscr{Y}$, however, this difficulty can be obviated, to a certain extent, with relative ease. Indeed, consider the following differential equation in a Banach space

$$\frac{df(t, x, y)}{dt} = \mathscr{Y} f(t, x, y)$$

with $\mathscr{Y} = \mathscr{A} + \sum_{i=1}^{k+m} \zeta_i \Delta_{u_i} \mathscr{A} + \sum_{j=1}^{k+r} \beta_j \Delta_{v_j}$. Let us now introduce the operator

$$D := \left(I + \sum_{i=1}^{k+m} \zeta_i \Delta_{u_i} \right)^{-\frac{1}{2}},$$

which is correctly defined for $\zeta_i > -1$, and rewrite the equation using this notation:

$$\frac{df(t, x, y)}{dt} = \left(D^{-2} \mathscr{A} + \sum_{j=1}^{k+r} \beta_j \Delta_{v_j} \right) f(t, x, y),$$

which is equivalent to

$$D^{-1} \frac{dDf(t, x, y)}{dt} = \left(D^{-1} D^{-1} \mathscr{A} D^{-1} + \sum_{j=1}^{k+r} \beta_j \Delta_{v_j} D^{-1} \right) Df(t, x, y).$$

By applying D to both parts of the equation above, we obtain

$$\frac{dDf(t, x, y)}{dt} = \left(D^{-1} \mathscr{A} D^{-1} + \sum_{j=1}^{k+r} \beta_j D \Delta_{v_j} D^{-1} \right) Df(t, x, y).$$

Since the operators D and Δ_{v_j} commute, the expression above is equivalent to

$$\frac{dg(t,x,y)}{dt} = \left(D^{-1}\mathscr{A}D^{-1} + \sum_{j=1}^{k+r} \beta_j \Delta_{v_j}\right) g(t,x,y),$$

where $g := Df$. We have thus rewritten the original equation in such a way that the previously non-self-adjoint operator $\mathscr{Y}$ is replaced with the self-adjoint operator

$$\mathscr{Y}' := D^{-1}\mathscr{A}D^{-1} + \sum_{j=1}^{k+r} \beta_j \Delta_{v_j},$$

and a one-to-one correspondence between the solutions f to the starting equation and the solutions g to the new equation can be established through the formula $g = Df$. Therefore, when it comes to analysing Cauchy problems, the operator $\mathscr{Y}$ can, for all intents and purposes, be considered self-adjoint.

We introduce the Laplace generating functions of the random variables $\mu_t(y)$ and μ_t for $z \geqslant 0$:

$$F(z;t,x,y) := \mathsf{E}_x e^{-z\mu_t(y)}, \qquad F(z;t,x) := \mathsf{E}_x e^{-z\mu_t}.$$

where E_x is the mean on condition $\mu_0(\cdot) = \delta_x(\cdot)$.

Theorem 2. *Let the operator A have the form* (1). *The functions $F(z;t,x)$ and $F(z;t,x,y)$ are continuously differentiable with respect to t uniformly with respect to $x, y \in \mathbf{Z}^d$ for all $0 \leqslant z \leqslant \infty$. They satisfy the inequalities $0 \leqslant F(z;t,x), F(z;t,x,y) \leqslant 1$ and are the solutions to the following Cauchy problems in $l^\infty(\mathbf{Z}^d)$*

$$\frac{dF(z;t,\cdot)}{dt} = AF(z;t,\cdot) + \sum_{j=1}^{k+r} \Delta_{v_j} f_j\left(F(z;t,\cdot)\right) \tag{8}$$

with the initial condition $F(z;0,\cdot) = e^{-z}$ and

$$\frac{dF(z;t,\cdot,y)}{dt} = AF(z;t,\cdot,y) + \sum_{j=1}^{k+r} \Delta_{v_j} f_j\left(F(z;t,\cdot,y)\right) \tag{9}$$

with the initial condition $F(z;0,\cdot,y) = e^{-z\delta_y(\cdot)}$.

Theorem 2 allows us to advance from analysing the BRW at hand to considering the corresponding Cauchy problem in a Banach space instead. Note that, contrary to the single branching source case examined in [11], there is not one but several terms $\Delta_{v_j} f_j(F)$ in the right-hand side of Eqs. (8) and (9), $j = 1, 2, \ldots, N$.

Theorem 3. *The moments $m_n(t,\cdot,y)\in l^2\left(\mathbf{Z}^d\right)$ and $m_n(t,\cdot)\in l^\infty\left(\mathbf{Z}^d\right)$ satisfy the following differential equations in the corresponding Banach spaces for all natural $n\geqslant 1$:*

$$\frac{dm_1}{dt}=\mathscr{Y}m_1, \tag{10}$$

$$\frac{dm_n}{dt}=\mathscr{Y}m_n+\sum_{j=1}^{k+r}\Delta_{v_j}g_n^{(j)}(m_1,\ldots,m_{n-1}),\qquad n\geqslant 2, \tag{11}$$

the initial values being $m_n(0,\cdot,y)=\delta_y(\cdot)$ and $m_n(0,\cdot)\equiv 1$ respectively. Here $\mathscr{Y}m_n$ stands for $\mathscr{Y}m_n(t,\cdot,y)$ or $\mathscr{Y}m_n(t,\cdot)$ respectively, and

$$g_n^{(j)}(m_1,\ldots,m_{n-1}):=\sum_{q=2}^{n}\frac{\beta_j^{(q)}}{q!}\sum_{\substack{i_1,\ldots,i_q>0\\ i_1+\cdots+i_q=n}}\frac{n!}{i_1!\cdots i_q!}m_{i_1}\cdots m_{i_q}. \tag{12}$$

Theorem 3 will later be used in the proof of Theorem 8 to help determine the asymptotic behaviour of the moments as $t\to\infty$.

Theorem 4. *The moments $m_1(t,x,\cdot)\in l^2\left(\mathbf{Z}^d\right)$ satisfy the following Cauchy problem in $l^2\left(\mathbf{Z}^d\right)$:*

$$\frac{dm_1(t,x,\cdot)}{dt}=\mathscr{Y}m_1(t,x,\cdot),\qquad m_1(0,x,\cdot)=\delta_x(\cdot).$$

This theorem allows us to obtain different differential equations by making use of the BRW symmetry.

Theorem 5. *The moment $m_1(t,x,y)$ satisfies both integral equations*

$$m_1(t,x,y)=p(t,x,y)+\sum_{j=1}^{k+r}\beta_j\int_0^t p(t-s,x,v_j)m_1(t-s,v_j,y)ds,$$

$$m_1(t,x,y)=p(t,x,y)+\sum_{j=1}^{k+r}\beta_j\int_0^t p(t-s,v_j,y)m_1(t-s,x,v_j)ds.$$

The moment $m_1(t,x)$ satisfies both integral equations

$$m_1(t,x)=1+\sum_{j=1}^{k+r}\beta_j\int_0^t p(t-s,x,v_j)m_1(s,v_j)ds, \tag{13}$$

$$m_1(t,x)=1+\sum_{j=1}^{k+r}\beta_j\int_0^t m_1(s,x,v_j)ds. \tag{14}$$

For $k>1$ the moments $m_k(t,x,y)$ and $m_k(t,x)$ satisfy the equations

$$\begin{aligned} m_k(t,x,y) &= m_1(t,x,y) \\ &\quad + \sum_{j=1}^{k+r} \int_0^t m_1(t-s,x,v_j) g_k^{(j)} \left(m_1(s,v_j,y),\ldots,m_{k-1}(s,v_j,y)\right) ds, \\ m_k(t,x) &= m_1(t,x) \\ &\quad + \sum_{j=1}^{k+r} \int_0^t m_1(t-s,x,v_j) g_k^{(j)} \left(m_1(s,v_j),\ldots,m_{k-1}(s,v_j)\right) ds. \end{aligned}$$

This theorem allows us to make transition from differential equations to integral equations. It is later used to prove Theorem 8.

Theorems 3–5 are a generalization to the case BRW$/r/k/m$ of Lemma 1.2.1, Theorem 1.3.1 and Theorem 1.4.1 from [11], proved there for BRW$/1/0/0$. The proofs of Theorems 3–5 differ only in technical details from the proofs of the above statements from [11] and are therefore omitted here.

3 Properties of the Operator $\mathscr{Y}$

We call a BRW supercritical if $\mu_t(y)$ and μ_t grow exponentially. As was mentioned in Introduction, one of the main results of this work is the numbers of particles limit behavior (5), from which it follows that the BRW with several branching sources with arbitrary intensities is supercritical if the operator $\mathscr{Y}$ has a positive eigenvalue λ. For this reason we devote this section to studying the spectral properties of the operator $\mathscr{Y}$.

We mention a statement proved in [11, Lemma 3.1.1].

Lemma 1. *The spectrum $\sigma(\mathscr{A})$ of the operator $\mathscr{A}$ is included in the half-line $(-\infty,0]$. Also, since the operator $\sum_{j=1}^{N}\beta_j\Delta_{v_j}$ is compact, $\sigma_{ess}(\mathscr{Y})=\sigma(\mathscr{A})\subset(-\infty,0]$, where $\sigma_{ess}(\mathscr{Y})$ denotes the essential spectrum [6] of the operator $\mathscr{Y}$.*

The following theorem provides a criterion of there being a positive eigenvalue in the spectrum of the operator $\mathscr{Y}$.

Theorem 6. *A number $\lambda>0$ is an eigenvalue and $f\in l^2(\mathbf{Z}^d)$ is the corresponding eigenvector of the operator $\mathscr{Y}$ if and only if the system of linear equations*

$$f(u_i)=\frac{1}{1+\zeta_i}\left(\lambda\sum_{j=1}^{k+m}\zeta_j f(u_j) I_{u_j-u_i}(\lambda)+\sum_{j=1}^{k+r}\beta_j f(v_j) I_{v_j-u_i}(\lambda)\right) \tag{15}$$

for $i=1,\ldots,k+m$, and

$$f(v_i)=\left(\lambda\sum_{j=1}^{k+m}\zeta_j f(u_j) I_{u_j-v_i}(\lambda)+\sum_{j=1}^{k+r}\beta_j f(v_j) I_{v_j-v_i}(\lambda)\right) \tag{16}$$

for $i=1,\dots,k+r$, with respect to the variables $f(u_j)$ and $f(v_j)$, where

$$I_x(\lambda):=G_\lambda(x,0)=\frac{1}{(2\pi)^d}\int_{[-\pi,\pi]^d}\frac{e^{-i(\theta,x)}}{\lambda-\phi(\theta)}d\theta,\qquad x\in\mathbf{Z}^d,$$

has a non-trivial solution.

Proof. For $\lambda>0$ to be an eigenvalue of the operator $\mathscr{Y}$ it is necessary and sufficient that there be a non-zero element $f\in l^2\left(\mathbf{Z}^d\right)$ that satisfies the equation

$$(\mathscr{Y}-\lambda I)f=\left(\mathscr{A}+\sum_{i=1}^{k+m}\zeta_i\Delta_{u_i}\mathscr{A}+\sum_{j=1}^{k+r}\beta_j\Delta_{v_j}-\lambda I\right)f=0.$$

Obviously, the solution sets of such an equation for the operators $\mathscr{Y}$ and $C^{-1}\mathscr{Y}C$ are the same for any operator C; let us set $C:=\left(I+\sum_{i=1}^{k+m}\zeta_i\Delta_{u_i}\right)^{\frac{1}{2}}$, which is correctly defined since $\zeta_i>-1$ for all i. Thus the equation above can be rewritten as follows:

$$\left(\mathscr{A}+\sum_{i=1}^{k+m}\zeta_i\mathscr{A}\Delta_{u_i}+\sum_{j=1}^{k+r}\beta_j\Delta_{v_j}-\lambda I\right)f=0$$

Since $(\Delta_{v_j}f)(x):=f(x)\delta_{v_j}(x)=f(v_j)\delta_{v_j}(x)$ and $(\mathscr{A}\Delta_{u_i}f)(x):=f(x)A\delta_{u_j}(x)$, the preceding expression can be rewritten as follows:

$$(\mathscr{A}f)(x)+\sum_{j=1}^{k+m}\zeta_jf(u_j)A\delta_{u_j}(x)+\sum_{j=1}^{k+r}\beta_jf(v_j)\delta_{v_j}(x)=\lambda f(x),\qquad x\in\mathbf{Z}^d.$$

We apply Fourier transform to this equality and obtain

$$(\widetilde{\mathscr{A}f})(\theta)+\sum_{j=1}^{k+m}\zeta_jf(u_j)\widetilde{\mathscr{A}\delta_{u_j}}(\theta)+\sum_{j=1}^{k+r}\beta_jf(v_j)e^{i(\theta,v_j)}=\lambda\tilde{f}(\theta),\tag{17}$$

for $\theta\in[-\pi,\pi]^d$. The Fourier transform $\widetilde{\mathscr{A}f}$ of $(\mathscr{A}f)(x)$ is of the form $\phi\tilde{f}$, where $\tilde{f}$ is the Fourier transform of f, and $\phi(\theta)$ is defined by the equality (7), see [11, Lemma 3.1.1]. With this in mind, and making use of the fact that, by the definition of the Fourier transform,

$$\widetilde{\mathscr{A}\delta_{u_j}}(\theta)=\phi(\theta)\widetilde{\delta_{u_j}}(\theta)=\phi(\theta)\sum_{x\in\mathbf{Z}^d}\delta_{u_j}(x)e^{i(x,\theta)}=\phi(\theta)e^{i(u_j,\theta)},$$

we rewrite equality (17) as

$$\phi(\theta)\tilde{f}(\theta)+\sum_{j=1}^{k+m}\zeta_jf(u_j)\phi(\theta)e^{i(u_j,\theta)}+\sum_{j=1}^{k+r}\beta_jf(v_j)e^{i(\theta,v_j)}=\lambda\tilde{f}(\theta),$$

or

$$\tilde{f}(\theta)=\frac{1}{\lambda-\phi(\theta)}\left[\sum_{j=1}^{k+m}\zeta_j f(u_j)\phi(\theta)e^{i(u_j,\theta)}+\sum_{j=1}^{k+r}\beta_j f(v_j)e^{i(\theta,v_j)}\right], \quad (18)$$

where $\theta\in[-\pi,\pi]^d$. Since $\lambda>0$ and $\phi(\theta)\leqslant 0$, $\int_{[-\pi,\pi]^d}|\lambda-\phi(\theta)|^{-2}d\theta<\infty$, which allows us to apply the inverse Fourier transform to equality (18): as

$$\begin{aligned}
&\Phi^{-1}\left[\frac{1}{\lambda-\phi(\theta)}\sum_{j=1}^{k+m}\zeta_j f(u_j)\phi(\theta)e^{i(u_j,\theta)}\right]\\
&=\Phi^{-1}\left[-\sum_{j=1}^{k+m}\zeta_j f(u_j)e^{i(u_i,\theta)}+\frac{\lambda}{\lambda-\phi(\theta)}\sum_{j=1}^{k+m}\zeta_j f(u_j)e^{i(u_j,\theta)}\right]\\
&=-\sum_{j=1}^{k+m}\zeta_j f(u_j)\frac{1}{(2\pi)^d}\int_{[-\pi,\pi]^d}e^{-i(\theta,u_j-x)}d\theta+\lambda\sum_{j=1}^{k+m}\zeta_j f(u_j)I_{u_j-x}(\lambda)\\
&=-\sum_{j=1}^{k+m}\zeta_j f(u_j)\mathbf{I}[x=u_j]+\lambda\sum_{j=1}^{k+m}\zeta_j f(u_j)I_{u_j-x}(\lambda)
\end{aligned}$$

we obtain

$$\begin{aligned}
&f(x)+\sum_{j=1}^{k+m}\zeta_j f(u_j)\mathbf{I}[x=u_j]\\
&\quad=\lambda\sum_{j=1}^{k+m}\zeta_j f(u_j)I_{u_j-x}(\lambda)\phi(\theta)e^{i(u_j,\theta)}+\sum_{j=1}^{k+r}\beta_j I_{v_j-x}(\lambda)f(v_j).
\end{aligned} \quad (19)$$

By choosing $x=u_i$, where $i=1,\dots,k+m$, or $x=v_i$, where $i=1,\dots,k+r$, we can rewrite (19) as follows:

$$f(u_i)=\frac{1}{1+\zeta_i}\left(\lambda\sum_{j=1}^{k+m}\zeta_j f(u_j)I_{u_j-u_i}(\lambda)+\sum_{j=1}^{k+r}\beta_j f(v_j)I_{v_j-u_i}(\lambda)\right),$$

$$f(v_i)=\lambda\sum_{j=1}^{k+m}\zeta_j f(u_j)I_{u_j-v_i}(\lambda)+\sum_{j=1}^{k+r}\beta_j f(v_j)I_{v_j-v_i}(\lambda).$$

We note that any solution of system (15) completely defines $f(x)$ on the entirety of its domain by formula (19), which proves the theorem. □

Let among $k+r$ sources, in which branching occurs, in $s\leq k+r$ sources intensities are $\beta_i>0$, $i=0,\dots,s$, and in $k+r-s$ sources intensities are $\beta_i\leq 0$, $i=s+1,\dots,k+r$. We represent the operator $\mathscr{Y}$ defined by (6) as follows:

$$\mathscr{Y}=\mathscr{A}+\sum_{i=1}^{k+m}\zeta_i\Delta_{u_i}\mathscr{A}+\sum_{i=1}^{s}\beta_i\Delta_{v_i}+\sum_{i=s+1}^{k+r}\beta_i\Delta_{v_i}.$$

Define operator

$$\mathscr{B} := \lambda I - \mathscr{A} - \sum_{i=1}^{k+m} \zeta_i \Delta_{u_i} \mathscr{A} - \sum_{i=s+1}^{k+r} \beta_i \Delta_{v_i},$$

then the eigenvector h corresponding to the eigenvalue λ of $\mathscr{Y}$ satisfies the equation

$$\mathscr{B}h = \sum_{i=1}^{s} \beta_i \delta_{v_i} \langle \delta_{v_i}, h \rangle.$$

Note that $\langle \mathscr{A}x, x \rangle \leq 0$. Besides, $\beta_i < 0$ for $i = s+1, \ldots, k+r$, and therefore $\langle \sum_{i=s+1}^{k+r} \beta_i \Delta_{v_i} x, x \rangle \leq 0$. Hence, the operator $\mathscr{B}$ is reversible. The problem of existence of positive eigenvalues of the operator $\mathscr{Y}$ is converted to the question of the existence of nonzero solutions for the equation $h = \mathscr{B}^{-1} \sum_{j=1}^{s} \beta_j \delta_{v_j} \langle \delta_{v_j}, h \rangle$, which, after introducing auxiliary variables $q_i = \langle \delta_{v_i}, h \rangle$ and scalar multiplication on the left of this equality by δ_{v_i} reduces to a finite system of equations

$$q_i = \sum_{j=1}^{s} \beta_j \langle \delta_{v_i}, \mathscr{B}^{-1} \delta_{v_j} \rangle q_j, \qquad i = 1, 2, \ldots, s. \tag{20}$$

Denote matrix $B^{(\lambda)}$:

$$B_{i,j}^{(\lambda)} := \beta_j \langle \delta_{v_i}, \mathscr{B}^{-1} \delta_{v_j} \rangle, \; i, j = 1, \ldots, s. \tag{21}$$

So the matrix representation (20) has the following form

$$q = B^{(\lambda)} q, \tag{22}$$

and the problem on positive eigenvalues for $\mathscr{Y}$ is reduced to the question of for which $\lambda > 0$ the number 1 is the matrix $B^{(\lambda)}$ eigenvalue.

Theorem 7. *Let $\lambda_0 > 0$ be the largest eigenvalue of the operator $\mathscr{Y}$. Then λ_0 is a simple eigenvalue of $\mathscr{Y}$, and 1 is the largest eigenvalue of the matrix $B^{(\lambda_0)}$.*

Proof. Denote $\zeta := \max(0, \max_i(\zeta_i)) \geq 0$ and note that the elements of the operator

$$\tilde{\mathscr{A}} := \mathscr{A} + \sum_{i=1}^{k+m} \zeta_i \Delta_{u_i} \mathscr{A} - \mathbf{a}(0,0)(\zeta + 1) I$$

are non-negative. It follows from Schur's test [4] that in each of the spaces $l^p(\mathbf{Z}^d)$ for the operator norm $\tilde{\mathscr{A}}$ there is an estimation

$$\|\tilde{\mathscr{A}}\|_p \leq -\mathbf{a}(0,0)(\zeta + 1). \tag{23}$$

Operator $\mathscr{B}$ can be represented as follows:

$$\mathscr{B} = \lambda I - \mathbf{a}(0,0)(\zeta+1)I - \sum_{i=s+1}^{k+r} \beta_i \Delta_{v_i} - \tilde{\mathscr{A}} = \mathscr{F}_\lambda - \tilde{\mathscr{A}},$$

where the operator

$$\mathscr{F}_\lambda = \lambda I - \mathbf{a}(0,0)(\zeta+1)I - \sum_{i=s+1}^{k+r} \beta_i \Delta_{v_i}$$

is diagonal with all its diagonal elements no less than $-\mathbf{a}(0,0)(\zeta+1)+\lambda > 0$. Then

$$\|\mathscr{F}_\lambda^{-1}\|_p \leq \frac{1}{-\mathbf{a}(0,0)(\zeta+1)+\lambda}. \tag{24}$$

Then $\mathscr{B}$ can be represented in the following form $\mathscr{B} = \mathscr{F}_\lambda\left(I - \mathscr{F}_\lambda^{-1}\tilde{\mathscr{A}}\right)$ and therefore

$$\mathscr{B}^{-1} = \left(I - \mathscr{F}_\lambda^{-1}\tilde{\mathscr{A}}\right)^{-1} \mathscr{F}_\lambda^{-1}. \tag{25}$$

Here by virtue of (23) and (24) the operator norm of $\mathscr{F}_\lambda^{-1}\tilde{\mathscr{A}}$ is less than one:

$$\|\mathscr{F}_\lambda^{-1}\tilde{\mathscr{A}}\|_p \leq \frac{-\mathbf{a}(0,0)(\zeta+1)}{-\mathbf{a}(0,0)(\zeta+1)+\lambda} < 1,$$

and therefore the operator $\left(I - \mathscr{F}_\lambda^{-1}\tilde{\mathscr{A}}\right)^{-1}$ can be represented as a series:

$$\left(I - \mathscr{F}_\lambda^{-1}\tilde{\mathscr{A}}\right)^{-1} = \sum_{n=0}^{\infty} \left(\mathscr{F}_\lambda^{-1}\tilde{\mathscr{A}}\right)^n.$$

Hence, by virtue of (25)

$$\mathscr{B}^{-1} = \sum_{n=0}^{\infty} \left(\mathscr{F}_\lambda^{-1}\tilde{\mathscr{A}}\right)^n \mathscr{F}_\lambda^{-1}. \tag{26}$$

Note that the right-hand side (26) is the sum of the products of operators (infinite matrices) with non-negative elements. Therefore, each element of the operator (infinite matrix) $\mathscr{B}^{-1}$ is non-negative.

Let us prove that each element of the operator $\mathscr{B}^{-1}$ is strictly positive. For the proof, we use the fact that our random walk is irreducible. Note that, since the random walk under the action of the operator $\mathscr{A}$ is irreducible, for any pair $x, y \in \mathbf{Z}^d$ there exists $n \geq 1$ and the set of points

$$u_0, u_1, \ldots, u_n \in \mathbf{Z}^d, \qquad u_0 = x,\ u_n = y, \tag{27}$$

such that $a(u_1-u_0)a(u_2-u_1)\cdots a(u_n-u_{n-1}) > 0$, whence follows

$$\overline{a}(u_1-u_0)\overline{a}(u_2-u_1)\cdots\overline{a}(u_n-u_{n-1}) > 0. \tag{28}$$

Note that the elements of the infinite matrix $\mathscr{B}^{-1}$ are indexed by pairs of points $x, y \in \mathbf{Z}^d$. In addition, the element with indices (x, y) of the matrix $\left(\mathscr{F}_\lambda^{-1}\tilde{\mathscr{A}}\right)^n$ in (26) is a sum of the form

$$\sum_{u_0=x,\ u_n=y} \overline{a}(u_1 - u_0)\overline{a}(u_2 - u_1)\cdots\overline{a}(u_n - u_{n-1})f_{u_0,u_1,\ldots,u_n} \tag{29}$$

taken over all possible "chains" of n elements, satisfying (27), in which positive factors $f_{u_0,u_1,\ldots,u_n}$ are formed due to the presence of the diagonal matrix $\mathscr{F}_\lambda^{-1}$ of $\left(\mathscr{F}_\lambda^{-1}\tilde{\mathscr{A}}\right)^n$. But by virtue of (28) at least one term in the sum (29) is strictly positive, while the rest are non-negative. Hence, the entire sum is also strictly positive, which implies that all elements of the operator (infinite matrix) $\mathscr{B}^{-1}$ are strictly positive. Since the elements of $\mathscr{B}^{-1}$ (see (21)) are positive, the matrix $B^{(\lambda)}$ is positive.

The right side of (25) contains the operators $\mathscr{F}_\lambda^{-1}$, whose elements, monotonically decreasing in $\lambda > 0$, tend to zero as $\lambda \to \infty$. Since in this case all multiplied and added operators (infinite matrices) are positive, then all their elements in this case will also decrease monotonically in $\lambda > 0$ and tend to zero as $\lambda \to \infty$.

We first show that if λ_0 is the operator $\mathscr{Y}$ largest eigenvalue, then the largest (absolute) eigenvalue of the matrix $B^{(\lambda_0)}$ is 1. Indeed, assume it is not the case.

It follows from (22) that $\lambda_0 > 0$ is an eigenvalue of $\mathscr{Y}$ if and only if 1 is an eigenvalue of $B^{(\lambda_0)}$. All elements of $B^{(\lambda_0)}$ are strictly positive. Consequently, by the Perron-Frobenius theorem, see [5, Theorem 8.4.4], the matrix $B^{(\lambda_0)}$ has a strictly positive eigenvalue that is strictly greater (by absolute value) than any other of its eigenvalues.

Let us denote the dominant eigenvalue of $B^{(\lambda_0)}$ by $\gamma(\lambda_0)$. Since we assumed that 1 is not the largest eigenvalue of $B^{(\lambda_0)}$, then $\gamma(\lambda_0) > 1$. Given that with respect to λ the functions $I_{x_i-x_j}(\lambda)$ are continuous, then all elements of $B^{(\lambda)}$ and all eigenvalues of $B^{(\lambda)}$ are continuous functions with respect to λ. All matrix $B^{(\lambda)}$ eigenvalues tend to zero as $\lambda \to \infty$, because for all i and j $I_{x_i-x_j}(\lambda) \to 0$ as $\lambda \to \infty$. Hence there is such a $\hat{\lambda} > \lambda_0$ that $\gamma(\hat{\lambda}) = 1$. Then, as was shown earlier, this $\hat{\lambda}$ has to be an eigenvalue of $\mathscr{Y}$, which contradicts the initial assumption that λ_0 is the largest eigenvalue of the operator $\mathscr{Y}$.

We have just proved that 1 is the largest eigenvalue of $B^{(\lambda_0)}$; then we obtain from the Perron-Frobenius theorem the simplicity of this eigenvalue. Therefore, to complete the proof, we have to show the simplicity of the eigenvalue λ_0 of $\mathscr{Y}$.

Suppose it is not, and λ_0 is not simple. In this case, there are at least two linearly independent eigenvectors f_1 and f_2 corresponding to the eigenvalue λ_0. Therefore, we can again applying the equality (19), notice that the linear independence of the vectors f_1 and f_2 is equivalent to the linear independence of the vectors

$$\hat{f}_i := (f_i(u_1), \ldots, f_i(u_N)), \qquad i = 1, 2.$$

From the definition of $B^{(\lambda)}$ and Theorem 6 it follows that vectors $\hat{f}_1$ and $\hat{f}_2$ satisfy $\left(B^{(\lambda_0)} - I\right) f = 0$. It contradicts the simplicity of eigenvalue 1 of $B^{(\lambda_0)}$. This completes the proof. □

Lemma 2. *Let $\mathscr{Y}$ be a self-adjoint continuous operator on a separable Hilbert space E, the spectrum of which is a disjoint union of two sets: fist one is a finite (counting multiplicity) set of isolated eigenvalues $\lambda_i > 0$ and second one is the remaining part of the spectrum which is included in $[-s, 0]$, $s > 0$. Then the solution $m(t)$ of the Cauchy problem*

$$\frac{dm(t)}{dt} = \mathscr{Y} m(t), \qquad m(0) = m_0,$$

satisfies the condition

$$\lim_{t\to\infty} e^{-\lambda_0 t} m(t) = C\left(m_0\right),$$

where $\lambda_0 = \max_i \lambda_i$.

Proof. Let us denote by V_{λ_i} the finite-dimensional eigenspace of $\mathscr{Y}$ that corresponds to the eigenvalue λ_i.

We consider the projection P_i of $\mathscr{Y}$ onto V_{λ_i}, see [6]. Let

$$x_i(t) := P_i m(t),$$
$$v(t) := \left(I - \sum_i P_i\right) m(t) = m(t) - \sum_i x_i(t).$$

All spectral operators P_i and $(I - \sum P_i)$ commute with $\mathscr{Y}$, see [6]. Therefore

$$\frac{dx_i(t)}{dt} = P_i \mathscr{Y} m(t) = \mathscr{Y} x_i(t)$$
$$\frac{dv(t)}{dt} = \left(I - \sum P_i\right) \mathscr{Y} m(t) = \left(I - \sum P_i\right) \mathscr{Y} \left(I - \sum P_i\right) v(t).$$

As $x_i(t) \in V_{\lambda_i}$, we can see that $\mathscr{Y} x_i(t) = \lambda_i x_i(t)$, from which it follows that $x_i(t) = e^{\lambda_i t} x_i(0)$. Since the spectrum of the operator

$$\mathscr{Y}_0 := \left(I - \sum P_i\right) \mathscr{Y} \left(I - \sum P_i\right)$$

is included into the spectrum of operator $\mathscr{Y}$ and $\mathscr{Y}_0$ has no isolated eigenvalues λ_i, it is included into $[-s, 0]$. From this for all $t \geqslant 0$ we obtain $|v(t)| \leqslant |v(0)|$, see [11, Lemma 3.3.5]. Hence

$$m(t) = \sum_i e^{\lambda_i t} P_i m(0) + v(t), \tag{30}$$

and the proof is complete. □

Remark 1. Let λ_0 be the largest eigenvalue of $\mathscr{Y}$. Denote $P_0 m(0) = C(m_0)$ in (30). Then $C(m_0) \neq 0$ if and only if the orthogonal projection $P_0 m(0)$ of the initial value $m_0 = m(0)$ onto the corresponding to the eigenvalue λ_0 eigenspace is non-zero.

If the eigenvalue λ_0 of $\mathscr{Y}$ is simple and f is an eigenvector corresponding to λ_0, the projection P_0 is defined by the formula $P_0 x = \frac{(f,x)}{(f,f)} f$, where $(\cdot,\cdot)$ denots scalar product in the Hilbert space E. In cases when this λ_0 is not simple, describing the projection P_0 is a significantly more difficult task.

We remind the reader that we proved the simplicity of the largest eigenvalue of $\mathscr{Y}$ above allowing us to bypass this issue.

Theorem 8. *Let defined by* (6) *operator* $\mathscr{Y}$ *with the parameters* $\{\zeta_i\}_{i=1}^{k+r}$ *and* $\{\beta_i\}_{i=1}^{k+m}$, *has a finite number of positive eigenvalues (counting multiplicity). We denote by* λ_0 *the largest of them, and the corresponding to* λ_0 *normalized vector by* f. *Then for* $t \to \infty$ *and all* $n \in \mathbf{N}$ *the following statements hold:*

$$m_n(t,x,y) \sim C_n(x,y) e^{n\lambda_0 t}, \quad m_n(t,x) \sim C_n(x) e^{n\lambda_0 t}, \tag{31}$$

where

$$C_1(x,y) = f(y) f(x), \qquad C_1(x) = f(x) \frac{1}{\lambda_0} \sum_{j=1}^{k+r} \beta_j f(v_j),$$

and the functions $C_n(x,y)$ *and* $C_n(x) > 0$ *for* $n \geqslant 2$ *are defined as follows:*

$$C_n(x,y) = \sum_{j=1}^{k+r} g_n^{(j)} \left(C_1(v_j,y), \ldots, C_{n-1}(v_j,y)\right) D_n^{(j)}(x),$$

$$C_n(x) = \sum_{j=1}^{k+r} g_n^{(j)} \left(C_1(v_j), \ldots, C_{n-1}(v_j)\right) D_n^{(j)}(x),$$

where $D_n^{(j)}(x)$ *are certain functions satisfying the estimate* $|D_n^{(j)}(x)| \leqslant \frac{2}{n\lambda_0}$ *for* $n \geqslant n_*$ *and some* $n_* \in \mathbf{N}$ *and* $g_n^{(j)}$ *are the functions defined in* (12).

Proof. For $n \in \mathbf{N}$ we consider the functions $\nu_n := m_n(t,x,y) e^{-n\lambda_0 t}$. From Theorem 3 (see Eqs. (10) and (11) for m_n) we obtain the following equations for ν_n:

$$\frac{d\nu_1}{dt} = \mathscr{Y} \nu_1 - \lambda_0 \nu_1,$$

$$\frac{d\nu_n}{dt} = \mathscr{Y} \nu_n - n\lambda_0 \nu_n + \sum_{j=1}^{k+r} \Delta_{v_j} g_n^{(j)} \left(\nu_1, \ldots, \nu_{n-1}\right), \qquad n \geqslant 2,$$

the initial values being $\nu_n(0,\cdot,y) = \delta_y(\cdot), n \in \mathbf{N}$.

Since λ_0 is the largest eigenvalue of $\mathscr{Y}$, the spectrum of $\mathscr{Y}_n := \mathscr{Y} - n\lambda_0 I$ for $n \geqslant 2$ is included into $(-\infty, -(n-1)\lambda_0]$. As it was shown, for example, in [11,

p. 58], that if the spectrum of a self-adjoint continuous operator $\widetilde{\mathscr{Y}}$ on a Hilbert space is included into $(-\infty, -s], s > 0$, and also $f(t) \to f_*$ as $t \to \infty$, then the solution of the equation

$$\frac{d\nu}{dt} = \widetilde{\mathscr{Y}}\nu + f(t)$$

satisfies $\nu(t) \to -\widetilde{\mathscr{Y}}^{-1} f_*$ condition. For this reason for $n \geqslant 2$ we obtain

$$\begin{aligned} C_n(x, y) = \lim_{t\to\infty} \nu_n &= -\sum_{j=1}^{k+r} \left(\mathscr{Y}_n^{-1} \Delta_{v_j} g_n^{(j)}(C_1(\cdot, y), \ldots, C_{n-1}(\cdot, y)) \right)(x) \\ &= -\sum_{j=1}^{k+r} g_n^{(j)}(C_1(v_j, y), \ldots, C_{n-1}(v_j, y))(\mathscr{Y}_n^{-1}\delta_{v_j}(\cdot))(x)). \end{aligned}$$

Now we prove the existence of such a natural number n_* that for all $n \geqslant n_*$ the estimates

$$D_n^{(j)}(x) := |(\mathscr{Y}_n^{-1}\delta_{v_j}(\cdot))(x)| \leqslant \frac{2}{n\lambda_0}$$

hold. We evaluate the operator $\mathscr{Y}_n^{-1}$ norm. For this, let us consider two vectors u and x such that $u = \mathscr{Y}_n x = \mathscr{Y} x - n\lambda_0 x$. Then $\|u\| \geqslant n\lambda_0\|x\| - \|\mathscr{Y} x\| \geqslant (n\lambda_0 - \|\mathscr{Y}\|)\|x\|$, hence $\|\mathscr{Y}_n^{-1} u\| = \|x\| \leqslant \|u\|/(n\lambda_0 - \|\mathscr{Y}\|)$, and for all $n \geqslant n_* = 2\lambda_0^{-1}\|\mathscr{Y}\|$ the estimate $\|\mathscr{Y}_n^{-1}\| \leqslant \frac{2}{n\lambda_0}$ holds. From this we conclude that

$$|(\mathscr{Y}_n^{-1}\delta_{v_j}(\cdot))(x)| \leqslant \|\mathscr{Y}_n^{-1}\delta_{v_j}(\cdot)\| \leqslant \|\mathscr{Y}_n^{-1}\|\|\delta_{v_j}(\cdot)\| \leqslant \frac{2}{n\lambda_0}, \qquad n \geqslant n_*.$$

Now we have to estimate the particle number moments asymptotic behaviour. It follows from (14) that as $t \to \infty$ the following asymptotic equivalences hold:

$$m_1(t, x) \sim \sum_{j=1}^{k+r} \beta_j \int_0^t m_1(s, x, v_j)\, ds \sim \sum_{j=1}^{k+r} \frac{\beta_j}{\lambda_0} m_1(t, x, v_j). \tag{32}$$

The function $m_1(t, x, 0)$ exhibits exponential growth as $t \to \infty$ and $m_1(t, x)$ will display the same behaviour.

We can now infer the asymptotic behaviour of the higher moments $m_n(t, x)$ for $n \geqslant 2$ from Eqs. (11) in a similar way to how it was done above for $m_n(t, x, y)$.

We proceed to prove the equalities for $C_1(x, y)$ and $C_1(x)$. The eigenvalue λ_0 is simple by Corollary 7 and it follows, according to Remark 1, that

$$C_1(x, y) = \lim_{t\to\infty} e^{-\lambda_0 t} m_1(t, x, y) = Pm_0 = (m_1(0, x, y), f)\, f(x).$$

But $m_1(0, x, y) = \delta_y(x)$, hence

$$C_1(x, y) = (m_1(0, x, y), f)\, f(x) = f(y) f(x).$$

We also obtain from (32) that

$$C_1(x) = \frac{1}{\lambda_0} \sum_{j=1}^{k+r} \beta_j C_1(x, v_j) = f(x) \frac{1}{\lambda_0} \sum_{j=1}^{k+r} \beta_j f(v_j),$$

which concludes the proof. □

Corollary 1. *$C_n(x,y) = \psi^n(y)C_n(x)$, where $\psi(y) = \frac{\lambda_0 f(y)}{\sum_{j=1}^{k+r}\beta_j f(v_j)}$.*

Proof. We prove the corollary by induction on n. For $n = 1$ the induction basis holds due to Theorem 8. Let us now deal with the induction step: according to Theorem 8,

$$C_{n+1}(x,y) = \sum_{j=1}^{k+r} g_{n+1}^{(j)}\left(C_1(v_j,y),\dots,C_n(v_j,y)\right) D_{n+1}^{(j)}(x),$$

$$C_{n+1}(x) = \sum_{j=1}^{k+r} g_{n+1}^{(j)}\left(C_1(v_j),\dots,C_n(v_j)\right) D_{n+1}^{(j)}(x);$$

therefore, it suffices to prove that for all j the equalities

$$g_{n+1}^{(j)}\left(C_1(v_j,y),\dots,C_n(v_j,y)\right) = \psi^{n+1}(y) g_{n+1}^{(j)}\left(C_1(v_j),\dots,C_n(v_j)\right)$$

hold. As a consequence of the definition and hypothesis of induction,

$$\begin{aligned} &g_{n+1}^{(j)}\left(C_1(v_j,y),\dots,C_n(v_j,y)\right) \\ &= \sum_{r=2}^{n+1}\frac{\beta_j^{(r)}}{r!}\sum_{\substack{i_1,\dots,i_r>0\\ i_1+\dots+i_r=n+1}}\frac{n!}{i_1!\cdots i_r!}C_{i_1}(v_j,y)\cdots C_{i_r}(v_j,y) \\ &= \psi^{n+1}(y)\sum_{r=2}^{n+1}\frac{\beta_j^{(r)}}{r!}\sum_{\substack{i_1,\dots,i_r>0\\ i_1+\dots+i_r=n+1}}\frac{n!}{i_1!\cdots i_r!}C_{i_1}(v_j)\cdots C_{i_r}(v_j), \end{aligned}$$

which proves the corollary. □

4 Proof of Theorem 1

Let us introduce the function

$$f(n,r) := \sum_{\substack{i_1,\dots,i_r>0\\ i_1+\dots+i_r=n}} i_1^{i_1}\cdots i_r^{i_r}, \qquad 1 \leqslant r \leqslant n.$$

The following auxiliary lemma is proved in [7, Lemma 9].

Lemma 3. *There is such a constant $C > 0$ that $f(n,r) < C\frac{n^n}{r^{r-1}}$ for all $n \geqslant r \geqslant 2$.*

We now turn to proving Theorem 1.

Proof. Let us define the functions

$$m(n,x,y) := \lim_{t\to\infty}\frac{m_n(t,x,y)}{m_1^n(t,x,y)} = \frac{C_n(x,y)}{C_1^n(x,y)},$$

$$m(n,x) := \lim_{t\to\infty} \frac{m_n(t,x)}{m_1^n(t,x)} = \frac{C_n(x)}{C_1^n(x)};$$

as follows from Theorem 8 and $G_\lambda(x,y)$ being positive, these definitions are sound. Corollary 1 yields

$$m(n,x,y) = m(n,x) = \frac{C_n(x)}{C_1^n(x)} = \frac{C_n(x,y)}{C_1^n(x,y)}.$$

From the above equalities and the asymptotic equivalences (31) we have Theorem 1 statements in terms of convergence of moments of the random variables $\xi(y) = \psi(y)\xi$ and ξ.

The distributions of the limit random variables $\xi(y)$ and ξ to be uniquely determined by their moments if, as was shown in [11], the Carleman condition

$$\sum_{n=1}^{\infty} m(n,x,y)^{-1/2n} = \infty, \qquad \sum_{n=1}^{\infty} m(n,x)^{-1/2n} = \infty \tag{33}$$

We establish below that the series for the $m(n,x)$ diverges and that, therefore, said moments define the random variable ξ uniquely; the statement concerning $\xi(y)$ and its moments can be proved in much the same manner.

Since $\beta_j^{(r)} = O(r!r^{r-1})$, there is such a constant D that for all $r \geqslant 2$ and $j = 1,\dots,k+r$ the inequality $\beta_j^{(r)} < Dr!r^{r-1}$ holds. Without loss of generality we assume that for all n

$$C_n(x) \leqslant \max_{j=1,\dots,k+r} C_n(v_j) = C_n(v_1).$$

Let $\gamma := 2NCDE\frac{\lambda_0\beta_2}{2}C_1^2(v_1)$, where C is defined in Lemma 3, and the constant E is such that $C_n(v_1) \leqslant \gamma^{n-1}n!n^n$ for $n \leqslant \max\{n_*, 2\}$, where n_* is defined in Theorem 8.

From this point on, the proof follows to the scheme of proof of [7, Th. 1] and is included only for readability.

Let us show by induction that

$$C_n(x) \leqslant C_n(v_1) \leqslant \gamma^{n-1}n!n^n.$$

The induction basis for $n = 1$ is valid due to the C choice. In order to prove the step of induction, we will show that

$$C_{n+1}(x) \leqslant C_{n+1}(v_1) \leqslant \gamma^n(n+1)!(n+1)^{n+1}.$$

It follows from $C_{n+1}(v_1)$ formula and the estimate for $D_n^{(j)}(x)$ from Theorem 8 that

$$C_{n+1}(v_1) \leqslant \sum_{j=1}^{N}\sum_{r=2}^{n+1} \frac{\beta_r^{(j)}}{r!} \sum_{\substack{i_1,\dots,i_r>0\\ i_1+\dots+i_r=n+1}} \frac{(n+1)!}{i_1!\cdots i_r!} C_{i_1}(v_1)\cdots C_{i_r}(v_1)\frac{2}{\lambda_0(n+1)}.$$

By the induction hypothesis

$$\frac{(n+1)!}{i_1!\cdots i_r!}C_{i_1}(0)\cdots C_{i_r}(0)\leqslant \gamma^{n+1-r}(n+1)!i_1^{i_1}\cdots i_r^{i_r};$$

which, added to the fact that $\beta_j^{(r)} < Dr!r^{r-1}$ and $\gamma^{n+1-r}\leqslant\gamma^{n-1}$, yields

$$\begin{aligned}\sum_{j=1}^{N}\sum_{r=2}^{n+1}\frac{\beta_j^{(r)}}{r!}&\sum_{\substack{i_1,\ldots,i_r>0\\ i_1+\cdots+i_r=n+1}}\frac{(n+1)!}{i_1!\cdots i_r!}C_{i_1}(v_1)\cdots C_{i_r}(v_1)\\ &\leqslant N\gamma^{n-1}D(n+1)!\sum_{r=2}^{n+1}r^{r-1}\sum_{\substack{i_1,\ldots,i_r>0\\ i_1+\cdots+i_r=n+1}}i_1^{i_1}\cdots i_r^{i_r}\\ &=N\gamma^{n-1}D(n+1)!\sum_{r=2}^{n+1}r^{r-1}f(n+1,r).\end{aligned}$$

We infer from Lemma 3 that

$$\begin{aligned}N\gamma^{n-1}D(n+1)!\sum_{r=2}^{n+1}r^{r-1}f(n+1,r)&\leqslant N\gamma^{n-1}(n+1)!DC\sum_{r=2}^{n+1}(n+1)^{n+1}\\ &\leqslant N\gamma^{n-1}DC(n+1)!(n+1)^{n+2}.\end{aligned}$$

Hence, by referring to the γ definition we obtain

$$C_{n+1}(x)\leqslant\gamma^n(n+1)!(n+1)^{n+1},$$

which completes the proof of the step of induction.

Since $n!\leqslant\left(\frac{n+1}{2}\right)^n$, $C_n(x)\leqslant\frac{\gamma^n}{2^n}(n+1)^{2n}$. Thus,

$$m(n,x)=\frac{C_n(x)}{C_1^n(x)}\leqslant\left(\frac{\gamma}{2C_1(x)}\right)^n(n+1)^{2n},$$

from which we obtain that

$$\sum_{n=1}^{\infty}m(n,x)^{-1/2n}\geqslant\sqrt{\frac{2C_1(x)}{\gamma}}\sum_{n=1}^{\infty}\frac{1}{n+1}=\infty.$$

The condition (33) is satisfied, and the corresponding Stieltjes moment problem for the moments $m(n,x)$ has a unique solution [8, Th. 1.11]. Hence, statements (5) are valid in terms of convergence in distribution and Theorem 1 is proved. □

Acknowledgment. The research was supported by the Russian Foundation for the Basic Research (RFBR), project No. 20-01-00487.

References

1. Albeverio, S., Bogachev, L.V., Yarovaya, E.B.: Asymptotics of branching symmetric random walk on the lattice with a single source. C. R. Acad. Sci. Paris Sér. I Math. **326**(8), 975–980 (1998). https://doi.org/10.1016/S0764-4442(98)80125-0
2. Bogachev, L.V., Yarovaya, E.B.: A limit theorem for a supercritical branching random walk on $\mathbf{Z}^d$ with a single source. Russian Math. Surv. **53**(5), 1086–1088 (1998). https://doi.org/10.1070/rm1998v053n05ABEH000077
3. Gikhman, I.I., Skorokhod, A.V.: The Theory of Stochastic Processes. II. Classics in Mathematics. Springer, Berlin (2004).Translated from the Russian by S. Kotz, Reprint of the 1975 edition
4. Halmos, P.R.: A Hilbert Space Problem Book. Graduate Texts in Mathematics, vol. 19, 2nd edn. Springer, New York (1982). https://doi.org/10.1007/978-1-4684-9330-6Encyclopedia of Mathematics and its Applications, 17
5. Horn, R.A., Johnson, C.R.: Matrix Analysis. Cambridge University Press, Cambridge (1985). https://doi.org/10.1017/CBO9780511810817
6. Kato, T.: Perturbation theory for linear operators. Die Grundlehren der mathematischen Wissenschaften, Band 132, Springer, New York (1966). https://doi.org/10.1007/978-3-662-12678-3
7. Khristolyubov, I.I., Yarovaya, E.B.: A limit theorem for a supercritical branching walk with sources of varying intensity. Teor. Veroyatn. Primen. **64**(3), 456–480 (2019). https://doi.org/10.1137/S0040585X97T989556
8. Shohat, J.A., Tamarkin, J.D.: The Problem of Moments. American Mathematical Society Mathematical Surveys, vol. I. American Mathematical Society, New York (1943)
9. Vatutin, V.A., Topchiĭ, V.A., Yarovaya, E.B.: Catalytic branching random walks and queueing systems with a random number of independent servers. Teor. Ĭmovīr. Mat. Stat. **69**, 1–15 (2003)
10. Yarovaya, E.: Spectral asymptotics of a supercritical branching random walk. Theory Probab. Appl. **62**(3), 413–431 (2018). https://doi.org/10.1137/S0040585X97T98871X
11. Yarovaya, E.B.: Branching random walks in a heterogeneous environment. Center of Applied Investigations of the Faculty of Mechanics and Mathematics of the Moscow State University, Moscow (2007). (in Russian)
12. Yarovaya, E.B.: Spectral properties of evolutionary operators in branching random walk models. Math. Notes **92**(1–2), 115–131 (2012). Translation of Mat. Zametki 92 (2012), no. 1, 123–140. https://doi.org/10.1134/S0001434612070139
13. Yarovaya, E.B.: Branching random walks with several sources. Math. Popul. Stud. **20**(1), 14–26 (2013). https://doi.org/10.1080/08898480.2013.748571
14. Yarovaya, E.: Positive discrete spectrum of the evolutionary operator of supercritical branching walks with heavy tails. Methodol. Comput. Appl. Probab. **19**(4), 1151–1167 (2016). https://doi.org/10.1007/s11009-016-9492-9

Efficient Improved Estimation Method for Non-Gaussian Regression from Discrete Data

Evgeny Pchelintsev[1(✉)] and Serguei Pergamenshchikov[2]

[1] Tomsk State University, Tomsk, Russian Federation
evgen-pch@yandex.ru
[2] Université de Rouen Normandie, Rouen, France
https://persona.tsu.ru/Home/UserProfile/393

Abstract. We study a robust adaptive nonparametric estimation problem for periodic functions observed in discrete fixed time moments with non-Gaussian Ornstein–Uhlenbeck noises. For this problem we develop a model selection method, based on the shrinkage (improved) weighted least squares estimates. We found constructive sufficient conditions for the observations frequency under which sharp oracle inequalities for the robust risks are obtained. Moreover, on the basis of the obtained oracle inequalities we establish for the proposed model selection procedures the robust efficiency property in adaptive setting. Then, we apply the constructed model selection procedures to estimation problems in Big Data models in continuous time. Finally, we provide Monte - Carlo simulations confirming the obtained theoretical results.

Keywords: Nonparametric regression · Non-Gaussian Ornstein–Uhlenbeck process · Discrete observations · Improved model selection method · Sharp oracle inequality · Asymptotic efficiency

1 Introduction

In this paper we consider the following nonparametric regression model in continuous time

$$\mathrm{d}y_t = S(t)\mathrm{d}t + \mathrm{d}\xi_t, \quad 0 \le t \le T, \tag{1}$$

where S is an unknown 1-periodic $\mathbb{R} \to \mathbb{R}$ function from $\mathbf{L}_2[0,1]$, the duration of observations T is integer and $(\xi_t)_{t\ge 0}$ is defined by a Ornstein – Uhlenbeck – Lévy defined as

$$\mathrm{d}\xi_t = a\xi_t \mathrm{d}t + \mathrm{d}u_t, \quad u_t = \varrho_1\, w_t + \varrho_2\, z_t, \quad \xi_0 = 0. \tag{2}$$

Here $(w_t)_{t\ge 0}$ is a standard Brownian motion, z_t is a pure jump Lévy process defined through the stochastic integral with respect to the compensated jump measure $\mu(\mathrm{d}s, \mathrm{d}x)$ with deterministic compensator $\widetilde{\mu}(\mathrm{d}s\,\mathrm{d}x) = \mathrm{d}s\Pi(\mathrm{d}x)$, i.e.

$$z_t = x * (\mu - \widetilde{\mu})_t = \int_0^t \int_{\mathbb{R}_*} v\,(\mu - \widetilde{\mu})(\mathrm{d}s\,\mathrm{d}v) \quad \text{and} \quad \mathbb{R}_* = \mathbb{R}\setminus\{0\},$$

A. N. Shiryaev et al. (Eds.): ICSM-5 2020, PROMS 371, pp. 164–177, 2021.
https://doi.org/10.1007/978-3-030-83266-7_12

$\Pi(\cdot)$ is the Lévy measure on $\mathbb{R}_*$, (see, for example in [2]), such that

$$\int_{\mathbb{R}_*} z^2\,\Pi(\mathrm{d}z) = 1 \quad \text{and} \quad \int_{\mathbb{R}_*} z^8\,\Pi(\mathrm{d}z) < \infty.$$

We assume that the unknown parameters $a \le 0$, ϱ_1 and ϱ_2 are such that

$$-\,a_{max} \le a \le 0, \quad 0 < \underline{\varrho} \le \varrho_1^2 \quad \text{and} \quad \sigma_Q = \varrho_1^2 + \varrho_2^2 \le \varsigma^*. \tag{3}$$

Moreover, we assume that the bounds a_{max}, $\underline{\varrho}$ and ς^* are functions of T, i.e. $a_{max} = a_{max}(T)$, $\underline{\varrho} = \underline{\varrho}_T$ and $\varsigma^* = \varsigma^*_T$, for which for any $\epsilon > 0$

$$\lim_{T\to\infty} \frac{a_{max}(T) + \varsigma^*_T}{T^\epsilon} = 0 \quad \text{and} \quad \liminf_{T\to\infty} T^\epsilon\, \underline{\varrho}_T > 0. \tag{4}$$

We denote by $\mathcal{Q}_T$ the family of all distributions of process (1)–(2) on the Skorokhod space $\mathbf{D}[0,n]$ satisfying the conditions (3)–(4). It should be noted that the process (2) is conditionally-Gaussian square integrated semimartingale with respect to σ-algebra $\mathcal{G} = \sigma\{z_t,\, t \ge 0\}$ which is generated by the jump process $(z_t)_{t\ge 0}$.

The problem is to estimate the unknown function S in the model (1) on the basis of observations

$$(y_{t_j})_{0\le j\le n}, \quad t_j = j\varDelta \quad \text{and} \quad \varDelta = \frac{1}{p}, \tag{5}$$

where $n = Tp$ and the observations frequency p is some fixed integer number. For this problem we use the quadratic risk, which for any estimate $\widehat{S}$, is defined as

$$\mathcal{R}_Q(\widehat{S}, S) := \mathbf{E}_{Q,S}\, \|\widehat{S} - S\|^2 \quad \text{and} \quad \|f\|^2 := \int_0^1 f^2(t)\mathrm{d}t, \tag{6}$$

where $\mathbf{E}_{Q,S}$ stands for the expectation with respect to the distribution $\mathbf{P}_{Q,S}$ of the process (1) with a fixed distribution Q of the noise $(\xi_t)_{0\le t\le n}$ and a given function S. Moreover, in the case when the distribution Q is unknown we use also the robust risk

$$\mathcal{R}^*_T(\widehat{S}, S) = \sup_{Q\in\mathcal{Q}_T} \mathcal{R}_Q(\widehat{S}, S). \tag{7}$$

Note that if $(\xi_t)_{t\ge 0}$ is a Brownian motion, then we obtain the well known white noise model (see, for example, [7] and [13]). Later, to take into account the dependence structure in the papers [6] and [10] it was proposed to use the Ornstein – Uhlenbeck noise processes, so called color Gaussian noises. Then, to study the estimation problem for non-Gaussian observations (1) in the papers [9,11] and [12] it was introduced impulse noises defined through the compound Poisson processes with unknown impulse distributions. However, compound Poisson processes can describe the impulse influence of only one fixed frequency and, therefore, such models are too restricted for practical applications. In this paper we consider more general pulse noises described by the Ornstein – Uhlenbeck – Lévy processes.

Our main goal in this paper is to develop improved estimation methods for the incomplete observations, i.e. when the process (1) is available for observations only in the fixed time moments (5). To this end we propose adaptive model selection method based on the improved weighted least squares estimates. For nonparametric estimation problem such approach was proposed in [15] for Lévy regression model.

2 Improved Estimation Method

First, we chose the trigonometric basis $(\phi_j)_{j\geq 1}$ in $\mathbf{L}_2[0,1]$, i.e. $\phi_1 \equiv 1$ and for $j \geq 2$

$$\phi_j(x) = \sqrt{2} \begin{cases} \cos(2\pi[j/2]x) & \text{for even } \; j; \\ \sin(2\pi[j/2]x) & \text{for odd } \; j, \end{cases} \tag{8}$$

where $[a]$ denotes the integer part of a. Note that if p is odd, then for any $1 \leq i,j \leq p$

$$(\phi_i, \phi_j)_p = \frac{1}{p}\sum_{l=1}^{p} \phi_i(t_l)\phi_j(t_l) = \mathbf{1}_{\{i=j\}}. \tag{9}$$

We use this basis to represent the function S on the lattice $\mathcal{T}_p = \{t_1, ..., t_p\}$ in the Fourier expansion form

$$S(t) = \sum_{j=1}^{p} \theta_j \phi_j(t) \quad \text{and} \quad \theta_j = (S, \phi_j)_p := \frac{1}{p}\sum_{k=1}^{p} S(t_k)\phi_j(t_k).$$

The coefficients θ_j can be estimated from the discrete data (5) as

$$\widehat{\theta}_j = \frac{1}{T}\int_0^T \psi_j(t)\, \mathrm{d}y_t \quad \text{and} \quad \psi_j(t) = \sum_{k=1}^{n} \phi_j(t_k)\mathbf{1}_{(t_{k-1}, t_k]}(t).$$

We note that the system of the functions $\{\psi_j\}_{1\leq j\leq p}$ is orthonormal in $\mathbf{L}_2[0,1]$. Now we set weighted least squares estimates for $S(t)$ as

$$\widehat{S}_\gamma(t) = \sum_{j=1}^{p} \gamma(j)\widehat{\theta}_j \psi_j(t) \tag{10}$$

with weights $\gamma = (\gamma(j))_{1\leq j\leq p}$ from a finite set $\Gamma \subset [0,\, 1]^p$. Now for the weight coefficients we introduce the following size characteristics

$$\nu = \#(\Gamma) \quad \text{and} \quad \nu_* = \max_{\gamma\in\Gamma} \sum_{j=1}^{p} \gamma(j),$$

where $\#(\Gamma)$ is the number of the vectors γ in Γ.

Definition 1. *Function* $\mathbf{g}(T)$ *is called slowly increasing as* $T \to \infty$*, if for any* $\epsilon > 0$

$$\lim_{T\to\infty} T^{-\epsilon}\,\mathbf{g}_T = 0.$$

$\mathbf{H}_1$) *For any vector* $\gamma \in \Gamma$ *there exists some fixed integer* $7 \le d = d(\gamma) \le p$ *such that their first* d *components are equal to one, i.e.* $\gamma(j) = 1$ for $1 \le j \le d$. *Moreover, we assume that the parameters* ν and ν_* *are functions of* T*, i.e.* $\nu = \nu(T)$ and $\nu_* = \nu_*(T)$*, and the functions* $\nu(T)$ *and* $T^{-1/3}\nu_*(T)$ *are slowly increasing as* $T \to \infty$.

Using this condition, we define the shrinkage weighted least squares estimates for S

$$S^*_\gamma(t) = \sum_{j=1}^{p} \gamma(j)\theta^*_j \psi_j(t), \quad \theta^*_j = \left(1 - \frac{\mathbf{c}_T}{\sqrt{\sum_{j=1}^{d} \widehat{\theta}^2_j}} \mathbf{1}_{\{1\le j\le d\}}\right) \widehat{\theta}_j, \tag{11}$$

where

$$\mathbf{c}_T = \frac{\varrho_T (d-6)}{2\left(\mathbf{r} + \sqrt{2d\varsigma^*/T}\right) T}$$

and the radius $\mathbf{r} > 0$ may be dependent of T, i.e. $\mathbf{r} = \mathbf{r}_T$ as a slowly increasing function for $T \to \infty$. To compare the estimates (10) and (11) we set

$$d_0 = \inf\{d \ge 7 \,:\, 5 + \ln d \le \check{a} d\} \quad \text{and} \quad \check{a} = \frac{1 - e^{-a_{max}}}{4a_{max}}.$$

Now we can compare the estimators (10) and (11) in mean square accuracy sense.

Theorem 1. *Assume that the condition* $\mathbf{H}_1$) *holds with* $d \ge d_0$. *Then for any* $p \ge d$ *and* $T \ge 3$

$$\sup_{Q\in\mathcal{Q}_T} \sup_{\|S\|\le \mathbf{r}} \left(\mathcal{R}_Q(S^*_\gamma, S) - \mathcal{R}_Q(\widehat{S}_\gamma, S)\right) < -\mathbf{c}_T^2. \tag{12}$$

Remark 1. The inequality (12) means that non-asymptotically, uniformly in $p \ge d$ the estimate (11) outperforms in square accuracy the estimate (10). Such estimators are called improved. Note that firstly for parametric regression models in continuous time similar estimators were proposed in [14] and [12]. Later, for Lévy models in nonparametric setting these methods were developed in [15].

3 Adaptive Model Selection Procedure

To obtain a good estimate from the class (11), we have to choose a weight vector $\gamma \in \Gamma$. The best way is to minimize the empirical squared error

$$\mathrm{Err}_p(\gamma) = \|S^*_\gamma - S\|^2$$

with respect to γ. Since this quantity depends on the unknown function S and, hence, depends on the unknown Fourier coefficients $(\theta_j)_{j\geq 1}$, the weight coefficients $(\gamma_j)_{j\geq 1}$ cannot be found by minimizing one. Then, one needs to replace the corresponding terms by their estimators. For this change in the empirical squared error, one has to pay some penalty. Thus, one comes to the cost function of the form

$$J_p(\gamma) = \sum_{j=1}^{p} \gamma^2(j)(\theta^*_j)^2 - 2\sum_{j=1}^{p} \gamma(j)\left(\theta^*_j\,\widehat{\theta}_j - \frac{\widehat{\sigma}_T}{T}\right) + \rho\,\widehat{P}_T(\gamma). \tag{13}$$

Here ρ is some positive penalty coefficient, $\widehat{P}_T(\gamma)$ is the penalty term is defined as

$$\widehat{P}_T(\gamma) = \frac{\widehat{\sigma}_T}{T}\sum_{j=1}^{p} \gamma^2(j),$$

where $\widehat{\sigma}_T$ is the estimate for the variance σ_Q which is chosen for $\sqrt{T} \leq p \leq T$ in the following form

$$\widehat{\sigma}_T = \frac{T}{p}\sum_{j=[\sqrt{T}]+1}^{p} \widehat{\theta}^2_j. \tag{14}$$

The substituting the weight coefficients, minimizing the cost function (13), in (11) leads to the improved model selection procedure, i.e.

$$S^* = S^*_{\gamma^*} \quad \text{and} \quad \gamma^* = \mathrm{argmin}_{\gamma\in\Gamma}\, J_p(\gamma). \tag{15}$$

It will be noted that γ^* exists because Γ is a finite set. If the minimizing sequence γ^* is not unique, one can take any minimizer. Unlike Pinsker's approach [16], here we do not use the regularity property of the unknown function to find the weights sequence γ^*, i.e. the procedure (15) is adaptive.

Now we study non-asymptotic property of the estimate (15). To this end we assume that

$\mathbf{H}_2$) *The observation frequency p is a function of T, i.e. $p = p(T)$ such that $\sqrt{T} \leq p \leq T$ and for any $\epsilon > 0$*

$$\lim_{T\to\infty} T^{\epsilon - 5/6} p = \infty.$$

First, we study the estimate (14).

Proposition 1. *Assume that the conditions $\mathbf{H}_1$) and $\mathbf{H}_2$) hold and the unknown function S has the square integrated derivative $\dot{S}$. Then for $T \geq 3$ and $\sqrt{T} < p \leq T$*

$$\mathbf{E}_{Q,S}|\widehat{\sigma}_T - \sigma_Q| \leq \mathbf{K}_T T^{-1/3}\left(1 + \|\dot{S}\|^2\right), \tag{16}$$

where the term $\mathbf{K}_T > 0$ is slowly increasing as $T \to \infty$.

Using this Proposition, we come to the following sharp oracle inequality for the robust risk of proposed improved model selection procedure.

Theorem 2. *Assume that the conditions* $\mathbf{H}_1) - \mathbf{H}_2)$ *hold and the function* S *has the square integrable derivative* $\dot{S}$. *Then for any* $T \geq 3$ *and* $0 < \rho < 1/2$ *the robust risk* (7) *of estimate* (15) *satisfies the following sharp oracle inequality*

$$\mathcal{R}_T^*(S^*, S) \leq \frac{1+5\rho}{1-\rho} \min_{\gamma \in \Gamma} \mathcal{R}_T^*(S_\gamma^*, S) + \frac{1}{\rho T} \mathbf{U}_T (1 + \|\dot{S}\|^2),$$

where the rest term $\mathbf{U}_T$ *is slowly increasing as* $T \to \infty$.

We use the condition $\mathbf{H}_1)$ to construct the special set Γ of weight vectors $(\gamma(j))_{j\geq 1}$ as it is proposed in [4] and [5] for which we will study the asymptotic properties of the model selection procedure (15). For this we consider the following grid

$$\mathcal{A}_T = \{1, \dots, \mathbf{k}\} \times \{r_1, \dots, r_m\},$$

where $r_i = i\delta$, $i = \overline{1, m}$ with $m = [1/\delta^2]$. We assume that the parameters $\mathbf{k} \geq 1$ and $0 < \delta \leq 1$ are functions of T, i.e. $\mathbf{k} = \mathbf{k}_T$ and $\delta = \delta(T)$, such that

$$\lim_{T\to\infty} \left(\frac{1}{\mathbf{k}_T} + \frac{\mathbf{k}_T}{\ln T} \right) = 0 \quad \text{and} \quad \lim_{T\to\infty} \left(\delta(T) + \frac{1}{T^\epsilon \delta(T)} \right) = 0$$

for any $\epsilon > 0$. One can take, for example,

$$\delta(T) = \frac{1}{\ln(T+1)} \quad \text{and} \quad \mathbf{k}(T) = k_0 + \sqrt{\ln(T+1)},$$

where $k_0 \geq 0$ is a fixed constant. For $\alpha = (\beta, r) \in \mathcal{A}_T$ we define the weights $\gamma_\alpha = (\gamma_\alpha(j))_{j\geq 1}$ as

$$\gamma_\alpha(j) = \mathbf{1}_{\{1\leq j \leq j_*(\alpha)\}} + \left(1 - (j/\omega_\alpha)^\beta\right) \mathbf{1}_{\{j_*(\alpha) < j \leq \omega_\alpha\}},$$

where $j_*(\alpha) = \omega_\alpha / \ln(T+1)$,

$$\omega_\alpha = \left(\frac{(\beta+1)(2\beta+1)}{\pi^{2\beta}\beta} \, r \, v_T \right)^{1/(2\beta+1)} \quad \text{and} \quad v_T = T/\varsigma^*.$$

Finally, we set

$$\Gamma = \{\gamma_\alpha, \, \alpha \in \mathcal{A}_T\}. \tag{17}$$

Remark 2. It should be noted, that in this case the condition $\mathbf{H}_1)$ holds true with $d = [j_*(\alpha)]$ (see, for example, [15]). Therefore, the model selection procedure (15) with the coefficients (17) satisfies the oracle inequality obtained in Theorem 2.

4 Asymptotic Efficiency

To study the efficiency properties we use the approach proposed by Pinsker in [16], i.e. we assume that the unknown function S belongs to the functional Sobolev ball $W_{k,\mathbf{r}}$ defined as

$$W_{k,\mathbf{r}} = \left\{ f \in \mathcal{C}^{(k)}_{per}[0,1] \,:\, \sum_{j=0}^{k} \|f^{(j)}\|^2 \le \mathbf{r} \right\}, \tag{18}$$

where $\mathbf{r} > 0$ and $k \ge 1$ are some unknown parameters, $\mathcal{C}^{k}_{per}[0,1]$ is the space of k times differentiable 1-periodic $\mathbb{R} \to \mathbb{R}$ functions such that for any $0 \le i \le k-1$ the periodic boundary conditions are satisfied, i.e. $f^{(i)}(0) = f^{(i)}(1)$. It should be noted that the ball $W_{k,\mathbf{r}}$ can be represented as an ellipse in $\mathbb{R}^{\infty}$ through the Fourier representation in $\mathbf{L}_2[0,1]$ for S, i.e.

$$S = \sum_{j=1}^{\infty} \tau_j \phi_j \quad \text{and} \quad \tau_j = \int_0^1 S(t)\phi_j(t)\mathrm{d}t.$$

In this case we can represent the ball (18)

$$W_{k,\mathbf{r}} = \left\{ f \in \mathbf{L}_2[0,1] \,:\, \sum_{j \ge 1} \mathbf{a}_j \tau_j^2 \le \mathbf{r} \right\}, \tag{19}$$

where $\mathbf{a}_j = \sum_{i=0}^{k} \|\phi^{(i)}\|^2 = \sum_{i=0}^{k} (2\pi[j/2])^{2i}$.

To compare the model selection procedure (15) with all possible estimation methods we denote by Σ_T the set of all estimators $\widehat{S}_T$ based on the observations $(y_{t_j})_{0 \le j \le n}$. According to the Pinsker method, firstly one needs to find a lower bound for risks. To this end, we set

$$l_k(\mathbf{r}) \;=\; ((2k+1)\mathbf{r})^{1/(2k+1)} \left(\frac{k}{(\pi(k+1))} \right)^{2k/(2k+1)}. \tag{20}$$

Using this coefficient we obtain the following lower bound.

Theorem 3. *The robust risks* (7) *are bounded from below as*

$$\liminf_{T \to \infty} \upsilon_T^{2k/(2k+1)} \inf_{\widehat{S}_T \in \Sigma_T} \sup_{S \in W_{k,\mathbf{r}}} \mathcal{R}^*_T(\widehat{S}_T, S) \ge l_k(\mathbf{r}), \tag{21}$$

where $\upsilon_T = T/\varsigma^*$.

Remark 3. The lower bound (21) is obtained on the basis of the Van - Trees inequality obtained in [15] for non-Gaussian Lévy processes.

To obtain the upper bound we need the following condition.

$\mathbf{H}_3$) *The parameter* ρ *in the cost function* (13) *is a function of* T, i.e. $\rho = \rho(T)$, *such that* $\lim_{T\to\infty} \rho(T) = 0$ *and*

$$\lim_{T \to \infty} T^{\epsilon} \rho(T) = +\infty$$

for any $\epsilon > 0$

Theorem 4. *Assume that the conditions* $\mathbf{H}_2) - \mathbf{H}_3)$ *hold. Then the model selection procedure* (15) *constructed through the weights* (17) *has the following upper bound*

$$\limsup_{T\to\infty} v_T^{2k/(2k+1)} \sup_{S\in W_{k,\mathbf{r}}} \mathcal{R}_T^*(S^*,S) \le l_k(\mathbf{r}).$$

It is clear that these theorems imply the following efficient property.

Theorem 5. *Assume that the conditions of Theorems 3 and 4 hold. Then the procedure* (15) *is asymptotically efficient, i.e.*

$$\lim_{T\to\infty} v_T^{2k/(2k+1)} \sup_{S\in W_{k,\mathbf{r}}} \mathcal{R}_T^*(S^*,S) = l_k(\mathbf{r})$$

and

$$\lim_{T\to\infty} \frac{\inf_{\widehat{S}_T\in\Sigma_T} \sup_{S\in W_{k,\mathbf{r}}} \mathcal{R}_T^*(\widehat{S}_T,S)}{\sup_{S\in W_{k,\mathbf{r}}} \mathcal{R}_T^*(S^*,S)} = 1. \tag{22}$$

Remark 4. Note that the parameter (20) defining the lower bound (21) is the well-known Pinsker constant, obtained in [16] for the model (1) with the Gaussian white noise process $(\xi_t)_{t\ge 0}$. For general semimartingale models the lower bound is the same as for the white noise model, but generally the normalization coefficient is not the same. In this case the convergence rate is given by $\left(T/\varsigma_T^*\right)^{-2k/(2k+1)}$ while in the white noise model the convergence rate is $(T)^{-2k/(2k+1)}$. So, if the upper variance threshold ς_T^* tends to zero, the convergence rate is better than the classical one, if it tends to infinity, it is worse and, if it is a constant, the rate is the same.

Remark 5. It should be noted that the efficiency property (22) is shown for the procedure (15) without using the Sobolev regularity parameters $\mathbf{r}$ and k, i.e. this procedure is efficient in adaptive setting.

5 Statistical Analysis for the Big Data Model

Now we apply our results for the high dimensional model (1), i.e. we consider this model with the parametric function

$$S(t) = \sum_{j=1}^{q} \beta_j \mathbf{u}_j(t), \tag{23}$$

where the parameter dimension q more than number of observations given in (5), i.e. $q > n$, the functions $(\mathbf{u}_j)_{1\le j\le q}$ are known and orthonormal in $\mathbf{L}_2[0,1]$. In this case we use the estimator (11) to estimate the vector of parameters $\beta = (\beta_j)_{1\le j\le q}$ as

$$\beta_\gamma^* = (\beta_{\gamma,j}^*)_{1\le j\le q} \quad \text{and} \quad \beta_{\gamma,j}^* = (\mathbf{u}_j, S_\gamma^*).$$

Moreover, we use the model selection procedure (15) as

$$\beta^* = (\beta^*_j)_{1\le j\le q} \quad \text{and} \quad \beta^*_j = (\mathbf{u}_j, S^*). \tag{24}$$

It is clear that

$$|\beta^*_\gamma - \beta|^2_q = \sum_{j=1}^{q} (\beta^*_{\gamma,j} - \beta_j)^2 = \|S^*_\gamma - S\|^2$$

and

$$|\beta^* - \beta|^2_q = \|S^* - S\|^2.$$

Therefore, Theorem 2 implies

Theorem 6. *Assume that conditions* $\mathbf{H}_1)$ - $\mathbf{H}_2)$ *hold and the function* (23) *has the square integrable derivative* $\dot{S}$. *Then for any* $T \ge 3$ *and* $0 < \rho < 1/2$

$$\sup_{Q\in\mathcal{Q}_T} \mathbf{E}_{Q,\beta}|\beta^* - \beta|^2_q \le \frac{1+5\rho}{1-\rho} \min_{\gamma\in\Gamma} \sup_{Q\in\mathcal{Q}_T} \mathbf{E}_{Q,\beta}|\beta^* - \beta|^2_q + \frac{1}{\rho T}\mathbf{U}_T(1 + \|\dot{S}\|^2),$$

where the term $\mathbf{U}_T$ *is slowly increasing as* $T \to \infty$.

Theorems 3 and 4 imply the efficiency property for the estimate (24) based on the model selection procedure (15) constructed on the weight coefficients (17) and the penalty threshold satisfying the condition $\mathbf{H}_3)$.

Theorem 7. *Assume that the conditions* $\mathbf{H}_2)$ – $\mathbf{H}_3)$ *hold. Then the estimate* (24) *is asymptotically efficient, i.e.*

$$\lim_{T\to\infty} v_T^{2k/(2k+1)} \sup_{S\in W_{k,\mathbf{r}}} \sup_{Q\in\mathcal{Q}_T} \mathbf{E}_{Q,\beta}|\beta^* - \beta|^2_q = l_k(\mathbf{r}) \tag{25}$$

and

$$\lim_{T\to\infty} \frac{\inf_{\widehat{\beta}_T\in\Xi_T} \sup_{S\in W_{k,\mathbf{r}}} \sup_{Q\in\mathcal{Q}_T} \mathbf{E}_{Q,\beta}|\widehat{\beta}_T - \beta|^2_q}{\sup_{S\in W_{k,\mathbf{r}}} \sup_{Q\in\mathcal{Q}_T} \mathbf{E}_{Q,\beta}|\beta^* - \beta|^2_q} = 1,$$

where Ξ_T *is the set of all possible estimators for the vector* β.

Remark 6. In the estimator (15) doesn't use the dimension q in (23). Moreover, it can be equal to $+\infty$. In this case it is impossible to use neither LASSO method nor Dantzig selector which are usually applied to similar models (see, for example, [17] and [1]). It should be emphasized also that the efficiency property (25) is shown without using any sparse conditions for the parameters $\beta = (\beta_j)_{1\le j\le q}$ usually assumed for such problems (see, for example, [3]).

6 Monte-Carlo Simulations

In this section we give the results of numerical simulations to assess the performance and improvement of the proposed model selection procedure (15). We simulate the model (1) with 1-periodic functions S of the forms

$$S_1(t) = t\,\sin(2\pi t) + t^2(1-t)\cos(4\pi t) \tag{26}$$

and

$$S_2(t) = \sum_{j=1}^{+\infty} \frac{1}{1+j^3} \sin(2\pi j t) \qquad (27)$$

on [0, 1] and the Ornstein – Uhlenbeck – Lévy noise process ξ_t is defined as

$$\mathrm{d}\xi_t = -\xi_t \mathrm{d}t + 0.5\,\mathrm{d}w_t + 0.5\,\mathrm{d}z_t, \quad z_t = \sum_{j=1}^{N_t} Y_j,$$

where N_t is a homogeneous Poisson process of intensity $\lambda = 1$ and $(Y_j)_{j\geq 1}$ is i.i.d. $\mathcal{N}(0,\,1)$ sequence (see, for example, [11]). We use the model selection procedure (15) constructed through the weights (17) in which $\mathbf{k} = 100 + \sqrt{\ln(T+1)}$,

$$r_i = \frac{i}{\ln(T+1)}, \quad m = [\ln^2(T+1)], \quad \rho = \frac{1}{(3+\ln T)^2}$$

and $\varsigma^* = 0.5$ We define the empirical risk as

$$\mathcal{R}(S^*,\,S) = \frac{1}{p}\sum_{j=1}^{p} \widehat{\mathbf{E}}\Delta_T^2(t_j) \quad \text{and} \quad \widehat{\mathbf{E}}\Delta_T^2(t) = \frac{1}{T}\sum_{l=1}^{N} \Delta_{T,l}^2(t),$$

where $\Delta_T(t) = S_T^*(t) - S(t)$ and $\Delta_{T,l}(t) = S_{T,l}^*(t) - S(t)$ is the deviation for the l-th replication. In this example we take $p = T/2$ and $N = 1000$.

Table 1. The sample quadratic risks for different optimal weights

T	200	500	1000	10000
$\mathcal{R}(S^*_{\gamma^*},\,S_1)$	2.8235	0.8454	0.0626	0.0024
$\mathcal{R}(\widehat{S}_{\widehat{\gamma}},\,S_1)$	6.0499	1.8992	0.4296	0.0419
$\mathcal{R}(\widehat{S}_{\widehat{\gamma}},\,S_1)/\mathcal{R}(S^*_{\gamma^*},\,S_1)$	2.1	2.2	6.9	17.7
$\mathcal{R}(S^*_{\gamma^*},\,S_2)$	2.3174	1.0199	0.0817	0.0015
$\mathcal{R}(\widehat{S}_{\widehat{\gamma}},\,S_2)$	7.1047	3.6592	0.8297	0.0299
$\mathcal{R}(\widehat{S}_{\widehat{\gamma}},\,S_2)/\mathcal{R}(S^*_{\gamma^*},\,S_2)$	3.1	3.6	10.2	19.9

Tables 1 and 2 give the sample risks for the improved estimate (15) and the model selection procedure based on the weighted least squares estimates (10) from [11] for different observation period T. One can observe that the improvement increases as T increases for the both models (26) and (27).

Remark 7. The figures show the behavior of the procedures (10) and (11) in the depending on the observation time T. The continuous lines are the functions (26) and (27), the dotted lines are the model selection procedures based on the least squares estimates $\widehat{S}$ and the dashed lines are the improved model selection

Table 2. The sample quadratic risks for the same optimal weights

T	200	500	1000	10000
$\mathcal{R}(S^*_{\widehat{\gamma}}, S_1)$	3.2017	0.9009	0.1284	0.0076
$\mathcal{R}(\widehat{S}_{\widehat{\gamma}}, S_1)$	6.0499	1.8992	0.4296	0.0419
$\mathcal{R}(\widehat{S}_{\widehat{\gamma}}, S_1)/\mathcal{R}(S^*_{\widehat{\gamma}}, S_1)$	1.9	2.1	3.3	5.5
$\mathcal{R}(S^*_{\widehat{\gamma}}, S_2)$	4.1586	1.9822	0.1032	0.0036
$\mathcal{R}(\widehat{S}_{\widehat{\gamma}}, S_2)$	7.1047	3.6592	0.8297	0.0299
$\mathcal{R}(\widehat{S}_{\widehat{\gamma}}, S_2)/\mathcal{R}(S^*_{\widehat{\gamma}}, S_2)$	1.7	1.8	8.0	8.3

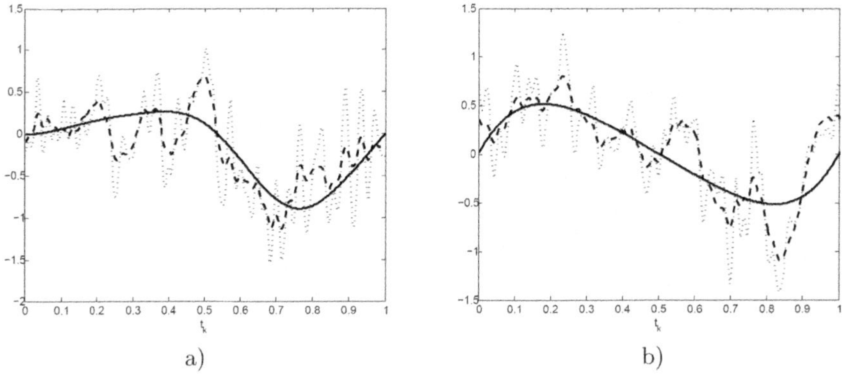

Fig. 1. Behavior of the regression functions and their estimates for $T = 200$ (a) – for the function S_1 and b) – for the function S_2).

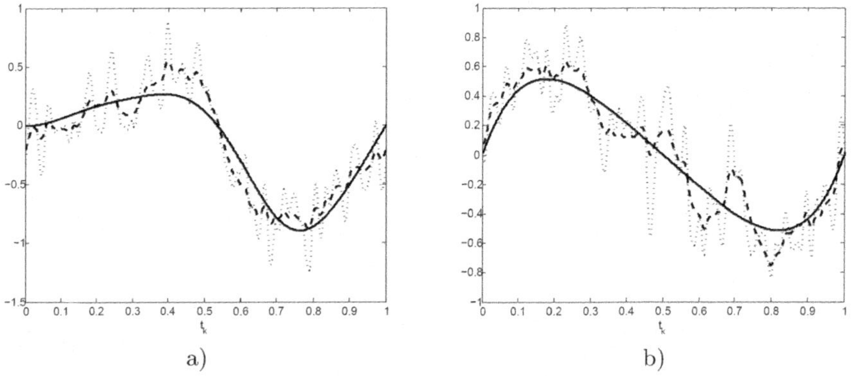

Fig. 2. Behavior of the regressions function and their estimates for $T = 500$ (a) – for the function S_1 and b) – for the function S_2).

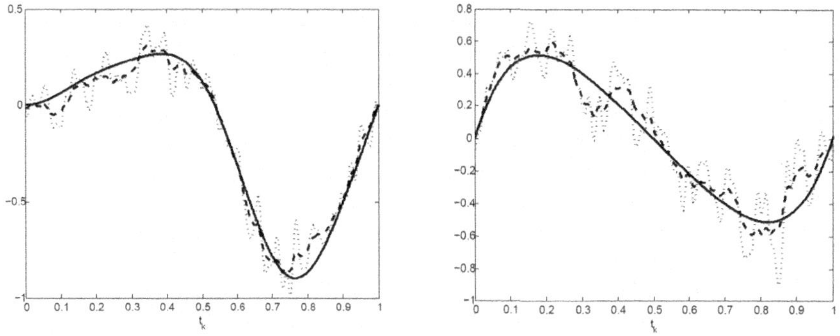

a) b)

Fig. 3. Behavior of the regression functions and their estimates for $T = 1000$ (a) – for the function S_1 and b) – for the function S_2).

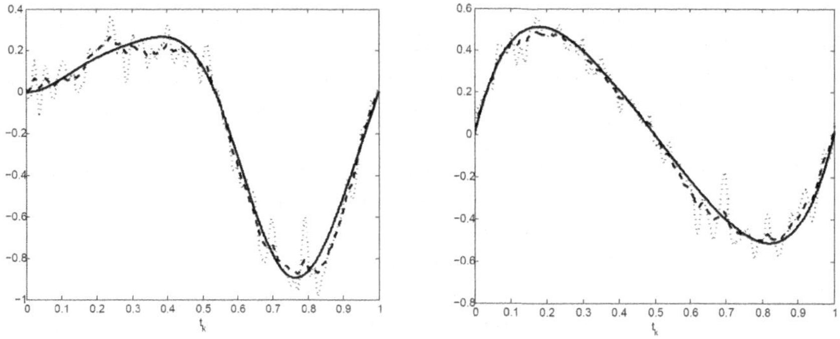

a) b)

Fig. 4. Behavior of the regression functions and their estimates for $T = 10000$ (a) – for the function S_1 and b) – for the function S_2).

procedures S^*. From the Table 2 for the same γ with various observations numbers T we can conclude that theoretical result on the improvement effect (12) is confirmed by the numerical simulations. Moreover, for the proposed shrinkage procedure, from the Table 1 and Figs. 1, 2, 3 and 4, one can be noted that the gain is significant for finite periods T.

7 Conclusion

In the conclusion we would like emphasized that in this paper we studied the following issues:

- we considered the nonparametric estimation problem for continuous time regression model (1) with the noise defined by non-Gaussian Ornstein–Uhlenbeck process with unknown distribution under the condition that this process can be observed only in the fixed discrete time moments (5);

- we proposed adaptive robust improved estimation method via model selection approach and we developed new analytical tools to provide the improvement effect for the non-asymptotic estimation accuracy. It turns out that in this case the accuracy improvement is much more significant than for parametric models, since according to the well-known James–Stein result [8] the accuracy improvement increases when dimension of the parameters increases. It should be noted, that for the parametric models this dimension is always fixed, while for the nonparametric models it tends to infinity, that is, it becomes arbitrarily large with an increase in the number of observations. Therefore, the gain from the application of improved methods is essentially increasing with respect to the parametric case;
- we found constructive conditions for the observation frequency under which we shown sharp non-asymptotic oracle inequalities for the robust risks (7). Then, through the obtained oracle inequalities we provide the efficiency property for the developed model selection methods in adaptive setting, i.e. when the regularity of regression function is unknown;
- we apply the developed model selection procedure to the estimation problem for the Big Data model in continuous time without using the parameter dimension and without assuming any sparse condition for the model parameters ;
- finally, we give Monte - Carlo simulations which confirm the obtained theoretical results.

Acknowledgments. This research was supported by RSF, project no. 20-61-47043.

References

1. Candés, E., Tao, T.: The Dantzig selector: statistical estimation when p is much larger than n. Ann. Stat. **35**, 2313–2351 (2007). https://doi.org/10.1214/009053606000001523
2. Cont, R., Tankov, P.: Financial Modelling with Jump Processes. Chapman & Hall, London (2004)
3. Fan, J., Fan, Y., Barut, E.: Adaptive robust variable selection. Ann. Stat. **42**, 321–351 (2014). https://doi.org/10.1214/13-AOS1191
4. Galtchouk, L., Pergamenshchikov, S.: Sharp non-asymptotic oracle inequalities for nonparametric heteroscedastic regression models. J. Nonparametr. Stat. **21**, 1–18 (2009). https://doi.org/10.1080/10485250802504096
5. Galtchouk, L., Pergamenshchikov, S.: Adaptive asymptotically efficient estimation in heteroscedastic nonparametric regression. J. Korean Stat. Soc. **38**, 305–322 (2009). https://doi.org/10.1016/J.JKSS.2008.12.001
6. Höpfner, R., Kutoyants, Y.A.: On LAN for parametrized continuous periodic signals in a time inhomogeneous diffusion. Stat. Decis. **27**, 309–326 (2009)
7. Ibragimov, I.A., Khasminskii, R.Z.: Statistical Estimation: Asymptotic Theory. Springer, New York (1981). https://doi.org/10.1007/978-1-4899-0027-2
8. James, W., Stein, C.: Estimation with quadratic loss. In: Proceedings of the Fourth Berkeley Symposium Mathematics, Statistics and Probability, vol. 1, pp. 361–380. University of California Press (1961)

9. Kassam, S.A.: Signal Detection in Non-Gaussian Noise. Springer, New York (1988). https://doi.org/10.1007/978-1-4612-3834-8
10. Konev, V.V., Pergamenshchikov, S.M.: General model selection estimation of a periodic regression with a Gaussian noise. Ann. Inst. Stat. Math. **62**, 1083–1111 (2010). https://doi.org/10.1007/s10463-008-0193-1
11. Konev, V.V., Pergamenshchikov, S.M.: Robust model selection for a semimartingale continuous time regression from discrete data. Stoch. Proc. Appl. **125**, 294–326 (2015). https://doi.org/10.1016/j.spa.2014.08.003
12. Konev, V., Pergamenshchikov, S., Pchelintsev, E.: Estimation of a regression with the pulse type noise from discrete data. Theory Probab. Appl. **58**, 442–457 (2014). https://doi.org/10.1137/s0040585x9798662x
13. Kutoyants, Y.A.: Identification of Dynamical Systems with Small Noise. Kluwer Academic Publishers Group, Berlin (1994)
14. Pchelintsev, E.: Improved estimation in a non-Gaussian parametric regression. Stat. Inference Stoch. Process. **16**, 15–28 (2013). https://doi.org/10.1007/s11203-013-9075-0
15. Pchelintsev, E.A., Pchelintsev, V.A., Pergamenshchikov, S.M.: Improved robust model selection methods for a Lévy nonparametric regression in continuous time. J. Nonparametr. Stat. **31**, 612–628 (2019). https://doi.org/10.1080/10485252.2019.1609672
16. Pinsker, M.S.: Optimal filtration of square integrable signals in Gaussian white noise. Probl. Transm. Inf. **17**, 120–133 (1981)
17. Tibshirani, R.: Regression shrinkage and selection via the Lasso. J. R. Stat. Soc. Ser. B. **58**, 267–288 (1996). https://doi.org/10.1111/J.2517-6161.1996.TB02080.X

Two–Sided Bounds for PDF's Maximum of a Sum of Weighted Chi-square Variables

Sergey G. Bobkov[1,2], Alexey A. Naumov[2], and Vladimir V. Ulyanov[2,3(✉)]

[1] University of Minnesota, Vincent Hall 228, 206 Church St SE, Minneapolis, MN 55455, USA
[2] Faculty of Computer Science, HSE University, 109028 Moscow, Russian Federation
[3] Faculty of Computational Mathematics and Cybernetics, Lomonosov Moscow State University, 119991 Moscow, Russian Federation
vulyanov@cs.msu.ru

Abstract. Two–sided bounds are constructed for a probability density function of a weighted sum of chi-square variables. Both cases of central and non-central chi-square variables are considered. The upper and lower bounds have the same dependence on the parameters of the sum and differ only in absolute constants. The estimates obtained will be useful, in particular, when comparing two Gaussian random elements in a Hilbert space and in multidimensional central limit theorems, including the infinite-dimensional case.

Keywords: Two–sided bounds · Weighted sum · Chi-square variable · Gaussian element

1 Introduction

In many statistical and probabilistic applications, we have to solve the problem of Gaussian comparison, that is, one has to evaluate how the probability of a ball under a Gaussian measure is affected, if the mean and the covariance operators of this Gaussian measure are slightly changed. In [1] we present particular examples motivating the results when such "large ball probability" problem naturally arises, including bootstrap validation, Bayesian inference and high-dimensional CLT, see also [2]. The tight non-asymptotic bounds for the Kolmogorov distance between the probabilities of two Gaussian elements to hit a ball in a Hilbert space have been derived in [1] and [3]. The key property of these bounds is that they are dimension-free and depend on the nuclear (Schatten-one) norm of the difference between the covariance operators of the elements and on the norm of the mean shift. The obtained bounds significantly improve the bound based on Pinsker's inequality via the Kullback–Leibler divergence. It was also established an anti-concentration bound for a squared norm $||Z - a||^2$, $a \in \mathbf{H}$, of a shifted Gaussian element Z with zero mean in a Hilbert space $\mathbf{H}$. The decisive role in

A. N. Shiryaev et al. (Eds.): ICSM-5 2020, PROMS 371, pp. 178–189, 2021.
https://doi.org/10.1007/978-3-030-83266-7_13

proving the results was played by the upper estimates for the maximum of the probability density function $g(x, a)$ of $||Z - a||^2$, see Theorem 2.6 in [1]:

$$\sup_{x \geq 0} g(x, a) \leq c\,(\varLambda_1 \varLambda_2)^{-1/4}, \tag{1}$$

where c is an absolute constant and

$$\varLambda_1 = \sum_{k=1}^{\infty} \lambda_k^2, \qquad \varLambda_2 = \sum_{k=2}^{\infty} \lambda_k^2$$

with $\lambda_1 \geq \lambda_2 \geq \ldots$ are the eigenvalues of a covariance operator $\varSigma$ of Z.

It is well known that $g(x, a)$ can be considered as a density function of a weighted sum of non-central χ^2 distributions. An explicit but cumbersome representation for $g(x, a)$ in finite dimensional space $\mathbf{H}$ is available (see, e.g., Sect. 18 in Johnson, Kotz and Balakrishnan [4]). However, it involves some special characteristics of the related Gaussian measure which makes it hard to use in specific situations. Our result (1) is much more transparent and provides sharp uniform upper bounds. Indeed, in the case $\mathbf{H} = \mathbf{R}^d$, $a = 0$, $\varSigma$ is the unit matrix, one has that the distribution of $||Z||^2$ is the standard χ^2 with d degrees of freedom and the maximum of its probability density function is proportional to $d^{-1/2}$. This is the same as what we get in (1).

At the same time, it was noted in [1] that obtaining lower estimates for $\sup_x g(x, a)$ remains an open problem. The latter problem was partially solved in [5], Theorem 1. However, it was done under additional conditions and we took into account the multiplicity of the largest eigenvalue.

In the present paper we get two–sided bounds for $\sup_x g(x, 0)$ in the finite-dimensional case $\mathbf{H} = \mathbf{R}^d$, see Theorem 1 below. The bounds are dimension-free, that is they do not depend on d. Thus, for the upper bounds (1), we obtain a new proof, which is of independent interest. And new lower bounds show the optimality of (1), since the upper and lower bounds differ only in absolute constants. Moreover, new two-sided bounds are constructed for $\sup_x g(x, a)$ with $a \neq 0$ in the finite-dimensional case $\mathbf{H} = \mathbf{R}^d$, see Theorem 2 below. Here we consider a typical situation, where λ_1 does not dominate the other coefficients.

2 Main Results

For independent standard normal random variables $Z_k \sim N(0, 1)$, consider the weighted sum

$$W_0 = \lambda_1 Z_1^2 + \cdots + \lambda_n Z_n^2, \qquad \lambda_1 \geq \cdots \geq \lambda_n > 0.$$

It has a continuous probability density function $p(x)$ on the positive half-axis. Define the functional

$$M(W_0) = \sup_x p(x).$$

Theorem 1. *Up to some absolute constants c_0 and c_1, we have*

$$c_0(A_1A_2)^{-1/4} \le M(W_0) \le c_1(A_1A_2)^{-1/4}, \tag{2}$$

where

$$A_1 = \sum_{k=1}^{n} \lambda_k^2, \qquad A_2 = \sum_{k=2}^{n} \lambda_k^2$$

and

$$c_0 = \frac{1}{4e^2\sqrt{2\pi}} > 0.013, \qquad c_1 = \frac{2}{\sqrt{\pi}} < 1.129.$$

Theorem 1 can be extended to more general weighted sums:

$$W_a = \lambda_1(Z_1 - a_1)^2 + \cdots + \lambda_n(Z_n - a_n)^2 \tag{3}$$

with parameters $\lambda_1 \ge \cdots \ge \lambda_n > 0$ and $a = (a_1, \ldots, a_n) \in \mathbf{R}^n$.

It has a continuous probability density function $p(x, a)$ on the positive half-axis $x > 0$. Define the functional

$$M(W_a) = \sup_x p(x, a).$$

Remark. It is known that for any non-centred Gaussian element Y in a Hilbert space, the random variable $||Y||^2$ is distributed as $\sum_{i=1}^{\infty} \lambda_i(Z_i - a_i)^2$ with some real a_i and λ_i such that

$$\lambda_1 \ge \lambda_2 \ge \cdots \ge 0 \quad \text{and} \quad \sum_{i=1}^{\infty} \lambda_i < \infty.$$

Therefore, the upper bounds for $M(W_a)$ immediately imply the upper bounds for the probability density function of $||Y||^2$.

Theorem 2. *If $\lambda_1^2 \le A_1/3$, then one has a two-sided bound*

$$\frac{1}{4\sqrt{3}} \frac{1}{\sqrt{A_1 + B_1}} \le M(W_a) \le \frac{2}{\sqrt{A_1 + B_1}},$$

where

$$A_1 = \sum_{k=1}^{n} \lambda_k^2, \qquad B_1 = \sum_{k=1}^{n} \lambda_k^2 a_k^2.$$

Moreover, the left inequality holds true without any assumption on λ_1^2.

Remark. In Theorem 2 we only consider a typical situation, where λ_1 does not dominate the other coefficients. Moreover, the condition $\lambda_1^2 \le A_1/3$ necessarily implies that $n \ge 3$. If this condition is violated, the behaviour of $M(W_a)$ should be studied separately.

3 Auxiliary Results

For the lower bounds in the theorems, one may apply the following lemma, which goes back to the work by Statulyavichus [6], see also Proposition 2.1 in [7].

Lemma 1. *Let η be a random variable with $M(\eta)$ denoting the maximum of its probability density function. Then one has*

$$M^2(\eta)\,\mathrm{Var}(\eta) \geq \frac{1}{12}. \tag{4}$$

Moreover, the equality in (4) is attained for the uniform distribution on any finite interval.

Remark. There are multidimensional extensions of (4), see e.g. [8,9] and Section III in [10].

Proof. Without loss of generality we may assume that $M(\eta) = 1$.

Put $H(x) = \mathbf{P}(|\eta - \mathbf{E}\eta| \geq x), \quad x \geq 0$.

Then, $H(0) = 1$ and $H'(x) \geq -2$, which gives $H(x) \geq 1 - 2x$, so

$$\begin{aligned}\mathrm{Var}(\eta) &= 2\int_0^\infty xH(x)\,dx \geq 2\int_0^{1/2} xH(x)\,dx \\ &\geq 2\int_0^{1/2} x(1-2x)\,dx = \frac{1}{12}.\end{aligned}$$

Lemma is proved.

The following lemma will give the lower bound in Theorem 2.

Lemma 2. *For the random variable W_a defined in (3), the maximum $M(W_a)$ of its probability density function satisfies*

$$M(W_a) \geq \frac{1}{4\sqrt{3}}\,\frac{1}{\sqrt{A_1 + B_1}}, \tag{5}$$

where

$$A_1 = \sum_{k=1}^n \lambda_k^2, \qquad B_1 = \sum_{k=1}^n \lambda_k^2 a_k^2.$$

Proof. Given $Z \sim N(0,1)$ and $b \in \mathbf{R}$, we have

$$\mathbf{E}\,(Z-b)^2 = 1 + b^2, \qquad \mathbf{E}\,(Z-b)^4 = 3 + 6b^2 + b^4,$$

so that $\mathrm{Var}((Z-b)^2) = 2 + 4b^2$. It follows that

$$\mathrm{Var}(W_a) = \sum_{k=1}^n \lambda_k^2\,(2 + 4a_k^2) = 2A_1 + 4B_1 \leq 4(A_1 + B_1).$$

Applying (4) with $\eta = W_a$, we arrive at (5).

Lemma is proved.

The proofs of the upper bounds in the theorems are based on the following lemma.

Lemma 3. *Let*

$$\alpha_1^2 + \cdots + \alpha_n^2 = 1.$$

If $\alpha_k^2 \leq 1/m$ *for* $m = 1, 2, \dots$, *then the characteristic function* $f(t)$ *of the random variable*

$$W = \alpha_1 Z_1^2 + \cdots + \alpha_n Z_n^2$$

satisfies

$$|f(t)| \leq \frac{1}{(1 + 4t^2/m)^{m/4}}. \tag{6}$$

In particular, in the cases $m = 4$ *and* $m = 3$, W *has a bounded density with* $M(W) \leq 1/2$ *and* $M(W) < 0.723$ *respectively.*

Proof. Necessarily $n \geq m$. The characteristic function has the form

$$f(t) = \prod_{k=1}^{n} (1 - 2\alpha_k it)^{-1/2},$$

so

$$-\log |f(t)| = \frac{1}{4} \sum_{k=1}^{n} \log(1 + 4\alpha_k^2 t^2).$$

First, let us describe the argument in the simplest case $m = 1$. For a fixed t, consider the concave function

$$V(b_1, \dots, b_n) = \sum_{k=1}^{n} \log(1 + 4b_k t^2)$$

on the simplex

$$Q_1 = \Big\{(b_1, \dots, b_n) : b_k \geq 0, \ b_1 + \cdots + b_n = 1\Big\}.$$

It has n extreme points $b^k = (0, \dots, 0, 1, 0, \dots, 0)$. Hence

$$\min_{b \in Q_1} V(b) = V(b^k) = \log(1 + 4t^2),$$

that is, $|f(t)| \leq (1 + 4t^2)^{-1/4}$, which corresponds to (6) for $m = 1$.

If $m = 2$, we consider the same function V on the convex set

$$Q_2 = \Big\{(b_1, \dots, b_n) : 0 \leq b_k \leq \frac{1}{2}, \ b_1 + \cdots + b_n = 1\Big\},$$

which is just the intersection of the cube $[0, \frac{1}{2}]^n$ with the hyperplane. It has $n(n-1)/2$ extreme points

$$b^{kj}, \quad 1 \leq k < j \leq n,$$

with coordinates $1/2$ on the j-th and k-th places and with zero elsewhere. Indeed, suppose that a point

$$b = (b_1, \dots, b_n) \in Q_2$$

has at least two non-zero coordinates $0 < b_k, b_j < 1/2$ for some $k < j$. Let x be the point with coordinates

$$x_l = b_l \quad \text{for} \quad l \neq k, j, \ \ x_k = b_k + \varepsilon, \quad \text{and} \quad x_j = b_j - \varepsilon,$$

and similarly, let y be the point such that

$$y_l = b_l \quad \text{for} \quad l \neq k, j, \ \ y_k = b_k - \varepsilon, \quad \text{and} \quad y_j = b_j + \varepsilon.$$

If $\varepsilon > 0$ is small enough, then both x and y lie in Q_2, while

$$b = (x + y)/2, \quad x \neq y.$$

Hence such b cannot be an extreme point. Equivalently, any extreme point b of Q_2 is of the form

$$b^{kj}, \quad 1 \leq k < j \leq n.$$

Therefore, we conclude that

$$\min_{b \in Q_2} V(b) = V(b^{kj}) = 2\log(1 + 2t^2),$$

which is the first desired claim.

In the general case, consider the function V on the convex set

$$Q_m = \Big\{(b_1, \dots, b_n) : 0 \leq b_k \leq \frac{1}{m}, \ b_1 + \dots + b_n = 1\Big\}.$$

By a similar argument, any extreme point b of Q_m has zero for all coordinates except for m places where the coordinates are equal to $1/m$. Therefore,

$$\min_{b \in Q_m} V(b) = V\Big(\frac{1}{m}, \dots, \frac{1}{m}, 0, \dots, 0\Big) = m\log(1 + 4t^2/m),$$

and we are done.

In case $m = 4$, using the inversion formula, we get

$$M(W) \leq \frac{1}{2\pi} \int_{-\infty}^{\infty} |f(t)|\, dt \leq \frac{1}{2\pi} \int_{-\infty}^{\infty} \frac{1}{1+t^2}\, dt = \frac{1}{2}.$$

Similarly, in the case $m = 3$,

$$M(W) \leq \frac{1}{2\pi} \int_{-\infty}^{\infty} \frac{1}{(1 + \frac{4}{3}\, t^2)^{3/4}}\, dt < 0.723.$$

Lemma is proved.

4 Proofs of Main Results

Proof of Theorem 1. In the following we shall write W instead of W_0.

If $n = 1$, then the distribution function and the probability density function of $W = \lambda_1 Z_1^2$ are given by

$$F(x) = 2\Phi\Big(\sqrt{\frac{x}{\lambda_1}}\Big) - 1, \quad p(x) = \frac{1}{\sqrt{2\pi\lambda_1}}\, e^{-x/(2\lambda_1)} \qquad (x > 0),$$

respectively. Therefore, p is unbounded near zero, so that $M(W) = \infty$. This is consistent with (2), in which case $A_1 = \lambda_1^2$ and $A_2 = 0$.

If $n = 2$, the density $p(x)$ is described as the convolution

$$p(x) \ = \ \frac{1}{2\pi\sqrt{\lambda_1\lambda_2}} \int_0^1 \frac{1}{\sqrt{(1-t)t}} \exp\Big\{ -\frac{x}{2}\Big[\frac{1-t}{\lambda_1} + \frac{t}{\lambda_2}\Big]\Big\}\, dt \qquad (x > 0). \quad (7)$$

Hence, p is decreasing and attains maximum at $x = 0$:

$$M(W) = \frac{1}{2\pi\sqrt{\lambda_1\lambda_2}} \int_0^1 \frac{1}{\sqrt{(1-t)t}}\, dt = \frac{1}{2\sqrt{\lambda_1\lambda_2}}.$$

Since $A_1 = \lambda_1^2 + \lambda_2^2$ and $A_2 = \lambda_2^2$, we conclude, using the assumption $\lambda_1 \geq \lambda_2$, that

$$\frac{1}{2}\,(A_1A_2)^{-1/4} \leq M(W) \leq \frac{1}{2^{3/4}}\,(A_1A_2)^{-1/4}.$$

As for the case $n \geq 3$, the density p is vanishing at zero and attains maximum at some point $x > 0$.

The further proof of Theorem 1 is based on the following observations and Lemma 3.

By homogeneity of (2), we may assume that $A_1 = 1$.

If $\lambda_1 \leq 1/2$, then all $\lambda_k^2 \leq 1/4$, so that $M(W) \leq 1/2$, by Lemma 3. Hence, the inequality of the form

$$M(W) \leq \frac{1}{2}\,(A_1A_2)^{-1/4}$$

holds true.

Now, let $\lambda_1 \geq 1/2$, so that $A_2 \leq 3/4$. Write

$$W = \lambda_1 Z_1^2 + \sqrt{A_2}\,\xi, \qquad \xi = \sum_{k=2}^n \alpha_k Z_k^2, \quad \alpha_k = \frac{\lambda_k}{\sqrt{A_2}}.$$

By construction, $\alpha_2^2 + \dots + \alpha_n^2 = 1$.

Case 1: $\lambda_2 \geq \sqrt{A_2}/2$. Since the function $M(W)$ may only decrease when adding an independent random variable to W, we get using (7) that

$$M(W) \leq M(\lambda_1 Z_1^2 + \lambda_2 Z_2^2) = \frac{1}{2\sqrt{\lambda_1\lambda_2}} \leq c\,(A_1A_2)^{-1/4},$$

where the last inequality holds with $c = 1$. This gives the upper bound in (2) with constant 1.

Case 2: $\lambda_2 \le \sqrt{A_2}/2$. It implies that $n \ge 5$ and all $\alpha_k^2 \le 1/4$ for $k > 1$. By Lemma 3 with $m = 4$, the random variable ξ has the probability density function q bounded by $1/2$. The distribution function of W may be written as

$$\mathbf{P}\{W \le x\} = \int_0^{x/\sqrt{A_2}} \mathbf{P}\Big\{|Z_1| \le \frac{1}{\sqrt{\lambda_1}}\,(x - y\sqrt{A_2})^{1/2}\Big\}\, q(y)\, dy, \quad x > 0,$$

and its density has the form

$$p(x) = \frac{1}{\sqrt{2\pi\lambda_1}} \int_0^{x/\sqrt{A_2}} \frac{1}{\sqrt{x - y\sqrt{A_2}}}\, e^{-(x - y\sqrt{A_2})/(2\lambda_1)}\, q(y)\, dy.$$

Equivalently,

$$p(x\sqrt{A_2}) = \frac{1}{\sqrt{2\pi\lambda_1}}\, A_2^{-1/4} \int_0^x \frac{1}{\sqrt{x-y}}\, e^{-(x-y)\sqrt{A_2}/(2\lambda_1)}\, q(y)\, dy. \tag{8}$$

Since $\lambda_1 \ge 1/2$, we immediately obtain that

$$M(W) \le A_2^{-1/4}\, \frac{1}{\sqrt{\pi}}\, \sup_{x>0} \int_0^x \frac{1}{\sqrt{x-y}}\, q(y) dy.$$

But, using $q \le 1/2$, we get

$$\begin{aligned}\int_0^x \frac{1}{\sqrt{x-y}}\, q(y) dy &= \int_{0<y<x,\ x-y<1} \frac{1}{\sqrt{x-y}}\, q(y) dy \\ &\quad + \int_{0<y<x,\ x-y>1} \frac{1}{\sqrt{x-y}}\, q(y) dy \\ &\le \frac{1}{2} \int_0^1 \frac{1}{\sqrt{z}}\, dz + 1 \,=\, 2.\end{aligned}$$

Thus,

$$M(W) \le 2A_2^{-1/4}\, \frac{1}{\sqrt{\pi}}.$$

Combining the obtained upper bounds for $M(W)$ in all cases we get the upper bound in (2).

For the lower bound, one may apply the inequality (4) in Lemma 1. Thus, we obtain that

$$M(W) \ge \frac{1}{2\sqrt{6}}$$

due to the assumption $A_1 = 1$ and the property $\mathrm{Var}(Z_1^2) = 2$.

If $\lambda_1^2 \le 1/2$, we have $A_2 \ge 1/2$. Hence,

$$M(W) \ge \frac{1}{2\sqrt{6}} \ge c_0\, (A_1 A_2)^{-1/4}, \tag{9}$$

where the last inequality holds true with

$$c_0 = \frac{1}{2^{5/4}\sqrt{6}} \geq 0.171.$$

In case $\lambda_1^2 \geq \frac{1}{2}$, we have $A_2 \leq 1/2$. Returning to the formula (8), let us choose $x = \mathbf{E}\xi + 2$ and restrict the integration to the interval

$$\Delta : \max(\mathbf{E}\xi - 2, 0) < y < \mathbf{E}\xi + 2.$$

On this interval necessarily

$$x - y \leq 4.$$

Therefore, (8) yields

$$M(W) \geq \frac{A_2^{-1/4}}{2\sqrt{2\pi\lambda_1}} \cdot e^{-2\sqrt{A_2}/\lambda_1}\, \mathbf{P}\{\xi \in \Delta\}.$$

Here,

$$\frac{A_2}{\lambda_1^2} = \frac{1}{\lambda_1^2} - 1 \leq 1,$$

and we get

$$M(W) \geq \frac{A_2^{-1/4}}{2\sqrt{2\pi}} \cdot e^{-2}\, \mathbf{P}\{\xi \in \Delta\}.$$

Now, recall that $\xi \geq 0$ and $\mathrm{Var}(\xi) = 2\,(\alpha_2^2 + \cdots + \alpha_n^2) = 2$. Hence, by Chebyshev's inequality,

$$\mathbf{P}\{|\xi - \mathbf{E}\xi| \geq 2\} \leq \frac{1}{4}\,\mathrm{Var}(\xi) = \frac{1}{2}.$$

That is, $\mathbf{P}\{\xi \in \Delta\} \geq 1/2$, and thus

$$M(W) \geq \frac{(A_1 A_2)^{-1/4}}{4\sqrt{2\pi}}\, e^{-2} \geq 0.013 \cdot (A_1 A_2)^{-1/4}.$$

Theorem 1 is proved.

Proof of Theorem 2. In the following we shall write W instead of W_a.

The lower bound in Theorem 2 immediately follows from (5) in Lemma 2 without any assumption on λ_1^2.

Our next aim is to reverse this bound up to a numerical factor under suitable natural assumptions.

Without loss of generality, let $A_1 = 1$. Our basic condition will be that $\lambda_1^2 \leq 1/3$, similarly to the first part of the proof of Theorem 1. Note that if $\lambda_1^2 \leq 1/3$ then necessarily $n \geq 3$.

As easy to check, for $Z \sim N(0,1)$ and $a \in \mathbf{R}$,

$$\mathbf{E}\, e^{it\,(Z-a)^2} = \frac{1}{\sqrt{1-2it}}\, \exp\Big\{a^2\, \frac{it}{1-it}\Big\}, \qquad t \in \mathbf{R},$$

so that

$$\Big|\mathbf{E}\, e^{it\,(Z-a)^2}\Big| = \frac{1}{(1+4t^2)^{1/4}}\,\exp\Big\{-2a^2\,\frac{t^2}{1+4t^2}\Big\}.$$

Hence, the characteristic function $f(t)$ of W satisfies

$$-\log|f(t)| = \frac{1}{4}\sum_{k=1}^n \log(1+4\lambda_k^2 t^2) + 2\sum_{k=1}^n a_k^2\,\frac{\lambda_k^2 t^2}{1+4\lambda_k^2 t^2}.$$

Since $\lambda_1^2 \leq \frac{1}{3}$, by the monotonicity, all $\lambda_k^2 \leq \frac{1}{3}$ as well. But, as we have already observed, under the conditions

$$0 \leq b_k \leq \frac{1}{3}, \quad b_1 + \dots + b_k = 1,$$

and for any fixed value $t \in \mathbf{R}$, the function

$$\psi(b_1,\dots,b_n) = \sum_{k=1}^n \log(1+4b_k t^2)$$

is minimized for the vector with coordinates

$$b_1 = b_2 = b_3 = \frac{1}{3} \quad \text{and} \quad b_k = 0 \quad \text{for} \quad k > 3.$$

Hence,

$$\psi(b_1,\dots,b_n) \geq 3\,\log(1+4t^2/3) \geq 3\,\log(1+t^2).$$

Therefore, one may conclude that

$$|f(t)| \leq \frac{1}{(1+t^2)^{3/4}}\,\exp\Big\{-2\sum_{k=1}^n a_k^2\,\frac{\lambda_k^2 t^2}{1+4\lambda_k^2 t^2}\Big\}. \tag{10}$$

It is time to involve the inversion formula which yields the upper bound

$$M(W) \leq \frac{1}{\pi}\int_0^\infty |f(t)|\,dt. \tag{11}$$

In the interval

$$0 < t < T = \frac{1}{2\lambda_1},$$

we have $\lambda_k^2 t^2 \leq 1/4$ for all k, and the bound (8) is simplified to

$$|f(t)| \leq \frac{1}{(1+t^2)^{3/4}}\,e^{-B_1 t^2}.$$

This gives

$$\int_0^T |f(t)|\,dt \leq I(B_1) \equiv \int_0^\infty \frac{1}{(1+t^2)^{3/4}}\,e^{-B_1 t^2}\,dt.$$

If $B_1 \le 1$,

$$I(B_1) \le \int_0^\infty \frac{1}{(1+t^2)^{3/4}}\,dt < 3,$$

while for $B_1 \ge 1$,

$$I(B_1) \le \int_0^\infty e^{-B_1 t^2}\,dt = \frac{\sqrt{\pi}}{2\sqrt{B_1}} < \frac{1}{\sqrt{B_1}}.$$

The two estimates can be united by

$$I(B_1) \le \frac{3\sqrt{2}}{\sqrt{1+B_1}}.$$

To perform the integration over the half-axis $t \ge T$, a different argument is needed. Put $p_k = a_k^2 \lambda_k^2 / B_1$, so that $p_k \ge 0$ and $p_1 + \dots + p_k = 1$. By Jensen's inequality applied to the convex function $V(x) = 1/(1+x)$ for $x \ge 0$ with points $x_k = 4\lambda_k^2 t^2$, we have

$$\begin{aligned}
\sum_{k=1}^n a_k^2 \frac{\lambda_k^2 t^2}{1+4\lambda_k^2 t^2} &= B_1 t^2 \sum_{k=1}^n p_k V(x_k) \\
&\ge B_1 t^2\, V(p_1 x_1 + \dots p_n x_n) \\
&= \frac{B_1 t^2}{1 + \frac{4t^2}{B_1} \sum_{k=1}^n a_k^2 \lambda_k^4} \ge \frac{B_1 t^2}{1 + \frac{4t^2}{3B_1} \sum_{k=1}^n a_k^2 \lambda_k^2} = \frac{B_1 t^2}{1 + \frac{4}{3} t^2},
\end{aligned}$$

where we used the property $\lambda_k^2 \le 1/3$. Moreover, since

$$t^2 \ge \frac{1}{(2\lambda_1)^2} \ge \frac{3}{4},$$

necessarily

$$\frac{t^2}{1 + \frac{4}{3} t^2} \ge \frac{3}{8}.$$

Hence, from (10) we get

$$|f(t)| \le \frac{1}{(1+t^2)^{3/4}}\, e^{-3B_1/4}, \quad t \ge T,$$

and

$$\int_T^\infty |f(t)|\,dt \le e^{-3B_1/4} \int_{\sqrt{3}/2}^\infty \frac{1}{(1+t^2)^{3/4}}\,dt < 1.68\, e^{-3B_1/4} < \frac{1.85}{\sqrt{1+B_1}}.$$

Combining the two estimates together for different regions of integration with $(3\sqrt{2} + 1.85)/\pi < 1.94$, the bound (11) leads to

$$M(W) < \frac{2}{\sqrt{A_1 + B_1}}.$$

Thus, this inequality, together with Lemma 2, completes the proof of the theorem.

Acknowledgements. Theorem 1 has been obtained under support of the Ministry of Education and Science of the Russian Federation as part of the program of the Moscow Center for Fundamental and Applied Mathematics under the agreement № 075-15-2019-1621. Theorem 2 was proved within the framework of the HSE University Basic Research Program. Research of S. Bobkov was supported by the NSF grant DMS-1855575.

References

1. Götze, F., Naumov, A.A., Spokoiny, V.G., Ulyanov, V.V.: Large ball probabilities, Gaussian comparison and anti-concentration. Bernoulli **25**(4A), 2538–2563 (2019). https://doi.org/10.3150/18-BEJ1062
2. Prokhorov, Y., Ulyanov, V.: Some approximation problems in statistics and probability. In: Limit Theorems in Probability, Statistics and Number Theory. Springer Proceedings in Mathematics and Statistics, vol. 42, pp. 235–249. Springer, Heidelberg (2013). https://doi.org/10.1007/978-3-642-36068-8_11
3. Naumov, A.A., Spokoiny, V.G., Tavyrikov, Y.E., Ulyanov, V.V.: Nonasymptotic estimates for the closeness of Gaussian measures on balls. Doklady Math. **98**(2), 490–493 (2018). https://doi.org/10.1134/S1064562418060248
4. Johnson, N., Kotz, S., Balakrishnan, N.: Continuous Univariate Distributions, vol. 1. Wiley, New York (1994)
5. Christoph, G., Prokhorov, Y.V., Ulyanov, V.V.: On distribution of quadratic forms in Gaussian random variables. Theory Probab. Appl. **40**(2), 250–260 (1996). https://doi.org/10.1137/1140028
6. Statulyavichus, V.A.: Limit theorems for densities and asymptotic expansions for distributions of sums of independent random variables. Theory Probab. Appl. **10**(4), 582–595 (1965). https://doi.org/10.1137/1110074
7. Bobkov, S.G., Chistyakov, G.P.: On concentration functions of random variables. J. Theor. Probab. **28**(3), 976–988 (2013). https://doi.org/10.1007/s10959-013-0504-1
8. Ball, K.: Logarithmically concave functions and sections of convex sets in R^n. Studia Math. **88**(1), 69–84 (1988)
9. Hensley, D.: Slicing convex bodies-bounds for slice area in terms of the body's covariance. Proc. Am. Math. Soc. **79**(4), 619–625 (1980). https://doi.org/10.2307/2042510
10. Bobkov, S., Madiman, M.: The entropy per coordinate of a random vector is highly constrained under convexity conditions. IEEE Trans. Inf. Theory **57**(8), 4940–4954 (2011). https://doi.org/10.1109/TIT.2011.2158475

On the Chromatic Number of a Random 3-Uniform Hypergraph

Yury A. Demidovich[1(✉)] and Dmitry A. Shabanov[1,2,3]

[1] Moscow Institute of Physics and Technology (National Research University), Dolgoprudnyi, Moscow Region, Russia
{demidovich.yua,dmitry.shabanov}@phystech.edu
[2] Lomonosov Moscow State University, Moscow, Russian Federation
[3] HSE University, Moscow, Russian Federation

Abstract. This paper is devoted to the problem concerning the chromatic number of a random 3-uniform hypergraph. We consider the binomial model $H(n, 3, p)$ and show that if $p = p(n)$ decreases fast enough then the chromatic number of $H(n, 3, p)$ is concentrated in 2 or 3 consecutive values which can be found explicitly as functions of n and p. This result is derived as an application of the solution of an extremal problem for doubly stochastic matrices.

Keywords: Random hypergraphs · Colorings · Second moment method · Doubly stochastic matrices

1 Introduction

The theory of random graphs and hypergraphs was always in the focus of study in probabilistic combinatorics. Recall that a *hypergraph* H is a pair of sets $H = (V, E)$, where V is a finite set whose elements are called *vertices*, and E is a family of subsets of V that are called edges of the hypergraph. If every edge consists of k vertices then a hypergraph is called *k-uniform*. An *r-coloring* of a vertex set is an arbitrary mapping $f : V \to \{1, \dots, r\}$. It is said to be *proper* if no edge is monochromatic. The *chromatic number* $\chi(H)$ of a hypergraph H is the minimum number of colors required for a proper coloring of H.

One of main stochastic models of random hypergraphs is the well-known binomial model of a random k-uniform hypergraph $H(n, k, p)$, which can be viewed as the Bernoulli scheme on k-subsets of an n-element set: every subset is included into $H(n, k, p)$ as an edge independently with probability p. We study the asymptotic behaviour of the chromatic number of $H(n, k, p)$ for large n, when k is fixed, and $p = p(n)$ is a function of n.

1.1 Related Work

The chromatic numbers of random graphs and hypergraphs have been intensively studied since the 1970s. For known results concerning $\chi(H(n, k, p))$ in the graph

A. N. Shiryaev et al. (Eds.): ICSM-5 2020, PROMS 371, pp. 190–203, 2021.
https://doi.org/10.1007/978-3-030-83266-7_14

case, $k=2$, the reader is referred to the recent paper [6]. In the current paper we concentrate only on the case $k \geq 3$. The asymptotics of the chromatic number of $H(n,k,p)$ in the dense case, when the expected number of edges is much larger than the number of vertices, i.e. when $pn^{k-1} \to +\infty$, was obtained by Shamir with coauthors [8,10] and by Krivelevich, Sudakov [7]. But we know much more about the limit distribution of $\chi(H(n,k,p))$ in the sparse case, when the expected number of edges is a linear function of n, i.e. $p = cn/\binom{n}{k}$ and $c>0$ does not depend on n. Dyer, Frieze and Greenhill [5] proved that in this case $\chi(H(n,k,p))$ is concentrated in two consecutive numbers, moreover, for some values, there is a concentration in exactly one number. For given $c>0$, let us denote $r_c = \min\{r \in \mathbb{N} : c < r^{k-1}\ln r\}$. Clearly, $c \in [(r_c-1)^{k-1}\ln(r_c-1), r_c^{k-1}\ln r_c)$. The authors of [5] established that

– if $c > r_c^{k-1}\ln r_c - \frac{1}{2}\ln r_c$ then

$$\mathsf{P}(\chi(H(n,k,p)) = r_c + 1) \to 1 \text{ as } n \to \infty;$$

– if $c < r_c^{k-1}\ln r_c - \frac{r_c-1}{r_c}(1+\ln r_c) - O\left(k^2 r_c^{1-k}\ln r_c\right)$ then

$$\mathsf{P}(\chi(H(n,k,p)) = r_c) \to 1 \text{ as } n \to \infty;$$

– if $c \in [r_c^{k-1}\ln r_c - \frac{r_c-1}{r_c}(1+\ln r_c) - O\left(k^2 r_c^{1-k}\ln r_c\right), r_c^{k-1}\ln r_c - \frac{1}{2}\ln r_c]$ then

$$\mathsf{P}(\chi(H(n,k,p)) \in \{r_c, r_c+1\}) \to 1 \text{ as } n \to \infty. \tag{1}$$

So, in many cases we obtain the exact limit distribution of the chromatic number. Later, the bounds in the third ambiguous case (1) were improved by Ayre, Coja–Oghlan and Greenhill [2] and by Shabanov [9]. They proved that up to the value $r_c^{k-1}\ln r_c - \frac{1}{2}\ln r_c - O(1)$ we still have the chromatic number equal to r_c.

The non-sparse case when $pn^{k-1} \to +\infty$ is not studied so well. Krivelevich and Sudakov showed that if additionally $p \to 0$ then

$$\chi(G(n,p)) \cdot \left(\frac{(k-1)d}{k\ln d}\right)^{-\frac{1}{k-1}} \xrightarrow{\mathsf{P}} 1 \text{ as } n \to +\infty, \tag{2}$$

where $d = p\binom{n-1}{k-1}$. But they did not investigate the concentration effect. The authors of the current paper study the chromatic number of the random hypergraph $H(n,k,p)$ for $k \geq 4$ [4] and proved the following theorem.

Theorem 1 ([4]). *Let $k \geq 4$ and $\varepsilon > 0$ be fixed. Denote $r_p = r_p(n) = \min\{r \in \mathbb{N} : c < r^{k-1}\ln r\}$ and $c = c(n) = p\binom{n}{k}\frac{1}{n}$. Suppose also that $c \leq n^{\frac{k-1}{2k+4}-\gamma}$ for some positive fixed γ, but $c \to +\infty$ as $n \to \infty$. Then we have the following concentration values for the chromatic number of $H(n,k,p)$:*

1. if $c \leq r_p^{k-1}\ln r_p - \frac{1}{2}\ln r_p - \frac{r_p-1}{r_p} - O\left(\frac{k^2\ln r_p}{r_p^{k/3-1}}\right)$ then

$$\mathsf{P}\left(\chi(H(n,k,p)) \in \{r_p, r_p+1\}\right) \longrightarrow 1 \text{ as } n \to \infty;$$

2. *if* $c > r_p^{k-1}\ln r_p - \frac{1}{2}\ln r_p + \varepsilon$ *for some fixed positive* $\varepsilon > 0$ *then*

$$\mathsf{P}\left(\chi(H(n,k,p)) \in \{r_p+1, r_p+2\}\right) \longrightarrow 1 \text{ as } n \to \infty.$$

3. *finally, if*

$$c \in \left(r_p^{k-1}\ln r_p - \frac{1}{2}\ln r_p - \frac{r_p - 1}{r_p} - O\left(\frac{k^2 \ln r_p}{r_p^{k/3-1}}\right), r_p^{k-1}\ln r_p - \frac{1}{2}\ln r_p + \varepsilon\right]$$

then

$$\mathsf{P}\left(\chi(H(n,k,p)) \in \{r_p, r_p+1, r_p+2\}\right) \longrightarrow 1 \text{ as } n \to \infty.$$

So, we see almost the same picture as in the sparse case but every time we have one more value.

1.2 Extremal Problem for Doubly Stochastic Matrices

The key ingredient of the proof of Theorem 1 is some result concerning the doubly stochastic matrices. Suppose that $r \geq 3$ is an integer. Let $\mathcal{M}_r$ denote the set of $r \times r$ real-valued matrices $M = (m_{ij}, i, j = 1, \ldots, r)$ with nonnegative elements satisfying the following conditions:

$$\sum_{i=1}^{r} m_{ij} = \frac{1}{r}, \text{ for any } j = 1, \ldots, r; \; \sum_{j=1}^{r} m_{ij} = \frac{1}{r}, \text{ for any } i = 1, \ldots, r. \tag{3}$$

So, for any $M \in \mathcal{M}_r$, the matrix $r \cdot M$ is doubly stochastic. Now, denote the following functions

$$\mathcal{H}_r(M) = -\sum_{i,j=1}^{r} m_{ij}\ln(r \cdot m_{ij}); \; \mathcal{E}_{r,k}(M) = \ln\left(1 - \frac{2}{r^{k-1}} + \sum_{i,j=1}^{r} m_{ij}^k\right). \tag{4}$$

Denote for $c > 0$, $\mathcal{G}_{c,r,k}(M) = \mathcal{H}_r(M) + c \cdot \mathcal{E}_{r,k}(M)$. It is known that if $c = c(r,k)$ is not too large then $\mathcal{G}_{c,r,k}(M)$ reaches its maximal value at the matrix J_r which has all entries equal to $1/r^2$. The first result of this type was obtained by Achlioptas and Naor in the breakthrough paper [1] for the graph case $k = 2$. Recently, it was improved by Kargaltsev, Shabanov and Shaikheeva [6]. For $k \geq 4$, Shabanov [9] proved the following.

Theorem 2 ([9]). *There exists an absolute constant* d *such that if* $k \geq 4$, $\max(r,k) > d$ *and*

$$c < r^{k-1}\ln r - \frac{1}{2}\ln r - \frac{r-1}{r} - O(k^2 r^{1-k/3}\ln r) \tag{5}$$

then for any $M \in \mathcal{M}_r$, $\mathcal{G}_{c,r,k}(M) \leq \mathcal{G}_{c,r,k}(J_r)$.

The aim of our work was to generalize Theorems 1 and 2 to the missed case $k = 3$.

1.3 New Results

The first new result of the paper provides the solution for the extremal problem concerning $\mathcal{G}_{c,r,k}$ in the case $k = 3$.

Theorem 3. *There exists an absolute constant r_0 such that if $r \geq r_0$ and*

$$c < r^2 \ln r - \frac{1}{2} \ln r - 1 - r^{-1/6} \tag{6}$$

then for any $M \in \mathcal{M}_r$, $\mathcal{G}_{c,r,3}(M) \leq \mathcal{G}_{c,r,3}(J_r)$.

Note that the obtained result is best possible in the following sense: if for some fixed $\varepsilon > 0$, it holds that $c > r^2 \ln r - \frac{1}{2} \ln r - 1 + \varepsilon$ then for any large enough r, there is $M \in \mathcal{M}_r$ such that $\mathcal{G}_{c,r,3}(M) > \mathcal{G}_{c,r,3}(J_r)$.

Theorem 3 and the second moment method allow us to estimate the chromatic number of the random 3-uniform hypergraph from above when $p = p(n)$ does not decrease too slowly.

Theorem 4. *Let $0 < \gamma < 1/5$ be fixed. Denote $c = c(n) = p\binom{n}{k}\frac{1}{n}$ and $r_p = r_p(n) = \min\{r \in \mathbb{N} : c < r^{k-1} \ln r\}$. Suppose that $c \leq n^{\frac{1}{5}-\gamma}$ and $c \to \infty$ as $n \to \infty$. If*

$$c < r_p^2 \ln r_p - \frac{1}{2} \ln r_p - 1 - r_p^{-1/6}, \tag{7}$$

then

$$\mathsf{P}\left(\chi\left(H\left(n, 3, p\right)\right) \leq r_p + 1\}\right) \longrightarrow 1 \text{ as } n \to \infty.$$

Together with a theorem from [4] (see Theorem 1 in [4]) our second theorem extends Theorem 1 to the missed case $k = 3$. For $c \leq n^{\frac{1}{5}-\gamma}$, we obtain the following values of the chromatic number of a random 3-uniform hypergraph:

1. if $c \leq r_p^2 \ln r_p - \frac{1}{2} \ln r_p - 1 - r_p^{-1/6}$ then
$$\mathsf{P}\left(\chi(H(n,3,p)) \in \{r_p, r_p + 1\}\right) \longrightarrow 1 \text{ as } n \to \infty;$$
2. if $c > r_p^2 \ln r_p - \frac{1}{2} \ln r_p + \varepsilon$ for some fixed positive $\varepsilon > 0$ then
$$\mathsf{P}\left(\chi(H(n,3,p)) \in \{r_p + 1, r_p + 2\}\right) \longrightarrow 1 \text{ as } n \to \infty.$$
3. if $c \in \left(r_p^2 \ln r_p - \frac{1}{2} \ln r_p - 1 - r_p^{-1/6}, r_p^2 \ln r_p - \frac{1}{2} \ln r_p + \varepsilon\right]$ then
$$\mathsf{P}\left(\chi(H(n,3,p)) \in \{r_p, r_p + 1, r_p + 2\}\right) \longrightarrow 1 \text{ as } n \to \infty.$$

In the next section we will prove Theorem 3.

2 Proof of Theorem 3

Note that (3) implies that the total sum of m_{ij} is equal to 1. Since $\mathcal{G}_{c,r,3}(J_r) = \ln r + c \cdot \ln\left(1 - \frac{1}{r^2}\right)^2$, we have

$$\mathcal{G}_{c,r,3}(J_r) - \mathcal{G}_{c,r,3}(M) = \mathcal{H}_r(J_r) - \mathcal{H}_r(M) - c\left(\mathcal{E}_{r,3}(M) - \mathcal{E}_{r,3}(J_r)\right)$$

$$= \ln r + \sum_{i,j=1}^{r} m_{ij} \ln(r \cdot m_{ij}) + c\left(\ln\left(1 - \frac{2}{r^2} + \sum_{i,j=1}^{r} m_{ij}^3\right) - \ln\left(1 - \frac{1}{r^2}\right)^2\right)$$

$$= \sum_{i,j=1}^{r} m_{ij} \ln(r^2 \cdot m_{ij}) - c \cdot \ln\left(1 + \frac{\sum_{i,j=1}^{r} m_{ij}^3 - r^{-4}}{\left(1 - \frac{1}{r^2}\right)^2}\right). \qquad (8)$$

We need to show that this value is nonnegative for any $M \in \mathcal{M}_r$. In fact, we prove a more precise statement and show that there exist some function $a = a(r) > 0$ such that given the condition (6) the following inequality holds for any $M \in \mathcal{M}_r$,

$$\mathcal{G}_{c,r,3}(J_r) - \mathcal{G}_{c,r,3}(M) \geq a(r) \cdot \sum_{i,j=1}^{r} \left(m_{ij} - \frac{1}{r^2}\right)^2. \qquad (9)$$

Our proof strategy follows the proof of Theorem 2 from [9], however we need to make some changes that allow to extend the result to the case $k = 3$.

2.1 Row Functions

Let us denote $\varepsilon_{ij} = m_{ij} - 1/r^2$. Due to (3), for any $i, j = 1, \ldots, r$, we have

$$\varepsilon_{ij} \in \left[-\frac{1}{r^2}, \frac{1}{r} - \frac{1}{r^2}\right], \quad \sum_{j'=1}^{r} \varepsilon_{ij'} = 0, \quad \sum_{i'=1}^{r} \varepsilon_{i'j} = 0. \qquad (10)$$

Let us also define the following "row" functions: for any $i = 1, \ldots, r$,

$$H_i(M) = \sum_{j=1}^{r} m_{ij} \ln(r^2 \cdot m_{ij}) = \sum_{j=1}^{r} \left(\frac{1}{r^2} + r^2 \varepsilon_{ij}\right) \ln(1 + r^2 \varepsilon_{ij}),$$

$$E_i(M) = \frac{\sum_{j=1}^{r} m_{ij}^3 - r^{-5}}{\left(1 - \frac{1}{r^2}\right)^2} = \left(1 - \frac{1}{r^2}\right)^{-2} \left(\frac{3}{r^2} \sum_{j=1}^{r} \varepsilon_{ij}^2 + \sum_{j=1}^{r} \varepsilon_{ij}^3\right). \qquad (11)$$

Clearly,

$$\mathcal{H}_r(J_r) - \mathcal{H}_r(M) = \sum_{i=1}^{r} H_i(M), \quad \mathcal{E}_{r,3}(M) - \mathcal{E}_{r,3}(J_r) \leq \sum_{i=1}^{r} E_i(M). \qquad (12)$$

Now we are going to estimate the differences $H_i(M) - c \cdot E_i(M)$, $i = 1, \ldots, r$, in various cases. The value $H_i(M) - c \cdot E_i(M)$ depends only on the i-th row of the matrix M. The classification of rows is the following. The row $M_i = (m_{ij}; j = 1, \ldots, r)$ is said to be

1. *central* if
$$\max_{j=1,\ldots,r} m_{ij} < \frac{1}{r} - \frac{1}{r\sqrt{\ln r}};$$
2. *good* if
$$\max_{j=1,\ldots,r} m_{ij} \in \left[\frac{1}{r} - \frac{1}{r\sqrt{\ln r}}, \frac{1}{r} - r^{-11/4}\right];$$
3. *bad* if
$$\max_{j=1,\ldots,r} m_{ij} > \frac{1}{r} - r^{-11/4}.$$

Now let us consider these three types of rows separately. Throughout the paper we use the estimates from [9] whenever it is possible. We also assume that r is large enough.

2.2 Central Rows

Proposition 1. *For any central row M_i,*

$$H_i(M) - c \cdot E_i(M) \geq \frac{r^2}{4} \sum_{j:\varepsilon_{ij}<0} \varepsilon_{ij}^2 + \left(2r\sqrt{\ln r} + O\left(r \ln \ln r\right)\right) \sum_{j:\varepsilon_{ij}\geq 0} \varepsilon_{ij}^2. \quad (13)$$

Proof. First, let us estimate the value $c \cdot E_i(M)$. Since every $\varepsilon_{ij} < \frac{1}{r} - \frac{1}{r^2} - \frac{1}{r\sqrt{\ln r}}$ and $c < r^2 \ln r$, we have

$$\begin{aligned} c \cdot E_i(M) &= c \cdot \left(1 - \frac{1}{r^2}\right)^{-2} \left(\frac{3}{r^2}\sum_{j=1}^{r} \varepsilon_{ij}^2 + \sum_{j=1}^{r} \varepsilon_{ij}^3\right) \\ &\leq r^2 \ln r \left(1 - \frac{1}{r^2}\right)^{-2} \left(\frac{3}{r^2} \sum_{j:\varepsilon_{ij}<0} \varepsilon_{ij}^2 + \left(\frac{3}{r^2} + \frac{1}{r} - \frac{1}{r^2} - \frac{1}{r\sqrt{\ln r}}\right) \sum_{j:\varepsilon_{ij}>0} \varepsilon_{ij}^2\right) \\ &\leq 4 \ln r \sum_{j:\varepsilon_{ij}<0} \varepsilon_{ij}^2 + \left(r \ln r - r\sqrt{\ln r} + O(\ln r)\right) \sum_{j:\varepsilon_{ij}>0} \varepsilon_{ij}^2. \quad (14) \end{aligned}$$

Now proceed to H_i. We need to estimate the value of $\left(\frac{1}{r^2} + \varepsilon_{ij}\right) \ln(1 + r^2\varepsilon_{ij})$ from below. In [9] it was proved that

1. if $\varepsilon_{ij} < 0$ then (see (34) in [9])
$$\left(\frac{1}{r^2} + \varepsilon_{ij}\right) \ln(1 + r^2\varepsilon_{ij}) \geq \varepsilon_{ij} + \frac{r^2}{2}\varepsilon_{ij}^2; \quad (15)$$
2. if $\varepsilon_{ij} \geq 0$ and $\varepsilon_{ij} \leq \frac{1}{r \ln r} - \frac{1}{r^2}$ then (see (34) in [9])
$$\left(\frac{1}{r^2} + \varepsilon_{ij}\right) \ln(1 + r^2\varepsilon_{ij}) \geq \varepsilon_{ij} + \frac{3r \ln r}{2(1 + 2\ln r/r)}\varepsilon_{ij}^2. \quad (16)$$

Assume that $\varepsilon_{ij} > \frac{1}{r\ln r} - \frac{1}{r^2}$. Then

$$\left(\frac{1}{r^2}+\varepsilon_{ij}\right)\ln(1+r^2\varepsilon_{ij}) \geq \varepsilon_{ij}\ln\left(\frac{r}{\ln r}\right) = \varepsilon_{ij}+\varepsilon_{ij}(\ln r - \ln\ln r - 1)$$
$$\geq \varepsilon_{ij} + r\varepsilon_{ij}^2\left(1-\frac{1}{r}-\frac{1}{\sqrt{\ln r}}\right)^{-1}(\ln r - \ln\ln r - 1)$$
$$= \varepsilon_{ij} + r\ln r\cdot\varepsilon_{ij}^2\left(1+\frac{1}{\sqrt{\ln r}}+O\left(\frac{\ln\ln r}{\ln r}\right)\right). \tag{17}$$

The obtained bounds (15), (16), (17) imply that for all large enough r,

$$H_i(M) \geq \frac{r^2}{2}\sum_{j:\varepsilon_{ij}<0}\varepsilon_{ij}^2 + \left(r\ln r + r\sqrt{\ln r} + O\left(r\ln\ln r\right)\right)\sum_{j:\varepsilon_{ij}\geq 0}\varepsilon_{ij}^2. \tag{18}$$

Together (14) and (18) provide the required estimate:

$$H_i(M) - c\cdot E_i(M) \geq \frac{r^2}{4}\sum_{j:\varepsilon_{ij}<0}\varepsilon_{ij}^2 + \left(2r\sqrt{\ln r} + O\left(r\ln\ln r\right)\right)\sum_{j:\varepsilon_{ij}\geq 0}\varepsilon_{ij}^2.$$

2.3 Good Rows

For good or bad row M_i, its maximal element is very close to $\frac{1}{r}$. So, it is convenient to define the value

$$m_i = \frac{1}{r} - \max_{j=1,\ldots,r} m_{ij}. \tag{19}$$

The inequality (28) from [9] estimates the value $H_i(M)$ in terms of the value m_i as follows:

$$H_i(M) \geq \frac{\ln r}{r} + m_i\ln m_i + m_i\ln\left(\frac{r}{r-1}\right) - m_i. \tag{20}$$

Note that these bounds hold for any row. We will use it very often in the remaining proof.

Proposition 2. *For any good row M_i,*

$$H_i(M) - c\cdot E_i(M) \geq \frac{1}{4}r^{-11/4}\ln r. \tag{21}$$

Proof. Let us estimate $c\cdot E_i(M)$. For a good row, we have $m_i \in [r^{-11/4}, 1/r\sqrt{\ln r}]$, so $m_i = o(r^{-1})$ and $m_i = \omega(r^{-3})$. Suppose that $m_{ij_0} = 1/r - m_i$ is the maximal element of M_i. Then (3) implies that $\sum_{j\neq j_0} m_{ij} = 1/r - m_{ij_0} = m_i$. Thus,

$$
\begin{aligned}
c \cdot E_i(M) &= c \cdot \left(1 - \frac{1}{r^2}\right)^{-2} \left(\sum_{j=1}^{r} m_{ij}^3 - r^{-5}\right) \leq c \cdot \left(1 - \frac{1}{r^2}\right)^{-2} \sum_{j=1}^{r} m_{ij}^3 \\
&= c \cdot \left(1 - \frac{1}{r^2}\right)^{-2} \left(\left(\frac{1}{r} - m_i\right)^3 + \sum_{j \neq j_0}^{r} m_{ij}^3\right) \\
&\leq c \cdot \left(1 - \frac{1}{r^2}\right)^{-2} \left(\frac{1}{r^3} - \frac{3m_i}{r^2} + \frac{3m_i^2}{r}\right).
\end{aligned}
$$

Here we use the fact that $\sum_{j \neq j_0}^{r} m_{ij}^3 \leq m_i^3$. Since $m_i = o(r)$ and $c < r^2 \ln r$ we obtain that

$$
\begin{aligned}
c \cdot E_i(M) &\leq c \cdot \left(\frac{1}{r^3} - \frac{3m_i}{r^2} + \frac{3m_i^2}{r} + O\left(\frac{1}{r^5}\right)\right) \\
&\leq \frac{\ln r}{r} - 3m_i \ln r(1 + o(1)) + O\left(\frac{\ln r}{r^3}\right).
\end{aligned} \tag{22}
$$

The general estimate (20) and the condition $m_i \geq r^{-11/4}$ imply that

$$
H_i(M) \geq \frac{\ln r}{r} + m_i \ln m_i(1 + o(1)) \geq \frac{\ln r}{r} - \frac{11}{4} m_i \ln r(1 + o(1)). \tag{23}
$$

The bounds (22) and (23) provide the required inequality:

$$
\begin{aligned}
H_i(M) - c \cdot E_i(M) &\geq \frac{1}{4} m_i \ln r(1 + o(1)) + O\left(\frac{\ln r}{r^3}\right) \\
&\geq \frac{1}{8} m_i \ln r \geq \frac{1}{4} r^{-11/4} \ln r.
\end{aligned}
$$

2.4 Bad Rows

Now it is time to deal with bad rows. Recall that in every bad row M_i there is an index j_0 such that $m_{ij_0} = \max_{j=1,\ldots,r} > \frac{1}{r} - r^{-11/4}$. The main problem here is that in this case the difference $H_i(M) - c \cdot E_i(M)$ can be negative. For $k \geq 4$, this negative value can be compensated by the bounds (13), (21) for central and good rows, if there is at least one non-bad row (see [9]). So, it remains to consider the case when all the rows are bad. Unfortunately, this is not the way for $k = 3$. Here we had to consider all the bad rows simultaneously.

Let $D \subset \{1, \ldots, r\}$ denote the set of indices of the bad rows in M. Introduce the following values:

$$
H_D(M) = \sum_{i \in D} H_i(M), \quad E_D(M) = \ln\left(1 + \sum_{i \in D} E_i(M)\right). \tag{24}
$$

The following statement estimates their difference.

Proposition 3. *Under the condition* (6) *the following inequality holds:*

$$H_D(M) - c \cdot E_D(M) \geq -\frac{|D| \ln r}{2r^3} + \frac{|D|^2 \ln r}{2r^4} + \frac{|D|}{r^{19/6}} + O\left(r^{-3}\right). \tag{25}$$

Proof. For $H_i(M)$, $i \in D$, we have the bound (20). So, it remains to estimate $c \cdot \mathcal{E}_D(M)$. Again, for any $i \in D$, we consider the maximal element of M_i,

$$m_{ij_0(i)} = \max_{j=1,\ldots,r} m_{ij} = \frac{1}{r} - m_i,$$

where $m_i \in [0, r^{-11/4}]$. Using (11) we get

$$\begin{aligned}\sum_{i \in B} E_i(M) &= \left(1 - \frac{1}{r^2}\right)^{-2} \sum_{i \in D} \left(\sum_{j=1}^{r} m_{ij}^3 - \frac{1}{r^5}\right) \\ &= \left(1 - \frac{1}{r}\right)^{-2} \sum_{i \in D} \left(\left(\frac{1}{r} - m_i\right)^3 + \sum_{j \neq j_0(i)} m_{ij}^3 - \frac{1}{r^5}\right).\end{aligned}$$

Note that $\sum_{j \neq j_0(i)} m_{ij}^3 \leq m_i^3 = O(r^{-33/4})$. Therefore,

$$\begin{aligned}\sum_{i \in B} \mathcal{E}_i(M) &= \left(1 - \frac{1}{r^2}\right)^{-2} \sum_{i \in D} \left(\frac{1}{r^3} - \frac{3m_i}{r^2} + \frac{3m_i^2}{r} - \frac{1}{r^5} + O(r^{-33/4})\right) \\ &= \sum_{i \in D} \left(\frac{1}{r^3} - \frac{3m_i}{r^2} - \frac{1}{r^5} + O(r^{-13/2})\right)\left(1 + \frac{2}{r^2} + O(r^{-4})\right).\end{aligned}$$

Now, we have

$$\left(\frac{1}{r^3} - \frac{3m_i}{r^2} - \frac{1}{r^5} + O(r^{-13/2})\right)\left(\frac{2}{r^2} + O(r^{-4})\right) = \frac{2}{r^5} + O(r^{-27/4}).$$

Consequently,

$$\begin{aligned}\sum_{i \in D} E_i(M) &= \sum_{i \in D} \left(\frac{1}{r^3} - \frac{3m_i}{r^2} + \frac{1}{r^5} + O(r^{-13/2})\right) \\ &= \frac{|D|}{r^3} - \frac{3}{r^2} \sum_{i \in D} m_i + \frac{|D|}{r^5} + O(|D| r^{-13/2}).\end{aligned} \tag{26}$$

Now, we want to estimate the square of this expression. Since $|D| \leq r$, the last three summands have the order $O(r^{-15/4})$. Therefore, (26) implies that

$$\left(\sum_{i \in D} E_i(M)\right)^2 = \frac{|D|^2}{r^6} + O(r^{-23/4}), \quad \left(\sum_{i \in D} E_i(M)\right)^3 = O(r^{-6}). \tag{27}$$

Now we are ready to estimate $c \cdot E_D(M)$. Using (26), (27) and applying Taylor expansion for the logarithm, we obtain

$$c \cdot E_D(M) = c \cdot \ln\left(1 + \sum_{i \in D} E_i(M)\right)$$
$$= c \cdot \left(\sum_{i \in D} E_i(M) - \frac{1}{2}\left(\sum_{i \in D} E_i(M)\right)^2 + O\left(\left(\sum_{i \in D} E_i(M)\right)^3\right)\right)$$
$$= c \cdot \left(\frac{|D|}{r^3} - \frac{3}{r^2}\sum_{i \in D} m_i + \frac{|D|}{r^5} + O(|D| r^{-13/2}) - \frac{|D|^2}{2r^6} + O(r^{-23/4}) + O(r^{-6})\right)$$
$$= c \cdot \left(\frac{|D|}{r^3} + \frac{|D|}{r^5} - \frac{|D|^2}{2r^6} - \frac{3}{r^2}\sum_{i \in D} m_i + O(r^{-11/2})\right).$$

The condition (6) states that $c < r^2 \ln r - \frac{1}{2}\ln r - 1 - r^{-1/6}$. Thus,

$$c \cdot E_D(M) < \left(r^2 \ln r - \frac{1}{2}\ln r - 1 - r^{-1/6}\right)$$
$$\times \left(\frac{|D|}{r^3} + \frac{|D|}{r^5} - \frac{|D|^2}{2r^6} - \frac{3}{r^2}\sum_{i \in D} m_i + O(r^{-11/2})\right)$$
$$= \frac{|D|\ln r}{r} + \frac{|D|\ln r}{r^3} - \frac{|D|^2 \ln r}{2r^4} - (3\ln r)\sum_{i \in D} m_i + O\left(\frac{\ln r}{r^{7/2}}\right)$$
$$- \frac{|D|\ln r}{2r^3} - \frac{|D|}{r^3} - \frac{|D|}{r^{19/6}} + O\left(\ln r \cdot r^{-15/4}\right)$$
$$= \frac{|D|\ln r}{r} + \frac{|D|\ln r}{2r^3} - \frac{|D|^2 \ln r}{2r^4} - \frac{|D|}{r^3} - \frac{|D|}{r^{19/6}} - (3\ln r)\sum_{i \in D} m_i + O\left(\frac{\ln r}{r^{7/2}}\right). \tag{28}$$

Let us complete the proof. Due to (20) we have the following lower bound for $H_D(M)$:

$$H_D(M) \geq \frac{|D|\ln r}{r} + \sum_{i \in D}\left[m_i \ln m_i + m_i \ln\left(\frac{r}{r-1}\right) - m_i\right].$$

Using (28), we obtain that

$$H_D(M) - c \cdot E_D(M) \geq -\frac{|D|\ln r}{2r^3} + \frac{|D|^2 \ln r}{2r^4} + \frac{|D|}{r^3} + \frac{|D|}{r^{19/6}} + O\left(\frac{\ln r}{r^{7/2}}\right)$$
$$+ \sum_{i \in D}\left[m_i \ln m_i + m_i \ln\left(\frac{r}{r-1}\right) - m_i + 3 m_i \ln r\right].$$

The function $f(x) = x \ln x + x \ln\left(\frac{r}{r-1}\right) - x + 3x \ln r$ is minimized when $x = (r-1)/r^4 \in [0, r^{-11/4})$. So, the minimal value is attained when $m_i = (r-1)/r^4$ for any $i \in D$. Hence,

$$\sum_{i\in D}\left[m_i\ln m_i + m_i\ln\left(\frac{r}{r-1}\right) - m_i + 3m_i\ln r\right]$$
$$\geq \sum_{i\in D}\left(\frac{r-1}{r^4}\left(\ln\left(\frac{r-1}{r^4}\right) + \ln\left(\frac{r}{r-1}\right) - 1 + 3\ln r\right)\right)$$
$$= -\sum_{i\in D}\frac{r-1}{r^4} = -\frac{(r-1)|D|}{r^4} = -\frac{|D|}{r^3} + O(r^{-3}).$$

This finally implies the required inequality

$$H_D(M) - c\cdot E_D(M) \geq -\frac{|D|\ln r}{2r^3} + \frac{|D|^2\ln r}{2r^4} + \frac{|D|}{r^{19/6}} + O\left(r^{-3}\right).$$

2.5 Completion of the Proof

It remains to summarize the obtained information. Now everything depends on the number of bad rows $|D|$ in the matrix M. Let $C, G \subset \{1,\ldots,r\}$ denote the set of indices of central and good rows, respectively. Recall (see (8), (11), (24)) that

$$\mathcal{G}_{c,r,3}(J_r) - \mathcal{G}_{c,r,3}(M) = \sum_{i=1}^{r} H_i(M) - c\cdot\ln\left(1 + \sum_{i=1}^{r} E_i(M)\right)$$
$$\geq \sum_{i\in C\cup G}(H_i(M) - c\cdot E_i(M)) + H_D(M) - c\cdot E_D(M). \tag{29}$$

Note that it is sufficient to show that $\mathcal{G}_{c,r,3}(J_r) - \mathcal{G}_{c,r,3}(M) \geq b$ for some $b = b(r) > 0$. This also implies the required inequality (9), because $\sum_{i,j=1}^{r}(m_{ij} - r^{-2})^2 < 1$.

Let us consider the following four cases.

1. If $|D| = 0$ then (9) follows from (13) and (21).
2. If $|D| \geq r - \frac{r^{5/6}}{\ln r}$ then (25) implies that the total contribution of bad rows is positive. Indeed, for large enough r,

$$H_D(M) - c\cdot E_D(M) \geq \frac{|D|\ln r}{2r^3}\left(\frac{|D|}{r} - 1\right) + \frac{|D|}{r^{19/6}} + O\left(r^{-3}\right)$$
$$\geq \frac{|D|\ln r}{2r^3}\left(-\frac{1}{r^{1/6}\ln r}\right) + \frac{|D|}{r^{19/6}} + O\left(r^{-3}\right) = \frac{|D|}{2r^{19/6}} + O\left(r^{-3}\right) > \frac{1}{3}r^{-13/6}.$$

   Hence, again (9) follows from (13) and (21).
3. Suppose $|D| < r - \frac{r^{5/6}}{\ln r}$, but $|G| \geq r^{4/5}$. Then

$$H_D(M) - c\cdot E_D(M) \geq -\frac{|D|\ln r}{2r^3} + O\left(r^{-3}\right) \geq -\frac{\ln r}{2r^2} + O\left(r^{-3}\right) > -\frac{\ln r}{r^2}.$$

The inequality (29) and the obtained bounds (13), (21) imply that for large enough r,

$$\mathcal{G}_{c,r,3}(J_r) - \mathcal{G}_{c,r,3}(M) \geq \sum_{i\in G} (H_i(M) - c \cdot E_i(M)) + H_D(M) - c \cdot E_D(M)$$
$$\geq r^{4/5}\frac{1}{4}r^{-11/4}\ln r - \frac{\ln r}{r^2} \geq \frac{1}{5}r^{-39/20}\ln r.$$

4. It remains to consider the situation when $|G| < r^{4/5}$ and $0 < |D| < r - \frac{r^{5/6}}{2\ln r}$. In this case there is at least $\frac{r^{5/6}}{\ln r} - r^{4/5}$ central rows in M. Suppose that $i_1, i_2 \in D$ are two indices corresponding to bad rows. Recall that any bad row has an element greater than $r^{-1} - r^{-11/4}$. Suppose that $m_{i_1 j_1}$ and $m_{i_2 j_2}$ are both greater than $1/r - r^{-11/4}$. Then it is straightforward to verify that the double stochastic property (3) implies that $j_1 \neq j_2$. So, the maximal elements of bad rows should be in different columns of matrix M. Without loss of generality we may assume that these elements are diagonal, i.e. for any $i \in D$,

$$m_{ii} = \max_{j=1,\ldots,r} m_{ij} > \frac{1}{r} - r^{-11/4}.$$

Recall the notation $\varepsilon_{ij} = m_{ij} - \frac{1}{r^2}$. If $j \in D$ then due to (10) we obtain that

$$\sum_{i\in C} \varepsilon_{ij} = -\varepsilon_{jj} - \sum_{i\in G\cup D; i\neq j} \varepsilon_{ij}.$$

We know that $\varepsilon_{jj} = m_{jj} - \frac{1}{r^2} \geq \frac{1}{r} - \frac{1}{r^2} - r^{-11/4}$ and any other element is at least $-r^{-2}$. Hence,

$$\sum_{i\in C} \varepsilon_{ij} \leq -\frac{1}{r} + \frac{1}{r^2} + r^{-11/4} + \frac{1}{r^2}\left(|G| + |D|\right).$$

In our case $|G| < r^{4/5}$ and $|D| < r - \frac{r^{5/6}}{\ln r}$, so, we get

$$\sum_{i\in C} \varepsilon_{ij} \leq -\frac{1}{r} + \frac{1}{r^2} + r^{-11/4} + \frac{1}{r^2}\left(r^{4/5} + r - \frac{r^{5/6}}{\ln r}\right)$$
$$= -\frac{r^{-7/6}}{\ln r}(1 + o(1)) < -\frac{r^{-7/6}}{2\ln r} < 0.$$

Hence, the sum over all negative summands is also less than $-\frac{r^{-7/6}}{2\ln r}$:

$$\sum_{i\in C:\varepsilon_{ij}<0} \varepsilon_{ij} \leq -\frac{r^{-7/6}}{2\ln r}.$$

By Cauchy–Schwarz inequality

$$\sum_{i\in C:\varepsilon_{ij}<0} \varepsilon_{ij}^2 \geq \frac{1}{r}\left(\frac{r^{-7/6}}{2\ln r}\right)^2 = \frac{r^{-10/3}}{4(\ln r)^2}. \tag{30}$$

Finally, by using (13), (21), (25), (29) and (30) we establish the required estimate

$$\begin{aligned}
\mathcal{G}_{c,r,3}(J_r) - \mathcal{G}_{c,r,3}(M) &\geq \sum_{i\in C} (H_i(M) - c\cdot E_i(M)) + H_B(M) - c\cdot E_B(M) \\
&\geq \sum_{i\in C} \frac{r^2}{4} \sum_{j:\varepsilon_{ij}<0} \varepsilon_{ij}^2 - \frac{|D|\ln r}{2r^3} + \frac{|D|^2\ln r}{2r^4} + \frac{|D|}{r^{19/6}} + O\left(r^{-3}\right) \\
&\geq \frac{r^2}{4} \sum_{i\in C} \sum_{j:\varepsilon_{ij}<0} \varepsilon_{ij}^2 - \frac{|D|\ln r}{2r^3} + O\left(r^{-3}\right) \\
&\geq \frac{r^2}{4} \sum_{i\in C} \sum_{j\in D:\varepsilon_{ij}<0} \varepsilon_{ij}^2 - \frac{|D|\ln r}{2r^3} + O\left(r^{-3}\right) \\
&= \frac{r^2}{4} \sum_{j\in D} \sum_{i\in C:\varepsilon_{ij}<0} \varepsilon_{ij}^2 - \frac{|D|\ln r}{2r^3} + O\left(r^{-3}\right) \\
&\geq \frac{r^2}{4}|D| \cdot \frac{r^{-10/3}}{4(\ln r)^2} - \frac{|D|\ln r}{2r^3} + O\left(r^{-3}\right) \\
&= |D| \cdot \frac{r^{-4/3}}{16(\ln r)^2}(1+o(1)) + O\left(r^{-3}\right).
\end{aligned}$$

Since $|D| \geq 1$, the obtained value is at least $\frac{r^{-4/3}}{16\ln^2 r}(1+o(1))$.

Theorem 3 is proved.

3 Sketch of the Proof of Theorem 4

In the last section we give a short sketch of the proof of Theorem 4. We just follow the proof of Theorem 1 which can be found in [4]. The general scheme was first developed by Coja-Oghlan, Panagiotou and Steger [3] in the case of graphs.

First of all, we have to estimate from below the probability that the chromatic number of the random hypergraph does not exceed r_p. By using the second moment method and Theorem 3 we prove the following lemma.

Lemma 1. *Suppose $pn^2 \to +\infty$ and $p \to 0$ as $n \to +\infty$. If the condition (7) holds then for all large enough n,*

$$\mathsf{P}\left(\chi(H(n,3,p)) \leq r_p\right) \geq n^{-2r_p^2}.$$

Lemma 1 helps to estimate the proportion of vertices of our hypergraph that can be properly colored with r_p colors. Let V_n denote the set of vertices of $H(n,3,p)$. The following statement is true.

Lemma 2. *Suppose that the conditions of Lemma 1 hold. Then with probability tending to 1, there exists a vertex subset U_0 with size at most $2r_p\sqrt{n\ln n}$ such that the chromatic number of the subhypergraph induced by $H(n,3,p)$ on $V_n \setminus U_0$ does not exceed r_p.*

Finally, we need to estimate the number of edges in any small induced sub-hypergraph in $H(n, k, p)$. Here we prove the following.

Lemma 3. *Suppose the conditions of Theorem 4 hold. Suppose that fixed* $\delta = \delta(\gamma) > 0$ *satisfies the inequality*

$$\delta < \frac{25\gamma}{18 + 60\gamma}.$$

Then with probability tending to 1, any vertex subset U *in* $H(n, 3, p)$ *with size at most* $r_p\sqrt{n}(\ln n)$ *has at most* $\left(\frac{2}{3} - \delta\right)|U|$ *edges inside.*

Theorem 4 is easily deduced from Lemmas 1–3. So, we know that with probability tending to 1, almost whole hypergraph can be properly colored with r_p colors. The remained small vertex subset U has size at most $2r_p\sqrt{n \ln n}$. Therefore, there is a small number of edges inside U.

Now we can increase this set U in such a way that there are no edges in the set of neighbors of the extended set U' and the size of U' is still less than $r_p\sqrt{n}(\ln n)$. By Lemma 3 the set U' can be properly colored with colors $\{1, 2\}$, its neighborhood W—with reserved color $r_p + 1$ and the remaining subset $V_n \setminus (U \cup W)$—with colors $\{1, 2, \ldots, r_p\}$. Clearly, this is a proper coloring of $H(n, 3, p)$ with $r_p + 1$ colors. Theorem 4 is proved.

Acknowledgments. Research of Yu.A. Demidovich was supported by the RFBR, project number 19-31-90016. Research of D.A. Shabanov was supported by the grant of the President of Russian Federation no. MD-1562.2020.1

References

1. Achlioptas, D., Naor, A.: The two possible values of the chromatic number of a random graph. Ann. Math. **162**(3), 1335–1351 (2005)
2. Ayre, P., Coja-Oghlan, A., Greenhill, C.: Hypergraph coloring up to condensation. Random Struct. Algorithms **54**(4), 615–652 (2019)
3. Coja-Oghlan, A., Panagiotou, K., Steger, A.: On the chromatic number of random graphs. J. Combin. Theory Ser. B **98**, 980–993 (2008)
4. Demidovich, Y.A., Shabanov, D.A.: On the chromatic number of random hypergraphs. Dokl. Math. **102**(2), 380–383 (2020)
5. Dyer, M., Frieze, A., Greenhill, C.: On the chromatic number of a random hypergraph. J. Combin. Theory Ser. B **113**, 68–122 (2015)
6. Kargaltsev, S., Shabanov, D., Shaikheeva, T.: Two values of the chromatic number of a sparse random graph. Acta Math. Univ. Comenian. **88**(3), 849–854 (2019)
7. Krivelevich, M., Sudakov, B.: The chromatic numbers of random hypergraphs. Random Struct. Algorithms **12**(4), 381–403 (1998)
8. Schmidt-Pruzan, J., Shamir, E., Upfal, E.: Random hypergraph coloring algorithms and the weak chromatic number. J. Graph Theory **8**, 347–362 (1985)
9. Shabanov, D.A.: Estimating the r-colorability threshold for a random hypergraph. Discret. Appl. Math. **282**, 168–183 (2020)
10. Shamir, E.: Chromatic number of random hypergraphs and associated graphs. Adv. Comput. Res. **5**, 127–142 (1989)

On Asymptotic Power of the New Test for Equality of Two Distributions

Viatcheslav Melas(✉) and Dmitrii Salnikov

Department of Mathematics, St. Petersburg State University, St. Petersburg, Russia

Abstract. The paper introduces a new test for equality of two distributions in a class of models. We proved analytically and by stochastic simulation that the test possesses high efficiency. For the case of normal and Cauchy distributions that differ only by shift the asymptotic power of the test appears to be approximately the same as for the Wilcoxon-Mann-Whitney, the Kolmogorov-Smirnov and the Anderson-Darling tests. But if the distributions differ by scale parameters the power of the new test is considerably better.

Keywords: Test for equality of two distributions · Asymptotic power · Cauchy distribution · Normal distribution

1 Formulation of the Problem

Let us consider the classical problem of testing hypothesis on the equality of two distributions

$$H_0 \ : \ F_1 = F_2 \tag{1}$$

against the alternative

$$H_1 \ : \ F_1 \neq F_2 \tag{2}$$

in the case of two independent samples $X = (X_1, \ldots, X_n)$ and $Y = (Y_1, \ldots, Y_m)$ with the distributions functions F_1 and F_2 respectively.

It is well known (see e.g. [1]) that in the case when both distributions differ only by the means and are normal the classical Student test has a few optimal properties. If the distributions are not normal but still differs only by means a widely popular Wilcoxon-Mann-Whitney (WMW) U-statistic is often used instead. However, it can be shown that if two normal populations differ only in variances, the power of WMW test is very low. If distributions are arbitrary there are some universal techniques such as tests by Kolmogorov-Smirnov and Cramer-von Mises (see [2]) and the Anderson-Darling test (see [3]) that can be applied but in many cases these tests can be not powerful.

Recently, Zech and Aslan [4] suggested the test based on U-statistics with the logarithmic kernel and provided its numerical justification for one and many dimensional cases in comparison with a few alternative techniques. However, to the best authors knowledge there are no analytical results about its asymptotic power. Here we introduce a similar but different test and provide a few analytical results on its power.

A. N. Shiryaev et al. (Eds.): ICSM-5 2020, PROMS 371, pp. 204–214, 2021.
https://doi.org/10.1007/978-3-030-83266-7_15

2 The New Test and Its Statistical Motivation

Assume that the distribution functions F_1 and F_2 belongs to the class of distribution functions of random variables ξ, such that

$$E[\ln^2(1+\xi^2)] < \infty. \tag{3}$$

Many distributions and, in particular, the Cauchy distribution have this property.

Among all distributions with given left hand side of (3) the Cauchy's one has the maximum entropy.

Consider the following test

$$\Phi_A = -\frac{1}{n^2}\sum_{1\le i<j\le n} g(X_i - X_j), \Phi_B = -\frac{1}{m^2}\sum_{1\le i<j\le m} g(Y_i - Y_j), \tag{4}$$

$$\Phi_{AB} = -\frac{1}{nm}\sum_{i=1}^{n}\sum_{j=1}^{m} g(X_i - Y_j), \Phi_{nm} = \Phi_{AB} - \Phi_A - \Phi_B, \tag{5}$$

where

$$g(u) = -\ln(1+|u|^2),$$

$g(x)$ is under a constant term precision the logarithm of the density of the standard Cauchy distribution. (Note that Zech and Aslan (2005) took $g(u) = -\ln(|u|)$).

We would like to have a test that is appropriate for the case where the basic distribution belongs to a rather general class of distributions and the alternative distribution differs only by shift and scale transformations.

In particular, we consider the class of distributions satisfying (3), but the approach can be generalized for other classes of distributions.

Consider the class of distributions given by the property (3). Note that if the parameters are known the test based on likelihood ratio is the most powerful among tests with given parameters.

The test suggested above can be considered as an approximation of logarithm of this ratio for the Cauchy distribution. We suppose that it will be very efficient for all distributions with property (3).

3 The Analytical Study of Asymptotic Power

Let us consider the case of two distributions having the property (3) and, in particular, the two that differ only by a shift. To simplify notations assume that $m = n$. The case $m \neq n$ is similar. Now the criterion (4)–(5) assumes the form

$$T_n = \Phi_{nn} = \frac{1}{n^2}\sum_{i,j=1}^{n} \ln(1+(X_i - Y_j)^2) - \frac{1}{n^2}\sum_{1\le i<j\le n} \ln(1+(X_i - X_j)^2) \tag{6}$$

$$-\frac{1}{n^2}\sum_{1\le i<j\le n}\ln(1+(Y_i-Y_j)^2). \tag{7}$$

Denote by $C(u,v)$ the Cauchy distribution with the density function

$$v/(\pi(v^2+(x-u)^2)).$$

Let $f(x)$ denotes the density of F_1. Denote

$$J_h=\int_R -g(x-y-|h|/\sqrt{n})f(x)f(y)dxdy,$$

where $g(u)=-\ln(1+|u|^2)$.

By expending the function $g(u)=\ln(1+|u|^2)$ into the Taylor series we obtain that for arbitrary density function $f(x)$ there exists the finite limit

$$J^*(h)=lim_{n\to\infty}n(J_h-J_0) \tag{8}$$

and it is equal to

$$(1/2)h^2\int_R g''_\theta(x-y-\theta)f(x)f(y)dxdy|_{\theta=0}.$$

(Note that the differentiation under integral is justified since the derivative $g''_\theta(x-y-\theta)|_{\theta=0}$ is less than 2.) That is

$$J^*(h)=h^2\int_R\frac{1-(x-y)^2}{(1+(x-y)^2)^2}f(x)f(y)dxdy.$$

Denote

$$\bar{b}=\sqrt{J^*(h)/h^2}.$$

The basic analytical result of the present paper is the following

Theorem 1. *Consider the problem of testing hypothesis on the equality of two distributions (1)–(2) where both functions have the property (3). Then*

(i) under the condition $n\to\infty$ the distribution function of nT_n converges under H_0 to that of the random variable

$$(aL)^2, \tag{9}$$

where L has the normal distribution with zero expectation and variance equal to 1, $a>0$ is some number.

(ii) Let $F_1(x)=F(x), F_2=F(x+\theta)$, where F is an arbitrary distribution function that is symmetric around a point and possess property (3), $\theta=h/\sqrt{n}, h$ is an arbitrary given number. Then the distribution function of nT_n converges under H_1 to that of the random variable

$$(aL+b)^2,$$

where $b = 0$ for the case of H_0 and $b = \bar{b}h$ for H_1. In this case the power of the criterion T_n with significance α is asymptotically equal to that is given by the formula

$$Pr\{L \geq z_{1-\alpha/2} - \bar{b}h/a\} + Pr\{L \leq -z_{1-\alpha/2} - \bar{b}h/a\},$$

where $z_{1-\alpha/2}$ is such that

$$Pr\{L \geq z_{1-\alpha/2}\} = \alpha/2.$$

If $F_1 = C(\nu, 1), F_2 = C(\nu + \theta, 1)$ then $b = h/3$.

Note that the analytical presentation for the coefficient a is a difficult problem that is not solved up to now. However this coefficient can be easily found by stochastic simulation. In the case of Cauchy distribution we found a heuristic formula $3a^2 = J_0$, that means $a = \sqrt{(2/3)\ln 3}$. This formula provide a very exact approximation for empirical power (see Tables 1, 2 and 3 in the next section).

Thus in the case of Cauchy distributions with scale parameter equal to 1 the power of the criterion T_n with significance α is approximately equal to

$$Pr\{L \geq z_{1-\alpha/2} - (1/\sqrt{6\ln 3})h\} + Pr\{L \leq -z_{1-\alpha/2} - (1/\sqrt{6\ ln3})h\}.$$

The proof of the theorem is given in the Appendix.

4 Simulation Results

In this section we present numerical results of the efficiency of new criterion in comparison with a few alternative criteria.

At the Tables 1, 2, 3, 4, 5, 6, 7, 8, 9, 10, 11 and 12 results for cases $n = 100$, 500, 1000 and different values of h with $\alpha = 0.05$ are given for normal and Cauchy distributions that differ either by shift or by scale parameters. The critical values were calculated in two ways: by simulation of the initial distribution and by random permutations (we used 800 random permutation in all cases). It worth to be noted that the results are very similar. Since the permutation technique is more universal, it can be recommended for practical applications.

Note that in all these cases when the distributions differ only in the shift parameters the power of T_n and that of the Wilcoxon-Mann-Whitney, the Kolmogorov-Smirnov and the Anderson-Darling tests were approximately equal to each other. It can be pointed out also that if the variances are not standard but are known we should simply make the corresponding normalisation. But for the cases where the distributions differ in scale parameters the Wilcoxon-Mann-Whitney is not appropriate at all and the power of the Kolmogorov-Smirnov and the Anderson-Darling tests is considerably lower.

Table 1. Cauchy distribution, $X \sim C(0,1)$, $Y \sim C(h/\sqrt{n},1)$, $n = 100$

h	$T_n, perm$	T_n, sim	$formulae$	$wilcox.test$	$ks.test$	$ad.test$
1	6.4	6.3	6.8	6.6	6.3	7.1
2	10.1	10.6	12.2	11.9	11.1	11.6
3	19.6	20.3	21.5	20.5	20.2	20.7
5	50.9	50.5	49.5	48.5	53.1	52.2
7	82	82.3	77.8	77.2	83.6	80.7
9	96.7	96.8	93.9	91.5	96.5	95.2

Table 2. Cauchy distribution, $X \sim C(0,1)$, $Y \sim C(h/\sqrt{n},1)$, $n = 500$

h	$T_n, perm$	T_n, sim	$formula$	$wilcox.test$	$ks.test$	$ad.test$
1	5.8	6.1	6.8	6.4	6.4	7.1
2	11.6	11.6	12.2	12.6	13.9	12.2
3	21	21.8	21.5	22.2	24.3	22.8
5	50.9	51	49.5	48	57.9	50.3
7	82.2	82.4	77.8	75.6	85.9	81.1
9	96.2	96.5	93.9	93.2	97.2	96.0

Table 3. Cauchy distribution, $X \sim C(0,1)$, $Y \sim C(h/\sqrt{n},1)$, $n = 1000$

h	$T_n, perm$	T_n, sim	$formula$	$wilcox.test$	$ks.test$	$ad.test$
1	6.3	6	6.8	6.8	8.1	6.8
2	11.4	11.9	12.2	12.9	13.4	12.9
3	21	20.9	21.5	22.8	26.2	22.2
5	53.6	53.6	49.5	50.8	59.6	54.2
7	84	84.5	77.8	79.5	87.6	84.4
9	96.6	96.6	93.9	93.2	98.3	96.3

Table 4. Cauchy distribution, $X \sim C(0,1)$, $Y \sim C(0,1+h/\sqrt{n})$, $n = 100$

h	$T_n, perm$	T_n, sim	$wilcox.test$	$ks.test$	$ad.test$
2	10.6	11.9	5.4	5.4	6.9
4	27.6	29.8	5.5	8.7	11.3
6	49.4	53.6	5.5	15.9	22.2
8	68.8	73.5	5.5	25	37.7
10	84.2	87.1	5.2	36.4	55.4

Table 5. Cauchy distribution, $X \sim C(0,1)$, $Y \sim C(0, 1 + h/\sqrt{n})$, $n = 500$

h	$T_n, perm$	T_n, sim	$wilcox.test$	$ks.test$	$ad.test$
2	9.4	10	4.5	6.3	6.2
4	28.5	30.6	4.8	14	12.3
6	54.5	56.5	5	26.1	29.7
8	79.5	80.5	5.2	43.3	51.0
10	93	94	5.2	62.2	74.2

Table 6. Cauchy distribution, $X \sim C(0,1)$, $Y \sim C(0, 1 + h/\sqrt{n})$, $n = 1000$

h	$T_n, perm$	T_n, sim	$wilcox.test$	$ks.test$	$ad.test$
2	10.2	10.5	5	7.6	7.3
4	32.4	33.8	5.2	13.8	14.9
6	61.1	62.8	5.2	27.9	32.8
8	84.8	85.6	5.2	47.4	59.7
10	96.1	97.1	5.4	67.9	82.8

Table 7. Normal distribution, $X \sim N(0,1)$, $Y \sim N(h/\sqrt{n}, 1)$, $n = 100$

h	$T_n, perm$	T_n, sim	$wilcox.test$	$ks.test$	$ad.test$
1	11.1	11.3	12.5	9.5	12.2
2	29.3	29	31.1	20.5	29.6
3	52.4	53.4	55.8	42	55
4	77.5	77.5	80.6	64.9	78.9
5	91.9	92.5	93.1	84.7	93.1

Table 8. Normal distribution, $X \sim N(0,1)$, $Y \sim N(h/\sqrt{n}, 1)$, $n = 500$

h	$T_n, perm$	T_n, sim	$wilcox.test$	$ks.test$	$ad.test$
1	9.2	8.9	9.6	8.3	9.0
2	23.9	23.9	26.3	20.6	25.4
3	47.3	48.9	51.7	41.4	49.7
4	75.3	75.1	77.8	66.9	76.9
5	91.1	91	92.8	86.1	92.6

Table 9. Normal distribution, $X \sim N(0, 1)$, $Y \sim N(h/\sqrt{n}, 1)$, $n = 1000$

h	$T_n, perm$	T_n, sim	$wilcox.test$	$ks.test$	$ad.test$
1	11	11.3	11.5	10	11.6
2	26.4	27.4	28.5	22	27.7
3	51.3	51.6	54.2	44.6	52.9
4	76.7	77	79.3	68.9	77.9
5	91.6	91.2	92.7	86.6	92.1

Table 10. Normal distribution, $X \sim N(0, 1)$, $Y \sim N(0, 1 + h/\sqrt{n})$, $n = 100$

h	$T_n, perm$	T_n, sim	$wilcox.test$	$ks.test$	$ad.test$
1	8.1	8.7	6.4	5.3	7.3
2	15	17.4	6.3	7.2	12.7
3	30.5	34.2	6.6	10.7	24.0
4	50.6	57.1	6.7	16.7	39.9
5	70.8	76.7	6.5	24.8	59.9

Table 11. Normal distribution, $X \sim N(0, 1)$, $Y \sim N(0, 1 + h/\sqrt{n})$, $n = 500$

h	$T_n, perm$	T_n, sim	$wilcox.test$	$ks.test$	$ad.test$
1	8.3	8.4	5	7.4	7.7
2	15.4	16.7	5.1	10.3	12.8
3	33.2	34.7	5.4	16.4	28.3
4	60	63.3	5.6	25.3	52.6
5	83.1	86.3	5.5	40.4	78.1

Table 12. Normal distribution, $X \sim N(0, 1)$, $Y \sim N(0, 1 + h/\sqrt{n})$, $n = 1000$

h	$T_n, perm$	T_n, sim	$wilcox.test$	$ks.test$	$ad.test$
1	6.7	6.9	5.4	6	6.7
2	15.1	16.4	5.5	9.9	13.1
3	33.2	36	5.4	16.1	30.6
4	62.2	64	5.6	27.5	56.8
5	84.6	86.6	5.4	43.6	81.1

5 Conclusion

In this paper we suggested a new test for equality of two distributions. In a wide class of distributions it was proved that the limiting distribution is the square of a Normal distribution. It allows to find asymptotic power analytically for the case of distributions that differ only by shift up to unknown parameter that can be found by stochastic simulation. The high efficiency of the test was confirmed by stochastic simulations.

Acknowledgments. The authors are indebted to professor Yakov Nikitin for the help in calculating the integrals. Work of Viatcheslav Melas was supported by RFBR (grant N 20-01-00096).

6 Appendix

Proof of Theorem 1. Let us consider the test (4)–(5) with the function $g(u) = -u^2$ that is the logarithm of the density of the standard Normal distribution.

Lemma 1. *For $g(x) = x^2$ the following identity holds*

$$\Phi_{nn} = (\bar{x} - \bar{y})^2$$

where

$$\bar{x} = (\sum_{i=1}^{n} X_i)/n, \bar{y} = (\sum_{i=1}^{n} Y_i)/n.$$

Denote

$$Z = (X, Y) = (X_1, \ldots, X_n, Y_1, \ldots, Y_n), V(Z) = \frac{1}{2} \sum_{i=1}^{2n} \sum_{j=1}^{2n} (Z_i - Z_j)^2.$$

The proof follows from the known formula [see e.g. [5], p. 296]

$$\frac{1}{n(n-1)} \sum_{1 \le i < j \le n} (X_i - X_j)^2 = \frac{1}{(n-1)} \sum_{i=1}^{n} (X_i - \bar{x})^2 \tag{10}$$

and the obvious identity

$$\sum_{i=1}^{2n} \sum_{j=1}^{2n} (Z_i - Z_j)^2 = \sum_{i,j=1}^{n} (X_i - X_j)^2 + \sum_{i,j=1}^{n} (Y_i - Y_j)^2 + 2 \sum_{i=1}^{n} \sum_{j=1}^{n} (X_i - Y_j)^2, \tag{11}$$

by direct but non trivial calculations.

Really, let us use the standard notation

$$S_x^2 = \frac{1}{(n-1)} \sum_{i=1}^{n} (X_i - \bar{x})^2$$

And S_y^2 and S_z^2 will be understood in the similar way. Denote

$$S_{xy} = \frac{1}{n^2}\sum_{i=1}^{n}\sum_{j=1}^{n}(X_i - Y_j)^2.$$

Note that due to formula (10) for X replaced by Z

$$V(Z) = 2n[\sum_{i=1}^{n}(X_i-(\bar{x}+\bar{y})/2)^2+\sum_{j=1}^{n}(Y_i-(\bar{x}+\bar{y})/2)^2] = 2n(n-1)(S_x^2+S_y^2)+n^2(\bar{x}-\bar{y})^2. \tag{12}$$

From (10) and (11) we obtain

$$n^2 S_{xy} = V(Z) - n(n-1)(S_x^2 + S_y^2). \tag{13}$$

Therefore

$$S_{xy} = \frac{1}{n}(n-1)(S_x^2 + S_y^2) + (\bar{x} - \bar{y})^2,$$

and we obtain

$$\Phi_{nn} = S_{xy} - \frac{1}{n}(n-1)(S_x^2 + S_y^2) = (\bar{x} - \bar{y})^2.$$

Thus Lemma 1 is proved. It follows from this lemma, that the criterion Φ_{nn} in this case is equivalent to the criterion $(\bar{x} - \bar{y})^2$.

Let us turn to the proof of the theorem.

Assume that either H_0 or H_1 holds. Then due to the law of large numbers for $U-$statistics [5] each of the sums

$$\Phi_{AB} = \frac{1}{n^2}\sum_{i,j=1}^{n} \ln(1 + (X_i - Y_j)^2),$$

$$\Phi_A + \Phi_B = \frac{1}{n^2}\sum_{1\leq i<j\leq n} \ln(1 + (X_i - X_j)^2) + \frac{1}{n^2}\sum_{1\leq i<j\leq n} \ln(1 + (Y_i - Y_j)^2)$$

tends to J_0.

Moreover,

$$\Phi_{AB} = J_0 + o(n^2),$$

$$\Phi_A + \Phi_B = J_0(1 - \frac{1}{n}) + o(n^2).$$

Note that

$$nT_n = n[\Phi_{AB} - J_0] - n[\Phi_A - \frac{1}{2}J_0] - n[\Phi_B - \frac{1}{2}J_0].$$

Let us apply the limit theorem for U-statistics (see Theorem 7.1 [5]) to each of the three terms in brackets. We obtain that nT_n tends to a random variable with a finite variance. Note that the conditions of the limit theorem are fulfilled for distributions F_1 with the property (3).

Note that $0 \le \ln(1+x^2) \le x^2$. By this reason Φ_{AB} is between 0 and S_{xy}. Due to theorem about the mean it is equal to $c_n S_{xy}$, $0 < c_n < 1$ and c_n tends to a constant c with $n \to \infty$. In a similar way, $\Phi_A + \Phi_B = c_{1n}(\frac{n}{n-1}(S_x^2 + S_y)^2)$ and c_{1n} tends to c_1 while $c_1 = c$.

Let C be an arbitrary positive number,

$$\tilde{X} = (\tilde{X}_1, \ldots, \tilde{X}_n), \;\; \tilde{Y} = (\tilde{Y}_1, \ldots, \tilde{Y}_n),$$

where $\tilde{X}_i = X_i$, if $|X_i| \le C$ and $\tilde{X}_i = C$ if $X_i > 0$, $\tilde{X}_i = -C$ if $X_i < 0$ otherwise. And $\tilde{Y}_i$ are determined similarly.

Consider the function

$$n\{\frac{1}{n^2}\sum_{i,j=1}^{n} \ln(1+(\tilde{X}_i - \tilde{Y}_j)^2 - \frac{1}{n^2}\sum_{i<j} \ln(1+(\tilde{X}_i - \tilde{X}_j)^2) - \tag{14}$$

$$\frac{1}{n^2}\sum_{i<j} \ln(1+(\tilde{Y}_i - \tilde{Y}_j)^2)\}. \tag{15}$$

Due to the presentations for Φ_{AB}, Φ_A and Φ_B derived above it can be checked that there exists a value t_n that depends on $\tilde{X}$ and $\tilde{Y}$ and numbers B_n such that it is equal to

$$t(\sum_{i=1}^{n} \tilde{X}_i/\sqrt{n} - \sum_{i=1}^{n} \tilde{Y}_i/\sqrt{n})^2 + B_n, \tag{16}$$

and B_n is $o(1)$.

Consider expression (14)–(15). Note that for distributions F_1 and F_2 satisfying (3) with $\tilde{X}_i$ and $\tilde{Y}_i$ replaced by X_i and Y_i, respectively, its variance is bounded from above due to that nT_n tends to a random variable with a finite variance. Therefore the expression (14)–(15) tends with $n \to \infty$ to a random variable with a finite variance for arbitrary C. Passing to the limit with $n \to \infty$ we obtain due to the central limit theorem that (16) has the limit distribution of the form (9), where L has the standard normal distribution. Since C is arbitrary we obtain that the limiting distribution has the required form.

For determining b in the part (ii) of the theorem we now can use the equality

$$(aL + b)^2 = \lim_{n\to\infty} nT_n, \tag{17}$$

that follows from the equality between (14)-(15) and (16). If H_0 take place we obviously have $b = 0$. In the case when H_1 take place EnT_n is asymptotically equivalent to

$$(n(J_h - J_0))^2 + En\hat{T_n}$$

where $\hat{T_n}$ received from T_n by replacing Y_i by $Y_i - b/\sqrt{n}$, $i = 1, \ldots, n$ and we obtain by passing to the limit with $n \to \infty$ that

$$b = \bar{b}h, \bar{b} = \sqrt{J^*(h)/h^2}.$$

And the asymptotic behaviour of the power announced in (ii) follows from the asymptotic normality of $\sqrt{nT_n}$. In order to calculate $\bar{b}$ in the case when F_1 is the standard Cachy distribution the following result is crucial.

Lemma 2. *If X and Y are independent random variables with the distribution $C(0,1)$, then*

$$E\ln(1+(X-Y)^2)=\ln 9,\quad E\ln(1+(X-Y-\theta)^2)-\ln 9=ln(1+\theta^2/9).$$

In order to prove this Lemma we need the following integrals

$$\int_R \frac{\ln(1+(x-y)^2)}{\pi(1+y^2)}dy=\ln(4+x^2),$$

$$\int_R \frac{\ln(4+x^2)}{\pi(1+x^2)}dx=\ln 9,$$

([6] 4.296.2 and 4.295.7.)

$$\int_R \frac{\ln(4+(x+\theta)^2)}{\pi(x^2+1)}dx=\ln(9+\theta^2),$$

[see [7], formula (2.6.14.19)]. Using these integrals we obtain

$$E\ln(1+(X-Y-\theta)^2)-\ln 9=2\int_R\int_R \frac{\ln(1+(x-y-\theta)^2)}{\pi^2(1+x^2)(1+y^2)}dxdy-\ln 9$$

$$=\int_R \frac{\ln(4+(y+\theta)^2)}{\pi(1+y^2)}dy-\ln 9=\ln(9+\theta^2)-\ln 9=\ln(1+\theta^2/9).$$

Submitting here $\theta=0$ we obtain both formulas of the Lemma. Note that $\theta^2=nh^2$ and

$$\lim_{n\to\infty} n\ln(1+\theta^2/9)=(1/9)h^2.$$

Therefore we obtain $\bar{b}=1/3$ that completes the proof of the theorem.

References

1. Lehmann, E.: Testing Statistical Hypotheses, Probability and Statistics Series. Wiley, Hoboken (1986)
2. Buening, H.: Kolmogorov-Smirnov and Cramer-von Mises type two-sample tests with various weight functions. Commun. Stat.-Simul. Comput. **30**, 847–865 (2001)
3. Anderson, T.W.: Anderson-Darling tests of goodness-of-fit. In: Lovric, M. (ed.) International Encyclopedia of Statistical Science. Springer, Heidelberg (2011). https://doi.org/10.1007/978-3-642-04898-2_118
4. Zech, G., Aslan, B.: New test for the multivariate two-sample problem based on the concept of minimum energy. J. Stat. Comput. Simul. **75**(2), 109–119 (2005)
5. Hoeffding, W.: A class of statistics with asymptotically normal distribution. Ann. Math. Stat. **19**, 293–325 (1948)
6. Gradshteyn, I.S., Ryzhik, I.M.: Table of Integrals, Series and Products. 7th edn. Amsterdam, Boston, Heidelberg, London
7. Prudnikov, A.P., Brychkov, Y.A., Marichev, O.I.: Integrals and Series. Elementary Functions, Nauka, Moscow (1981). [in Russian]

Random Dimension Low Sample Size Asymptotics

Gerd Christoph[1] and Vladimir V. Ulyanov[2,3(✉)]

[1] Department of Mathematics, University of Magdeburg, Magdeburg, Germany
gerd.christoph@ovgu.de
[2] Faculty of Computational Mathematics and Cybernetics,
Lomonosov Moscow State University, Moscow, Russian Federation
vulyanov@cs.msu.ru
[3] Faculty of Computer Science, HSE University,
109028 Moscow, Russian Federation

Abstract. A first investigation of high-dimensional low-sample-size (HDLSS) asymptotics, Hall, Marron and Neeman (2005) discovered a surprisingly rigid geometric structure. A sample of size k taken from the standard m-dimensional normal distribution is for large m close to the vertices of the k-dimensional simplex in m-dimensional vector space. It follows from the analysis of three geometric statistics: the length of an observation, the distance between any two independent observations and the angle between these vectors. We generalize and refine the results constructing the second order Chebyshev-Edgeworth expansions under assumption that the data dimension is random and different scaling factors are chosen.

Keywords: HDLSS data · Chebyshev-Edgeworth expansions · Random dimension · Student's t-distribution · Laplace approximation

1 Three Geometric Statistics of Gaussian Vectors

We continue to study properties of high-dimensional Gaussian random vectors. In our earlier papers Christoph, Prokhorov and Ulyanov [8] and Bobkov, Naumov and Ulyanov [5] two-sided bounds were constructed for a probability density function of the distance of a Gaussian random element Y with zero mean from a point a in a Hilbert space $\mathbb{H}$. We get new results for basic geometric statistics connected with high-dimensional random normal vectors.

Let $\mathbf{X}_1 = (X_{1,1}, ..., X_{1,m})^T, ..., \mathbf{X}_k = (X_{k,1} ..., X_{k,m})^T$ be a random sample.

In a high-dimension low-sample-size (HDLSS) data it is assumed that dimension m tends to infinity and sample size k is fixed.

One of the first investigation of HDLSS data was done in Hall, Marron and Neeman (2005) [14]. It became the basis of research in high-dimensional mathematical statistics. See a recent survey on HDLSS asymptotics and its applications in Aoshima et al. [1]. Further development see e.g. in Fujikoshi, Ulyanov

A. N. Shiryaev et al. (Eds.): ICSM-5 2020, PROMS 371, pp. 215–228, 2021.
https://doi.org/10.1007/978-3-030-83266-7_16

and Shimizu [12] when both m and k may tend to infinity. This is an important framework of the current data analysis called *Big data*. In [14] it was discovered a surprisingly rigid geometric structure. A sample of size k taken from the standard m-dimensional normal distribution is close for large m to the vertices of the k-dimensional simplex in $\mathbb{R}^m$. It follows from the analysis of three geometric statistics:

the **length** $||\mathbf{X}_i||_m$ of an observation,

the **distance** $||\mathbf{X}_i - \mathbf{X}_j||_m$ between any two independent observations, and the **angle** $\theta_m = \text{ang}(\mathbf{X}_i, \mathbf{X}_j)$ between these vectors.

We generalize and refine the results constructing the second order Chebyshev-Edgeworth expansions under assumption that the data dimension is random and different scaling factors are chosen.

In case of $\dim \mathbb{H} < \infty$ we consider a sample of size k when the dimension of the observations is a random variable N_n with values in $\mathbb{N}_+ = \{1, 2, \ldots\}$.

The present work continues our investigations in Christoph and Ulyanov [9] on these three geometric statistics of Gaussian vectors with randomly distributed dimension N_n which depends on parameter $n \in \mathbb{N}_+$ and $N_n \to \infty$ in probability as $n \to \infty$. Let the vectors $\mathbf{X}_1, ..., \mathbf{X}_k$ and $N_1, N_2, ...$ be defined on one and the same probability space and it is assumed that they are independent. If $T_m := T_m(\mathbf{X}_1, ..., \mathbf{X}_k)$ is some statistic of the vectors $\mathbf{X}_1, ..., \mathbf{X}_k$ with *non-random dimension* $m \in \mathbb{N}_+$ then the random variable $T_{N_n} = T_{N_n}(\omega)$ is defined as:

$$T_{N_n}(\omega) := T_{N_n(\omega)}(\mathbf{X}_1(\omega), ..., \mathbf{X}_k(\omega)), \quad \omega \in \Omega \quad \text{and} \quad n \in \mathbb{N}_+.$$

Therefore, the statistics T_{N_n} based on statistics T_m are constructed from the sample $\{\mathbf{X}_1, ..., \mathbf{X}_k\}$, where these vectors have the dimension N_n.

In [9], the distribution function of the normalized angle $\theta_m = \text{ang}(\mathbf{X}_i, \mathbf{X}_j)$ was approximated by a second order Chebyshev-Edgeworth expansion with a bound $\leq C m^{-2}$ for all $m \in \mathbb{N}_+$. Furthermore, the fixed dimension m of the Gaussian vectors was substituted by a random number N_n and expansions for statistics θ_{N_n} were proved.

A natural question arises whether similar results hold for the length $||\mathbf{X}_i||_{N_n}$ and the distance $||\mathbf{X}_i - \mathbf{X}_j||_{N_n}$ of Gaussian vectors with random dimension N_n.

Two cases of random dimensions (or random sample sizes) N_n are considered as e.g. in Bening, Galieva and Korolev [2], Christoph, Monakhov and Ulyanov [7] and Christoph and Ulyanov [9]:

i) The random dimension $N_n = N_n(r) \in \mathbb{N}_+$ has negative binomial distribution displaced by 1 with probability of success $1/n$, positive parameter $r > 0$ and probabilities

$$\mathbb{P}(N_n(r) = j) = \frac{\Gamma(j + r - 1)}{\Gamma(j)\,\Gamma(r)} \left(\frac{1}{n}\right)^r \left(1 - \frac{1}{n}\right)^{j-1}, \; j \in \mathbb{N}_+. \tag{1}$$

ii) The random dimension $N_n = N_n(s) \in \mathbb{N}_+$ is discrete Pareto-like distributed with parameters $n \in \mathbb{N}_+$, $s > 0$ and distribution function

$$\mathbb{P}(N_n(s) \le k) = \left(\frac{k}{s+k}\right)^n \quad \text{where} \quad N_n(s) = \max_{1\le j\le n} Y_j(s), \tag{2}$$

and $Y(s), Y_1(s), Y_2(s), ...,$ are independent discrete Pareto II distributed random variables with the common distribution

$$\mathbb{P}\big(Y(s) \le k\big) = \frac{k}{s+k} \quad \text{and} \quad \mathbb{P}(Y(s) = k) = \frac{s}{(s+k)\,(s+k-1)}, \; k \in \mathbb{N}_+. \tag{3}$$

The discrete $Y(s)$ on integers is the discretized continuous Pareto II (Lomax) random variable, see Buddana and Kozubowski [6].

Both cases of random dimensions of the Gaussian vectors are also interesting because $\mathbb{E}N_n(r) = r(n-1)+1 < \infty$ and $\mathbb{E}N_n(s) = \infty$, which has an influence on the normalization factors.

The rest of the paper is organized as follows: In Sect. 2, Chebyshev-Edgeworth expansions are proved for the geometric statistics of Gaussian vectors with fixed dimension m. Section 3 presents the transfer theorem for results with fixed sample size (in our case the dimension of the vectors) m to those with random sample size N_n. The main results are given in Sects. 4 and 5 when the random sample size is negative binomial $N_n(r)$ or discrete Pareto-like $N_n(s)$ distributed, respectively. In Sect. 6 the main results are proved.

2 Approximation for Geometric Statistics of m-Dimensional Normal Vectors

Let $\mathbf{X}_i = (X_{i,1}, ..., X_{i,m})^T, ..., \mathbf{X}_j = (X_{j,1} ..., X_{j,m})^T$ be m-dimensional vectors chosen from a sample $\{\mathbf{X}_1, ..., \mathbf{X}_k\}$ of normal distribution $\mathcal{N}(\mathbf{0}_m, \mathrm{I}_m)$ with mean vectors $\mathbb{E}\mathbf{X}_k = \mathbf{0}_m$ and covariance matrix I_m for $1 \le i < j \le k \le m$.

The **length** of the vector $\mathbf{X}_j$ is defined by the Euclidean distance $||\cdot||_m$:

$$||\mathbf{X}_i||_m = S_m^{1/2} \quad \text{with} \quad S_m = \sum\nolimits_{k=1}^{m} X_{i,k}^2\,. \tag{4}$$

and similarly the **distance** $||\mathbf{X}_i - \mathbf{X}_j||_m$ between any two independent vectors

$$||\mathbf{X}_i - \mathbf{X}_i||_m = \sum\nolimits_{k=1}^{m} (X_{i,k} - X_{j,k})^2\,. \tag{5}$$

The distribution of distance $||\mathbf{X}_i - \mathbf{X}_j||_m$ is closely linked to the distribution of length $||\mathbf{X}_i||_m$, since $(X_{i,k} - X_{j,k})/\sqrt{2}$ has also standard normal distribution $\Phi(x)$. Therefore

$$\mathbb{P}(||\mathbf{X}_i - \mathbf{X}_j||_m/\sqrt{2} \le x) = \mathbb{P}(||\mathbf{X}_i||_m \le x). \tag{6}$$

The **angle** $\theta_m = \text{ang}(\mathbf{X}_i, \mathbf{X}_j)$ between these two independent vectors with vertex at the origin and the **sample correlation coefficient** $R_m(\mathbf{X}_i, \mathbf{X}_j)$ are connected by:

$$\cos\theta_m = \frac{||\mathbf{X}_i||_m^2 + ||\mathbf{X}_j||_m^2 - ||\mathbf{X}_i - \mathbf{X}_j||_m^2}{2\,||\mathbf{X}_i||_m\,||\mathbf{X}_j||_m} = R_m(\mathbf{X}_i, \mathbf{X}_j) = R_m. \quad (7)$$

Hall, Marron and Neeman [14] showed

- for the length $||\mathbf{X}_i||_m = \sqrt{m} + \mathcal{O}_p(1)$,
- for the distance $||\mathbf{X}_i - \mathbf{X}_j||_m = \sqrt{2m} + \mathcal{O}_p(1)$ with $i \neq j$ and
- for the θ_m = angle $\text{ang}(\mathbf{X}_i, \mathbf{X}_j) = \frac{1}{2}\pi + \mathcal{O}_p(m^{-1/2})$ with $i \neq j$,

where $1 \leq i < j \leq k \leq m$ and $\mathcal{O}_p$ refers to the stochastic boundedness.

The length of the vector $\mathbf{X}_i$ drawn from an m-dimensional normal distribution $\mathcal{N}(\mathbf{0}, \mathrm{I}_m)$ is defined in (4) as $||\mathbf{X}_i||_m = S_m^{1/2}$, where the statistics S_m as a sum of the squares of m independent standard normal random variables has **chi-square distribution** with m degrees of freedom and

$$V_m = \frac{S_m - m}{\sqrt{2\,m}} \quad (8)$$

is asymptotically standard normally distributed. With the two-term Chebyshev-Edgeworth expansions in the central limit theorem for the distribution function of V_m, the following inequality results for all $m \in \mathbb{N}$

$$\left| P\Big(V_m \leq x\Big) - \Phi(x) - \varphi(x)\left(\frac{\lambda_3\,H_2(x)}{6\,\sqrt{m}} + \frac{\lambda_3^2\,H_5(x)}{72\,m} + \frac{\lambda_4\,H_3(x)}{24\,m}\right)\right| \leq \frac{C}{m^{3/2}}$$

where $H_2(x) = x^2 - 1$, $H_3(x) = x^3 - 3x$, $H_5(x) = x^5 - 10x^3 + 15x$ are the Chebyshev-Hermite polynomials, skewness $\lambda_3 = \sqrt{8}$ and excess kurtosis $\lambda_4 = 12$ of S_1, see Petrov [19, Sec. 5.7, Theorem 5.18].

Then $S_m = m(1 + \sqrt{2/m}\,V_m)$ and Tayor expansion of $(1+u)^{1/2}$ lead to

$$||\mathbf{X}_i||_m = S_m^{1/2} = \sqrt{m}\left(1 + \frac{1}{\sqrt{2\,m}}\,V_m - \frac{1}{4\,m}\,V_m^2 + \frac{\sqrt{2}}{8\,m^{3/2}}V_m^3 + \ldots\right) \quad (9)$$

Define the statistics

$$Z_m = \sqrt{2}\Big(\frac{||\mathbf{X}_i||_m}{\sqrt{m}} - 1\Big) \quad \text{and} \quad Z_m^* = \sqrt{2}\Big(\frac{||\mathbf{X}_i - \mathbf{X}_j||_m}{\sqrt{2\,m}} - 1\Big), \quad (10)$$

then (6) results in

$$P\Big(\sqrt{m}\,Z_m \leq x\Big) = P\Big(\sqrt{m}\,Z_m^* \leq x\Big). \quad (11)$$

It follows from (9) that the statistic $T_1 = \sqrt{m}Z_m$ holds

$$T_1 = \sqrt{m}Z_m = V_m - \frac{\sqrt{2}}{4\sqrt{m}}\,V_m^2 + \frac{\sqrt{1}}{4\,m}\,V_m^3 + \ldots \quad (12)$$

Following the sketch of the proof in Kawaguchi, Ulyanov and Fujikoshi [16, Theorem 1] (The coefficients in the polynomial $l_2(x)$ are incorrect.) and calculating the characteristic function $f_{T_1}(t)$, we obtain

$$f_{T_1}(t) = \mathbb{E}\left[e^{itV_m}\left(1 - \frac{\sqrt{2}(it)}{4\sqrt{m}}V_m^2 + \frac{(it)}{4\,m}V_m^3 + \frac{(it)^2}{16\,m}V_m^4 + \mathcal{O}_p(m^{-3/2})\right)\right]$$

$$= e^{-t^2/2}\left(1 - \frac{\sqrt{2}((it)^3 + 3(it))}{12\sqrt{m}} + \frac{(it)^6 - 6(it)^4 - 9(it)^2}{144\,m})\right) + \mathcal{O}(m^{-3/2}). \quad (13)$$

This results in the related expansion of the corresponding distribution function:

Proposition 1. *Let* $\mathbf{X}_i$ *be a vector drawn from an* m*-dimensional normal distribution* $\mathcal{N}(\mathbf{0}_m, \mathrm{I}_m)$. *Then with the asymptotic expansion for the distribution of normalized length* $Z_m = \sqrt{2}\Big(\frac{||\mathbf{X}_i||_m}{\sqrt{m}} - 1\Big)$ *we obtain the following inequality for all* $m \in \mathbb{N}$*:*

$$\left|P\Big(\sqrt{m}\,Z_m \le x\Big) - \Phi(x) - \varphi(x)\left(\frac{x^2-4}{6\sqrt{2\,m}} + \frac{x^5 - 16x^3 + 24x}{144\,m}\right)\right| \le \frac{C}{m^{3/2}}. \quad (14)$$

Corollary 1. *Let* $\mathbf{X}_i$ *and* $\mathbf{X}_j$, $i \ne j$ *be independent random vectors with an* m*-dimensional normal distribution* $\mathcal{N}(\mathbf{0}_m, \mathrm{I}_m)$. *Due to (11), distribution function of the normalized distance* $Z_m^* = \sqrt{2}\Big(\frac{||\mathbf{X}_i - \mathbf{X}_j||_m}{\sqrt{2\,m}} - 1\Big)$ *has the same asymptotic expansion as the distribution of normalized length* Z_m *and inequality (14) with replacing* Z_m *by* Z_m^*.

Second order Chebyshev-Edgeworth expansion of the angle $\theta_m = \mathrm{ang}(\mathbf{X}_i, \mathbf{X}_j)$ between independent vectors $\mathbf{X}_i$ and $\mathbf{X}_j$ with vertex at the origin and the corresponding sample correlation coefficient $R_m(\mathbf{X}_i, \mathbf{X}_j)$ with computable error bounds of approximation are shown in Christoph and Ulyanov [9, Section 2], using results of Konishi [17, Sect. 4], Johnson, Kotz and Balakrishnan [15, Chap. 32], Christoph, Ulyanov and Fujikoshi [11]:

$$\sup_x \left|P\Big(\sqrt{m}\;R_m \le x\Big) - \Phi(x) - \frac{x^3 - 5x}{4\,m}\varphi(x)\right| \le \frac{B_1}{m^2} \quad (15)$$

and

$$\sup_x \left|P\left(\sqrt{m}(\theta_m - \frac{\pi}{2}) \le x\right) - \Phi(x) - \frac{x^3 - 15x}{12\,m}\varphi(x)\right| \le \frac{B_2}{m^2}. \quad (16)$$

The estimates (15) and (16) were used in Christoph and Ulyanov [9] to obtain second order approximations the statistics R_{N_n} and $\Theta_{N_n} = \theta_{N_n} - \pi/2$ when the non-random dimension m of the vectors is replaced be a random dimension N_n, where the random dimension $N_n \to \infty$ in probability when the parameter $n \to \infty$.

Analogous results for the statistics $||\mathbf{X}_i||_m$ and $||\mathbf{X}_i - \mathbf{X}_j||_m$ are proven in Sects. 4 and 5 below, when the non-random dimension m is replaced be a random dimension N_n.

3 Auxiliary Proposition

In this section, expansions for the distribution function of statistics T_{N_n} obtained from samples with random sample size (here with random dimension N_n of the considered vectors $\mathbf{X}_i$) are obtained. These depend directly on the expansions concerning statistics T_m based on non-random samples size m and expansions regarding the random sample size N_n.

First we formulate the conditions determining expansions for the statistic T_m with $\mathbb{E}T_m = 0$ and the normalized random dimension N_n:

Assumption A: *Given $\gamma \in \{-1/2, 0, 1/2\}$, $a > 1$, $C_1 > 0$ and differentiable functions $f_1(x), f_2(x)$ with bounded derivatives $f_1'(x), f_2'(x)$ such that*

$$\sup_x \left| \mathbb{P}\big(m^\gamma T_m \le x\big) - \Phi(x) - \frac{f_1(x)}{\sqrt{m}} - \frac{f_2(x)}{m} \right| \le \frac{C_1}{m^a} \quad \text{for all} \quad m \in \mathbb{N}. \tag{17}$$

Remark 1. Statistics satisfying Assumption A are shown in (14), (15) and (16).

Assumption B: *Given constants $b > 0$ and $C_2 > 0$, real numbers g_n with $0 < g_n \uparrow \infty$ if $n \to \infty$, a distribution function $H(y)$ with $H(0+) = 0$ and a function $h_2(y)$of bounded variation that*

$$\sup_{y \ge 0} \left| \mathbb{P}\left(\frac{N_n}{g_n} \le y\right) - H(y) - \frac{h_2(y)\, \mathbb{I}_{\{b>1\}}(b)}{n} \right| \le \frac{C_2}{n^b} \quad \text{for all} \quad n \ge 1. \tag{18}$$

where $\mathbb{I}_A(x) = \begin{cases} 1, & x \in A \\ 0, & x \notin A \end{cases}$ *defines the indicator function of a set $A \subset \mathbb{R}$.*

Remark 2. The random dimensions $N_n(r)$ and $N_n(s)$ given in (1) and (2), respectively, fulfill Assumption B as shown in [9, Propositions 1 and 2], see (29) and (39) below.

Proposition 2. *Let $\gamma \in \{1/2,\, 0,\, -1/2\}$ and both Assumption A and B as well as the following requirements on $H(.)$ and $h_2(.)$ are fulfilled*

$$\left.\begin{array}{lr} i: \ H(1/g_n) \le c_1\, g_n^{-b} & for\, b > 0, \\ ii: \ \int_0^{1/g_n} y^{-1/2} dH(y) \le c_2\, g_n^{-b+1/2} & for\, b > 1/2, \\ iii: \ \int_0^{1/g_n} y^{-1} dH(y) \le c_3\, g_n^{-b+1} & for\, b > 1, \end{array}\right\} \tag{19}$$

$$\left.\begin{array}{lr} i: \ h_2(0) = 0, \quad and \quad |h_2(1/g_n)| \le c_4\, n\, g_n^{-b} & for\, b > 1, \\ ii: \ \int_0^{1/g_n} y^{-1} |h_2(y)| dy \le c_5\, n\, g_n^{-b} & for\, b > 1, \end{array}\right\} \tag{20}$$

where b is the convergence rate in (18). Then for all $n \ge 1$ is valid:

$$\sup_{x \in \mathbb{R}} \left| \mathbb{P}\Big(g_n^\gamma T_{N_n} \le x\Big) - G_{n,2}(x) \right| \le C_1\, \mathbb{E}\left(N_n^{-a}\right) + (C_3 D_n + C_4)\, n^{-b} + I_n, \tag{21}$$

where with $a > 1, b > 0, f_1(z), f_2(z), h_2(y)$ are given in (17) and (18)

$$G_{n,2}(x) = \left\{ \begin{array}{ll} \int_0^\infty \Phi(x\,y^\gamma)\mathrm{d}H(y), & 0 < b \leq 1/2, \\ \int_0^\infty \Big(\Phi(xy^\gamma) + \dfrac{f_1(x\,y^\gamma)}{\sqrt{g_n y}}\Big)\mathrm{d}H(y) =: G_{n,1}(x), & 1/2 < b \leq 1, \\ G_{n,1}(x) + \int_0^\infty \dfrac{f_2(x\,y^\gamma)}{g_n y}\mathrm{d}H(y) + \int_0^\infty \dfrac{\Phi(x\,y^\gamma)}{n}\mathrm{d}h_2(y), & b > 1, \end{array} \right\} \quad (22)$$

$$D_n = \sup_x \int_{1/g_n}^\infty \left| \frac{\partial}{\partial y}\left(\Phi(xy^\gamma) + \frac{f_1(xy^\gamma)}{\sqrt{g_n y}} + \frac{f_2(xy^\gamma)}{y g_n}\right)\right| \mathrm{d}y, \quad (23)$$

$$I_n = \sup_x \left(|I_1(x,n)| + |I_2(x,n)|\right), \quad (24)$$

$$I_1(x,n) = \int_{1/g_n}^\infty \Big(\frac{f_1(x\,y^\gamma)\,\mathbb{I}_{(0,1/2]}(b)}{\sqrt{g_n y}} + \frac{f_2(x\,y^\gamma)}{g_n\,y}\Big)\mathrm{d}H(y), \qquad b \leq 1, \quad (25)$$

and

$$I_2(x,n) = \int_{1/g_n}^\infty \Big(\frac{f_1(x\,y^\gamma)}{n\,\sqrt{g_n y}} + \frac{f_2(x\,y^\gamma)}{n\,g_n y}\Big)\mathrm{d}h_2(y), \qquad b > 1. \quad (26)$$

The constants C_1, C_3, C_4 are independent of n.

Proof. The proof is based on the statement in [2, Theorem 3.1] for $\gamma \geq 0$. Since in Theorems 1 and 2 in the present paper as well as in Christoph and Ulyanov [9, Theorems 1 and 2] the case $\gamma = -1/2$ is also considered, therefore the proof was adapted to $\gamma \in \{1/2,\, 0,\, -1/2\}$ in [9]. The conditions (19) and (20) guarantee integration range $(0, \infty)$ of the integrals in (22). The approximation function $G_{n,2}(x)$ in (22) is now a polynomial in $g_n^{-1/2}$ and $n^{-1/2}$. Present Proposition 2 differs from Theorems 1 and 2 in [9] only by the term $f_1(xy^\gamma)\,(g_n y)^{-1/2}$ and the added condition (19ii) to estimate this term. Therefore here the details are omitted. □

Remark 3. The domain $[1/g_n, \infty)$ of integration depends on g_n in (23), (25) and (26). Some of the integrals in (25) and (26) could tend to infinity with $1/g_n \to 0$ as $n \to \infty$ and thus worsen the convergence rates of the corresponding terms. See (47) in Sect. 6.

In the next two sections we consider the statistics Z_m and Z_m^* defined in (10) and the cases when the random dimension N_n is given in either (1) or (2). We use Proposition 2 when the limit distributions of scaled statistics Z_{N_n} are scale mixtures $G_\gamma(x) = \int_0^\infty \Phi(x\,y^\gamma)\mathrm{d}H(y)$ with $\gamma \in \{1/2,\, 0,\, -1/2\}$ that can be expressed in terms of the well-known distributions. We obtain non-asymptotic results for the statistics Z_{N_n} and $Z_{N_n}^*$, using second order approximations the statistics Z_m and Z_m^* given in (14) as well as for the random sample size N_n. In both cases the jumps of the distribution function of the random sample size N_n only affect the function $h_2(y)$ in formula (18).

4 The Random Dimension $N_n(r)$ is Negative Binomial Distributed

The negative binomial distributed dimension $N_n(r)$ has probability mass function (1)) and $g_n = \mathbb{E}(N_n(r)) = r\,(n-1)+1$. Schluter and Trede [21] (Sect. 2.1) underline the advantage of this distribution compared to the Poisson distribution for counting processes. They showed in a general unifying framework

$$\lim_{n\to\infty} \sup_y |\mathbb{P}(N_n(r)/g_n \le y) - G_{r,r}(y)| = 0, \tag{27}$$

where $G_{r,r}(y)$ is the Gamma distribution function with the identical shape and scale parameters $r > 0$ and density

$$g_{r,r}(y) = \frac{r^r}{\Gamma(r)}\, y^{r-1}\, e^{-ry}\, \mathbb{I}_{(0\,,\,\infty)}(y) \quad \text{for all} \quad y \in \mathbb{R}. \tag{28}$$

Statement (27) was proved earlier in Bening and Korolev [3, Lemma 2.2].

In [9, Proposition 1] the following inequality was proved for $r > 0$:

$$\sup_{y\ge 0}\left|\mathbb{P}\left(\frac{N_n(r)}{g_n} \le y\right) - G_{r,r}(y) - \frac{h_{2;r}(y)\,\mathbb{I}_{\{r>1\}}(r)}{n}\right| \le \frac{C_2(r)}{n^{\min\{r,2\}}}, \tag{29}$$

$$\text{where} \quad h_{2;r}(y) = \frac{1}{2\,r}\, g_{r,r}(y)\,\big((y-1)(2-r) + 2Q_1(g_n\, y)\big) \quad \text{for} \quad r > 1,$$

$$Q_1(y) = 1/2 - (y - [y]) \quad \text{and} \quad [y] \quad \text{is the integer part of a value } y. \tag{30}$$

Both Bening, Galieva and Korolev [2] and Gavrilenko, Zubov and Korolev [13] showed the rate of convergence in (29) for $r \le 1$. In Christoph, Monakhov and Ulyanov [7, Theorem 1] the Chebyshev-Edgeworth expansion (29) for $r > 1$ is proved.

Remark 4. The random dimension $N_n(r)$ satisfies Assumption 2 of the Transfer Propositions 2 with $g_n = \mathbb{E}N_n(r)$, $H(y) = G_{r,r}(y)$, $h_2(y) = h_{2;r}(y)$ and $b = 2$.

In (21), negative moment $\mathbb{E}(N_n(r))^{-a}$ is required where m^{-a} is rate of convergence of Chebyshev-Edgeworth expansion for T_m in (17). Negative moments $\mathbb{E}(N_n(r))^{-a}$ fulfill the estimate:

$$\mathbb{E}\big(N_n(r)\big)^{-a} \le C(a,r)\begin{cases} n^{-\min\{r,\,a\}}, r \ne a \\ \ln(n)\, n^{-a}, r = a \end{cases} \quad \text{for all} \quad r > 0 \quad \text{and} \quad a > 0. \tag{31}$$

For $r = a$ the factor $\ln n$ cannot be removed. In Christoph, Ulyanov and Bening [10, Corollary 4.2] leading terms for the negative moments of $\mathbb{E}\big(N_n(r)\big)^{-p}$ were derived for any $p > 0$ that lead to (31).

The expansions of the length of the vector Z_m in (14) as well as of the sample correlation coefficient R_n in (15) and the angle θ_m in (16) have as limit distribution the standard normal distribution $\Phi(x)$. Therefore, with $g_n = \mathbb{E}N_n(r)$ and $\gamma \in \{1/2, 0, -1/2\}$, limit distributions for

$$\mathbb{P}\Big(g_n^\gamma (N_n(r))^{1/2-\gamma} Z_{N_n(r)} \le x\Big) \quad \text{are} \quad G_\gamma(x,r) = \int_0^\infty \Phi(x\, y^\gamma) \mathrm{d}G_{r,r}(y).$$

These scale mixtures distributions $G_\gamma(x,r)$ are calculated in Christoph and Ulyanov [9, Theorems 3–5]. We apply Proposition 2 to the statistics

$$T_{N_n(r)} = N_n(r)^{1/2-\gamma}\, Z_{N_n(r)} \quad \text{with the normalizing factor} \quad g_n^\gamma = \mathbb{E}(N_n(r))^\gamma.$$

The limit distributions are:

- for $\gamma = 1/2$ and $r > 0$ the **Student's t-distribution** $S_{2\,r}(x)$ with density

$$s_{2\,r}(x) = \frac{\Gamma(r+1/2)}{\sqrt{2\,r\pi}\,\Gamma(r)} \left(1 + \frac{x^2}{2\,r}\right)^{-(r+1/2)}, \quad x \in \mathbb{R}, \tag{32}$$

- for $\gamma = 0$ the **normal law** $\Phi(x)$,
- for $\gamma = -1/2$ and $r = 2$ the **generalized Laplace distributions** $L_2(x)$ with density $l_2(x)$:

$$L_2(x) = \frac{1}{2} + \frac{1}{2}\,\mathrm{sign}(x)\,(1 - (1+|x|)\,e^{-2\,|x|}) \quad \text{and} \quad l_2(x) = \left(\frac{1}{2} + |x|\right)\, e^{-2\,|x|}.$$

For arbitrary $r > 0$ Macdonald functions $K_{r-1/2}(x)$ occur in the density $l_r(x)$, which can be calculated in closed form for integer values of r.

The standard Laplace density with variance 1 is $l_1(x) = \dfrac{1}{\sqrt{2}}\,\mathrm{e}^{-\sqrt{2}\,|x|}$.

Theorem 1. *Let Z_m and $N_n(r)$ with $r > 0$ be defined by (10) and (1), respectively. Suppose that (14) is satisfied for Z_m and (29) for $N_n(r)$. Then the following statements hold for all $n \in \mathbb{N}_+$:*

(i) **Student's t approximation** *using scaling factor $\sqrt{\mathbb{E}N_n(r)}$ by $Z_{N_n(r)}$*

$$\sup_x \left|\mathbb{P}\left(\sqrt{g_n}\, Z_{N_n(r)} \le x\right) - S_{2r;n}(x)\right| \le C_r \begin{cases} n^{-\min\{r,3/2\}}, & r \ne 3/2, \\ \ln(n)\, n^{-3/2}, & r = 3/2, \end{cases} \tag{33}$$

where

$$S_{2r;n}(x) = S_{2r}(x) + s_{2r}(x) \left(\frac{\sqrt{2}\,((2r-5)x^2 - 8r)}{12\,(2r-1)\sqrt{g_n}}\, \mathbb{I}_{\{r>1/2\}}(r) \right.$$
$$\left. + \frac{96r^2x + (-64r^2+128r)x^3 + (4r^2-32r+39)x^5}{(x^2+2r)(2r-1)\,g_n} \mathbb{I}_{\{r>1\}}(r) \right), \tag{34}$$

(ii) **Normal approximation** *with random scaling factor* $N_n(r)$ *by* $Z_{N_n(r)}$

$$\sup_x \left|\mathbb{P}(\sqrt{N_n(r)}\, Z_{N_n(r)} \le x) - \Phi_{n,2}(x)\right| \le C_r \begin{cases} n^{-\min\{r,3/2\}}, & r \ne 3/2, \\ \ln(n)\, n^{-3/2}, & r = 3/2, \end{cases} \tag{35}$$

where

$$\Phi_{n,2}(x) = \Phi(x) + \frac{\sqrt{2}\, r\, \Gamma(r-1/2)}{12\, \Gamma(r)\, \sqrt{g_n}} (x^2 - 4)\varphi(x)\, \mathbb{I}_{\{r>1/2\}}(r) + \frac{x^5 - 16\, x^3 + 24\, x}{144\, g_n} \left(\frac{r}{r-1} \mathbb{I}_{\{r>1\}}(r) + \ln n\, \mathbb{I}_{\{r=1\}}(r) \right). \tag{36}$$

(iii) **Generalized Laplace approximation** *if* $r = 2$ *with mixed scaling factor* $g_n^{-1/2}\, N_n(2)$ *by* $Z_{N_n(2)}$

$$\sup_x \left|\mathbb{P}\left(g_n^{-1/2}\, N_n(2)\, Z_{N_n(2)} \le x\right) - L_{n;2}(x)\right| \le C_2\, n^{-3/2} \tag{37}$$

where

$$L_{n;2}(x) = L_2(x) - \frac{1}{3\sqrt{g_n}} \left(\frac{1}{\sqrt{2}} + \sqrt{2}|x| - x^2 \right) e^{-2|x|} + \frac{1}{33\, g_n} (12\sqrt{2}\, x - 15|x|\, x + 2\, x^3)\, e^{-2|x|}. \tag{38}$$

5 The Random Dimension $N_n(s)$ is Discrete Pareto-Like Distributed

The Pareto-like distributed dimension $N_n(s)$ has probability mass function (2) and $\mathbb{E}(N_n(s)) = \infty$. Hence $g_n = n$ is chosen as normalizing sequence for $N_n(s)$.

Bening and Korolev [4, Sect. 4.3] showed that for integer $s \ge 1$

$$\lim_{n\to\infty} \sup_{y>0} |\mathbb{P}(N_n(s) \le n\, y) - H_s(y)| = 0.$$

where $H_s(y) = e^{-s/y}\mathbb{I}_{(0,\infty)}(y)$ is the continuous distribution function of the inverse exponential $W(s) = 1/V(s)$ with exponentially distributed $V(s)$ having rate parameter $s > 0$. As $\mathbb{P}(N_n(s) \le y)$, so $H_s(y)$ is heavy tailed with shape parameter 1 and $\mathbb{E}W(s) = \infty$.

Lyamin [18] proved a bound $|\mathbb{P}(N_n(s) \le n\, y) - H_s(y)| \le C/n$ and $C < 0.37$ for integer $s \ge 1$.

In [9, Proposition 2] the following results are presented for $s > 0$:

$$\sup_{y>0} \left| \mathbb{P}\left(\frac{N_n(s)}{n} \le y\right) - H_s(y) - \frac{h_{2;s}(y)}{n} \right| \le \frac{C_3(s)}{n^2}, \quad \text{for all} \quad n \in \mathbb{N}_+, \tag{39}$$

with $H_s(y) = e^{-s/y}$ and $h_{2;s}(y) = s\, e^{-s/y}\left(s - 1 + 2Q_1(n\, y)\right)/(2\, y^2)$ for $y > 0$, where $Q_1(y)$ is defined in (30). Moreover

$$\mathbb{E}\big(N_n(s)\big)^{-p} \le C(p)\, n^{-\min\{p,2\}}, \tag{40}$$

where for $0 < p \le 2$ the order of the bound is optimal.

The Chebyshev-Edgeworth expansion (39) is proved in Christoph, Monakhov and Ulyanov [7, Theorem 4]. The leading terms for the negative moments $\mathbb{E}\big(N_n(s)\big)^{-p}$ were derived in Christoph, Ulyanov and Bening [10, Corollary 5.2] that lead to (40).

Remark 5. The random dimension $N_n(s)$ satisfies Assumption 2 of the Transfer Propositions 2 with $H_s(y) = \mathrm{e}^{-s/y}$, $h_2(y) = h_{2;s}(y)$, $g_n = n$ and $b = 2$.

With $g_n = n$ and $\gamma \in \{1/2,\, 0,\, -1/2\}$, the limit distributions for

$$\mathbb{P}\left(n^{\gamma} N_n(s)^{1/2-\gamma} Z_{N_n(s)} \leq x\right) \quad \text{are now} \quad G_{\gamma}(x,s) = \int_0^{\infty} \Phi(x\, y^{\gamma}) \mathrm{d}H_s(y).$$

These scale mixtures distributions $G_{\gamma}(x,s)$ are calculated in Christoph and Ulyanov [9, Theorems 6–8]. We apply Proposition 2 to statistics

$$T_{N_n(s)} = N_n(s)^{1/2-\gamma}\, Z_{N_n(s)} \quad \text{with the normalizing factor } n^{\gamma}.$$

The limit distributions are:

- for $\gamma = 1/2$ **Laplace distributions** $L_{1/\sqrt{s}}(x)$ with density

$$l_{1/\sqrt{s}}(x) = \sqrt{s/2}\, \mathrm{e}^{-\sqrt{2\,s}|x|},$$

- for $\gamma = 0$ the **standard normal law** $\Phi(x)$ and
- for $\gamma = -1/2$ the **scaled Student's t-distribution** $S_2^*(x; \sqrt{s})$ with density

$$s_2^*(x; \sqrt{s}) = \frac{1}{2\sqrt{2\,s}} \Big(1 + \frac{x^2}{2\,s}\Big)^{-3/2}.$$

Theorem 2. *Let Z_m and $N_n(s)$ with $s > 0$ be defined by (10) and (2), respectively. Suppose that (14) is satisfied for Z_m and (39) for $N_n(s)$. Then the following statements hold for all $n \in \mathbb{N}_+$:*

(i) **Laplace approximation** *with non-random scaling factor n^{γ} by $Z_{N_n(s)}$:*

$$\sup_x \left|\mathbb{P}\left(\sqrt{n}\, Z_{N_n(s)} \leq x\right) - L_{1/\sqrt{s};n}(x)\right| \leq C_s\, n^{-3/2} \tag{41}$$

where

$$L_{1/\sqrt{s};n}(x) = L_{1/\sqrt{s}}(x) + l_{1/\sqrt{s}}(x) \left(\frac{\sqrt{2}}{12\, s\sqrt{n}}\, (sx^2 - 2\,(1 + \sqrt{2\,s}\;|x|) + \frac{s}{72\, n} \left(\frac{x^3\,|x|}{\sqrt{2\,s}} - \frac{8\,x^2}{s} + \frac{6\,x}{s^2}\,(1 + \sqrt{2\,s}\;|x|) \right)\right) \tag{42}$$

,

(ii) **Normal approximation** *with random scaling factor* $\sqrt{N_n(s)}$ *by* $Z_{N_n(r)}$*:*

$$\sup_x \left| \mathbb{P}\left(\sqrt{N_n(s)}\, Z_{N_n(s)} \le x\right) - \Phi_{n,2}(x)\right| \le C_s\, n^{-3/2}, \tag{43}$$

where

$$\Phi_{n,2}(x) = \Phi(x) + \varphi(x)\left(\frac{\sqrt{2}\,\pi(x^2-4)}{24\sqrt{n}} + \frac{x^5 - 16x^3 + 24x}{144\, s\, n}\right) \tag{44}$$

(iii) **Scaled Student's t-distribution** *with mixed scaling factor by* $Z_{N_n(s)}$

$$\sup_x \left| \mathbb{P}\left(n^{-1/2}\, N_n(s)\, Z_{N_n(s)} \le x\right) - S^*_{n;2}(x)\right| \le C_s\, n^{-3/2}, \tag{45}$$

where

$$S^*_{n;2}(x;\sqrt{s}) = S^*_2(x;\sqrt{s}) + s^*_2(x;\sqrt{s})\left(-\frac{\sqrt{2}\,(x^2+8\,s)}{12(2\,s+x^2)\sqrt{n}}\right.$$
$$\left. + \frac{1}{144\,n}\left(\frac{105x^5}{(2\,s+x^2)^3} + \frac{240x^3}{(2\,s+x^2)^2} + \frac{72x}{2\,s+x^2}\right)\right). \tag{46}$$

6 Proofs of Main Results

Proof. The proofs of Theorems 1 and 2 are based on Proposition 2. The structure of the functions f_1, f_2 and $h_2$in Assumptions A and B is similar to the structure of the corresponding functions in Conditions 1 and 2 in [9]. Therefore, the estimates of the term D_n and of the integrals $I_1(x,n)$ and $I_2(x,n)$ in (23), (25) and (24) as well as the validity of (19) and (20) in Proposition 2 when $H(y)$ is $G_{r,r}(y)$ or $H_s(y)$ can be shown analogously to the proofs for Lemmas 1, 2 or 4 in [9]. In Remark 3 above it was pointed out that the integrals in (25) and (25) can degrade the convergence rate. Let $r < 1$. With $|f_2(x\,y^\gamma| \le c^*$ we get

$$\int_{1/g_n}^{\infty} \frac{|f_2(x\,y^\gamma)|}{g_n\, y} \mathrm{d}G_{r,r}(y) \le \frac{c^* r^r}{\Gamma(r)\, g_n} \int_{1/g_n}^{\infty} y^{r-2}\mathrm{d}y \le \frac{c^* r^r}{(1-r)\Gamma(r)}\, g_n^{-r}. \tag{47}$$

The additional term $f_1(xy^\gamma)\,(g_n y)^{-1/2}$ in (17) in Assumption A is to be estimated with condition (19ii).

Moreover, the bounds for $\mathbb{E}(N_n)^{-3/2}$ follow from (31) and (40), since $a = 3/2$ in Assumption A, considering the approximation (14).

The integrals in (22) in Proposition 2 are still to be calculated. Similar integrals are calculated in great detail in the proofs of Theorems 3–8 in [9]. To obtain (34), we compute the integrals with Formula 2.3.3.1 in Prudnikov et al. [20]

$$M_\alpha(x) = \frac{r^r}{\Gamma(r)\sqrt{2\pi}} \int_0^\infty y^{\alpha-1} \mathrm{e}^{-(r+x^2/2)y} dy = \frac{\Gamma(\alpha)\, r^{r-\alpha}}{\Gamma(r)\sqrt{2\,\pi}} \left(1 + x^2/(2r)\right)^{-\alpha} \tag{48}$$

for $\alpha = r - 1/2,\ r\ +\ 1/2,\ r\ +\ 3/2$ and $p = r + x^2/2$.

Lemma 2 in [9] and $\int_0^\infty y^{-1} dG_{r,r}(y) = r/(r-1)$ for $r > 1$ lead to (36).

To show (38) we use Formula 2.3.16.2 in [20] with $n = 0, 1$ and Formula 2.3.16.3 in [20] with $n = 1, 2$ and $p = 2$ and $q = x^2/2$.

To obtain (42), we calculate the integrals again with Formula 2.3.16.3 in [20], with $p = x^2/2 > 0$, $q = s > 0$, $n = 0, 1, 2$.

Lemma 4 in [9] and $\int_0^\infty y^{-a-1} \mathrm{e}^{-s/y} dy = s^{-a} \Gamma(a)$ for $a = 3/2$, 2 lead to (44).

Finally, in $\int_0^\infty f_k(x/y^\gamma) y^{-2-k/2} \mathrm{e}^{-s/y} \mathrm{d}y$ we use the substitution $s/y = u$ to obtain, with (48), the terms in (46). □

Acknowledgements. Theorem 1 has been obtained under support of the Ministry of Education and Science of the Russian Federation as part of the program of the Moscow Center for Fundamental and Applied Mathematics under the agreement № 075-15-2019-1621. Theorem 2 was proved within the framework of the HSE University Basic Research Program.

References

1. Aoshima, M., Shen, D., Shen, H., Yata, K., Zhou, Y.-H., Marron, J.S.: A survey of high dimension low sample size asymptotics. Aust. N. Z. J. Stat. **60**(1), 4–19 (2018). https://doi.org/10.1111/anzs.12212
2. Bening, V.E., Galieva, N.K., Korolev, V.Y.: Asymptotic expansions for the distribution functions of statistics constructed from samples with random sizes [in Russian]. Inf. Appl. IPI RAN **7**(2), 75–83 (2013)
3. Bening, V.E., Korolev, V.Y.: On the use of Student's distribution in problems of probability theory and mathematical statistics. Theory Probab. Appl. **49**(3), 377–391 (2005)
4. Bening, V.E., Korolev, V.Y.: Some statistical problems related to the Laplace distribution [in Russian]. Inf. Appl. IPI RAN **2**(2), 19–34 (2008)
5. Bobkov, S.G., Naumov, A.A., Ulyanov V.V.: Two-sided inequalities for the density function's maximum of weighted sum of chi-square variables. arXiv:2012.10747v1 (2020). https://arxiv.org/pdf/2012.10747.pdf
6. Buddana, A., Kozubowski, T.J.: Discrete Pareto distributions. Econ. Qual. Control **29**(2), 143–156 (2014)
7. Christoph, G., Monakhov, M.M., Ulyanov, V.V.: Second-order Chebyshev-Edgeworth and Cornish-Fisher expansions for distributions of statistics constructed with respect to samples of random size. J. Math. Sci. (N.Y.) **244**(5), 811–839 (2020). Translated from Zapiski Nauchnykh Seminarov POMI, 466, Veroyatnost i Statistika. 26, 167–207 (2017)
8. Christoph, G., Prokhorov, Yu., Ulyanov, V.: On distribution of quadratic forms in Gaussian random variables. Theory Prob. Appl. **40**(2), 250–260 (1996)
9. Christoph, G., Ulyanov, V.V.: Second order expansions for high-dimension low-sample-size data statistics in random setting. Mathematics **8**(7), 1151 (2020)
10. Christoph, G., Ulyanov, V.V., Bening, V.E.: Second order expansions for sample median with random sample size. arXiv:1905.07765v2 (2020). https://arxiv.org/pdf/1905.07765.pdf

11. Christoph, G., Ulyanov, V.V., Fujikoshi, Y.: Accurate approximation of correlation coefficients by short Edgeworth-Chebyshev expansion and its statistical applications. In: Shiryaev, A.N., Varadhan, S.R.S., Presman, E.L. (eds.) Prokhorov and Contemporary Probability Theory. In Honor of Yuri V. Prokhorov. Springer Proceedings in Mathematics & Statistics, vol. 33, pp. 239–260. Springer, Heidelberg (2013). https://doi.org/10.1007/978-3-642-33549-5_13
12. Fujikoshi, Y., Ulyanov, V.V., Shimizu, R.: Multivariate Statistics. High-Dimensional and Large-Sample Approximations. Wiley Series in Probability and Statistics. Wiley, Hoboken (2010)
13. Gavrilenko, S.V., Zubov, V.N., Korolev, V.Y.: The rate of convergence of the distributions of regular statistics constructed from samples with negatively binomially distributed random sizes to the Student distribution. J. Math. Sci. (N.Y.) **220**(6), 701–713 (2017)
14. Hall, P., Marron, J.S., Neeman, A.: Geometric representation of high dimension, low sample size data. J. R. Stat. Soc. Ser. **67**, 427–444 (2005)
15. Johnson, N.L., Kotz, S., Balakrishnan, N.: Continuous Univariate Distributions, vol. 2, 2nd edn. Wiley, New York (1995)
16. Kawaguchi, Y., Ulyanov, V.V., Fujikoshi, Y.: Asymptotic distributions of basic statistics in geometric representation for high-dimensional data and their error bounds (Russian). Inf. Appl. **4**, 12–17 (2010)
17. Konishi, S.: Asymptotic expansions for the distributions of functions of a correlation matrix. J. Multivar. Anal. **9**, 259–266 (1979)
18. Lyamin, O.O.: On the rate of convergence of the distributions of certain statistics to the Laplace distribution. Mosc. Univ. Comput. Math. Cybern. **34**(3), 126–134 (2010)
19. Petrov, V.V.: Limit Theorems of Probability Theory. Sequences of Independent Random Variables. Clarendon Press, Oxford (1995)
20. Prudnikov, A.P., Brychkov, Y.A., Marichev, O.I.: Integrals and Series, Volume 1: Elementary Functions, 3rd edn. Gordon & Breach Science Publishers, New York (1992)
21. Schluter, C., Trede, M.: Weak convergence to the student and Laplace distributions. J. Appl. Probab. **53**(1), 121–129 (2016)

Limit of the Smallest Eigenvalue of a Sample Covariance Matrix for Spherical and Related Distributions

Pavel Yaskov(✉)

Steklov Mathematical Institute of RAS, Moscow, Russia
yaskov@mi-ras.ru

Abstract. For an isotropic random vector $\mathbf{x}_p$ in $\mathbb{R}^p$ and an independent of $\mathbf{x}_p$ random variable η, consider a sample covariance matrix associated with $\eta\mathbf{x}_p$. We show that with probability one, the empirical spectral distribution of this matrix has a weak limit μ and its smallest eigenvalue tends to the left edge of the support of μ, when p grows at the same rate as the sample size, quadratic forms of $\mathbf{x}_p$ satisfy a weak concentration property, and a uniform integrability condition holds for $\mathbf{x}_p$.

Keywords: Random matrices · Bai-Yin theorem

1 Introduction

This paper contributes to the recent literature on the smallest eigenvalues of sample covariance matrices. Under various dependence and moment conditions, lower bounds on the smallest eigenvalues are obtained in [1,2,8,10–12,14,17–19,21,22,24], among others.

In this paper, we consider sample covariance matrices of the form

$$\widehat{\Sigma}_n = \frac{1}{n}\sum_{k=1}^{n} \eta_k^2 \mathbf{x}_{pk}\mathbf{x}_{pk}^\top,$$

hereinafter $\{(\mathbf{x}_{pk}, \eta_k)\}_{k=1}^n$ are i.i.d. copies of $(\mathbf{x}_p, \eta)$, η is a random variable, and $\mathbf{x}_p$, $p \in \mathbb{N}$, is a random vector in $\mathbb{R}^p$ such that it is independent of η and $\mathbb{E}\mathbf{x}_p\mathbf{x}_p^\top = I_p$ for the identity matrix $I_p \in \mathbb{R}^{p\times p}$. Following [8] and [20], we impose uniform integrability conditions on $\mathbf{x}_p$ and assume that quadratic forms of $\mathbf{x}_p$ satisfy a weak concentration property (see Sect. 2 for details). As is shown in [26], a version of this property is a necessary and sufficient condition for the limiting spectral distribution of $\widehat{\Sigma}_n$ to be the Marchenko-Pastur distribution when $\eta \equiv 1$. Our conditions on $\mathbf{x}_p$ are general enough to cover many models of interest, in particular, $\mathbf{x}_p$ having either a centred log-concave distribution, or independent centred entries with uniformly integrable squares.

This work is supported by the Russian Science Foundation under grant 19-11-00290.

A. N. Shiryaev et al. (Eds.): ICSM-5 2020, PROMS 371, pp. 229–241, 2021.
https://doi.org/10.1007/978-3-030-83266-7_17

Due to the weak concentration property for quadratic forms, our setting can be considered as an extension of the case of spherical distributions corresponding to $\eta \mathbf{x}_p$ with $\mathbf{x}_p/\sqrt{p}$ uniformly distributed over the unit sphere S^{p-1} in $\mathbb{R}^p$.

Our main result is that with probability one, the empirical spectral distribution of $\widehat{\Sigma}_n$ has a nonrandom weak limit μ and the smallest eigenvalue of $\widehat{\Sigma}_n$ tends to the left edge of the support of μ when $n \to \infty$ and $p = p(n)$ is such that $p/n \to y \in (0, 1)$. These results are new, as we do not assume the existence of moments of order higher than 2. Our results improve the results of [8], where it is shown (for $\eta \equiv 1$) that the smallest eigenvalues converge in probability. In the proofs, we will use the method of Srivastava and Vershynin [17] (going back to [6]) with the modifications from [8] that allow to apply the weak concentration property for quadratic forms efficiently.

The paper is structured as follows. Section 2 presents main results. Proofs are given Sect. 3. An Appendix contains some technical lemmas.

2 Main Results

Let us introduce some notation and assumptions. Denote the spectral norm of $A \in \mathbb{C}^{p\times p}$ by $\|A\|$ and the smallest eigenvalue of symmetric $A \in \mathbb{R}^{p\times p}$ by $\lambda_{\min}(A)$. Set $\mathbb{C}^+ = \{z \in \mathbb{C} : \mathrm{Im}(z) > 0\}$. Let also $I(B) = 1$ if B holds and $I(B) = 0$ otherwise. Assume further that $\mathbb{R}^p$ is equipped with the standard Euclidean norm. For any $p \in \mathbb{N}$, we say that $\mathbf{x}_p$ is an isotropic random vector in $\mathbb{R}^p$ if $\mathbb{E}\mathbf{x}_p\mathbf{x}_p^\top = I_p$ for the identity matrix $I_p \in \mathbb{R}^{p\times p}$.

Our main assumptions are as follows.

(A1) For every sequence of orthogonal projectors $(\Pi_p)_{p=1}^\infty$ with $\Pi_p \in \mathbb{R}^{p\times p}$,

$$\mathbf{x}_p^\top \Pi_p \mathbf{x}_p = \mathrm{tr}(\Pi_p) + o_{\mathbb{P}}(p), \quad p \to \infty,$$

where $o_{\mathbb{P}}(\cdot)$ stands for $o(\cdot)$ in probability.

(A2) The family $\{(\mathbf{x}_p, v_p)^2 : v_p \in S^{p-1}, p \in \mathbb{N}\}$ is uniformly integrable, where S^{p-1} is the unit sphere in $\mathbb{R}^p$.

Assumption (A1) is a form of the weak concentration property for quadratic forms. Namely, we have the following proposition.

Proposition 1. *If* (A1) *holds and* $\mathbf{x}_p$ *is an isotropic random vector in* $\mathbb{R}^p$ *for all* $p \in \mathbb{N}$, *then*

$$\sup \mathbb{E}|\mathbf{x}_p^\top A_p \mathbf{x}_p - \mathrm{tr}(A_p)| = o(p), \quad p \to \infty,$$

where sup *is taken over symmetric positive semi-definite matrices* $A_p \in \mathbb{R}^{p\times p}$ *with* $\|A_p\| \leqslant 1$.

Assumption (A1) holds in any of the following cases:

(C1) each $\mathbf{x}_p$ has a centred isotropic log-concave distribution,

(C2) each $\mathbf{x}_p$ has independent entries $(X_{pk})_{k=1}^p$ such that $\mathbb{E}X_{pk} = 0$, $\mathbb{E}X_{pk}^2 = 1$, and the following Lindeberg condition holds:

$$\lim_{p\to\infty} \frac{1}{p} \sum_{k=1}^p \mathbb{E}X_{pk}^2 I(|X_{pk}| > \varepsilon\sqrt{p}) = o(p) \text{ for all } \varepsilon > 0.$$

This is shown in Lemma 2.5 of [15] for (C1) and in Proposition 2.1 of [25] for (C2). For more general classes of $\mathbf{x}_p$ satisfying (A1), we refer to [3,23,27].

For log-concave $\mathbf{x}_p$, (A2) also holds (see [8]). For $\mathbf{x}_p$ with independent entries, (A2) follows from the elementary proposition below.

Proposition 2. *If* (C2) *holds and* $\{X_{pk}^2 : p \in \mathbb{N}, 1 \leqslant k \leqslant p\}$ *is a uniformly integrable family, then* (A2) *holds.*

The main result of this paper can be stated as follows.

Theorem 1. *Let* $\mathbf{x}_p$ *be an isotropic random vector in* $\mathbb{R}^p$ *for all* $p \in \mathbb{N}$ *and let* η *be a random variable independent of* $\mathbf{x}_p$. *If* $p = p(n)$ *is such that* $p/n \to y \in (0,1)$ *and* (A1)–(A2) *hold, then with probability one,*

$$\lambda_{\min}(\widehat{\Sigma}_n) \to \max\{0, \sup_{s>0} \lambda(s)\}, \quad n \to \infty,$$

where $\lambda(s) = -ys^{-1} + \mathbb{E}\eta^2/(1+\eta^2 s)$, $s > 0$.

When $\eta \equiv 1$ and the entries of all $\mathbf{x}_p$ are i.i.d. centered random variables with unit variance, Theorem 1 reduces to the Bai-Yin theorem [4] (in the form of [19]) stating that with probability one,

$$\lambda_{\min}(\widehat{\Sigma}_n) \to (1-\sqrt{y})^2.$$

The proof of Theorem 1 consists of two parts. The first part is a lower bound on $\underline{\lim}\, \lambda_{\min}(\widehat{\Sigma}_n)$ that is derived using the method of Srivastava and Vershynin [17] with the modifications from [8] and [21]. The second part is an upper bound on $\overline{\lim}\, \lambda_{\min}(\widehat{\Sigma}_n)$, which follows from the theorem below.

Theorem 2. *Let* $\mathbf{x}_p$ *be an isotropic random vector in* $\mathbb{R}^p$ *for all* $p \in \mathbb{N}$ *and let* η *be a random variable independent of* $\mathbf{x}_p$. *If* (A1) *holds and* $p = p(n)$ *is such that* $p/n \to y > 0$, *then with probability one, the empirical spectral distribution*

$$\mu_n = \frac{1}{p} \sum_{k=1}^{p} \delta_{\lambda_k}$$

weakly converges to a probability measure μ, *whose Stieljes transform* $s = s(z)$, $z \in \mathbb{C}^+$, *is a unique solution in* $\mathbb{C}^+$ *of the equation* $z = -s^{-1} + y^{-1}\mathbb{E}\eta^2/(1+\eta^2 s)$. *Here* $\lambda_k = \lambda_k(n)$, $1 \leqslant k \leqslant p$, *are eigenvalues of* $n\widehat{\Sigma}_n/p$.

Theorem 2 extends Theorem 19.1.8 in [16] when $H^{(0)}$ is the null matrix, $c = 1/y$, and $\tau_\alpha = \eta_\alpha^2$. The latter uses a stronger version of (A1) with the convergence in L_2 instead of convergence in probability.

Notes and Comments. The lower edge of the support of μ from Theorem 2 is equal to $a = \max\{0, \sup\{\lambda(s)/y : s > 0\}\}$ for $\lambda = \lambda(s)$ given in Theorem 1 (see the proof of Lemma 3.2 in [22]). This implies that $\lambda_{\min}(\widehat{\Sigma}_n) \leqslant ay + o(1)$ a.s.

3 Proofs

Proof of Proposition 1. By (A1), if $\Pi_p \in \mathbb{R}^{p\times p}$ are orthogonal projectors for all $p \geqslant 1$, then for $Z_p = 1 + (\mathbf{x}_p^\top \Pi_p \mathbf{x}_p - \operatorname{tr}(\Pi_p))/p$, we have $\mathbb{E}Z_p = 1$, $Z_p \geqslant 0$ a.s., and $Z_p \xrightarrow{\mathbb{P}} 1$ as $p \to \infty$. Hence, $Z_p \to 1$ in L_1 for any sequence of orthogonal projectors $\{\Pi_p\}_{p=1}^\infty$. Thus,

$$\sup_{\Pi_p} \mathbb{E}|\mathbf{x}_p^\top \Pi_p \mathbf{x}_p - \operatorname{tr}(\Pi_p)| = o(p)$$

with sup taken over all orthogonal projectors $\Pi_p \in \mathbb{R}^{p\times p}$. Any non-zero diagonal matrix $D \in \mathbb{R}^{p\times p}$ with diagonal entries $\|D\| = \lambda_1 \geqslant \ldots \geqslant \lambda_p \geqslant 0$ can be written as

$$\frac{D}{\lambda_1} = \sum_{k=1}^{p} w_k D_k,$$

where $\lambda_{p+1} = 0$, $w_k = (\lambda_k - \lambda_{k+1})/\lambda_1 \geqslant 0$ are such that $\sum_{k=1}^p w_k = 1$, and each D_k is a diagonal matrix with diagonal entries in $1, \ldots, 1, 0, \ldots, 0$ (k ones). Hence, any (non-zero) symmetric positive semi-definite matrix A_p with $\|A_p\| \leqslant 1$ can be written as a convex combination of some orthogonal projectors up to a factor $\|A_p\|$. As a result, by the convexity of the L_1-norm,

$$\sup_{A_p}(\mathbb{E}|\mathbf{x}_p^\top A_p \mathbf{x}_p - \operatorname{tr}(A_p)|/\|A_p\|) \leqslant \sup_{\Pi_p} \mathbb{E}|\mathbf{x}_p^\top \Pi_p \mathbf{x}_p - \operatorname{tr}(\Pi_p)| = o(p),$$

where A_p as above. This finishes the proof of the lemma.

Proof of Proposition 2. Denote by $\mathcal{G}$ the class of convex non-decreasing functions $F : \mathbb{R}_+ \to \mathbb{R}_+$ such that $F(0) = 0$, $F(x)/x \to \infty$ as $x \to \infty$, and

$$\text{there is } c > 0 \text{ such that } F(2x) \leqslant cF(x) \text{ for all } x \geqslant 0. \tag{1}$$

By the lemma on page 770 in [13], the family $\mathcal{A} = \bigcup_{p\geqslant 1}\{X_{pk}^2\}_{k=1}^p$ is uniformly integrable iff there is $G \in \mathcal{G}$ such that $\sup\{\mathbb{E}G(X) : X \in \mathcal{A}\} = M < \infty$.

Note also if $F \in \mathcal{G}$, then $H(x) = F(x^2)$ also belongs to $\mathcal{G}$, since F is non-decreasing and convex, $f(x) = x^2$ is convex, and $F(4x^2) \leqslant c^2F(x^2)$ for all $x \geqslant 0$ and $c > 0$ from (1). In addition, $\bigl\{\sum_{j=1}^k v_j X_{pj}\bigr\}_{k=1}^p$ is a martingale for all $v = (v_1, \ldots, v_p) \in \mathbb{R}^p$. Therefore, by the Burkholder-Davis-Gundy inequality (see Theorem 1.1 in [7]), and the convexity of $G(x)$ and $G(x^2)$,

$$\mathbb{E}G((\mathbf{x}_p, v)^2) \leqslant C\mathbb{E}G\Bigl(\sum_{k=1}^p v_k^2 X_{pk}^2\Bigr) \leqslant C\sum_{k=1}^p v_k^2 \mathbb{E}G(X_{pk}^2) \leqslant CM$$

when $\sum_{k=1}^p v_k^2 = 1$, where $C > 0$ depends only on G. Thus, using again the lemma on page 770 in [13], we conclude that the family defined in (A2) is uniformly integrable.

Proof of Theorem 1. As follows from the remark after Theorem 2,

$$\lambda_{\min}(\widehat{\Sigma}_n) \leqslant \max\{0, \sup_{s>0} \lambda(s)\} + o(1) \quad \text{a.s.}$$

as $n \to \infty$. If $\sup_{s>0} \lambda(s) \leqslant 0$, then $\lambda_{\min}(\widehat{\Sigma}_n) \leqslant o(1)$ a.s. Since $\lambda_{\min}(\widehat{\Sigma}_n) \geqslant 0$, we conclude that $\lambda_{\min}(\widehat{\Sigma}_n) \to 0$ a.s.

Assume that $\sup_{s>0} \lambda(s) > 0$. The function $\lambda = \lambda(s)$, $s > 0$, is continuous (by the dominated convergence theorem) and such that $\lambda(s) \to 0$, $s \to \infty$, and

$$\lambda(s) = -\frac{y}{s} + \frac{1}{s}\mathbb{E}\frac{\eta^2}{s^{-1}+\eta^2} = -\frac{y}{s} + \frac{o(1)}{s} \to -\infty, \quad s \to 0+.$$

Hence, $\sup_{s>0} \lambda(s) = \lambda(\varphi)$ for some $\varphi > 0$.

We now describe the method of Srivastava and Vershynin [17] with the modifications proposed in [8]. Let $A_0 \in \mathbb{R}^{p\times p}$ be the zero matrix and

$$A_k = A_k(n) = \sum_{j=1}^{k} \eta_j^2 \mathbf{x}_j \mathbf{x}_j^\top, \quad k = 1, \dots, n,$$

where $\mathbf{x}_j$ is a short-hand for $\mathbf{x}_{pj}$. Further, let $m_A(l) = \operatorname{tr}(A - lI_p)^{-1}$. Let now $l_0 = -p/\varphi$ for φ given above. Then $m_{A_0}(l_0) = \varphi$.

Suppose for a moment that l_k, Δ_k, and Δ_k^R, $k = 1, \dots, n$, are such that: (a) $\Delta_k, \Delta_k^R \geqslant 0$; (b) $l_k = l_{k-1} + \Delta_k - \Delta_k^R$; (c) $\lambda_{\min}(A_{k-1}) > l_{k-1} + \Delta_k$; (d) $m_{A_k}(l_{k-1} + \Delta_k) \leqslant m_{A_{k-1}}(l_{k-1}) \leqslant \varphi$. Here and in what follows, Δ_k, Δ_k^R, and l_k may implicitly depend on p and n, but for brevity, we do not indicate this dependence in our notation. Explicit constructions will be given below.

Since $\Delta_k^R \geqslant 0$, then (a)–(c) imply that $\lambda_{\min}(A_k) > l_k$ and (d) implies that $m_{A_k}(l_k) \leqslant m_{A_{k-1}}(l_{k-1}) \leqslant \varphi$. As a result,

$$n\lambda_{\min}(\widehat{\Sigma}_n) = \lambda_{\min}(A_n) \geqslant l_n = l_0 + \sum_{k=1}^{n} \Delta_k - \sum_{k=1}^{n} \Delta_k^R$$

Following [8], we fix $\varepsilon \in (0,1)$ and set

$$\Delta_k^R = \min\{l \in \mathbb{Z}_+ : m_{A_k}(l_{k-1}+\Delta_k - l) - m_{A_k}(l_{k-1}+\Delta_k - l - 1) \leqslant (\varepsilon p)^{-1}\}. \quad (2)$$

By such definition, $m_{A_k}(l_k) - m_{A_k}(l_k - 1) \leqslant (\varepsilon p)^{-1}$ for $1 \leqslant k \leqslant n$. Further, we will also assume that $\varepsilon < \varphi^{-2}$ and, as a result, $m_{A_0}(l_0) - m_{A_0}(l_0 - 1) = \varphi^2/(p+\varphi) \leqslant (\varepsilon p)^{-1}$. We have the following upper bound on $\sum_{k=1}^n \Delta_k^R$.

Lemma 1. *If* (a)–(d) *hold and* $\{\Delta_k^R\}_{k=1}^n$ *are defined in* (2), *then*

$$\sum_{k=1}^{n} \Delta_k^R \leqslant \varepsilon\varphi p.$$

In addition, $\lambda_{\min}(A_k) - l_k \geqslant \sqrt{\varepsilon p} - 1$ *for all* $k = 1, \dots, n$.

The proof of Lemma 1 can be found in the Appendix. By the lemma (recall also $l_0 = -p/\varphi$),

$$\lambda_{\min}(\widehat{\Sigma}_n) \geqslant -\varphi^{-1}\frac{p}{n} + \frac{1}{n}\sum_{k=1}^{n} \Delta_k - \varepsilon\varphi\frac{p}{n}.$$

Denoting $\mathbb{E}_k = \mathbb{E}[\cdot|\mathcal{F}_k]$ for $\mathcal{F}_k = \sigma(\mathbf{x}_1, \ldots, \mathbf{x}_k)$, $1 \leqslant k \leqslant n$, and the trivial σ-algebra $\mathcal{F}_0$, we arrive at the bound

$$\lambda_{\min}(\widehat{\Sigma}_n) \geqslant -\frac{p}{n}\varphi^{-1} + \frac{1}{n}\sum_{k=1}^{n} \mathbb{E}_{k-1}\Delta_k + \frac{Z_n}{\sqrt{n}} - \varepsilon\varphi\frac{p}{n},$$

where

$$Z_n = \frac{1}{\sqrt{n}}\sum_{k=1}^{n}(\Delta_k - \mathbb{E}_{k-1}\Delta_k).$$

We need one more lemma (see Lemma 4.4 in [21]).

Lemma 2. *Let $(D_k)_{k=1}^n$ be a sequence of non-negative random variables adapted to a filtration $(\mathcal{F}_k)_{k=1}^n$ such that $\mathbb{E}(D_k^2|\mathcal{F}_{k-1}) \leqslant 1$ a.s., $k = 1, \ldots, n$, where $\mathcal{F}_0$ is the trivial σ-algebra. If*

$$Z = \frac{1}{\sqrt{n}}\sum_{k=1}^{n}(D_k - \mathbb{E}(D_k|\mathcal{F}_{k-1})),$$

then $\mathbb{P}(Z < -t) \leqslant \exp\{-t^2/2\}$ for all $t > 0$.

If we choose Δ_k being uniformly bounded by a positive constant, then, by Lemma 2, we will have that $\sum_{n=1}^{\infty} \mathbb{P}(Z_n < -\log n) < \infty$ and in view of the Borel-Cantelli lemma, $Z_n < -\log n$ infinitely often with zero probability. In addition, using $p/n = y + o(1)$, we will get that

$$\lambda_{\min}(\widehat{\Sigma}_n) \geqslant -\frac{y}{\varphi} + \frac{1}{n}\sum_{k=1}^{n} \mathbb{E}_{k-1}\Delta_k - \varepsilon\varphi y + o(1) \quad \text{a.s.}$$

To choose Δ_k, we need a version of Lemma 3.2 from [8] (for a proof, see the Appendix).

Lemma 3. *Let $A \in \mathbb{R}^{p\times p}$ be symmetric, $x \in \mathbb{R}^p$, $t, l \in \mathbb{R}$, $\varphi > 0$, and $\varepsilon \in (0,1)$. Let also*

$$Q(x,l) = x^\top(A - lI_p)^{-1}x \quad \text{and} \quad q(x,l) = \frac{x^\top(A - lI_p)^{-2}x}{\operatorname{tr}(A - lI_p)^{-2}}.$$

If $\operatorname{tr}(A^{-1}) \leqslant \varphi$, $\lambda_{\min}(A) \geqslant 2\varepsilon^{-2}$, and

$$\Delta = (1-\varepsilon)^2\frac{t^2 q(x,0) I(q(x,0) < \varepsilon^{-1})}{1 + t^2(1+2\varepsilon)\varphi} I(Q(x,\varepsilon^{-1}) < (1+2\varepsilon)\varphi),$$

then $\lambda_{\min}(A) > \Delta$ and $\operatorname{tr}(A + t^2xx^\top - \Delta I_p)^{-1} \leqslant \operatorname{tr}(A^{-1})$.

For $k = 1, \ldots, n$, define Δ_k and l_k inductively by $l_k = l_{k-1} + \Delta_k - \Delta_k^R$ with Δ_k^R from (2) and

$$\begin{aligned}\Delta_k =& (1-\varepsilon)^2\frac{\eta_k^2 q_k(\mathbf{x}_k, l_{k-1}) I(q_k(\mathbf{x}_k, l_{k-1}) < \varepsilon^{-1})}{1 + \eta_k^2(1+2\varepsilon)\varphi} \cdot \\ & \cdot I(Q_k(\mathbf{x}_k, l_{k-1} + \varepsilon^{-1}) < (1+2\varepsilon)\varphi),\end{aligned}$$

where (Q_k, q_k) are defined as (Q, q) in Lemma 3 with $A = A_{k-1}$. Taking now $p \geqslant \varepsilon^{-1}(2\varepsilon^{-2}+1)^2 + 2\varphi\varepsilon^{-2}$ and using Lemma 1 and 3, we will guarantee that

$$\lambda_{\min}(A_0) - l_0 = p/\varphi \geqslant 2\varepsilon^{-2} \quad \text{and} \quad \lambda_{\min}(A_k) - l_k \geqslant \sqrt{\varepsilon p} - 1 \geqslant 2\varepsilon^{-2},$$

$\lambda_{\min}(A_{k-1}) > l_{k-1} + \Delta_k$, and $m_{A_k}(l_k) \leqslant m_{A_{k-1}}(l_{k-1})$ for $k = 1, \ldots, n$.

To finish the proof, we need to bound $\mathbb{E}_{k-1}\Delta_k$ from below.

Lemma 4. *Under the conditions of Lemma 3, let $t = \eta$ and $x = \mathbf{x}_p$. Assume also that $m_A(0) - m_A(-1) \leqslant (\varepsilon p)^{-1}$. Then*

$$\mathbb{E}\Delta = (1-\varepsilon)^2 \mathbb{E}\frac{\eta^2}{1+\eta^2(1+2\varepsilon)\varphi}(1 - \mathbb{E}q(\mathbf{x}_p, 0)I(B))$$

and $\mathbb{P}(B) \leqslant \delta(\varepsilon, p)$, where for L_p given in Proposition 1,

$$\delta(\varepsilon, p) = \varepsilon + \frac{1}{\varepsilon\varphi}\left[2(\varepsilon^2 + (\varepsilon p)^{-1})(1+\varepsilon) + \frac{L_p}{\varepsilon^3 - (\varepsilon p)^{-1}}\right]$$

and $B = \{q(\mathbf{x}_p, 0) \geqslant \varepsilon^{-1}\} \cup \{Q(\mathbf{x}_p, \varepsilon^{-1}) \geqslant (1+2\varepsilon)\varphi\}$.

Lemma 5. *If* (A2) *holds, then*

$$M_p(\delta) = \sup \frac{1}{\operatorname{tr}(A_p)}\mathbb{E}(\mathbf{x}_p^\top A_p \mathbf{x}_p)I(B) \to 0, \quad \delta \to 0+,$$

uniformly in p, where sup *is taken over all nonzero symmetric positive semi-definite matrices $A_p \in \mathbb{R}^{p\times p}$ and all events B with $\mathbb{P}(B) \leqslant \delta$.*

Combining Lemma 4 and 5 yields

$$\frac{1}{n}\sum_{k=1}^{n} \mathbb{E}_{k-1}\Delta_k \geqslant (1-\varepsilon)^2 \mathbb{E}\frac{\eta^2}{1+\eta^2(1+2\varepsilon)\varphi}(1 - M_p(\delta_p))$$

and

$$\lambda_{\min}(\widehat{\Sigma}_n) \geqslant -\frac{y}{\varphi} + (1-\varepsilon)^2 \mathbb{E}\frac{\eta^2}{1+\eta^2(1+2\varepsilon)\varphi}(1 - M(\delta_p)) - \varepsilon\varphi y + o(1) \quad \text{a.s.},$$

where $\delta_p = \delta(\varepsilon, p)$ and $M(\delta) = \sup\{M_p(\delta) : p \geqslant 1\}$. Now, by Proposition 1, $\delta_p \to \delta(\varepsilon)$ as $p \to \infty$ for $\delta(\varepsilon) = \varepsilon + 2\varepsilon(1+\varepsilon)/\varphi$. As a result,

$$\varliminf_{n\to\infty} \lambda_{\min}(\widehat{\Sigma}_n) \geqslant -\frac{y}{\varphi} + (1-\varepsilon)^2 \mathbb{E}\frac{\eta^2}{1+\eta^2(1+2\varepsilon)\varphi}\big(1 - M(2\delta(\varepsilon))\big) - \varepsilon\varphi y \text{ a.s.}$$

for all $\varepsilon \in (0, \min\{\varphi^{-2}, 1\})$. Taking $\varepsilon \to 0+$, we get via Lemma 5 and the dominated convergence theorem that $M(2\delta(\varepsilon)) \to 0$ and with probability one,

$$\varliminf_{n\to\infty} \lambda_{\min}(\widehat{\Sigma}_n) \geqslant \lambda(\varphi) = -\frac{y}{\varphi} + \mathbb{E}\frac{\eta^2}{1+\eta^2\varphi}.$$

We have proved in the beginning that with probability one,

$$\overline{\lim_{n\to\infty}} \lambda_{\min}(\widehat{\Sigma}_n) \leqslant \lambda(\varphi).$$

As a result, $\mathbb{P}(\lambda_{\min}(\widehat{\Sigma}_n) \to \lambda(\varphi)) = 1$.

Proof of Theorem 2. If each $\mathbf{x}_p$ has the standard normal distribution, then the result follows from Theorem 7.2.2 in [16]. Consider now the general case. We will use the Stieltjes transform method. However, instead of Stieltjes transforms of the form $s_\nu(z) = \int_{\mathbb{R}} (\lambda - z)^{-1}\nu(d\lambda)$, $z \in \mathbb{C}^+$, for a finite measure ν on the Borel σ-algebra of $\mathbb{R}$, it is easier to work with Stieltjes transforms defined by

$$S_\nu(t) = \int_0^\infty \frac{\nu(d\lambda)}{\lambda + t}, \quad t > 0,$$

when the support of ν belongs to $\mathbb{R}_+$.

Let μ be the probability measure such that $s_\mu = s_\mu(z)$, $z \in \mathbb{C}^+$, is a unique solution in $\mathbb{C}^+$ of $z = -s^{-1} + y^{-1}\mathbb{E}\eta^2/(1+\eta^2 s)$. The existence of such solution and the corresponding probability measure μ is established in Lemma 7.2.4 in [16] (see also Step 3 of the proof of Theorem 4.3 on page 88 in [5]). In addition, $\mu(\mathbb{R}_+) = 1$, since μ can be obtained as a weak limit of probability measures with support in $\mathbb{R}_+$ (when any $\mathbf{x}_p$ has the standard normal distribution).

By continuity, $S_\mu = S_\mu(t)$, $t > 0$, is a unique positive solution of

$$-t = -S^{-1} + y^{-1}\mathbb{E}\eta^2/(1+\eta^2 S) \quad \text{or} \quad 1 = tS + \frac{1}{y}\mathbb{E}\frac{\eta^2 S}{1+\eta^2 S}. \tag{3}$$

In addition, by Lemma 3.1 in [26], if $\{\mu_n\}_{n\geqslant 1}$ are random probability measures with support in $\mathbb{R}_+$, then $\mathbb{P}(\mu_n \to \mu \text{ weakly}) = 1$ iff $\mathbb{P}(S_{\mu_n}(t) \to S_\mu(t)) = 1$ for all $t > 0$, hereinafter all limits with respect to n mean that n passes to infinity.

For $n \geqslant 1$, let μ_n be the empirical spectral distribution of $n\widehat{\Sigma}_n/p$. Then

$$S_n(t) = \frac{1}{p}\mathrm{tr}(n\widehat{\Sigma}_n/p + tI_p)^{-1} = \mathrm{tr}(n\widehat{\Sigma}_n + tpI_p)^{-1}, \quad t > 0,$$

is its Stieltjes transform. We will prove $\mathbb{P}(\mu_n \to \mu \text{ weakly}) = 1$ by showing that $S_n(t) \to S_\mu(t)$ a.s. for all $t > 0$. Applying the standard martingale argument (see Step 1 in the proof of Theorem 3.10 on page 54 in [5]) and the bound

$$w^\top(A + tI_p)^{-2}w/(1 + w^\top(A + tI_p)^{-1}w) \leqslant t^{-1}$$

valid for all symmetric positive semi-definite $A \in \mathbb{R}^{p\times p}$, $w \in \mathbb{R}^p$, and $t > 0$, we derive that

$$S_n(t) - \mathbb{E}S_n(t) \to 0 \quad \text{a.s.} \tag{4}$$

We finish the proof by checking that $\mathbb{E}S_{\mu_n}(t)$ converges to $S_\mu(t)$.

Let $(\mathbf{x}_p, \eta) = (\mathbf{x}_{p,n+1}, \eta_{n+1})$ be such that $\{(\mathbf{x}_{pk}, \eta_k)\}_{k=1}^{n+1}$ are i.i.d. Put

$$A_n = n\widehat{\Sigma}_n = \sum_{k=1}^n \eta_k^2 \mathbf{x}_{pk}\mathbf{x}_{pk}^\top \quad \text{and} \quad B_n = A_n + \eta^2\mathbf{x}_p\mathbf{x}_p^\top = \sum_{k=1}^{n+1} \eta_k^2 \mathbf{x}_{pk}\mathbf{x}_{pk}^\top.$$

We have

$$p = \operatorname{tr}\big((B_n + tpI_p)(B_n + tpI_p)^{-1}\big)$$
$$= \sum_{k=1}^{n+1} \eta_k^2 \mathbf{x}_{pk}^\top (B_n + tpI_p)^{-1}\mathbf{x}_{pk} + tp\operatorname{tr}(B_n + tpI_p)^{-1}.$$

Taking the expectation and using the exchangeability of $\{(\mathbf{x}_{pk}, \eta_k)\}_{k=1}^{n+1}$,

$$p = (n+1)\mathbb{E}(\eta^2 \mathbf{x}_p^\top (B_n + tpI_p)^{-1}\mathbf{x}_p) + tp\,\mathbb{E}\operatorname{tr}(B_n + tpI_p)^{-1}. \tag{5}$$

Recall the Sherman-Morrison formula

$$(A + ww^\top)^{-1} = A^{-1} - \frac{A^{-1}ww^\top A^{-1}}{1 + w^\top A^{-1} w}$$

valid for all symmetric positive definite $A \in \mathbb{R}^{p\times p}$ and $w \in \mathbb{R}^p$. The latter implies that

$$\operatorname{tr}(A + ww^\top)^{-1} = \operatorname{tr}(A^{-1}) - \frac{w^\top A^{-2} w}{1 + w^\top A^{-1} w} \text{ and } w^\top (A + ww^\top)^{-1} w = \frac{w^\top A^{-1} w}{1 + w^\top A^{-1} w}.$$

Therefore,

$$\mathbb{E}\operatorname{tr}(B_n + tpI_p)^{-1} = \mathbb{E}\operatorname{tr}(A_n + tpI_p)^{-1} + \mathbb{E}\frac{\eta^2 \mathbf{x}_p \operatorname{tr}(A_n + tpI_p)^{-2}\mathbf{x}_p}{1 + \eta^2 \mathbf{x}_p \operatorname{tr}(A_n + tpI_p)^{-1}\mathbf{x}_p}$$
$$= \mathbb{E}S_n(t) + O(1/p)$$

and $\mathbb{E}\operatorname{tr}(B_n + tpI_p)^{-1} \leqslant t^{-1}$. Further, we will show that

$$\mathbb{E}(\eta^2 \mathbf{x}_p^\top (B_n + tpI_p)^{-1}\mathbf{x}_p) = \mathbb{E}\frac{\eta^2 \mathbb{E}S_n(t)}{1 + \eta^2 \mathbb{E}S_n(t)} + o(1). \tag{6}$$

Suppose for a moment that (6) holds. As $p/n = y + o(1)$, (5) reduces to

$$\frac{1}{y}\mathbb{E}\frac{\eta^2 \mathbb{E}S_n(t)}{1 + \eta^2 \mathbb{E}S_n(t)} + t\mathbb{E}S_n(t) = 1 + o(1).$$

Note that $(\mathbb{E}S_n(t))_{n=1}^\infty$ is a bounded positive sequence. By (3), $S = S_\mu(t)$ is a unique solution in $\mathbb{R}_+$ of the limiting equation $y^{-1}\mathbb{E}(\eta^2 S)/(1 + \eta^2 S) + tS = 1$ when $t > 0$. As a result, any converging subsequence of $(\mathbb{E}S_n(t))_{n=1}^\infty$ tends to $S(t)$. Hence, $\mathbb{E}S_n(t) \to S_\mu(t)$. By (4), we conclude that $S_n(t) \to S_\mu(t)$ a.s. for all $t > 0$ and $\mu_n \to \mu$ weakly a.s.

To finish the proof, we need to check (6). By the Sherman-Morrison formula,

$$\eta^2 \mathbf{x}_p^\top (B_n + tpI_p)^{-1}\mathbf{x}_p = \eta^2 \mathbf{x}_p^\top (A_n + \eta^2 \mathbf{x}_p \mathbf{x}_p^\top + tpI_p)^{-1}\mathbf{x}_p$$
$$= \frac{\eta^2 \mathbf{x}_p^\top (A_n + tpI_p)^{-1}\mathbf{x}_p}{1 + \eta^2 \mathbf{x}_p^\top (A_n + tpI_p)^{-1}\mathbf{x}_p}.$$

By Proposition 1 and the independence of $\mathbf{x}_p$ and A_n, we get

$$\mathbf{x}_p^\top (A_n + tpI_p)^{-1}\mathbf{x}_p - S_n(t) \xrightarrow{\mathbb{P}} 0.$$

By (4), we also have $S_n(t) - \mathbb{E}S_n(t) \xrightarrow{\mathbb{P}} 0$. Hence,

$$\left|\frac{\eta^2\mathbf{x}_p^\top (A_n + tpI_p)^{-1}\mathbf{x}_p}{1+\eta^2\mathbf{x}_p^\top (A_n + tpI_p)^{-1}\mathbf{x}_p} - \frac{\eta^2\mathbb{E}S_n(t)}{1+\eta^2\mathbb{E}S_n(t)}\right| \leqslant$$

$$\leqslant \eta^2|\mathbf{x}_p^\top (A_n + tpI_p)^{-1}\mathbf{x}_p - \mathbb{E}S_n(t)| \xrightarrow{\mathbb{P}} 0.$$

Now, (6) follows from the last inequality the dominated convergence theorem. This finishes the proof of Theorem 2.

Appendix

Proof of Lemma 1. By the definition of $\varDelta_k^R$, for all $0 \leqslant l < \varDelta_k^R$,

$$m_{A_k}(l_{k-1} + \varDelta_k - l) - m_{A_k}(l_{k-1} + \varDelta_k - l - 1) > \frac{1}{\varepsilon p}.$$

Hence, using $l_k = l_{k-1} + \varDelta_k - \varDelta_k^R$ and (d) in the proof of Theorem 1, we get

$$\frac{\varDelta_k^R}{\varepsilon p} \leqslant \sum_{l=0}^{\varDelta_k^R - 1} (m_{A_k}(l_{k-1} + \varDelta_k - l) - m_{A_k}(l_{k-1} + \varDelta_k - l - 1)) =$$

$$= m_{A_k}(l_{k-1} + \varDelta_k) - m_{A_k}(l_k) \leqslant m_{A_{k-1}}(l_{k-1}) - m_{A_k}(l_k)$$

and

$$\sum_{k=1}^n \varDelta_k^R \leqslant \varepsilon p \sum_{k=1}^n (m_{A_{k-1}}(l_{k-1}) - m_{A_k}(l_k)) \leqslant \varepsilon p\, m_{A_0}(l_0) \leqslant \varepsilon \varphi p.$$

Let us prove the second inequality. By definition, $m_{A_k}(l) = \operatorname{tr}(A_k - lI_p)^{-1}$. Hence, it follows from $\lambda_{\min}(A_k) > l_{k-1} + \varDelta_k$, $l_k = l_{k-1} + \varDelta_k - \varDelta_k^R$, and the definition of $\varDelta_k^R$ that

$$0 \leqslant \frac{1}{\lambda_{\min}(A_k) - l_k} - \frac{1}{\lambda_{\min}(A_k) - l_k + 1} \leqslant m_{A_k}(l_k) - m_{A_k}(l_k - 1) \leqslant \frac{1}{\varepsilon p}.$$

Hence, $(\lambda_{\min}(A_k) - l_k)(\lambda_{\min}(A_k) - l_k + 1) \geqslant \varepsilon p$ and, as a result,

$$\lambda_{\min}(A_k) - l_k \geqslant \sqrt{\varepsilon p} - 1.$$

Lemma 1 is proved.

Proof of Lemma 3. We have $\operatorname{tr}(A^{-1}) \leqslant \varphi$ and $\varDelta \leqslant \varepsilon^{-1} < \lambda_{\min}(A)$. Due to Lemma 2.2 in [17], we will get that $\operatorname{tr}(A + t^2 xx^\top - \varDelta I_p)^{-1} \leqslant \operatorname{tr}(A^{-1})$ if we check that $\varDelta(1 + t^2 Q(x, \varDelta)) = \varDelta(1 + Q(tx, \varDelta)) \leqslant q(tx, \varDelta) = t^2 q(x, \varDelta)$.

It follows from $\Delta \leqslant \varepsilon^{-1} < \lambda_{\min}(A)$ that $Q(x,\Delta) \leqslant Q(x,\varepsilon^{-1})$ and

$$\frac{1+t^2Q(x,\Delta)}{1+t^2(1+2\varepsilon)\varphi} I(Q(x,\varepsilon^{-1}) < (1+2\varepsilon)\varphi) \leqslant 1.$$

Therefore, $\Delta(1+t^2Q(x,\Delta)) \leqslant (1-\varepsilon)^2 t^2 q(x,0)$. In addition,

$$q(x,0) = \frac{x^\top A^{-2}x}{\operatorname{tr}(A^{-2})} \leqslant \frac{x^\top (A-\Delta I_p)^{-2}x}{\operatorname{tr}(A^{-2})} = \frac{\operatorname{tr}(A-\Delta I_p)^{-2}}{\operatorname{tr}(A^{-2})} q(x,\Delta).$$

We only need to check that $(1-\varepsilon)^{-2}\operatorname{tr}(A^{-2}) \geqslant \operatorname{tr}(A-\Delta I_p)^{-2}$. Writing $A = \sum_{i=1}^p \lambda_i e_i e_i^\top$ for some $\lambda_i \in \mathbb{R}$ and orthonormal $e_i \in \mathbb{R}^p$, $i=1,\dots,p$, we rewrite the last inequality as

$$\frac{1}{(1-\varepsilon)^2}\sum_{i=1}^p \frac{1}{\lambda_i^2} \geqslant \sum_{i=1}^p \frac{1}{(\lambda_i-\Delta)^2}. \tag{7}$$

For any $\lambda \geqslant \varepsilon^{-2}$,

$$\frac{1}{(\lambda-\Delta)^2} \leqslant \frac{1}{(\lambda-\varepsilon^{-1})^2} = \frac{1}{(1-(\lambda\varepsilon)^{-1})^2\lambda^2} \leqslant \frac{1}{(1-\varepsilon)^2\lambda^2}.$$

Noting that $\lambda_i \geqslant \lambda_{\min}(A) \geqslant \varepsilon^{-2}$, we arrive at (7). This finishes the proof of $\Delta(1+t^2Q(x,\Delta)) \leqslant t^2 q(x,\Delta)$. Lemma 3 is proved.

Proof of Lemma 4. The first inequality follows from the definition of Δ and the independence of η and $\mathbf{x}_p$. Let us prove the second inequality.

Since $\lambda_{\min}(A) \geqslant 2/\varepsilon^2$ and $0<\varepsilon<1$, we have for $\lambda \geqslant \lambda_{\min}(A)$,

$$(1+\varepsilon)(\lambda - 1/\varepsilon) \geqslant \lambda \quad \text{or} \quad \lambda \geqslant \frac{1+\varepsilon}{\varepsilon^2}$$

and, as a result, $m_A(1/\varepsilon) \leqslant (1+\varepsilon)m_A(0) \leqslant (1+\varepsilon)\varphi$. Therefore,

$$\begin{aligned}\mathbb{P}(B) &\leqslant \mathbb{P}(q(\mathbf{x}_p,0) \geqslant \varepsilon^{-1}) + \mathbb{P}(Q(\mathbf{x}_p,\varepsilon^{-1}) \geqslant (1+2\varepsilon)\varphi)\\ &\leqslant \frac{\mathbb{E}q(\mathbf{x}_p,0)}{\varepsilon^{-1}} + \mathbb{P}(Q(\mathbf{x}_p,\varepsilon^{-1}) - m_A(1/\varepsilon) \geqslant \varepsilon\varphi)\\ &\leqslant \varepsilon + \frac{1}{\varepsilon\varphi}\mathbb{E}|Q(\mathbf{x}_p,\varepsilon^{-1}) - m_A(1/\varepsilon)|,\end{aligned}$$

where we take into account that $\mathbb{E}q(\mathbf{x}_p,0) = 1$ for isotropic $\mathbf{x}_p$. Writing $A = \sum_{i=1}^p \lambda_i e_i e_i^\top$ for some $\lambda_i \in \mathbb{R}$ and orthonormal $e_i \in \mathbb{R}^p$, $i=1,\dots,p$, we get

$$\mathbb{E}|Q(\mathbf{x}_p,\varepsilon^{-1}) - m_A(1/\varepsilon)| \leqslant R_1 + R_2,$$

where

$$R_1 = \sum_{i:\lambda_i \leqslant \varepsilon^3 p} \frac{\mathbb{E}(\mathbf{x}_p,e_i)^2+1}{\lambda_i - 1/\varepsilon}, \quad R_2 = \mathbb{E}\Big|\sum_{i:\lambda_i > \varepsilon^3 p} \frac{(\mathbf{x}_p,e_i)^2-1}{\lambda_i - 1/\varepsilon}\Big|.$$

Using inequalities $(1+\varepsilon)(\lambda_i - 1/\varepsilon) \geqslant \lambda_i$, we deduce that

$$R_1 \leqslant 2(1+\varepsilon) \sum_{i:\lambda_i \leqslant \varepsilon^3 p} \frac{1}{\lambda_i} \leqslant 2(1+\varepsilon) \sum_{i=1}^{p} \frac{\varepsilon^3 p + 1}{\lambda_i(\lambda_i + 1)} \leqslant 2(\varepsilon^2 + (\varepsilon p)^{-1})(1+\varepsilon),$$

$$R_2 \leqslant \frac{L_p}{\varepsilon^3 - (\varepsilon p)^{-1}},$$

where L_p is given in Proposition 1 and we have also used that

$$\sum_{i=1}^{p} \frac{1}{\lambda_i(\lambda_i + 1)} = m_A(0) - m_A(-1).$$

Combining the above bounds, we obtain the second inequality in Lemma 4.

Proof of Lemma 5. By (A2), the family $\mathcal{G} = \{(\mathbf{x}_p, v_p)^2 : v_p \in S^{p-1}, p \in \mathbb{N}\}$ is uniformly integrable. The same can be said about its convex hull (e.g., see Theorem 20 on page 23-II in [9]). For any nonzero symmetric positive semi-definite $A_p \in \mathbb{R}^{p\times p}$, $\mathbf{x}_p^\top A_p \mathbf{x}_p / \mathrm{tr}(A_p)$ belongs to the convex hull of $\mathcal{G}$ and the desired property follows from the standard properties of uniformly integrable families (e.g., see Theorem 19 on page 22-II in [9]).

References

1. Adamczak, R., Litvak, A.E., Pajor, A., Tomczak-Jaegermann, N.: Quantitative estimates of the convergence of the empirical covariance matrix in log-concave ensembles. J. Amer. Math. Soc. **23**(2), 535–561 (2010)
2. Adamczak, R., Litvak, A.E., Pajor, A., Tomczak-Jaegermann, N.: Sharp bounds on the rate of convergence of the empirical covariance matrix. C. R. Math. Acad. Sci. Paris **349**(3–4), 195–200 (2011)
3. Anatolyev, S., Yaskov, P.: Asymptotics of diagonal elements of projection matrices under many instruments/regressors. Economet. Theor. **33**(3), 717–738 (2017)
4. Bai, Z., Yin, Y.-O.: Limit of the smallest eigenvalue of a large-dimensional sample covariance matrix. Ann. Probab. **21**(3), 1275–1294 (1993)
5. Bai, Z., Silverstein, J.: Spectral Analysis of Large Dimensional Random Matrices, 2nd edn. Springer, New York (2010). https://doi.org/10.1007/978-1-4419-0661-8
6. Batson, J.D., Spielman, D.A., Srivastava, N.: Twice-Ramanujan sparsifiers. In: STOC'09-Proceedings of the 2009 ACM International Symposium on Theory of Computing, pp. 255–262. ACM, New York (2009)
7. Burkholder, D.L., Davis, B.J., Gundy, R.F.: Integral inequalities for convex functions of operators on martingales. In: Proceedings of the Sixth Berkeley Symposium on Mathematical Statistics and Probability, vol. 2: Probability Theory, pp. 223–240. Univ. of Calif. Press (1972)
8. Chafaï, D., Tikhomirov, K.: On the convergence of the extremal eigenvalues of empirical covariance matrices with dependence. Probab. Theory Relat. Fields **170**, 847–889 (2017). https://doi.org/10.1007/s00440-017-0778-9
9. Dellacherie, C., Meyer, P.-A.: Probabilities and Potential. Herman, Paris (1978)

10. van de Geer, S., Muro, A.: On higher order isotropy conditions and lower bounds for sparse quadratic forms. Elec. J. of Stat. **8**, 3031–3061 (2014)
11. Koltchinskii, V., Mendelson, S.: Bounding the smallest singular value of a random matrix without concentration. Int. Math. Res. Not. **23**, 12991–13008 (2015)
12. Lecue, G., Mendelson, S.: Sparse recovery under weak moment assumptions. J. Eur. Math. Soc. **19**, 881–904 (2017)
13. Meyer, P.-A.: Sur le lemme de la Valleé Poussin et un théoreme de Bismut. Séminaire de probabilités (Strasbourg) **12**, 770–774 (1978)
14. Oliveira, R.I.: The lower tail of random quadratic forms, with applications to ordinary least squares and restricted eigenvalue properties. Probab. Theory Relat. Fields **166**, 1175–1194 (2016). https://doi.org/10.1007/s00440-016-0738-9
15. Pajor, A., Pastur, L.: On the limiting empirical measure of eigenvalues of the sum of rank one matrices with log-concave distribution. Studia Math. **195**, 11–29 (2009)
16. Pastur, L., Shcherbina, M.: Eigenvalue distribution of large random matrices. In: Mathematical Surveys and Monographs, vol. 171. American Mathematical Society, Providence (2011)
17. Srivastava, N., Vershynin, R.: Covariance estimation for distributions with $2+\varepsilon$ moments. Ann. Probab. **41**, 3081–3111 (2013)
18. Tikhomirov, K.E.: The smallest singular value of random rectangular matrices with no moment assumptions on entries. Israel J. Math. **212**(1), 289–314 (2016). https://doi.org/10.1007/s11856-016-1287-8
19. Tikhomirov, K.: The limit of the smallest singular value of random matrices with i.i.d. entries. Adv. Math. **284**, 1–20 (2015)
20. Yaskov, P.A.: On the behaviour of the smallest eigenvalue of a high-dimensional sample covariance matrix. Russian Math. Surv. **68**(3), 569–570 (2013)
21. Yaskov, P.: Lower bounds on the smallest eigenvalue of a sample covariance matrix. Elect. Comm. in Probab. **19**, 1–10 (2014)
22. Yaskov, P.: Sharp lower bounds on the least singular value of a random matrix without the fourth moment condition. Elect. Comm. in Probab. **20**, 1–9 (2015)
23. Yaskov, P.: Variance inequalities for quadratic forms with applications. Math. Methods Statist. **24**(4), 309–319 (2015). https://doi.org/10.3103/S1066530715040055
24. Yaskov, P.: Controlling the least eigenvalue of a random Gram matrix. Linear Algebra Appl. **504**, 108–123 (2016)
25. Yaskov, P.: A short proof of the Marchenko-Pastur theorem. C.R. Math. **354**, 319–322 (2016)
26. Yaskov, P.: Necessary and sufficient conditions for the Marchenko-Pastur theorem. Electron. Commun. Probab. **21**, Article no. 73, 1–8 (2016)
27. Yaskov, P.: LLN for quadratic forms of long memory time series and its applications in random matrix theory. J. Theor. Probab. **31**(4), 2032–2055 (2018). https://doi.org/10.1007/s10959-017-0767-z

On Positive Recurrence of One-Dimensional Diffusions with Independent Switching

In Memory of Svetlana Anulova (19.10.1952–21.11.2020)

Alexander Veretennikov[1,2,3](✉)

[1] University of Leeds, Leeds, UK
ayv@iitp.ru
[2] National Research University Higher School of Economics, Moscow, Russian Federation
[3] Institute for Information Transmission Problems, Moscow, Russian Federation
https://www.hse.ru/en/org/persons/125432921

Abstract. Positive recurrence of one-dimensional diffusion with switching, with an additive Wiener process, and with one recurrent and one transient regime is established under suitable conditions on the drift in both regimes and on the intensities of switching. The approach is based on an embedded Markov chain with alternating jumps: one jump increases the average of the square norm of the process, while the next jump decreases it, and under suitable balance conditions this implies positive recurrence.

Keywords: 1D diffusion · Switching · Positive recurrence

1 Introduction

On a probability space $(\Omega, \mathcal{F}, (\mathcal{F}_t), \mathbb{P})$ with a one-dimensional $(\mathcal{F}_t)$-adapted Wiener process $W = (W_t)_{t\ge 0}$ on it, a one-dimensional SDE with switching is considered,

$$dX_t = b(X_t, Z_t)\,dt + dW_t, \quad t \ge 0, \quad X_0 = x,\ Z_0 = z,$$

where Z_t is a continuous-time Markov process on the state space $S = \{0, 1\}$ with (positive) intensities of respective transitions $\lambda_{01} =: \lambda_0$, & $\lambda_{10} =: \lambda_1$; the process Z is assumed to be independent of W and adapted to the filtration $(\mathcal{F}_t)$. We assume that these intensities are constants; this may be relaxed. Under the regime $Z = 0$ the process X is assumed positive recurrent, while under the regime $Z = 1$ its modulus may increase in the square mean with the rate comparable to the decrease rate under the regime $Z = 0$. This vague wording will be specified in the assumptions. Denote

$$b(x, 0) = b_-(x), \quad b(x, 1) = b_+(x).$$

A. N. Shiryaev et al. (Eds.): ICSM-5 2020, PROMS 371, pp. 242–252, 2021.
https://doi.org/10.1007/978-3-030-83266-7_18

The problem addressed in this paper is to find sufficient conditions for the positive recurrence (and, hence, for convergence to the stationary regime) for solutions of stochastic differential equations (SDEs) with switching in the case where not for all values of the modulating process the SDE is recurrent, and where it is recurrent, this property is assumed to be "not very strong". Earlier a similar problem was tackled in [2] in the exponential recurrent case; its method apparently does not work for the weaker polynomial recurrence. A new approach is offered. Other SDEs with switching were considered in [1,4,5,7], see also the references therein. Neither of these works address exactly the problem which is attacked in this paper: some of them tackled an exponential recurrence, some other study the problem of a simple recurrence versus transience.

2 Main Result: Positive Recurrence

The existence and pathwise uniqueness of the solution follows easily from [9], or from [6], or from [8], although, neither of these papers tackles the case with switching. The next theorem is the main result of the paper.

Theorem 1. *Let the drift b be bounded and let there exist $r_-, r_+, M > 0$ such that*

$$xb_-(x) \le -r_-, \quad xb_+(x) \le +r_+, \quad \forall\, |x| \ge M, \tag{1}$$

and

$$2r_- > 1 \quad \& \quad \kappa_1^{-1} := \frac{\lambda_0(2r_+ + 1)}{\lambda_1(2r_- - 1)} < 1. \tag{2}$$

Then the process (X, Z) is positive recurrent; moreover, there exists $C > 0$ such that for all M_1 large enough and all $x \in \mathbb{R}$

$$\mathbb{E}_x \tau_{M_1} \le C(x^2 + 1), \tag{3}$$

where

$$\tau_{M_1} := \inf(t \ge 0 : |X_t| \le M_1).$$

Moreover, the process (X_t, Z_t) has a unique invariant measure, and for each nonrandom initial condition x, z there is a convergence to this measure in total variation when $t \to \infty$.

3 Proof

Denote $\|b\| = \sup_x |b(x)|$. Let $M_1 \gg M$ (the value M_1 will be specified later); denote

$$T_0 := \inf(t \ge 0 : Z_t = 0),$$

and

$$0 \le T_0 < T_1 < T_2 < \ldots,$$

where each T_n is defined as the next moment of switch of the component Z; let

$$\tau := \inf(T_n \ge 0 : |X_{T_n}| \le M_1).$$

It suffices to evaluate from above the value $\mathbb{E}_x\tau$ because $\tau \ge \tau_{M_1}$. Let us choose $\epsilon > 0$ such that

$$\lambda_0(2r_+ + 1 + \epsilon) = q\lambda_1(2r_- - 1 - \epsilon) \tag{4}$$

with some $q < 1$ (see (2)). Note that for $|x| \le M$ there is nothing to prove; so assume $|x| > M$.

Lemma 1. *Under the assumptions of the theorem for any $\delta > 0$ there exists M_1 such that*

$$\max\left[\sup_{|x|>M_1} \mathbb{E}_x\left(\int_0^{T_1} 1(\inf_{0\le s\le t}|X_s|\le M)dt|Z_0=0\right),\right.$$
$$\left.\sup_{|x|>M_1} \mathbb{E}_x\left(\int_0^{T_0} 1(\inf_{0\le s\le t}|X_s|\le M)dt|Z_0=1\right)\right] < \delta. \tag{5}$$

Proof. Let X^i_t, $i = 0, 1$ denote the solution of the equation

$$dX^i_t = b(X^i_t, i)\, dt + dW_t, \quad t \ge 0, \quad X^i_0 = x.$$

Let $Z_0 = 0$; then $T_0 = 0$. The processes X and X^0 coincide a.s. on $[0, T_1]$ due to uniqueness of solution. Therefore, due to the independence of Z and W, and, hence, of Z and X^0, we obtain

$$\mathbb{E}_x\left(\int_0^{T_1} 1(\inf_{0\le s\le t}|X_s| \le M)dt|Z_0 = 0\right) = \mathbb{E}_x\int_0^{T_1} 1(\inf_{0\le s\le t}|X^0_s| \le M)dt$$
$$= \mathbb{E}_x\int_0^{\infty} 1(t < T_1)1(\inf_{0\le s\le t}|X^0_s| \le M)dt = \int_0^{\infty} \mathbb{E}_x 1(t < T_1)\mathbb{P}(\inf_{0\le s\le t}|X^0_s| \le M)dt$$
$$= \int_0^{\infty} \exp(-\lambda_0 t)\mathbb{P}(\inf_{0\le s\le t}|X^0_s| \le M)dt.$$

Let us take t such that

$$\int_t^{\infty} e^{-\lambda_0 s}ds < \delta/2.$$

Now, by virtue of the boundedness of b, it is possible to choose $M_1 > M$ such that for this value of t we have

$$t\,\mathbb{P}_x(\inf_{0\le s\le t}|X^0_s| \le M) < \delta/2.$$

The bound for the second term in (5) follows by using the process X^1 and the intensity λ_1 in the same way. QED

Lemma 2. *If M_1 is large enough, then under the assumptions of the theorem for any $|x| > M_1$ for any $k = 0, 1, \ldots$*

$$\mathbb{E}_x(X^2_{T_{2k+1}\wedge\tau}|Z_0 = 0, \mathcal{F}_{T_{2k}}) \le \mathbb{E}_x(X^2_{T_{2k}\wedge\tau}|Z_0 = 0, \mathcal{F}_{T_{2k}})$$
$$-1(\tau > T_{2k})\lambda_0^{-1}((2r_- - 1) - \epsilon), \tag{6}$$

$$\mathbb{E}_x(X^2_{T_{2k+2}\wedge\tau}|Z_0 = 1, \mathcal{F}_{T_{2k+1}}) \le \mathbb{E}_x(X^2_{T_{2k+1}\wedge\tau}|Z_0 = 1, \mathcal{F}_{T_{2k+1}})$$
$$+1(\tau > T_{2k+1})\lambda_1^{-1}((2r_- + 1) + \epsilon). \tag{7}$$

Proof. **1.** Recall that $T_0 = 0$ under the condition $Z_0 = 0$. We have,

$$T_{2k+1} = \inf(t > T_{2k} : Z_t = 1).$$

In other words, the moment T_{2k+1} may be treated as "T_1 after T_{2k}". Under $Z_0 = 0$ the process X_t coincides with X^0_t until the moment T_1. Hence, we have on $[0, T_1]$ by Ito's formula

$$dX^2_t - 2X_t dW_t = 2X_t b_-(X_t)dt + dt \le (-2r_- + 1)dt,$$

on the set $(|X_t| > M)$ due to the assumptions (1). Further, since $1(|X_t| > M) = 1 - 1(|X_t| \le M)$, we obtain

$$\int_0^{T_1\wedge\tau} 2X_t b_-(X_t)dt$$
$$= \int_0^{T_1\wedge\tau} 2X_t b_-(X_t)1(|X_t| > M)dt + \int_0^{T_1\wedge\tau} 2X_t b_-(X_t)1(|X_t| \le M)dt$$
$$\le -2r_- \int_0^{T_1\wedge\tau} 1(|X_t| > M)dt + \int_0^{T_1\wedge\tau} 2M\|b\|1(|X_t| \le M)dt$$
$$= -2r_- \int_0^{T_1\wedge\tau} 1dt + \int_0^{T_1\wedge\tau} (2M\|b\| + 2r_-)1(|X_t| \le M)dt$$
$$\le -2r_- \int_0^{T_1\wedge\tau} 1dt + (2M\|b\| + 2r_-) \int_0^{T_1\wedge\tau} 1(|X_t| \le M)dt.$$

Thus, always for $|x| > M_1$,

$$\mathbb{E}_x \int_0^{T_1\wedge\tau} 2X_t b_-(X_t)dt$$
$$\le -2r_- E \int_0^{T_1\wedge\tau} 1dt + (2M\|b\| + 2r_-)E_x \int_0^{T_1\wedge\tau} 1(|X_t| \le M)dt$$

$$= -2r_- \mathbb{E} \int_0^{T_1 \wedge \tau} 1dt + (2M\|b\| + 2r_-)\mathbb{E}_x \int_0^{T_1 \wedge \tau} 1(|X_t| \le M)dt$$

$$\le -2r_- \mathbb{E} \int_0^{T_1 \wedge \tau} 1dt + (2M\|b\| + 2r_-)\mathbb{E}_x \int_0^{T_1} 1(|X_t| \le M)dt$$

$$\le -2r_- \mathbb{E} \int_0^{T_1 \wedge \tau} 1dt + (2M\|b\| + 2r_-)\delta.$$

For our fixed $\epsilon > 0$ let us choose $\delta = \lambda_0^{-1}\epsilon/(2M\|b\| + 2r_-)$. Then, since $|x| > M_1$ implies $T_1 \wedge \tau = T_1$ on $(Z_0 = 0)$, we get

$$\mathbb{E}_x X^2_{T_1 \wedge \tau} - x^2 \le -(2r_- - 1)\mathbb{E}_x \int_0^{T_1} dt + \lambda_0^{-1}\epsilon = -\lambda_0^{-1}((2r_- - 1) - \epsilon).$$

Substituting here $X_{T_{2k}}$ instead of x and writing $\mathbb{E}_x(\cdot|\mathcal{F}_{T_{2k}})$ instead of $\mathbb{E}_x(\cdot)$, and multiplying by $1(\tau > T_{2k})$, we obtain the bound (6), as required.
2. The condition $Z_0 = 1$ implies the inequality $T_0 > 0$. We have,

$$T_{2k+2} = \inf(t > T_{2k+1} : Z_t = 0).$$

In other words, the moment T_{2k+2} may be treated as "T_0 after T_{2k+1}". Under $Z_0 = 1$ the process X_t coincides with X^1_t until the moment T_0. Hence, we have on $[0, T_0]$ by Ito's formula

$$dX_t^2 - 2X_t dW_t = 2X_t b_+(X_t)dt + dt \le (2r_+ + 1)dt,$$

on the set $(|X_t| > M)$ due to the assumptions (1). Further, since $1(|X_t| > M) = 1 - 1(|X_t| \le M)$, we obtain

$$\int_0^{T_0 \wedge \tau} 2X_t b_+(X_t)dt$$

$$= \int_0^{T_0 \wedge \tau} 2X_t b_+(X_t)1(|X_t| > M)dt + \int_0^{T_0 \wedge \tau} 2X_t b_+(X_t)1(|X_t| \le M)dt$$

$$\le 2r_+ \int_0^{T_0 \wedge \tau} 1(|X_t| > M)dt + \int_0^{T_0 \wedge \tau} 2M\|b\|1(|X_t| \le M)dt$$

$$= 2r_+ \int_0^{T_0 \wedge \tau} 1dt + \int_0^{T_1 \wedge \tau} (2M\|b\| - 2r_+)1(|X_t| \le M)dt$$

$$\le 2r_+ \int_0^{T_0 \wedge \tau} 1dt + 2M\|b\| \int_0^{T_0 \wedge \tau} 1(|X_t| \le M)dt.$$

Thus, always for $|x| > M_1$,

$$\mathbb{E}_x \int_0^{T_0 \wedge \tau} 2X_t b_+(X_t)dt$$

$$\le 2r_+ E\int_0^{T_0\wedge\tau} 1dt + 2M\|b\|E_x\int_0^{T_0\wedge\tau} 1(|X_t|\le M)dt$$

$$= 2r_+\mathbb{E}\int_0^{T_0\wedge\tau} 1dt + 2M\|b\|\mathbb{E}_x\int_0^{T_1\wedge\tau} 1(|X_t|\le M)dt$$

$$\le 2r_+\mathbb{E}\int_0^{T_0\wedge\tau} 1dt + 2M\|b\|\mathbb{E}_x\int_0^{T_0} 1(|X_t|\le M)dt$$

$$\le 2r_+\mathbb{E}\int_0^{T_0\wedge\tau} 1dt + 2M\|b\|\delta.$$

For our fixed $\epsilon > 0$ let us choose $\delta = \lambda_0^{-1}\epsilon/(2M\|b\|)$. Then, since $|x| > M_1$ implies $T_0\wedge\tau = T_0$ on $(Z_0 = 1)$, we get

$$\mathbb{E}_x X^2_{T_1\wedge\tau} - x^2 \le -(2r_- - 1)\mathbb{E}_x\int_0^{T_1} dt + \lambda_0^{-1}\epsilon = -\lambda_0^{-1}((2r_- - 1) - \epsilon).$$

Substituting here $X_{T_{2k+1}}$ instead of x and writing $\mathbb{E}_x(\cdot|\mathcal{F}_{T_{2k+1}})$ instead of $\mathbb{E}_x(\cdot)$, and multiplying by $1(\tau > T_{2k+1})$, we obtain the bound (7), as required. QED

Lemma 3. *If M_1 is large enough, then under the assumptions of the theorem for any $k = 0, 1, \dots$*

$$\mathbb{E}_x(X^2_{T_{2k+2}\wedge\tau}|Z_0 = 0, \mathcal{F}_{T_{2k+1}}) \le \mathbb{E}_x(X^2_{T_{2k+1}\wedge\tau}|Z_0 = 0, \mathcal{F}_{T_{2k+1}})$$
$$+1(\tau > T_{2k+1})\lambda_1^{-1}((2r_+ + 1) + \epsilon)), \tag{8}$$

and

$$\mathbb{E}_x(X^2_{T_{2k+1}\wedge\tau}|Z_0 = 1, \mathcal{F}_{T_{2k}}) \le \mathbb{E}_x(X^2_{T_{2k}\wedge\tau}|Z_0 = 1, \mathcal{F}_{T_{2k}})$$
$$-1(\tau > T_{2k})\lambda_0^{-1}((2r_+ - 1) - \epsilon)). \tag{9}$$

Proof. Let $Z_0 = 0$; recall that it implies $T_0 = 0$. If $\tau \le T_{2k+1}$, then (8) is trivial. Let $\tau > T_{2k+1}$. We will substitute x instead of $X_{T_{2k}}$ for a while, and will be using the solution X^1_t of the equation

$$dX^1_t = b(X^1_t, 1)\,dt + dW_t, \quad t\ge T_1, \quad X^1_{T_1} = X_{T_1}.$$

For M_1 large enough, since $|x|\wedge|X_{T_1}| > M_1$ implies $T_2\le\tau$, and due to the assumptions (1) we guarantee the bound

$$1(|X_{T_1}| > M_1)(\mathbb{E}_{X_{T_1}}X^2_{T_2\wedge\tau} - X^2_{T_1\wedge\tau})$$

$$\le 1(|X_{T_1}| > M_1)(\mathbb{E}_{X_{T_1}}(T_2 - T_1)((2r_+ + 1) + \epsilon))$$

$$= +1(|X_{T_1}| > M_1)(\lambda_1^{-1}((2r_+ + 1) + \epsilon))$$

in the same way as the bound (7) in the previous lemma. In particular, it follows that for $|x| > M_1$

$$(\mathbb{E}_{X_{T_1}} X^2_{T_2\wedge\tau} - X^2_{T_1\wedge\tau}) \le 1(|X_{T_1}| > M_1)(\mathbb{E}_{X_{T_1}}(T_2 - T_1)((2r_+ + 1) + \epsilon))$$

$$= +1(|X_{T_1}| > M_1)(\lambda_1^{-1}((2r_+ + 1) + \epsilon)),$$

since $|X_{T_1}| \le M_1$ implies $\tau \le T_1$ and $\mathbb{E}_{X_{T_1}} X^2_{T_2\wedge\tau} - X^2_{T_1\wedge\tau} = 0$. So, on the set $|x| > M_1$ we have,

$$\mathbb{E}_x(\mathbb{E}_{X_{T_1}} X^2_{T_2\wedge\tau} - X^2_{T_1\wedge\tau})$$

$$\le \mathbb{E}_x 1(|X_{T_1}| > M_1)(\lambda_1^{-1}((2r_+ + 1) + \epsilon)) \le \lambda_1^{-1}((2r_+ + 1) + \epsilon).$$

Now substituting back $X_{T_{2k}}$ in place of x and multiplying by $1(\tau > T_{2k+1})$, we obtain the inequality (8), as required.

For $Z_0 = 1$ we have $T_0 > 0$, and the bound (9) follows in a similar way. QED

Now we can complete the proof of the theorem. Consider the case $Z_0 = 0$ where $T_0 = 0$. Note that the bound (6) of the Lemma 2 together with the bound (8) of the Lemma 3 can be equivalently rewritten as follows:

$$\mathbb{E}_x X^2_{T_{2k+1}\wedge\tau} - \mathbb{E}_x X^2_{T_{2k}\wedge\tau} \le -((2r_- - 1) - \epsilon)\mathbb{E}_x(T_{2k+1}\wedge\tau - T_{2k}\wedge\tau), \quad (10)$$

and

$$\mathbb{E}_x X^2_{T_{2k}\wedge\tau} - \mathbb{E}_x X^2_{T_{2k-1}\wedge\tau} \le ((2r_+ + 1) + \epsilon)\mathbb{E}_x(T_{2k}\wedge\tau - T_{2k-1}\wedge\tau). \quad (11)$$

We have the identity

$$\tau \wedge T_n = T_0 + \sum_{m=0}^{n-1}((T_{m+1}\wedge\tau) - (T_m\wedge\tau)).$$

Therefore,

$$\mathbb{E}_x(\tau\wedge T_n) = \mathbb{E}_x T_0 + \mathbb{E}_x\sum_{m=0}^{n-1}((T_{m+1}\wedge\tau) - (T_m\wedge\tau)),$$

Since $T_n \uparrow \infty$, by virtue of the monotonic convergence in both parts and due to Fubini theorem we obtain,

$$\mathbb{E}_x\tau = \mathbb{E}_x T_0 + \sum_{m=0}^{\infty}\mathbb{E}_x((T_{m+1}\wedge\tau) - (T_m\wedge\tau)) \quad (12)$$

$$= \mathbb{E}_x T_0 + \sum_{k=0}^{\infty} \mathbb{E}_x((T_{2k+1} \wedge \tau) - (T_{2k} \wedge \tau))$$

$$+ \sum_{k=0}^{\infty} \mathbb{E}_x((T_{2k+2} \wedge \tau) - (T_{2k+1} \wedge \tau)).$$

Due to (10) and (11) we have,

$$\mathbb{E}_x(T_{2k+1} \wedge \tau - T_{2k} \wedge \tau) \le ((2r_- - 1) - \epsilon)^{-1} \left(\mathbb{E}_x X^2_{T_{2k+1}\wedge\tau} - \mathbb{E}_x X^2_{T_{2k}\wedge\tau}\right)$$

$$\mathbb{E}_x X^2_{T_{2m+2}\wedge\tau} - x^2$$

$$\le ((2r_+ + 1) + \epsilon) \sum_{k=0}^{m} \mathbb{E}_x(T_{2k+2} \wedge \tau - T_{2k+1} \wedge \tau)$$

$$-((2r_- - 1) - \epsilon) \sum_{k=0}^{m} \mathbb{E}_x(T_{2k+1} \wedge \tau - T_{2k} \wedge \tau)$$

$$= \sum_{k=0}^{m} (-((2r_- - 1) - \epsilon)(\mathbb{E}_x(T_{2k+1} \wedge \tau - T_{2k} \wedge \tau)$$

$$+((2r_+ + 1) + \epsilon)\mathbb{E}_x(T_{2k+2} \wedge \tau - T_{2k+1} \wedge \tau))\,.$$

By virtue of Fatou's lemma we get

$$x^2 \ge ((2r_- - 1) - \epsilon) \sum_{k=0}^{m} (\mathbb{E}_x(T_{2k+1} \wedge \tau - T_{2k} \wedge \tau)$$
$$-((2r_+ + 1) + \epsilon) \sum_{k=0}^{m} \mathbb{E}_x(T_{2k+2} \wedge \tau - T_{2k+1} \wedge \tau). \tag{13}$$

Note that $1(\tau > T_{2k+1}) \le 1(\tau > T_{2k})$. So, $\mathbb{P}(\tau > T_{2k}) \ge \mathbb{P}(\tau > T_{2k+1})$. Hence,

$$\lambda_0 \mathbb{E}_x(T_{2k+1} \wedge \tau - T_{2k} \wedge \tau) - \lambda_1 \mathbb{E}_x(T_{2k+2} \wedge \tau - T_{2k+1} \wedge \tau)$$

$$= \lambda_0 \mathbb{E}_x(T_{2k+1} \wedge \tau - T_{2k} \wedge \tau) 1(\tau \ge T_{2k})$$

$$-\lambda_1 \mathbb{E}_x (T_{2k+2}\wedge\tau - T_{2k+1}\wedge\tau) 1(\tau \geq T_{2k+1})$$

$$= \lambda_0 \mathbb{E}_x 1(\tau > T_{2k}) \mathbb{E}_{X_{T_{2k}}} (T_{2k+1}\wedge\tau - T_{2k}\wedge\tau)$$

$$-\lambda_1 \mathbb{E}_x 1(\tau > T_{2k+1}) \mathbb{E}_{X_{T_{2k+1}}} (T_{2k+2}\wedge\tau - T_{2k+1}\wedge\tau)$$

$$= \lambda_0 \mathbb{E}_x 1(\tau > T_{2k}) \lambda_0^{-1} - \lambda_1 \mathbb{E}_x 1(\tau > T_{2k+1}) \lambda_1^{-1}$$

$$= \mathbb{E}_x 1(\tau > T_{2k}) - \mathbb{E}_x 1(\tau > T_{2k+1}) \geq 0.$$

Thus,

$$\mathbb{E}_x (T_{2k+2}\wedge\tau - T_{2k+1}\wedge\tau) \leq \frac{\lambda_0}{\lambda_1} \mathbb{E}_x (T_{2k+1}\wedge\tau - T_{2k}\wedge\tau).$$

Therefore, we estimate

$$((2r_+ + 1) + \epsilon) \sum_{k=0}^{m} \mathbb{E}_x (T_{2k+2}\wedge\tau - T_{2k+1}\wedge\tau)$$

$$\leq ((2r_+ + 1) + \epsilon) \frac{\lambda_0}{\lambda_1} \sum_{k=0}^{m} \mathbb{E}_x (T_{2k+1}\wedge\tau - T_{2k}\wedge\tau)$$

$$= q((2r_- - 1) - \epsilon) \sum_{k=0}^{m} \mathbb{E}_x (T_{2k+1}\wedge\tau - T_{2k}\wedge\tau).$$

So, (13) implies that

$$x^2 \geq ((2r_- - 1) - \epsilon) \sum_{k=0}^{m} (\mathbb{E}_x (T_{2k+1}\wedge\tau - T_{2k}\wedge\tau)$$

$$-((2r_+ + 1) + \epsilon) \sum_{k=0}^{m} \mathbb{E}_x (T_{2k+2}\wedge\tau - T_{2k+1}\wedge\tau)$$

$$\geq (1-q)((2r_- - 1) - \epsilon) \sum_{k=0}^{m} (\mathbb{E}_x (T_{2k+1}\wedge\tau - T_{2k}\wedge\tau)$$

$$\geq \frac{1-q}{2} ((2r_- - 1) - \epsilon) \sum_{k=0}^{m} (\mathbb{E}_x (T_{2k+1}\wedge\tau - T_{2k}\wedge\tau)$$

$$+\frac{1-q}{2q}\left((2r_+ + 1) + \epsilon\right) \sum_{k=0}^{m} \mathbb{E}_x(T_{2k+2} \wedge \tau - T_{2k+1} \wedge \tau).$$

Denoting $c := \min\left(\frac{1-q}{2q}\left((2r_+ + 1) + \epsilon\right), \frac{1-q}{2}\left((2r_- - 1) - \epsilon\right)\right)$, we conclude that

$$x^2 \ge c \sum_{k=0}^{2m} \mathbb{E}_x(T_{k+1} \wedge \tau - T_k \wedge \tau).$$

So, as $m \uparrow \infty$, by the monotone convergence theorem we get the inequality

$$\sum_{k=0}^{\infty} \mathbb{E}_x(T_{k+1} \wedge \tau - T_k \wedge \tau) \le c^{-1} x^2.$$

Due to (12), it implies that (in the case $T_0 = 0$)

$$\mathbb{E}_x \tau \le c^{-1} x^2, \tag{14}$$

as required. Recall that this bound is established for $|x| > M_1$, while in the case of $|x| \le M_1$ the left hand side in this inequality is just zero.

In the case of $Z_0 = 1$ (and, hence, $T_0 > 0$), we have to add the value $\mathbb{E}_x T_0 = \lambda_1^{-1}$ to the right hand side of (14), which leads to the bound (3), as promised.

In turn, this bound implies existence of the invariant measure, see [3, Section 4.4]. Convergence to it in total variation follows due to the coupling method in a standard way. So, this measure is unique. The details and some extensions of this issue will be provided in another paper. QED

Acknowledgments. The author is deeply indebted to S. Anulova who made useful comments to the first versions of the manuscript and with whom the author started studying the theme of switched diffusions some years ago. This article is devoted to her memory. The article was prepared within the framework of the HSE University Basic Research Program in part which includes Lemmata 1 and 2; the Proof of Lemma 3 and Theorem 1 was funded by Russian Foundation for Basic Research grant 20-01-00575a.

References

1. Anulova, S.V., Veretennikov, A.Y.: Exponential convergence of degenerate hybrid stochastic systems with full dependence. In: Korolyuk, V., Limnios, N., Mishura, Y., Sakhno, L., Shevchenko, G. (eds.) Modern Stochastics and Applications. SOIA, vol. 90, pp. 159–174. Springer, Cham (2014). https://doi.org/10.1007/978-3-319-03512-3_10
2. Cloez, B., Hairer, M.: Exponential ergodicity for Markov processes with random switching. Bernoulli **21**(1), 505–536 (2015)

3. Khasminskii, R.Z.: Stochastic Stability of Differential Equations, 2nd edn. Springer, Berlin (2012). https://doi.org/10.1007/978-3-642-23280-0
4. Khasminskii, R.Z.: Stability of regime-switching stochastic differential equations. Probl. Inf. Transm. **48**, 259–270 (2012). https://doi.org/10.1134/S0032946012030064
5. Mao, X., Yin, G., Yuan, C.: Stabilization and destabilization of hybrid systems of stochastic differential equations. Automatica **43**, 264–273 (2007)
6. Nakao, S.: Comparison theorems for solutions of one-dimensional stochastic differential equations. In: Maruyama, G., Prokhorov, Y.V. (eds.) Proceedings of the Second Japan-USSR Symposium on Probability Theory. LNM, vol. 330, pp. 310–315. Springer, Heidelberg (1973). https://doi.org/10.1007/BFb0061496
7. Shao, J., Yuan, C.: Stability of regime-switching processes under perturbation of transition rate matrices. Nonlinear Anal. Hybrid Syst. **33**, 211–226 (2019)
8. Veretennikov, A.Y.: On strong solutions and explicit formulas for solutions of stochastic integral equations. Math. USSR-Sb. **39**(3), 387–403 (1981). https://doi.org/10.1070/SM1981v039n03ABEH001522
9. Zvonkin, A.K.: A transformation of the phase space of a diffusion process that removes the drift. Math. USSR-Sb. **22**(1), 129–149 (1974)

Applications of Stochastic Methods

On Two-Type Branching Random Walks and Their Applications for Genetic Modelling

Yulia Makarova(✉), Vladimir Kutsenko, and Elena Yarovaya

Lomonosov Moscow State University, Leninskie Gory 1, Moscow 119234, Russia
ykmakarova@gmail.com, vlakutsenko@yandex.ru, yarovaya@mech.math.msu.su

Abstract. We consider a model of the evolution of a population in the presence of epistatic lethal alleles. A model which describes the evolution of lethal and non-lethal alleles based on two-type branching random walks on multidimensional lattices is presented. We study this model in terms of subpopulations of particles generated by a single particle of each type located at every lattice point. The differential equations for the generating functions and factorial moments for the particle subpopulations are obtained. For the first moments we get explicit solutions for cases significant in the genetic context. The asymptotic behaviour for the first moments of particle distribution at lattice points is obtained for a random walk with finite variance of jumps.

Keywords: Branching random walks · Two-type processes · Multidimensional lattices · Generating functions · Population genetics · Epistasis · Lethal alleles

1 Introduction

In population genetics it is common to consider five main features of the evolution process: mutation, selection, population structure, gene transfer method, and drift, see [5]. To correctly model the evolution process, it is important to consider each of them. Therefore, in this Section, we briefly describe all these five features. We need to recall a few definitions. A locus is the physical location of a gene on a chromosome. Alleles are alternative forms or versions of a gene at a particular locus, see [7]. Consider a population of organisms with a single set of chromosomes, i.e., haploid organisms. Let these organisms differ only in one locus, denote alternative alleles in this locus by A and a. Let W_A and W_a be the average number of offsprings of an organism with the alleles A and a respectively. These values are sometimes referred to as fitness [5].

Now let us go back to the five basic concepts of evolution. A mutation is a persistent change in the genome. For humans, the probability of mutation of nucleotide per generation is approximately 10^{-8} (see [10]) which is small compared to the number of pairs of nucleotides in DNA (this number is equal to

A. N. Shiryaev et al. (Eds.): ICSM-5 2020, PROMS 371, pp. 255–268, 2021.
https://doi.org/10.1007/978-3-030-83266-7_19

$3 \cdot 10^9$). The selection mechanism reflects the change in the frequency $p_A = p_A(t)$ of the allele A in the population. Note that the frequency of the alternative allele is equal to $p_a = p_a(t) = 1 - p_A$. In the simplest case of an infinite population continuous-time selection can be described by a deterministic equation of the form $\frac{dp_A}{dt} = sp_A(1 - p_A)$ with initial condition $p_A(0) \in [0, 1]$ and $s = 1 - W_a/W_A$, see [5]. The structure of a population is primarily a configuration of the space in which this population is located. We should also note that introducing a spatial structure into the model of evolution is a complex and important problem. One of the classical models of spatial structure is the Fischer-Kolmogorov-Petrovskii-Piskunov equation $\frac{\partial p}{\partial t} = sp(1-p) + D\frac{\partial^2 p}{\partial x^2}$ which is the selection equation described earlier with one-dimensional shift according to diffusion coefficient $D \in \mathbb{R}$, see [9]. The fourth concept that we have to explain is the gene transfer method. In particular, the transfer method can be divided into asexual and sexual. In asexual reproduction, the offspring is an identical copy of the parent. During sexual reproduction, the offspring inherits half of the genetic material from each of the parents. One of the important characteristics of sexual reproduction is crossover. Crossover is the process of exchanging sections of homologous chromosomes. Finally drift is the phenomenon of random changes in allele frequencies in a population. For example, drift can be modelled by the discrete-time Bienaymé-Galton-Watson process, see [2].

We have finished reviewing the five main characteristics of the evolutionary process. Note that branching processes can describe all characteristics except the spatial structure. Mutations can be modeled by multi-type processes, selection can be modeled by different intensities of particle branching, and the stochastic nature of the process itself is responsible for drift. The method of gene transfer can also be described in the population structure, but in a slightly more complex way, see [8]. As we have discussed, the spatial structure of a population is essential. The stochastic process which combines the properties of a branching process and random walk is called branching random walk (BRW). In contrast to branching processes, BRWs seemingly have not been widely applied in evolutionary biology. The presented work partially fills this gap. In particular, we focus on the problem of missing heritability which is the fact that single genetic variations cannot account for much of the heritability of diseases, behaviours, and other phenotypes, see [15]. One of the possible candidates for the solution to this problem is epistatic interaction. Epistasis is the phenomenon when one gene masks or alters the effect of another one, see [7]. The study of such variants is important in the context of the missing heritability problem. Epistatic variants may carry a significant part of undiscovered heritability. In the present paper we show how consideration of BRW with several types of particles allowes us to describe the evolution of a population with carriers of epistatic lethal mutations.

The structure of paper is as follows. In Sect. 2 we describe the model of the evolution of epistatic lethal mutations and give its description in genetic terms. In Sect. 3 we present the model of two-type BRWs on multidimensional lattice $\mathbb{Z}^d$, $d \in \mathbb{N}$ and obtain differential equations for the generating functions of subpopulations generated by a single particle of each type. In Sect. 4 we get differential equations for the factorial moments of subpopulations. In Sect. 5 we study the solutions for the equations for the first moments.

2 Description of Genetic Models

In this Section we describe a model of the evolution of organisms with epistatically lethal mutations on chromosomes in genetic terms. Consider a diploid organism, i.e. organism with a double set of chromosomes. Consider two loci L_1 and L_2 on the same chromosome of a diploid organism. Combination of alleles on the same chromosome is sometimes called a haplotype. We assume that the organisms do not differ in loci except (L_1, L_2). Denote two various alleles in locus L_1 by L_1' and L_1'' and two various alleles in locus L_2 by L_2' and L_2''. Denote haplotype $L_1'L_2'$ as O, haplotype $L_1''L_2'$ as B, haplotype $L_1'L_2''$ as C, and haplotype $L_1''L_2''$ as D. Consider a population in which the haplotypes under consideration have reached equilibrium frequencies (p_O, p_B, p_C, p_D) for types O, B, C and D respectively, see [5]. In addition we assume that this distribution does not change in time. Note that the genotype of an organism in our assumptions and notations can be indicated by two labels x/y, where $x, y \in \{O, B, C, D\}$ and x/y is the same as y/x.

Lethal alleles (or lethals) are alleles that prevent survival (see [6]). In the present paper, we consider dominant and incomplete dominant epistatic lethals. We consider a dominant epistatic lethal allele as an allele that leads to nonviability of the organism if there is at least one copy of the D haplotype. We consider an incomplete dominant epistatic lethal as an allele that leads to nonviability of the organism if there are two copies of the D haplotype. Besides, this lethal can lead to a partial loss of reproductive function in carriers of one copy of the D haplotype.

Let us firstly describe the model of organism reproduction in terms of birth and death of particles with corresponding labels in case of alleles that do not affect fitness. Consider an organism (particle) with labels x/y, where $x, y \in \{O, B, C, D\}$. We assume that during small time dt the following transformations are possible:

1. a particle can die with the probability $\mu dt + o(dt)$;
2. a particle can produce copy of itself and a particle with label x_1/y_1 with the probability $\lambda b_{x_1/y_1} dt + o(dt)$, where

$$b_{x_1/y_1} = \frac{1}{2} p_{y_1} \mathbb{I}(x_1 \in \{x, y\}),$$

and $\mathbb{I}(\cdot)$ is indicator function and $x_1, y_1 \in \{O, B, C, D\}$. We assume that offsprings start their evolution processes independently of the others with the same birth and death intensities.

Remark 1. From a biological point of view, the branching process presented above represents the following process. In the process of transformation, the organism either dies or forms a pair with a random organism from the population. The offspring of this pair inherits the random haplotype from each parent. Let us assume that the second organism from the pair "returns" to the population and, from now on, we consider only the two remaining particles. Also we assume that there is no inbreeding and that recombination between loci L_1 and L_2 does not occur. Note that for simplicity of notation, we considered the case of binary splitting, i.e. particle can produce two offsprings. Although we can consider an arbitrary number of offsprings by introducing appropriate intensities. This more general model will be discussed in Sect. 3.

As we have already mentioned in the problems of evolutionary biology, it is important to consider the spatial structure. In our research the space is the multidimensional lattice $\mathbb{Z}^d$, $d \in \mathbb{N}$. A particle located at any point in a short time can go through the branching process described earlier, or move to another point on the lattice with a specified probability depending on both starting and finishing points. We will describe the random walk in Sect. 3 in details.

Let us examine how the BRW model reformulates in the case of dominant epistatic lethales. Particles with at least one D haplotype are lethal, and particles without D haplotype are all identical to each other. Therefore all particles labelled as x/y, $x, y \in \{O, B, C\}$ can be combined together and studied in the framework of single-type BRW. This problem is investigated under various initial conditions, branching source configurations, and conditions on the underlying walk, e.g. see [1,3,11,14].

However, from a biological point of view, the case of incomplete dominant epistatic lethales is more interesting. In this case, the particles without haplotype D are unaffected while particles with one copy of D have altered fitness. Moreover, two carriers of D haplotype can produce a particle with labels D/D, which immediately dies. Straightforward consideration of the particle reproduction scheme under these conditions allows one to notice that the particles can be divided into three classes concerning conditional probabilities of splitting. Let us denote particles with labels x/y, where $x, y \in \{O, B, C\}$ as type I particles, particles with labels $\{O/D, B/D, C/D\}$ as type II particles, and D/D as type III particles. Type I represents healthy organisms, type II represents carriers, and type III represents unviable organisms. In the condition of splitting type I particles can produce particles of type I with the probability $1 - p_D$ and particles of type II with the probability p_D. During splitting type II particles can produce particles of type I with the probability $\frac{1}{2}(1 - p_D)$, particles of type II with the probability $\frac{1}{2}$, and particles of type III with the probability $\frac{1}{2}p_D$. Thus, in this case, the branching intensities introduced earlier can be rewritten as follows. Note that we introduce additional notation $\beta(\cdot, \cdot)$ to facilitate the transition to the model in the next section. During a small time dt the following transformations for type I particle are possible:

1. a particle can die with the probability $\mu_1 dt + o(dt)$, $\mu_1 \geq 0$;
2. a particle can produce two newborn particles of type I with the probability $\beta_1(2, 0)dt + o(dt)$, $\beta_1(2, 0) = \lambda_1(1 - p_D) \geq 0$;

3. a particle can produce a particle of type I and a particle of type II with the probability $\beta_1(1,1)dt + o(dt)$, $\beta_1(1,1) = \lambda_1 p_D \geq 0$.

During small time dt the following transformations for type II particle are possible:

1. a particle can die with the probability $(\mu_2 + \frac{\lambda_2}{2} p_D)dt + o(dt)$, $\mu_2 \geq 0$, $\lambda_2 \geq 0$;
2. a particle can produce two newborn particles of type II with the probability $\beta_2(0,2)dt + o(dt)$, $\beta_2(0,2) = \frac{\lambda_2}{2}$;
3. a particle can produce a particle of type I and a particle of type II with the probability $\beta_2(1,1)dt + o(dt)$, $\beta_2(1,1) = \frac{\lambda_2}{2}(1 - p_D)$.

In this model, the number of particles of both types is of interest. The number of particles of the first type describes the number of healthy offsprings of organisms that were on $\mathbb{Z}^d$ at the initial time moment. The number of particles of the second type describes the number of offsprings who are carriers of the mutation. In Sect. 3 we introduce the formal definition of the model.

3 Two-Type BRWs on $\mathbb{Z}^d$

In this Section we consider continuous-time BRW with two types of particles on $\mathbb{Z}^d$, $d \in \mathbb{N}$. The objects of the study are subpopulations which can be represented as the following column-vectors:

$$n_1(t,x,y) = [n_{11}(t,x,y), n_{12}(t,x,y)]^T,$$
$$n_2(t,x,y) = [n_{21}(t,x,y), n_{22}(t,x,y)]^T.$$

Here $n_i(t,x,y)$, $i = 1,2$ is the vector of particles at the time moment $t > 0$ at the point $y \in \mathbb{Z}^d$, generated by a single particle of type i which at time moment $t = 0$ was at the site $x \in \mathbb{Z}^d$. Its components $n_{ij}(t,x,y)$, $j = 1,2$ are the numbers of particles at the point $y \in \mathbb{Z}^d$ of type j, generated by a single particle of type i at $x \in \mathbb{Z}^d$ at the moment $t = 0$. We assume that

$$n_{ij}(0,x,y) = \delta_i^j \delta_x(y), \tag{1}$$

where the first δ_l^m is the Kronecker function on $\mathbb{R}$, that is for l, $m \in \mathbb{R}$

$$\delta_l^m = \begin{cases} 1, & l = m; \\ 0, & l \neq m \end{cases}$$

and the second $\delta_u(v)$ is the Kronecker function on $\mathbb{Z}^d$, that is for u, $v \in \mathbb{Z}^d$

$$\delta_u(v) = \begin{cases} 1, & u = v; \\ 0, & u \neq v. \end{cases}$$

We will assume that the evolution of particles of each type consists of several opportunities. A particle of each type stays at some point on $\mathbb{Z}^d$ exponentially distributed time up to the first transformation. After that there can be the following transformations:

1. Firstly, a particle of type i, $i = 1, 2$, can die with the rate $\mu_i \geq 0$, so that particle can die with the probability $\mu_i dt + o(dt)$ during small time period dt;
2. Secondly, each particle of type i can produce new particles of both types. Denote by $\beta_i(k,l) \geq 0$, $k+l \geq 2$, the rate of a particle of type i to produce k particles of type $i = 1$ and l particles of type $i = 2$. Then we define the corresponding generating function of branching (without particle death) for $i = 1, 2$ (see [13]):
$$F_i(z_1, z_2) = \sum_{k+l\geq 2} z_1^k z_2^l \beta_i(k,l); \tag{2}$$
3. Finally, particles can jump between the points on the lattice. We assume that the probability of jump from a point $x \in \mathbb{Z}^d$ to a point $x+v \in \mathbb{Z}^d$ during the small time period dt is equal to $\varkappa_i a_i(x, x+v)dt + o(dt)$, $i = 1, 2$. Here $\varkappa_i > 0$ is the diffusion coefficient. In what follows we consider a symmetric random walk, that is the case when $a_i(x,y) = a_i(y,x)$. Moreover, random walk will be assumed to be homogeneous in space: $a_i(x, x+v) =: a_i(v)$ and irreducible, so that $span\{v : a_i(v) > 0\} = \mathbb{Z}^d$. Also $a_i(0) = -1$, $\sum_v a_i(v) = 0$. Then the migration operator has the form
$$\mathcal{L}_i\psi(x) = \varkappa_i \sum_v [\psi(x+v) - \psi(x)]a_i(v). \tag{3}$$

The aim of research is to study the behaviour of each subpopulation $n_i(t,x,y)$. Then given $z = (z_1, z_2)$, let us introduce the generating function
$$\Phi_i(t,x,y;z) = \mathsf{E} z_1^{n_{i1}(t,x,y)} z_2^{n_{i2}(t,x,y)}. \tag{4}$$

This generating function specifies the evolution of a single particle of type $i = 1, 2$. Let us consider what can happen to this particle (later, using this we can obtain a differential equation for the generating functions). Firstly, the initial particle at a point x can die with the probability $\mu_i dt + o(dt)$ (then the subpopulation of this particle will disappear). Secondly, this particle can produce k particles of type 1 and l particles of type 2 with the probability $\beta_i(k,l)dt + o(dt)$. Thirdly, the particle can jump with the probability $\varkappa_i a_i(v)dt + o(dt)$ from a point $x \in \mathbb{Z}^d$ to a point $x + v \in \mathbb{Z}^d$. Finally, there can happen nothing with a particle during time dt. From this we get the following Lemma.

Lemma 1. *The generating functions $\Phi_i(t,x,y;z)$, $i = 1, 2$, specified by (4) satisfy the differential equation*
$$\begin{aligned}\frac{\partial \Phi_i(t,x,y;z)}{\partial t} &= \mathcal{L}_i \Phi_i(t,x,y;z) + \mu_i(1 - \Phi_i(t,x,y;z)) \\ &+ F_i(\Phi_1(t,x,y;z), \Phi_2(t,x,y;z)) \\ &- \sum_{k+l\geq 2} \beta_i(k,l)\Phi_i(t,x,y;z);\end{aligned} \tag{5}$$
$$\Phi_i(0,x,y;z) = \begin{cases} 1, & x \neq y; \\ z_i, & x = y. \end{cases} \tag{6}$$

Proof. Given an $i = 1, 2$, consider the generating function $\Phi_i(\cdot, x, y; z)$ at the time moment $t + dt$:

$$\Phi_i(t+dt, x, \cdot; \cdot) = \Big(1 - \varkappa_i dt - \mu_i dt - \sum_{k+l\geq 2} \beta_i(k,l)dt\Big)\Phi_i(t, x, \cdot; \cdot) + \varkappa_i \sum_v \Phi_i(t, x+v, \cdot; \cdot) a_i(v)\, dt + \mu_i dt + \sum_{k+l\geq 2} \beta_i(k,l)\Phi_1^k(t, x, \cdot; \cdot)\Phi_2^l(t, x, \cdot; \cdot)\, dt + o(dt).$$

Then

$$\Phi_i(t+dt, x, \cdot; \cdot) - \Phi_i(t, x, \cdot; \cdot) = -\Big(\varkappa_i + \mu_i + \sum_{k+l\geq 2} \beta_i(k,l)\Big)\Phi_i(t, x, \cdot; \cdot)\, dt + \varkappa_i \sum_v \Phi_i(t, x+v, \cdot; \cdot) a_i(v)\, dt + \mu_i dt + \sum_{k+l\geq 2} \beta_i(k,l)\Phi_1^k(t, x, \cdot; \cdot)\Phi_2^l(t, x, \cdot; \cdot)\, dt + o(dt).$$

Therefore,

$$\frac{\partial \Phi_i(t, x, y; z)}{\partial t} = \mathcal{L}_i \Phi_i(t, x, y; z) + \mu_i(1 - \Phi_i(t, x, y; z)) + \sum_{k+l\geq 2} \beta_i(k,l)(\Phi_1^k(t, x, y; z)\Phi_2^l(t, x, y; z) - \Phi_i(t, x, y; z)).$$

Here, according to formula (2), we have

$$\sum_{k+l\geq 2} \beta_i(k,l)\Phi_1^k(t, x, y; z)\Phi_2^l(t, x, y; z) = F_i(\Phi_1(t, x, y; z), \Phi_2(t, x, y; z)),$$

and hence

$$\frac{\partial \Phi_i(t, x, y; z)}{\partial t} = \mathcal{L}_i \Phi_i(t, x, y; z) + \mu_i(1 - \Phi_i(t, x, y; z)) + F_i(\Phi_1(t, x, y; z), \Phi_2(t, x, y; z)) - \sum_{k+l\geq 2} \beta_i(k,l)\Phi_i(t, x, y; z)).$$

The initial condition for the latter equation follows from (1):

$$\Phi_i(0, x, y; z) = \mathsf{E} z_1^{n_{i1}(0,x,y)} z_2^{n_{i2}(0,x,y)} = \mathsf{E} z_1^{\delta_i^1 \delta_x(y)} z_2^{\delta_i^2 \delta_x(y)} = z_1^{\delta_i^1 \delta_x(y)} z_2^{\delta_i^2 \delta_x(y)} = z_i^{\delta_x(y)}.$$

So, we obtain the desired results (5), (6) which completes the proof of Lemma 1.

4 Factorial Moments

In BRWs one of the ways to study the behaviour of the process is to study the behaviour of its moments. Let us define for all $r \geq 1$ and $i, j = 1, 2$

$$m_{ij}^{(r)}(t,x,y) = \mathsf{E}\Big[n_{ij}(t,x,y)\big(n_ij(t,x,y)-1\big)\dots\big(n_{ij}(t,x,y)-r+1\big)\Big].$$

Here $m_{ij}^{(r)}(t,x,y)$ is the factorial moment for the subpopulation $n_{ij}(t,x,y)$ of the order r.

To obtain the differential equation for $m_{ij}^{(r)}(t,x,y)$, firstly, note that

$$\begin{aligned}\frac{\partial^r \Phi_i(t,x,y;z)}{\partial z_j^r} &= \frac{\partial^r \mathsf{E} z_1^{n_{i1}(t,x,y)} z_2^{n_{i2}(t,x,y)}}{\partial z_j^r} \\ &= \mathsf{E}\Big[n_{ij}(t,x,y)\big(n_{ij}(t,x,y)-1\big)\dots\big(n_{ij}(t,x,y)-r+1\big) \\ &\quad \times z_1^{n_{i1}(t,x,y)-r\delta_j^1} z_2^{n_{i2}(t,x,y)-r\delta_j^2}\Big]\end{aligned}$$

Then by fixing $z = (1, 1)$ we have

$$\frac{\partial^r \Phi_i(t,x,y;z)}{\partial z_j^r}\bigg|_{z=(1,1)} = m_{ij}^{(r)}(t,x,y).$$

Therefore, to obtain the differential equation for the factorial moment of the r-th order we differentiate r times both sides of Eq. (5) from Lemma 1 over z_j:

$$\begin{aligned}\frac{\partial^{r+1}\Phi_i(t,x,y;z)}{\partial z_j^r \partial t} &= \partial_{z_j^r}\Big(\mathcal{L}_i\Phi_i(t,x,y;z) + \mu_i(1-\Phi_i(t,x,y;z)) \\ &\quad + F_i(\Phi_1(t,x,y;z), \Phi_2(t,x,y;z)) \\ &\quad - \sum_{k+l\geq 2} \beta_i(k,l)\Phi_i(t,x,y;z)\Big).\end{aligned}$$

Taking here $z = (z_1, z_2)$ we obtain the representation of the left hand side of Eq. (5)

$$\frac{\partial^{r+1}\Phi_i(t,x,y;z)}{\partial z_j^r \partial t}\bigg|_{z=(1,1)} = \frac{\partial}{\partial t}\frac{\partial^r \Phi_i(t,x,y;z)}{\partial z_j^r}\bigg|_{z=(1,1)} = \frac{\partial m_{ij}^{(r)}(t,x,y)}{\partial t}.$$

while the right hand side of the same equation equals to

$$\begin{aligned}&\partial_{z_j^r}\Big(\mathcal{L}_i\Phi_i(t,x,y;z) + \mu_i(1-\Phi_i(t,x,y;z)) + \sum_{k+l\geq 2}\beta_i(k,l)\big(\Phi_1^k(t,x,y;z) \\ &\quad \times \Phi_2^l(t,x,y;z) - \Phi_i(t,x,y;z)\big)\Big)\Big|_{z=(1,1)} = \Big(\mathcal{L}_i\big(\partial_{z_j^r}\Phi_i(t,x,y;z)\big) \\ &\quad - \mu_i\big(\partial_{z_j^r}\Phi_i(t,x,y;z)\big) - \sum_{k+l\geq 2}\beta_i(k,l)\big(\partial_{z_j^r}\Phi_i(t,x,y;z)\big) + \sum_{k+l\geq 2}\beta_i(k,l) \\ &\quad \times \big(\partial_{z_j^r}(\Phi_1^k(t,x,y;z)\Phi_2^l(t,x,y;z))\big)\Big)\Big|_{z=(1,1)}.\end{aligned}$$

Apply the Leibniz's formula

$$\partial_{t^n}\big(f(t)g(t)\big)=\sum_{k=0}^{n}\binom{n}{k}\partial_{t^k}\big(f(t)\big)\partial_{t^{n-k}}\big(g(t)\big)$$

to the last term. Then we continue the previous relation and obtain

$$\Big(\mathcal{L}_i\big(\partial_{z_j^r}\Phi_i(t,x,y;z)\big)-\mu_i\big(\partial_{z_j^r}\Phi_i(t,x,y;z)\big)-\sum_{k+l\geq 2}\beta_i(k,l)\big(\partial_{z_j^r}\Phi_i(t,x,y;z)\big)$$
$$+\sum_{k+l\geq 2}\beta_i(k,l)\sum_{s=0}^{r}\binom{r}{s}\big(\partial_{z_j^s}\Phi_1^k(t,x,y;z)\big)\big(\partial_{z_j^{r-s}}\Phi_2^l(t,x,y;z)\big)\Big)\Big|_{z=(1,1)}.$$

To differentiate terms $\partial_{z_j^s}\Phi_i^l(t,x,y;z)$ we apply Faà di Bruno's formula (see [4])

$$\big(f(g(t))\big)^{(n)}=\sum_{k=1}^{n}\big(f(g(t))\big)^{(k)}B_{n,k}\Big(g'(t),\ldots,g^{(n-k+1)}(t)\Big),$$

where $B_{n,k}(x_1,\ldots,x_{n-k+1})$ is Bell polynomial which is defined as

$$B_{n,k}(x_1,\ldots,x_{n-k+1})=\sum\frac{n!}{j_1!\ldots j_{n-k+1}!}\Big(\frac{x_1}{1!}\Big)_1^j\cdots\Big(\frac{x_{n-k+1}}{(n-k+1)!}\Big)^{j_{n-k+1}},$$

where sum is taken for all sets of parameters $\{j_1,\ldots,j_{n-k+1}\}$, so that

$$j_1+\ldots j_{n-k+1}=k,\quad j_1+2j_2+\ldots(n-k+1)j_{n-k+1}=n.$$

Continuing the previous relation we have

$$\Big(\mathcal{L}_i\big(\partial_{z_j^r}\Phi_i(t,x,y;z)\big)-\Big(\sum_{k+l\geq 2}\beta_i(k,l)+\mu_i\Big)\big(\partial_{z_j^r}\Phi_i(t,x,y;z)\big)$$
$$+\sum_{k+l\geq 2}\beta_i(k,l)\Big[\big(\partial_{z_j^r}\Phi_1^k(t,x,y;z)\big)\Phi_2^l(t,x,y;z)+\Phi_1^k(t,x,y;z)\big(\partial_{z_j^r}\Phi_2^l(t,x,y;z)\big)\Big]$$
$$+\sum_{k+l\geq 2}\sum_{s=1}^{r-1}\binom{r}{s}\beta_i(k,l)\times\Big[\sum_{q=1}^{\min(r-s,k)}k(k-1)\ldots(k-q+1)\Phi_1^{k-q}(t,x,y;z)$$
$$\times B_{r-s,q}\big(\Phi_1'(t,x,y;z),\ldots,\Phi_1^{(r-s-q+1)(t,x,y;z)}\big)\Big]\times\Big[\sum_{p=1}^{\min(l,s)}l(l-1)\ldots(l-p+1)$$
$$\times B_{s,p}\big(\Phi_2'(t,x,y;z),\ldots,\Phi_2^{(s-l+1)}(t,x,y;z)\big)\Big]\Big)\Big|_{z=(1,1)}.$$

Let us define

$$\bar{x}=\begin{cases}1, & x=2;\\ 2, & x=1\end{cases}$$

then we obtain the differential equation for the factorial moment $m_{ij}^{(r)}(t,x,y)$:

$$\begin{aligned}\frac{\partial m_{ij}^{(r)}(t,x,y)}{\partial t} &= \mathcal{L}_i m_{ij}^{(r)}(t,x,y) + \Big(\sum_{k+l\geq 2}(k-1)\beta_i(k,l) - \mu_i\Big) m_{ij}^{(r)}(t,x,y)\\ &\quad + \sum_{k+l\geq 2} l m_{\bar{i}j}^{(r)}(t,x,y) + \sum_{k+l\geq 2}\sum_{s=1}^{r-1}\binom{r}{s}\beta_i(k,l)\\ &\quad \times \Big[\sum_{q=1}^{\min(r-s,k)} k(k-1)\dots(k-q+1)\\ &\quad \times B_{r-s,q}(m_{ij}^{(1)}(t,x,y),\dots,m_{ij}^{(r-s-q+1)}(t,x,y))\Big]\\ &\quad \times \Big[\sum_{p=1}^{\min(l,s)} l(l-1)\dots(l-p+1)\\ &\quad \times B_{s,p}(m_{\bar{i}j}^{(1)}(t,x,y),\dots,m_{\bar{i}j}^{(s-l+1)}(t,x,y))\Big]\end{aligned}$$

with initial condition

$$\begin{aligned}m_{ij}^{(r)}(0,x,y) &= \mathsf{E}\Big[n_{ij}(0,x,y)\big(n_{ij}(0,x,y)-1\big)\dots\big(n_{ij}(0,x,y)-r+1\big)\Big]\\ &= \mathsf{E}\Big[\delta_i^j\delta_x(y)\big(\delta_i^j\delta_x(y)-1\big)\dots\big(\delta_i^j\delta_x(y)-r+1\big)\Big]\\ &= \delta_1^r\delta_i^j\delta_x(y).\end{aligned}$$

Remark 2. BRWs with two types of particles (or more common models with arbitrary finite number of particle types) can generalise widely studied models with one type. To obtain models with one type we should assume in our model, for example, that intensities $\beta_1(k,l) = 0$ for all $k+l \geq 2$ and $l > 0$. In this case particles of the first type can produce only the offsprings of the first type. Such models were studied, for instance, in [12,14].

5 Solutions of Differential Equations for the First Moments

In this Section we consider the solutions of differential equations for the first moments $m_{ij}(t,x,y) = m_{ij}^{(1)}(t,x,y) = \mathsf{E}n_{ij}(t,x,y)$, $i,j = 1,2$, in case when generators of random walk for both particle types are equal, so that $\mathcal{L}_1 = \mathcal{L}_2 = \mathcal{L}$, where $\mathcal{L}$ acts by formula (3). We are going to omit the calculus as they are pretty huge and introduce only obtained results.

But firstly, we consider the following parabolic problem

$$\frac{\partial p(t,x,y)}{\partial t} = \mathcal{L}p(t,x,y), \qquad p(0,x,y) = \delta_x(y), \tag{7}$$

and introduce some designations:

$$b = \sum_{k+l\geq 2} l\beta_1(k,l), \qquad c = \sum_{k+l\geq 2} k\beta_2(k,l),$$
$$\beta_1 = \sum_{k+l\geq 2} (k-1)\beta_1(k,l), \qquad \beta_2 = \sum_{k+l\geq 2} (l-1)\beta_2(k,l).$$

We assume that $b < \infty$, $c < \infty$, and $\beta_i < \infty$, $i = 1, 2$, and consider the solutions for the first moments with regard to the values b and c. As $b \geq 0$ and $c \geq 0$ we consider three cases: 1. $b = 0$ and $c \geq 0$; 2. $b = 0$ and $b \geq 0$ and $c = 0$; 3. $b > 0$ and $c > 0$. The fact that the first two of these conditions intersect does not interfere with further considerations. In future formulae we assume that $p(t, x, y)$ is the solution of Cauchy problem (7).

1. *Case* $b = 0$, $c \geq 0$. Here we have that $\beta_1 = 0$ then

$$\left.\begin{aligned}
&m_{11}(t,x,y) = e^{-\mu_1 t} p(t,x,y);\\
&m_{21}(t,x,y) = \begin{cases} \frac{c\left(e^{-\mu_1 t} - e^{(\beta_2-\mu_2)t}\right)}{\mu_2-\beta_2-\mu_1} p(t,x,y), & \text{if } \mu_1 \neq \mu_2 - \beta_2,\\ cte^{-\mu_1 t} p(t,x,y), & \text{if } \mu_1 = \mu_2 - \beta_2; \end{cases}\\
&m_{12}(t,x,y) = 0;\\
&m_{22}(t,x,y) = e^{(\beta_2-\mu_2)t} p(t,x,y).
\end{aligned}\right\} \tag{8}$$

2. *Case* $b \geq 0, c = 0$. Here we have that $\beta_2 = 0$ then

$$\left.\begin{aligned}
&m_{11}(t,x,y) = e^{(\beta_1-\mu_1)t} p(t,x,y);\\
&m_{21}(t,x,y) = 0;\\
&m_{12}(t,x,y) = \begin{cases} \frac{b\left(e^{(\beta_1-\mu_1)t} - e^{-\mu_2 t}\right)}{\mu_2-\beta_2-\mu_1} p(t,x,y), & \text{if } \mu_1 - \beta_1 \neq \mu_2,\\ bte^{-\mu_2 t} p(t,x,y), & \text{if } \mu_1 - \beta_1 = \mu_2; \end{cases}\\
&m_{22}(t,x,y) = e^{-\mu_2 t} p(t,x,y).
\end{aligned}\right\} \tag{9}$$

3. *Case* $b > 0, c > 0$. Here

$$\left.\begin{aligned}
m_{11}(t,x,y) &= \frac{e^{C_1 t}}{2C_2}\Big((\beta_1 - \mu_1 - C_1 + C_2)e^{C_2 t}\\
&\quad + (C_1 + C_2 - \beta_1 + \mu_1)e^{-C_2 t}\Big) p(t,x,y);\\
m_{21}(t,x,y) &= \frac{ce^{C_1 t}}{2C_2}\left(e^{C_2 t} - e^{-C_2 t}\right) p(t,x,y);\\
m_{12}(t,x,y) &= \frac{be^{C_1 t}}{2C_2}\left(e^{C_2 t} - e^{-C_2 t}\right) p(t,x,y);\\
m_{22}(t,x,y) &= \frac{e^{C_1 t}}{2C_2}\Big((C_1 + C_2 - \beta_1 + \mu_1)e^{C_2 t}\\
&\quad + (C_1 - C_2 - \beta_1 + \mu_1)e^{-C_2 t}\Big) p(t,x,y),
\end{aligned}\right\} \tag{10}$$

where

$$C_1 = \frac{1}{2}\Big(\beta_1 + \beta_2 - \mu_1 - \mu_2\Big), \quad C_2 = \frac{1}{2}\Big[\big((\beta_1 - \beta_2) - (\mu_1 - \mu_2)\big)^2 + 4bc\Big]^{1/2}.$$

Remark 3. In this case we also can find the asymptotic behaviour for the first moments $m_{ij}(t,x,y)$, $i = 1,2$, $j = 1,2$, for $t \to \infty$. Their behaviour essentially depends on the properties of the underlying random walk. For example, if the condition is valid

$$\sum_{v \neq 0} a_i(v)|v|^2 < \infty, \quad i = 1,2, \tag{11}$$

where $|\cdot|$ denotes the vector norm, then underlying random walk has finite variance of jumps. It was obtained, e.g. see [14], that $p(t,x,y) \sim \gamma_d / t^{d/2}$, $t \to \infty$, where the constant $\gamma_d > 0$.

Let us discuss some obtained results under the assumption (11). Consider the case $b = 0$, $c \geq 0$, which describes the processes when particles of the first type cannot produce offsprings. Let us assume that $\mu_1 = \mu_2 - \beta_2 = 0$. It means that particles of the first type can only jump between lattice points. According to formula (8), as $t \to \infty$, we get

$$\begin{array}{lll} d = 1 & d = 2 & d \geq 3 \\ m_{ii}(t,x,y) \sim \dfrac{\gamma_1}{\sqrt{t}}, & m_{ii}(t,x,y) \sim \dfrac{\gamma_2}{t} & m_{ii}^{(1)}(t,x,y), \sim \dfrac{\gamma_d}{t^{d/2}}, \quad i = 1,2 \\ m_{21}(t,x,y) \sim c\gamma_1\sqrt{t}, & m_{21}(t,x,y) \sim c\gamma_2, & m_{21}^{(1)}(t,x,y) \sim \dfrac{c\gamma_d}{t^{d/2-1}}. \end{array}$$

Note that when $d = 2$ the value $m_{21}(t,x,y)$ tends to constant, so that we have steady state in terms of the first moments for subpopulations. In contrast to this case, when $d = 1$ mean nummber of particles of subpopulation $n_{21}(t,x,y)$ grows with the rate $\sqrt{t}$. However when $d \geq 3$ the same subpopulation degenerates at each lattice point. The same results can be obtained in case $b \geq 0$, $c = 0$, see (9). The only difference is that in this case particles of the second type can only jump between lattice points.

In case $b > 0$, $c > 0$ the parameter $C_2 > 0$ for all β_i, $\mu_i \in \mathbb{R}$, $i = 1,2$. From (10), as $t \to \infty$, we get the asymptotic behaviour of the first moments

$$\left.\begin{aligned} m_{11}(t,x,y) &\sim \frac{\gamma_d e^{(C_1+C_2)t}}{2C_2 t^{d/2}}\big(\beta_1 - \mu_1 - C_1 + C_2\big); \\ m_{21}(t,x,y) &\sim \frac{c\gamma_d e^{(C_1+C_2)t}}{2C_2 t^{d/2}}; \\ m_{12}(t,x,y) &\sim \frac{b\gamma_d e^{(C_1+C_2)t}}{2C_2 t^{d/2}}; \\ m_{22}(t,x,y) &\sim \frac{\gamma_d e^{(C_1+C_2)t}}{2C_2 t^{d/2}}\big(C_1 + C_2 - \beta_1 + \mu_1\big). \end{aligned}\right\}$$

The asymptotic behaviour of the moments depend on the value $C_1 + C_2$. If $C_1 + C_2 > 0$ when all subpopulations $n_{ij}(t,x,y)$, $i,j = 1,2$ will grow with exponential rate while if $C_1 + C_2 \le 0$ subpopulations will degenerate with either exponential rate ($C_1 + C_2 < 0$) or polynomial rate ($C_1 = -C_2$).

Subpopulations of particle offsprings for $i = 1, 2$, $j = 1, 2$ defined as

$$n_{ij,x}(t) := \sum_{y \in \mathbb{Z}^d} n_{ij}(t, x, y), \quad n_{ij,y}(t) := \sum_{x \in \mathbb{Z}^d} n_{ij}(t, x, y),$$

may be of particular interest for the models considered in Sect. 2.

Let $m_{ij,x}(t) = \mathsf{E} n_{ij,x}(t)$, $m_{ij,y}(t) = \mathsf{E} n_{ij,y}(t)$, $i = 1, 2$, $j = 1, 2$. From here we get

$$m_{ij,x}(t) = \sum_{y \in \mathbb{Z}^d} m_{ij}(t, x, y), \quad m_{ij,y}(t) = \sum_{x \in \mathbb{Z}^d} m_{ij}(t, x, y).$$

Hence, summing the left and right sides of the equations for $m_{ij}(t, x, y)$ for y or x in virtue of the equality $\sum_{y \in \mathbb{Z}^d} p(t, x, y) = \sum_{x \in \mathbb{Z}^d} p(t, x, y) = 1$, see, e.g. [14], we obtain the following results for the first moments that $m_{ij,x}(t) = m_{ij,y}(t) =: m_{ij}(t)$ for all $x, y \in \mathbb{Z}^d$, $i = 1, 2$, $j = 1, 2$. Due to the fact that the dependence on spatial coordinates in systems (8)–(10) is contained only in the function $p(t, x, y)$, so in consequence of the homogeneity of the branching medium and the initial conditions of the BRW model the right-hand side of the systems for $m_{ij}(t)$ is defined only by properties of the branching process at the sources.

Remark 4. Let us sum up the results from the genetic point of view. The case $\beta_2 = 0$ corresponds to the model of incomplete dominant epistatic lethals in the assumption that carriers of a single copy of the D haplotype (considered in Sect. 2) are viable, but cannot produce offsprings. In this case, the mean number of particles of the second type (carriers) located at $y \in \mathbb{Z}^d$ generated by a particles of the first type (healthy organisms) is determined by $m_{12}(t, x, y)$ from Eq. (9). The behaviour of $m_{12}(t, x, y)$ depends on ratio between mortality rate μ_2 and difference $\mu_1 - \beta_1$. Of particular interest is the case of $\mu_2 = \mu_1 - \beta_1$ in which the asymptotic behaviour of moments differs sharply depending on lattice dimension d. In the general case, when particles of the second type (carriers of haplotype D) are viable, the behaviour of the first moments of subpopulations is described by Eqs. (10). The asymptotic behaviour of the first moments $m_{ij}(t, x, y)$ and $m_{ij}(t)$ depends on parameters C_1, C_2 based on the birth and death intensities of both types of organisms.

Acknowledgments. The authors are grateful to Prof. S. Molchanov for useful discussions. The study was supported by the Russian Foundation for the Basic Research (RFBR), project No. 20-01-00487.

References

1. Bulinskaya, E.V.: Spread of a catalytic branching random walk on a multidimensional lattice. Stochast. Process. Appl. **128**(7), 2325–2340 (2015)
2. Dawson, D.A.: Introductory lectures on stochastic population systems. arXiv:1705.03781 (2017)

3. Ermakova, E., Mahmutova, P., Yarovaya, E.: Branching random walks and their applications for epidemic modelling. Stoch. Model. **35**(3), 300–317 (2019)
4. di Bruno, F.: Sullo sviloppo dell Funczioni. In: Annali di Scienze Mathematiche e Fisiche, pp. 479–480 (1855). (in Italian)
5. Gillespie, J.H.: Population Genetics: A Concise Guide. JHU Press, Baltimore (2005)
6. Gluecksohn-Waelsch, S.: Lethal genes and analysis of differentiation. Science **142**(3597), 1269–1276 (1963)
7. Hartl, D.L., Clark, A.G.: Principles of Population Genetics. Sinauer Associates, Sunderland (1997)
8. Karlin, S., Taylor, H.M.: A First Course in Stochastic Processes. Academic Press, Cambridge (2012)
9. Kolmogorov, A.N., Petrovskii, I.G., Piskunov, N.S.: A study of the diffusion equation with increase in the quality of matter, and its application to a biological problem. Bull. Moscow Univ. Math. Ser. A **1**(6), 1–25 (1937). (in Russian)
10. Kong, A., Frigge, M.L., Masson, G., Besenbacher, S., et al.: Rate of de novo Mutations and the importance of father's age to disease risk. Nature **488**(7412), 471–475 (2012)
11. Makarova, Y., Han, D., Molchanov, S., Yarovaya, E.: Branching random walks with immigration. Lyapunov stability. Markov Process. Relat. Fields **25**(4), 683–708 (2019)
12. Molchanov, S.A., Yarovaya, E.B.: Large deviations for a symmetric branching random walk on a multidimensional lattice. Proc. Steklov Inst. Math. **282**, 186–201 (2013). https://doi.org/10.1134/S0081543813060163
13. Sevastyanov, B.A.: Branching Processes. Nauka, Moscow (1971). (in Russian)
14. Yarovaya, E.B.: Branching random walks in a heterogeneous environment. Center of Applied Investigations of the Faculty of Mechanics and Mathematics of the Moscow State University, Moscow (2007). (in Russian)
15. Young, A.I.: Solving the missing heritability problem. PLoS Genet. **15**(6), e1008222 (2019)

Reconstruction of Multivariable Functions Under Uncertainty by Means of the Scheme of Metric Analysis

A. V. Kryanev[1,2], V. V. Ivanov[1,2], L. A. Sevastyanov[2,3(✉)], and D. K. Udumyan[4]

[1] National Research Nuclear University "MEPhI", Moscow, Russian Federation
avkryanev@mephi.ru
[2] Joint Institute for Nuclear Research (JINR), Dubna, Moscow, Russian Federation
[3] Peoples' Friendship University of Russia (RUDN University), Moscow, Russian Federation
[4] University of Miami, 1320 S. Dixie Hwy, Coral Gables, FL 33124, USA

Abstract. The problem of the reconstruction of a multivariable function whose values with chaotic errors are given at a finite number of points is considered in the paper. The problems of this kind arise when solving applied problems in various fields of research, including physics, engineering, economics, etc. We propose a new approach for solving this problem with the help of a metric analysis. The paper gives numerical two examples of the solution of the problem of the reconstruction of multivariable function, demonstrating the effectiveness of the proposed scheme. In the first example, the results of estimating the exact value of the function at the points where the values of the function with errors are known, in the second example, the results of reconstructing the physical characteristics of the core of a nuclear reactor are presented.

Keywords: Multivariable function · Reconstruction · Metric analysis

1 Metric Analysis Interpolation Scheme

The problem of the reconstruction of a multivariable function is a key problem to solving many applied tasks [1–7].

In this paper we propose the computational scheme for solving the problem of the reconstruction of multivariable function.

This approach uses information about the location of the points $\mathbf{X_1},...,\mathbf{X_n}$ of the function $F(\mathbf{X})$, $\mathbf{X} = (\mathbf{X_1}...,\mathbf{X_m})^T$ at which the values $Y_k, k = 1,...,n$ of the function are given.

The proposed in this paper scheme can be used even in those cases when the number n of points, in which the values of the function are given, is less than the number m of its arguments.

A. N. Shiryaev et al. (Eds.): ICSM-5 2020, PROMS 371, pp. 269–279, 2021.
https://doi.org/10.1007/978-3-030-83266-7_20

Interpolation schema presented in this paper is based on the metric analysis [5–7].

Consider the problem of determining the value of an unknown function

$$Y = F(\mathbf{X}), \tag{1}$$

for which its values $Y_k, k = 1, ..., n$ are given with chaotic errors in the knots $\mathbf{X_k} = (X_{k1}, ..., X_{km})^T$, $k = 1, ..., n$, and the desired value of the function (1) must be restored at the point $\mathbf{X}^*$.

According to the method of interpolation, based on metric analysis, the interpolation value Y^* is found as a solution of problems of minimum of the measure of metric uncertainty at the point $\mathbf{X}^* = (X_1^*, ..., X_m^*)^T$ [8–10].

$$\sigma^2_{ND}(Y^*; \mathbf{z}^*) = (W\mathbf{z}^*, \mathbf{z}^*), \tag{2}$$

where W is a matrix of the metric uncertainty and the interpolation value is determined by a linear combination

$$Y^* = \sum_{k=1}^{n} z_k^* \cdot Y_k, \quad \sum_{k=1}^{n} z_k^* = 1 \tag{3}$$

and is given by

$$Y^* = \frac{(W^{-1}\mathbf{Y}, \mathbf{1})}{(W^{-1}\mathbf{1}, \mathbf{1})}. \tag{4}$$

The matrix of the metric uncertainty W is defined by

$$W = \begin{pmatrix} \rho^2(\mathbf{X}_1, \mathbf{X}^*) & \dots & (\mathbf{X}_1, \mathbf{X}_n) \\ \vdots & \ddots & \vdots \\ (\mathbf{X}_n, \mathbf{X}_1) & \dots & \rho^2(\mathbf{X}_n, \mathbf{X}^*) \end{pmatrix} \tag{5}$$

where $\rho^2(\mathbf{X}_1, \mathbf{X}^*) = \sum_{l=1}^{m} V_l \cdot (X_{il} - X_l^*)^2$, $(\mathbf{X}_i, \mathbf{X}_j) = \sum_{l=1}^{m} (X_{il} - X_l^*) \cdot (X_{jl} - X_l^*)$, $i \neq j$, $i, j = 1, ..., n,$, where $V(l)$, $l = 1, \dots, m$, $\sum_{l=1}^{m} V_l = m$ are metric weights.

Theorem 1 (on convergence [9]). *The interpolation value of the function* Y^* *converges to the exact value* Y_k, *as* $\mathbf{X}^* \to \mathbf{X}_k$, $k = 1, ..., n$.

From Theorem 1 it follows that the interpolation function obtained using the metric analysis scheme is the continuous function of m variables.

When forming the matrix of metric uncertainty (5) its elements can be determined in some ways, in particular, the scalar product $(\mathbf{X}_i, \mathbf{X}_j)$ can be introduced in different ways. As a result, you can get different convergence rates of the interpolation function to the function values at the interpolation nodes. You can, for example, introduce the dot product as $(\mathbf{X}_i, \mathbf{X}_j) = \sum_{l=1}^{m} (X_{il} - X_l^*)^{\alpha} \cdot (X_{jl} - X_l^*)^{\alpha}$, $i \neq j$, $i, j = 1, ..., n,$, where $\alpha > 1$ is a parameter. In this case, the rate of convergence will be of order ρ^{α}. This approach is often convenient when the considered applied problem has a requirement on the rate of convergence.

Note that the presented multivariable function interpolation scheme, based on metric analysis (unlike many other interpolation schemes), does not use any general representation of the interpolated function. This scheme allows to calculate the interpolated values of the function at each given point $\mathbf{X}^*$ separately, taking into account the location of the point $\mathbf{X}^*$ in the m-dimensional space E_m with respect to the points $\mathbf{X}_k$, $k = 1, \ldots, n$ in which the values of the function are known.

The meaning of metric weights is that they take into account the degree of change of the function under study when its arguments change. In this case, the metric uncertainty matrix will take into account not only the geometric arrangement of points in the original geometric space, but also the different level of change in the function relative to different function arguments.

From $\sum_{l=1}^{m} V_l = m$ it follows that if $V_l > 1$ ($V_l < 1$), then this indicates a larger (smaller) level of change in the function when changing the l-th argument with respect to the same degree of sensitivity of the function to changes in arguments.

The interpolation scheme of multivariable functions of metric analysis allows one to consider the different level of sensitivity of changes in the function to changes in the arguments with the help of metric weights $V_l, l = 1, ..., m$. In our previous works to find the metric weights w_l the scheme, based on the successive elimination of arguments and taking into account the changes in the function, was used (see [8–11]).

The quality of solving the problems of separating deterministic and chaotic components for multivariable functions largely depends on the quality of the definition of metric weights.

2 Metric Analysis Reconstruction Scheme

Consider the problem of restoring the functional dependence $Y = F(X_1, ..., X_m) = F(\mathbf{X})$ in the presence of chaotic deviations from the exact values at given points.

In this paper we propose the more effective scheme for determining the metric weights V_l, $l = 1, ..., m$, based on a multifactor linear model of the relationship between the values of the function Y and its arguments X_l, $l = 1, ..., m$.

The new scheme for determining metric weights V_l, $l = 1, ..., m$ is based on the calculation of weighting factors u_l, $l = 1, ..., m$ of the linear regression model:

$$Y = u_0 + \sum_{l=1}^{m} u_l \cdot X_l + \varepsilon, \tag{6}$$

where ε is random noise.

According to the LSM, the parameter estimates $\mathbf{u} = (u_1, ..., u_m)^T$ of model (6) are given by:

$$\mathbf{u} = K_{\mathbf{X}}^{-1}\mathbf{cov}(Y, \mathbf{X}), \tag{7}$$

where $K_{\mathbf{X}}^{-1}$ is the inverse matrix, and the elements of the covariance matrix $K_{\mathbf{X}}$ are defined by:

$$cov(X_i, X_j) = \frac{1}{n-1} \sum_{k=1}^{n} (X_{ki} - \bar{X}_i) \cdot (X_{kj} - \bar{X}_j), \tag{8}$$

and vector components $\mathbf{cov}(Y, \mathbf{X}) = (cov(Y, X_1)...cov(Y, X_m))^T$ are calculated according to equations: $cov(Y, X_j) = \frac{1}{n-1} \sum_{k=1}^{n} (Y_k - \bar{Y}) \cdot (X_{kj} - \bar{X}_j)$, $\bar{X}_i = \frac{1}{n} \sum_{k=1}^{n} X_{ki}$, $\bar{Y} = \frac{1}{n} \sum_{k=1}^{n} Y_k$.

Then the values of the metric weights v_l, $l = 1, ..., m$ are calculated by the formulas:

$$v_l = \frac{|u_l|}{\sum_{j=1}^{m} |u_j|} \cdot m, \quad l = 1, ..., m. \tag{9}$$

Remark 1. Since the metric weights depend on the values of the function at the interpolation nodes, the interpolation values obtained by the metric analysis method, in the general case, will nonlinearly depend on the given values Y_k, $k = 1, ..., n$ of the function.

Thus, the above scheme for choosing a weighting metric makes it possible to identify the degree of influence of each of the arguments and takes into account the different degree of influence by moving to a new metric with corresponding unequal weights.

Let us consider the degenerate case when an argument (factor) is introduced in the function under study, on which the function does not depend. Then, when implementing the scheme for finding the weights of the new metric an unambiguous result will be obtained: the metric weight corresponding to the above factor will be zero and this factor will be automatically excluded from further consideration. Therefore, the above scheme for the transition to a metric with weights makes it possible to take into account, just as it is done in factor analysis, the influence of arguments on the change in the function and to exclude insignificant ones from them, lowering the dimension of the factor space.

Let us consider the degenerate case when an argument (factor) is introduced in the function under study, on which the function does not depend. In this case, an unambiguous result will be obtained: the metric weight corresponding to the above factor will be zero, and this factor will be automatically excluded from further consideration. The above scheme for the transition to a metric with weights makes it possible to take into account, just as it is done in factor analysis, the influence of factors on the change in the function and exclude insignificant ones from them, reducing the dimension of the factor space.

Remark 2. In the conditions of a strong correlation of part of the arguments $\mathbf{X}_k = (X_{k1}, ..., X_{km})^T$, $k = 1, ..., n$, and, thus, a singular or ill-conditioned matrix $K_{\mathbf{X}}$ it is necessary to carry out regularization by replacing the matrix $K_{\mathbf{X}}$ by the regularized matrix, for example, the matrix $K_{\mathbf{X},\beta} = K_{\mathbf{X}} + \beta * diag(K_{11}, ..., K_{mm})$, $\beta > 0$, where K_{ij}, $i, j = 1, ..., m$ is the regularized matrix, for example, the matrix $K_{\mathbf{X}}$.

Function values Y_k, $k = 1, ..., n$ are known with errors at points $\mathbf{X}_k = (X_{k1}, ..., X_{km})^T$, $k = 1, ..., n$.

Thus, we have the system of equations:

$$Y_k = Y_{kdet} + \varepsilon_k, \quad k = 1, ..., n, \tag{10}$$

where $\mathbf{Y}_{det} = (Y_{1det}, ..., Y_{ndet})^T$ is the sought vector of deterministic components (estimates of the values of the function) at points $\mathbf{X}_k = (X_{k1}, ..., X_{km})^T$, $k = 1, ..., n$ and $\varepsilon = (\varepsilon_1, ..., \varepsilon_n)^T$ is the vector of chaotic components.

For any point $\mathbf{X}^*$ we are looking for value Y_α

$$Y_\alpha = \sum_{i=1}^{n} z_i \cdot Y_i = (\mathbf{z}, \mathbf{Y}), \tag{11}$$

where vector $\mathbf{z}$ is the solution of the following problem of minimum of total uncertainty:

$$(W\mathbf{z}, \mathbf{z}) + \alpha \cdot (K_{\mathbf{Y}}\mathbf{z}, \mathbf{z}) - min, \quad (\mathbf{z}, \mathbf{1}) = 1, \tag{12}$$

$\alpha \geq 0$ is the smoothing parameter, $K_{\mathbf{Y}}$ is the covariance matrix of the vector of random components $\varepsilon = (\varepsilon_1, ..., \varepsilon_n)^T$, and the matrix of metric uncertainty W is calculated with respect to the point $\mathbf{X}^*$ (see (5)).

Expression $(W\mathbf{z}, \mathbf{z})$ represents the metric uncertainty in the restored value of the function, and expression $\alpha \cdot (K_{\mathbf{Y}}\mathbf{z}, \mathbf{z})$ represents the stochastic uncertainty of the recovered value.

The problem (12) can be solved by means of Lagrange multipliers.

The solution of problem (12) is defined by the equality:

$$Y_\alpha = \left((W + \alpha \cdot K_{\mathbf{Y}})^{-1}\mathbf{1}, \mathbf{Y}\right) / \left((W + \alpha \cdot K_{\mathbf{Y}})^{-1}\mathbf{1}, \mathbf{1}\right). \tag{13}$$

When $\alpha \to \infty$ the value Y_α for the point $\mathbf{X}_k$ is defined by the equality:

$$Y_\infty = \frac{(K_{\mathbf{Y}}^{-1}\mathbf{1}, \mathbf{Y})}{(K_{\mathbf{Y}}^{-1}\mathbf{1}, \mathbf{1})}. \tag{14}$$

When $\alpha \to +0$ the value Y^* for the point $\mathbf{X}_k$ is defined by the equality:

$$Y_0 = \frac{(W^{-1}\mathbf{1}, \mathbf{Y})}{(W^{-1}\mathbf{1}, \mathbf{1})}. \tag{15}$$

The value of Y^* in the point $\mathbf{X}^*$ is given by:

$$Y^* = Y_{\alpha_*}, \tag{16}$$

where α_* is found from the equality

$$\|\mathbf{Y} - \mathbf{Y}_{\alpha_*}\|^2 = n \cdot \sigma^2, \tag{17}$$

where $\mathbf{Y}_{\alpha_*} = (Y_{1\alpha_*}, ..., Y_{n\alpha_*})^T$, σ^2 is the mean value of variances of chaotic components ε_k, $k = 1, ..., n$.

Consider the restoration of the functional dependence, when the values of the function at each given point are known with different levels of errors. In such cases, reconstruction should be performed with the value of found using the generalized residual principle.

Let at points $\mathbf{X}_1, .., \mathbf{X}_n$ the values of the function Y_k, $k = 1, ..., n$ are given with errors, the variances of which are equal to $\sigma_1^2, ..., \sigma_1^2$, respectively.

In this case, the covariance matrix has a diagonal form:

$$K_{\mathbf{Y}} = diag\left(\sigma_1^2, ..., \sigma_1^2\right). \tag{18}$$

We introduce the residual functional

$$\delta(\alpha) = \frac{1}{n} \cdot \sum_{k=1}^{n} \frac{\left(Y_k^*(\alpha) - Y_k\right)^2}{\sigma_k^2} - 1, \tag{19}$$

where $Y_k^*(\alpha)$ are the reconstructed values at the nodes $\mathbf{X}_k$, where the values of the function are given.

To restore the original functional dependence a value of α_0 is chosen for which

$$\delta(\alpha_0) = 0, \tag{20}$$

By the found value of α_0 for points $\mathbf{X}_1, .., \mathbf{X}_n$ are the recovered values $Y_k^*(\alpha_0)$, $k = 1, ..., n$.

The quantity $Y_{ch}(\mathbf{X}_k) = Y_k - Y_k^*(\alpha_0)$ is the chaotic component of the function value at the point $\mathbf{X}_k$.

The actual finding of a suitable value of 0 according to the residual method is reduced to the sequential finding of the restored values $Y_k^*(\alpha)$, $k = 1, .., n$ for different α and the choice of such reconstructed values for which equality (20) is most accurately satisfied.

Since the parameter α is continuous, in the numerical implementation it is possible to find a suitable value of α_0 in such a way that equality (20) is fulfilled with a predetermined accuracy.

We note once again that interpolation and restoration of functional dependencies by the method of metric analysis does not imply setting the basis system of functions, and at each point where the interpolation or reconstructed value is calculated, its location relative to the interpolation nodes is individually taken into account.

The total measure of uncertainty $\sigma_{sum}^2(Y/\mathbf{X}^*)$ of the value of Y at the point $\mathbf{X}^*$ is determined by the equality

$$\sigma_{sum}^2\left(Y/\mathbf{X}^*\right) = (W\mathbf{z}, \mathbf{z}) + \alpha \cdot (K_{\mathbf{Y}}\mathbf{z}, \mathbf{z}) = (V_{sum}\mathbf{z}, \mathbf{z}), \tag{21}$$

where $V_{tot} = (W\mathbf{z}, \mathbf{z}) + \alpha \cdot (K_{\mathbf{Y}}\mathbf{z}, \mathbf{z})$ is a symmetric positive definite $(n \times n)$ – matrix.

The matrix V_{tot} from (21) will be called the matrix of the total uncertainty of the value of the function Y at the point $\mathbf{X}^*$.

Consider the inverse matrix V_{tot}^{-1}. The quantity $I(Y/X^*) = \left(V_{tot}^{-1}\mathbf{1}, \mathbf{1}\right) > 0$ will be called information about the value of the function at the point $\mathbf{X}^*$ from the values Y of the function, known with errors at points $\mathbf{X}_1, ..., \mathbf{X}_n$ [9].

3 Numerical Results

The numerical results for restored values (deterministic components) of the multivariable function, using a metric analysis for a test case, are given below.

The values of the function were restored at the same points in which the noisy values of the function were known. Chaotic values were generated using a normally distributed random variable $N(0, \sigma^2)$.

The following indicators for restoring the original function were calculated:

1. deterministic components Y_{kdet}, $k = 1, ..., n$;
2. chaotic components ε_k, $k = 1, ..., n$;
3. initial relative errors $\Delta_k = \left|\frac{Y_k^* - Y_k}{Y_k^*}\right|$, $k = 1, ..., n$;
4. mean initial error $\Delta = \frac{1}{n} \cdot \sum_{k=1}^{n} \Delta_k$;
5. relative recovery errors $\delta_k = \left|\frac{Y_k^* - Y_{kdet}}{Y_k^*}\right|$, $k = 1, ..., n$;
6. mean recovery error $\delta = \frac{1}{n} \cdot \sum_{k=1}^{n} \delta_k$.

Here Y_k^* are exact values of the function; Y_k are noisy function values, $Y_k = Y_{kdet} + \varepsilon_k$, $\varepsilon_k \sim N(0, \sigma^2)$ is a normally distributed random variable; Y_{kdet} are restored values of the deterministic component of our function.

Example. The function Y of four variables $\mathbf{x} = (x_1, x_2, x_3, x_4)^T$ has the form $Y = (V\mathbf{x}, \mathbf{x}) + (\mathbf{c}, \mathbf{x})$, where

$$V = \begin{pmatrix} 5 & 0 & 0 & 0 \\ 0 & 3 & 0 & 0 \\ 0 & 0 & 6 & 0 \\ 0 & 0 & 0 & 0.3 \end{pmatrix},$$

$\mathbf{c} = (-0.8, 2.0, 1.0, 0.5)^T$, and the values of arguments lie in the unit four-dimensional domain D: $0 \le x_i \le 1$, $i = 1, 2, 3, 4$.

The exact values of the function were calculated at randomly chosen 24 points ($n = 24$) in the domain D. The exact values of the function are given in the first column of Table 1. Then, using a normally distributed random variable with $\sigma = 1.2$, chaotic components $\varepsilon_k, k = 1, ..., 24$, were generated and added in sequence to 24 exact values of the function. Thus obtained noisy values of the original function are presented in the second column of Table 1. The third column of Table 1 shows the values of the relative errors.

The separation of the deterministic and chaotic components is realized according to the scheme presented above. The optimum value α_* of parameter α according to (20) was determined using equality

$$\alpha_* = argmin \left| \frac{1}{24} \sum_{k=1}^{24} (Y_k - Y_{k\alpha})^2 - 1.44 \right|.$$

Table 1. Values of selected deterministic and chaotic components

Exact value Y_k^*	Value with chaotic component Y_k	Initial relative error $\Delta_k = \left\lvert \frac{Y_k^* - Y_k}{Y_k^*} \right\rvert$	Restored value Y_{kdet}	Chaotic value ε_k	Relative recovery error $\delta_k = \left\lvert \frac{Y_k^* - Y_{kdet}}{Y_k^*} \right\rvert$
11.114	11.049	0.006	10.587	0.463	0.047
6.484	4.017	0.381	6.700	−2.684	0.033
4.574	3.866	0.155	4.722	−0.857	0.032
5.486	4.045	0.263	4.951	−0.907	0.098
3.552	4.164	0.172	3.624	0.540	0.020
9.581	9.211	0.039	9.41	−0.197	0.018
5.486	5.348	0.025	4.951	0.397	0.098
3.552	1.612	0.546	3.624	−2.012	0.020
9.581	8.543	0.108	9.41	−0.937	0.018
5.212	5.769	0.107	5.396	0.373	0.035
6.937	5.864	0.155	6.299	−0.435	0.092
6.233	6.452	0.035	6.691	−0.239	0.074
11.332	12.411	0.095	10.576	1.836	0.067
8.764	10.173	0.161	8.660	1.512	0.012
10.146	11.741	0.157	10.070	1.672	0.008
7.687	8.227	0.070	8.328	−0.100	0.083
7.466	6.238	0.165	6.963	−0.726	0.067
5.973	5.047	0.155	5.690	−0.643	0.047
9.560	12.300	0.287	9.762	2.539	0.021
12.935	13.551	0.048	12.733	0.819	0.016
5.866	6.488	0.106	5.770	0.718	0.016
7.524	7.166	0.048	8.010	−0.845	0.065
10.028	9.503	0.052	10.003	−0.500	0.003
6.283	5.531	0.120	6.093	−0.562	0.030

In Table 1 the values of selected deterministic and chaotic components are presented in the fourth and fifth columns respectively. Finally, the sixth column of Table 1 gives the relative error values of restored function.

The following indicators for deterministic component were obtained: $\alpha_* = 4.2$; $\Delta = 0.144$; $\delta = 0.077$.

Below there are the numerical results of reconstruction using the proposed schemes for metric analysis of physical indicators of the state of the active zone of the nuclear reactor using the accumulated data on the reactor operation.

The important physical indicator is considered: the macroscopic generation cross section in thermal group and the macroscopic cross section for scattering from the fast group to the thermal group, which depends on controlled physical parameters of the components of the metric analysis of physical indicators of the state of the active zone.

The numerical results of the reconstruction were compared with the known values of the indicators, the errors of the reconstruction values were calculated. They are presented in the two tables below.

Tables 2 and 3 show the results of reconstruction using the scheme for determining the metric weights presented in this paper.

Table 2. Macroscopic generation cross section in thermal group

Recovery performance indicator	Equal weight scheme			Scheme with weights		
Amount of points	6	12	18	6	12	18
Error share < 5%	97,1	99,3	100,0	98,2	100,0	99,6
Error share < 1%	81,3	89,7	89,3	83,1	91,9	91,9
Error share < 0,5%	62,5	74,3	76,1	64,7	77,6	75,0
Error share < 0,1%	24,6	36,8	34,6	33,8	38,2	37,1
Maximum error	13,11	5,57	4,82	11,05	3,70	5,43
Average error	0,777	0,491	0,486	0,689	0,389	0,398

Table 3. Macroscopic generation cross section for scattering from the fast group to thermal group

Recovery performance indicator	Equal weight scheme			Scheme with weights		
Amount of points	6	12	18	6	12	18
Error share < 5%	99,3	100,0	100,0	99,7	100,0	100,0
Error share < 1%	86,4	89,2	89,2	86,4	90,2	88,5
Error share < 0,5%	68,5	76,6	76,6	71,0	77,3	78,7
Error share < 0,1%	30,8	35,0	38,1	33,9	39,2	42,0
Maximum error	6,272	3,529	3,003	5,508	3,418	2,943
Average error	0,487	0,380	0,365	0,368	0,389	0,347

Tables 4 and 5 show for comparison the results of restoration using the sequential exclusion of arguments scheme (scheme no. 1) and the scheme presented in this paper (scheme no. 2).

4 Conclusion

Based on the metric analysis, the scheme and algorithm for solving the problem of the reconstruction of multivariable function, whose values are given at a

Table 4. Macroscopic cross section for scattering from thermal group to thermal group

Recovery performance indicator	Equal weight scheme			Scheme no 1			Scheme no 2		
Amount of points	6	12	18	6	12	18	6	12	18
Error share < 5%	100,0	100,0	100,0	100,0	100,0	100,0	100,0	100,0	100,0
Error share < 1%	99,1	100,0	100,0	98,7	100,0	100,0	99,7	100,0	100,0
Error share < 0,5%	96,9	99,7	100,0	95,3	99,4	100,0	96,2	100,0	100,0
Error share < 0,1%	66,4	81,4	88,1	71,4	83,6	88,1	75,5	93,7	97,2
Maximum error	1,561	0,747	0,360	1,795	0,692	0,410	1,374	0,362	0,156
Average error	0,116	0,058	0,041	0,114	0,055	0,041	0,096	0,034	0,027

Table 5. Macroscopic cross-section of fission in the thermal group

Recovery performance indicator	Equal weight scheme			Scheme no 1			Scheme no 2		
Amount of points	6	12	18	6	12	18	6	12	18
Error share < 5%	97,3	99,3	99,3	98,7	99,7	99,7	98,3	99,7	99,7
Error share < 1%	80,1	88,2	86,9	83,2	90,2	90,6	83,2	91,6	90,2
Error share < 0,5%	63,3	74,1	76,1	69,7	78,8	78,1	65,3	79,5	76,1
Error share < 0,1%	28,3	39,7	40,7	33,0	46,1	46,1	32,3	42,8	41,1
Maximum error	12,372	7,247	5,718	15,829	6,151	5,704	8,790	7,000	5,199
Average error	0,825	0,488	0,468	0,662	0,389	0,395	0,661	0,378	0,395

finite number of the points, are proposed. In this scheme, a priori information about the form of the functional dependence is not used (only information on the continuity of the reconstructed function is used). Numerical experiments on the reconstruction of multivariable function for a number of multivariable functions with the help of the proposed schemes have shown that the scheme allows one to the reconstruction of multivariable function even in the presence of a small number of points at which values of the being analyzed function are given. Moreover, the number of such points can be less than the number of arguments. For the examples presented in the paper, the mean errors of the reconstruction of multivariable considered functions were calculated. Numerical experiments on the reconstruction for a number of functions of many variables using the proposed scheme where conducted. The results of experiments have shown that the proposed scheme allows one to restore the values of function even in the situation when we have a small number of points. From a comparison of average errors Δ and δ for the example given by us one can see that the new scheme allows reducing relative errors δ_k twice.

Acknowledgments. The authors thank I. A. Malinkin for help with the numerical calculations presented in the paper.

References

1. Simonoff, J.S.: Smoothing Methods in Statistics, 2nd edn. Springer, New York (1998). https://doi.org/10.1007/978-1-4612-4026-6
2. Haykin, S.: Neural Networks: A Comprehensive Foundation. Prentice Hall, New York (1999)
3. Sendov, B., Andreev, A.: Approximation and interpolation theory. Handbook of Numerical Analysis, vol. 3, pp. 223–462. Elsevier (1994)
4. Masjukov, A., Masjukov, V.: A new fast iterative method for interpolation of multivariate scattered data. Comput. Methods Appl. Math. **5**, 1–18 (2005)
5. Taranik, V.A.: Application of the Lagrange interpolation polynomial for functions with many variables, vol. 8, no. 2(13), pp. 69–76. Science Rise, Technology Center (2015)
6. Andreasen, M.M.: Non-linear DSGE models and the central difference Kalman filter. J. Appl. Econometr. **28**(6), 929–955 (2013)
7. Antoniou, I., Ivanov, V.V., Zrelov, P.V.: Wavelet filtering of network traffic measurements. Physica A **324**, 733–753 (2003)
8. Kryanev, A.V., Lukin, G.V., Udumyan, D.K.: Metric analysis and applications. Numer. Methods Program. Adv. Comput. Sci. J. **10**, 408–414 (2009). (in Russian)
9. Kryanev, A.V., Lukin, G.V., Udumyan, D.K.: Metric Analysis and Data Processing, Science edn, Moscow (2012). (in Russian)
10. Kryanev, A.V., Udumyan, D.K., Lukin, G.V., Ivanov, V.V.: Metric analysis approach for interpolation and forecasting of time processes. Appl. Math. Sci. **8**(22), 1053–1060 (2014)
11. Kryanev, A.V., Ivanov, V.V., Malinkin, I.A., Sevastyanov, L.A., Udumyan, D.K.: Interpolation of multivariable functions by means of the nonlinear schema of metric analysis. In: Proceedings of International Conference "The Fourth Symposium on Methods of Nonlinear Mathematical Physics", Greece (2020). (in press)

Application of Deep Learning Methods for the Identification of Partially Observable Subgraphs Under the Conditions of a Priori Uncertainty and Stochastic Disturbances (Using the Example of the Problem of Recognizing Constellations)

V. A. Galkin(✉) and A. V. Makarenko

Institute of Control Sciences, Russian Academy of Sciences, ul. Profsoyuznaya 65, 117977 Moscow, Russia

Abstract. This paper demonstrates the effective capabilities of deep neural networks in solution of the problem of structural identification on graphs in conditions of a priori uncertainty, incomplete observability and stochastic disturbances which is also knows as subgraph detection or recovery. The problem of identification of observed constellations in a photo of the night sky was considered as a test. The solution with quality of 0.927 F_1 is obtained. In this work we synthesized original ResNet architecture of the convolution neural network with 26 trainable layers, 415 193 configurable parameters, carried out statistical analysis of structural characteristics of the dataset and adapted the standard binary cross entropy loss function, developed a special strategy for learning the neural network. Moreover, an adequate criterion of observability of the constellation in the image was formed. We also studied the influence of noise on the quality and stability of the received solutions.

Keywords: Deep learning · Graph identification · Stochastic disturbances

1 Introduction

Let's introduce an undirected graph

$$\Gamma^{\mathrm{S}} = \left\langle \mathrm{V}^{\mathrm{S}}, \mathrm{E}^{\mathrm{S}} \right\rangle, \quad \mathrm{V}^{\mathrm{S}} \neq \emptyset, \quad \mathrm{E}^{\mathrm{S}} \neq \emptyset, \quad \left|\mathrm{V}^{\mathrm{S}}\right| = K_{\mathrm{S}}^{\mathrm{V}} \ll \infty, \quad \left|\mathrm{E}^{\mathrm{S}}\right| = K_{\mathrm{S}}^{\mathrm{E}} \ll \infty, \tag{1}$$

with the nodes and edges set by sets of V^{S} and E^{S} respectively. According to this topology, the Γ^{S} graph does not contain multiple edges, has no loops, and has the only connected component. Each node and edge of the Γ^{S} graph is associated

A. N. Shiryaev et al. (Eds.): ICSM-5 2020, PROMS 371, pp. 280–291, 2021.
https://doi.org/10.1007/978-3-030-83266-7_21

with their property vectors: $\mathbf{P}^{\mathrm{V}}_{\mathrm{S}\,i}$ – of the i node, $i = \overline{1,\, K^{\mathrm{V}}_{\mathrm{S}}}$ and $\mathbf{P}^{\mathrm{E}}_{\mathrm{S}\,ij}$ – of the edge linking the i and j nodes, $i \neq j$, $j = \overline{1,\, K^{\mathrm{V}}_{\mathrm{S}}}$.

The solution of various structural identification tasks like detection or recovery of subgraphs on graphs like Γ^{S}, is relevant from both an applied and theoretical point of view. There are two classical groups of algorithms for searching and selection of subgraphs:

- In a completely observed graph: modularity optimization (Newman 2006), stochastic block models (Airoldi et al. 2008), spectral graph-partitioning (Newman 2013), clique percolation (Du et al. 2007), clustering (Chen and Saad 2010), and label propagation (Li et al. 2020).
- Under the conditions of incomplete observability of nodes and edges of the source graph: based on the incorporated additional information (Yang et al. 2013), the similarity of topological structures (Yan and Gregory 2012; Yan and Gregory 2011), and a hierarchical gamma process (Zhou 2015).

Recently, deep learning methods have been actively used in solution various problems on the graphs in the conditions of a priori uncertainty and stochastic disturbances (Goodfellow et al. 2016). It can be seen from scientific and information resources, that main focus is on tasks of evaluating certain characteristics of graphs, subgraphs and nodes (Lin et al. 2018; Zügner et al. 2018).

This work demonstrates the possibility of effective application of deep neural networks for the solution of structural identification problem on graphs in conditions of a priori uncertainty, incomplete observability and stochastic disturbances. The problem of identification of constellations in a photo of the night sky is considered as a test one. The priori uncertainty arises due to the lack of information about the time and coordinates of the survey and the direction of the optical axis. Incomplete observability could be formed by partial shielding stars by clouds. Stochastic disturbances form "false stars" due to the instrument noise of the photo registration equipment.

2 Statement of the Problem and Related Work

Let Γ^{N} graph similar to Γ^{S} graph is available for the observation. In general, the observed Γ^{N} graph has a non-zero intersection with Γ^{S} graph:

$$\Gamma^{\mathrm{S}} \cap \Gamma^{\mathrm{N}} \neq \emptyset, \quad \Gamma^{\mathrm{S}} \neq \Gamma^{\mathrm{N}}. \tag{2}$$

It's also possible to present the Γ^{N} graph as

$$\Gamma^{\mathrm{N}} = \Gamma^{\mathrm{N}}_{\circ} \cup \Gamma^{*}, \tag{3}$$

where $\Gamma^{\mathrm{N}}_{\circ} \approx \Gamma^{\mathrm{S}}_{\circ}$ addition or distribution the observable subgraph of the Γ^{S} graph, and the Γ^{*} graph represents the stochastic addition or distribution forming false nodes and edges. Characteristics of $\mathbf{P}^{\mathrm{V}}_{\mathrm{N}\circ\,i}$ and $\mathbf{P}^{\mathrm{E}}_{\mathrm{N}\circ\,ij}$ also include random misrepresentations according to their true identities $\mathbf{P}^{\mathrm{V}}_{\mathrm{S}\circ\,i}$ and $\mathbf{P}^{\mathrm{E}}_{\mathrm{S}\circ\,ij}$.

The final test problem can be formulated as highlighting the most effective assessment of the Γ^{N} graph from the observed $\Gamma^{\mathrm{S}}_{\mathrm{o}}$ graph:

$$L\Big[\Gamma^{\mathrm{N}}\big|(\mathrm{P}^{\mathrm{V}}_{\mathrm{N}}, \mathrm{P}^{\mathrm{E}}_{\mathrm{N}}), \Gamma^{\mathrm{So}}\big|(\mathrm{P}^{\mathrm{V}}_{\mathrm{So}}, \mathrm{P}^{\mathrm{E}}_{\mathrm{So}})\Big] \to \min, \tag{4}$$

where L describes an error function. And the solution of the problem (4) implies the identification of the nodes of the Γ^{S} graph which are part of Γ^{N}.

In the test task defined above, the nodes of Γ^{S} graph represent clusters of stars called constellations, the edges describe relations between constellations with common borders on the celestial sphere. Characteristics of the nodes stand for the corresponding astronomical properties of constellations and their borders. The sky with stars is the observed part of the celestial sphere in a horizontal coordinate system is fixed to a location on Earth. It's worth to note that the constellation "Serpens" is divided into two: "Serpens Caput" and "Serpens Cauda". In this way, the total number of constellations $K^{\mathrm{V}}_{\mathrm{S}}$ is equal to 89 and $K^{\mathrm{E}}_{\mathrm{S}} = 264$. The graph described above is shown in Fig. 1.

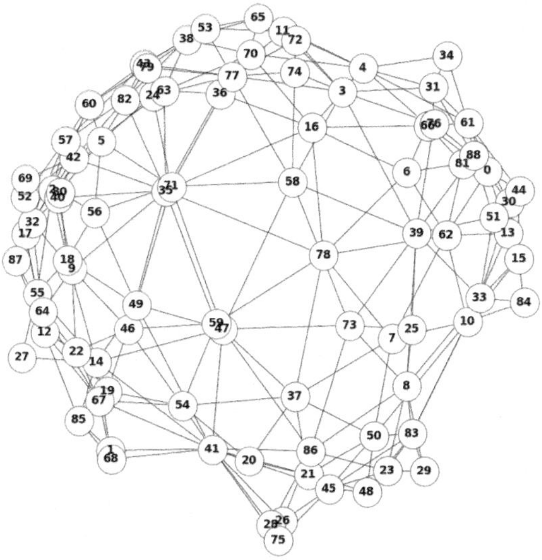

Fig. 1. Constellation graph Γ^{S}. Node number (starting with 0) – constellation numbers in alphabetical order.

In the stated statement, besides theoretical, the problem has a high applied importance for solving the problems of identification of stars and astronomical navigation.

Solution based on the comparison of templates is presented in work (Ji et al. 2015). Accuracy on the test images formed 74% with correct constellation detection share of about 92.8%, but the average recognition time is 85 s, which is not applicable in real time conditions.

In our work, we use approach of developing and training a deep convolutional neural network which is able to approximate unknown mapping of the observed sky to the constellation space.

3 Raw Data and Dataset Preparation

3.1 Images of the Starry Sky

In this task, the observed data are images of the starry sky, which are parts of the celestial sphere in the horizontal coordinate system. The horizontal coordinate system is connected to the Earth (or another celestial object) and participates in the Earth's own rotation. As a result, distant celestial bodies such as stars, move in circles with a period of time equal to the Earth's period of rotation and are motionless relative to each other, provided that their proper motions are not taken into account.

In order to create an astronomical dataset where the neural network is being trained and tested, we have developed our software based on the Tycho-2 star catalog. The dataset was created using an iterative algorithm based on Earth observer parameters (location of the observer, viewing angles and GMT). Stars with a magnitude value of no more than 6.5, which corresponds to stars visible to the eye, were used in the construction of images of the starry sky. Also, the spectral properties of stars were not taken into account, i.e. a photo of the starry sky is a binary image. This decision was made in order to solve the problem of identification constellations only by the geometric pattern of the observed starry.

Thus, a training and testing dataset was formed with a total size of 1 284 780 and 48 772 samples respectively. It should be noted that the samples presented do not overlap with the earth observer parameters to avoid leakage of data during the testing phase.

The characteristics of the images of the starry sky are as follows:

1. size 240×240 pixels;
2. the field of view angle $20^\circ \times 20^\circ$, as with modern star sensors;

3.2 Constellation Labels

To train the neural network, a marked dataset is essential, i.e. each input image requires an answer that contains the marks of the constellations depicted in the starry sky photo. In this task, an obvious and natural criterion for the observability of constellations can be formed, namely an entry criterion based on the current boundaries of the constellations. In other words, a constellation is observable if any part of the image is part of a constellation on a celestial sphere. An example of how this criterion is used is given in Figs. 2 and 3. Thus the resulting binary vector of length 89 is formed from the colors of the built mask.

However, as will be shown in the future, this observability criterion is not optimal and is not suitable for the problem solution. The adequate criterion for

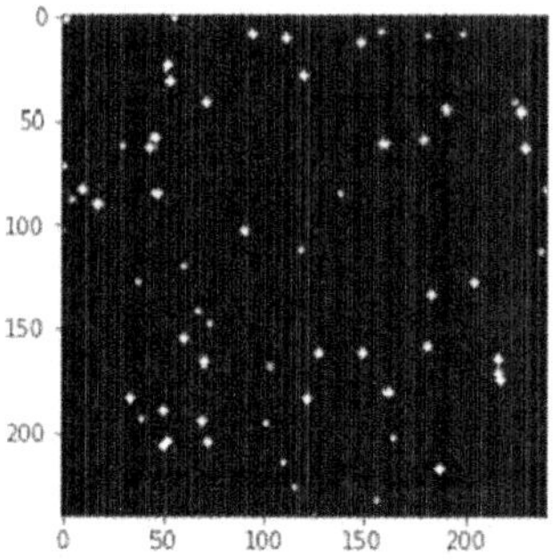

Fig. 2. Original image of the starry sky built in the developed software

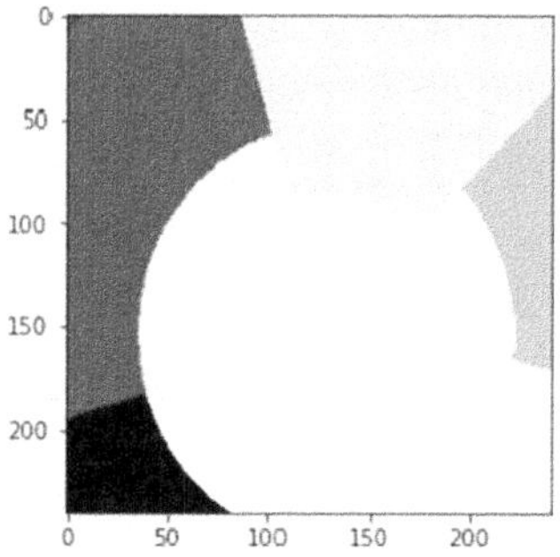

Fig. 3. A mask corresponding to Fig. 2, based on the current constellation boundaries.

the observability of constellations into images of the starry sky will be based on the observation of stars in a photo from the constellation. A comparative analysis of the two criteria, as well as the reasonableness of the choice of criteria, will be investigated during an exploratory data analysis.

4 Exploratory Data Analysis

In order to obtain a priori information on the generated dataset for further use in improving the solution, we will carry out an exploratory data analysis. The main tool for exploratory analysis of data is to empirically extract features, explore their relationship and identify anomalies. We have identified two types of features: the features of constellations, which are part of the node property vector, and the statistical characteristics of the dataset.

- The feature of constellations:
  1 number of bordering constellations;
  2 constellation area in the equatorial coordinate system (deterministic value);
  3 number of stars in the constellation;
- Statistical characteristics of the dataset:
  4 distribution of class labels by observability criterion based on modern constellation boundaries;
  5 relative area of the constellation in the data set - the percentage of peaks in all instances;
  6 distribution of class labels by observability criterion based on the entry of at least one star from constellations into the image of a starry sky;
  7 percentage of constellation labels in the entire dataset that are observed on the basis of modern boundaries but do not contain any of their own stars;

At the beginning of the exploratory analysis, a number of images were obtained for each constellation, where the constellation is observed and unobserved. By averaging all constellations, we obtained a ratio of 97% : 3%, where 3% is the percentage of samples where the constellation appears in the image, and 97% of pictures where the constellation does not appear in the image. Thus, there is a class imbalance in the data, which makes it difficult to solve the problem. It is therefore necessary to change the learning and evaluation strategy for the network. It should also be noted that it is impossible to balance the data by adding images and augmenting them, due to the intersection of classes in the starry sky images. The imbalance of classes must therefore be taken into account by weighting the loss function.

The next step is to study the number of boundary constellations in the selected constellation, in other words, the number of boundary constellations is the node degree of the graph of the starry sky. Median number of boundaries for all constellations is 6, but there are abnormal constellations where:

1. The number of neighbors is 2 or 3. These are rare constellations, the recognition of which is particularly difficult.
2. The number of bordering constellations is 10 and 14 ("Eridan", "Hydra" respectively). These constellations are located at the center of the equatorial coordinate system, have a large number of stars, a large area and are less difficult to recognize.

Here is the percentile distribution of stars within the constellations. The numerical values of this distribution are shown in Table 1.

Table 1. Percentile distribution of the number of stars

Percentile	0	5	25	50	75	95	100
The number of the star	17	23	42	67	145	224	276

The total number of stars in a constellation determines the amount of information available for recognition. So, the more stars, the more recognizable the constellation pattern is, and the number of stars also determines the stability of the recognition of the constellation. However, this fact is beyond the scope of this chapter and will be explained in the further talk. It should also be noted that the maximum number of stars in the image in the dataset is 152, that means that constellations with large numbers of stars never appear entirely in dataset photographs, making it difficult to form a pattern of these constellations in the latent space of the neural network.

One of the key areas of exploratory analysis is the adequate selection of constellation labelling criteria. For this step, we will calculate the following characteristic – the percentage of constellation labels that are observed on the basis of the criteria based on constellation boundaries, while containing at least one star of that constellation. Let us denote this characteristic for λ. The percentile distribution of this characteristic is given in Table 2. It can be seen that, due to the high sparseness of the stars in the image, the prevailing majority of labels formed on the basis of modern constellation boundaries do not contain their own stars at all, i.e. there will be no information for the neural network to make a decision. Therefore, the criterion of constellation observability by boundaries is not optimal and is not applicable in this task. The number of observable stars of the constellation must be taken into account when forming the labels.

Table 2. Percentile distribution of λ

Percentile	0	5	25	50	75	95	100
λ	14.686%	21.326%	29.473%	32.623%	44.041%	63.703%	75.714%

Finally, let's examine the relationship of the highlighted characteristics. To do this, let us consider the linear relationship between the features, namely, let us calculate correlation coefficients in pairs. The matrix of paired correlation:

$$C = \begin{pmatrix} 1 & 0.723 & 0.627 & 0.643 & 0.639 & 0.670 & 0.285 \\ 0.723 & 1 & 0.844 & 0.761 & 0.772 & 0.906 & 0.171 \\ 0.627 & 0.844 & 1 & 0.674 & 0.693 & 0.783 & 0.183 \\ 0.643 & 0.761 & 0.674 & 1 & 0.994 & 0.945 & 0.709 \\ 0.639 & 0.772 & 0.693 & 0.994 & 1 & 0.954 & 0.709 \\ 0.670 & 0.906 & 0.783 & 0.945 & 0.954 & 1 & 0.508 \\ 0.285 & 0.171 & 0.183 & 0.709 & 0.709 & 0.508 & 1 \end{pmatrix}.$$

Let's visualise the resulting pairwise correlations in a form of a weighted undirected graph and pairwise scatter plots. The corresponding graphs are shown on the Figs. 4 and 5.

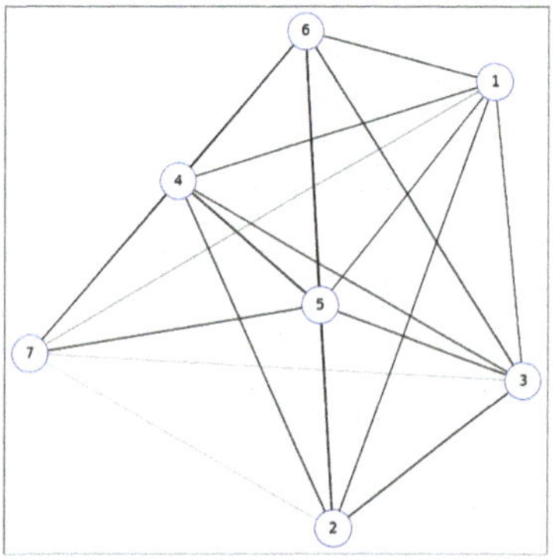

Fig. 4. Weighted feature correlation graph

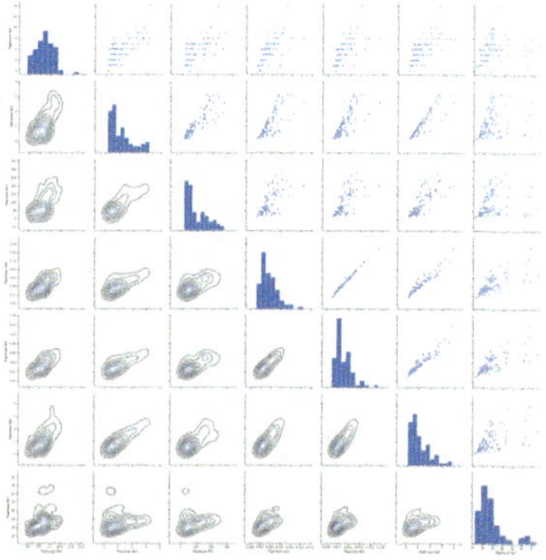

Fig. 5. Pair diagrams of feature scattering

The main advantage of the built up pairwise correlations is the following: the resulting dependency matrix can be substituted for the resulting neural network errors during training and the weight of the loss function can be adjusted so as to "knock down" the correlation between the error and the output so as to achieve an even distribution of the error, while the overall quality and stability of the neural network will increase.

5 Experiment Results

5.1 Neural Network Architecture

To solve the problem of recognition of constellations, the architecture of a deep convolution neural network was designed, which is a modification of the classic ResNet architecture (He et al. 2016). Software implementation of the neural network is carried out in the pytorch framework version 1.6.0. The depth of the network is equal to 26 trained layers. The size of the network input $1 \times 240 \times 240$ is a single-channel image, the output size of the resulting vector is 89. Activation function on the output layer is a sigmoid function, characterizing the probability of class appearance (observation of the constellation). The total number of configurable network parameters is 415 193. To estimate the quality of network approximation F_1 score was used (Goodfellow et al. 2016).

5.2 Reference Solution

In order to obtain a starting point for the solution of the task with which the comparison will be made in the future, we received a reference solution, which was carried out under simplified conditions.

Initially, binary cross entropy (Goodfellow et al. 2016) was used to learn the neural network, which was further weighted by the number of class samples (in which the constellation is observed) to neutralize class imbalances. The Adam optimizer was used for learning for 15 epochs with a standard learning speed and 256 mini-batch size. In this experiment, constellation labels were formed based on constellation boundaries. The results of testing the resulting solution are shown in Table 3.

Table 3. Accuracy of the reference solution on the test dataset.

F_1, min	F_1, median	F_1, max
0.234	0.450	0.799

The results presented in Table 3 show that the neural network is actually non-functional. Based on the fact that a deep neural network is a universal approximator (Cybenko 1989) the solution is possible. For this reason, we apply the information obtained from the exploratory analysis of the data, namely, change the observability criterion of the constellations in the image.

5.3 The Research of the Constellation Observability Criterion Based on the Entry of Stars from the Constellations

In this section let's learn the neural network under the same conditions as in Sect. 5.2, but change the condition for the formation of the resulting labels. A constellation is observed if the image contains at least k stars. In this criterion, parameter k characterizes the information threshold sufficient for the neural network to make a decision about the observability of the constellation.

In order to determine the minimum value of sufficient information, we will train the neural network, provided that the constellation is observed in the image of the starry sky, if the constellation:

1. includes at least 7 constellation stars;
2. includes at least 5 constellation stars;
3. includes at least 3 constellation stars;
4. includes at least 1 constellation stars;

The results of the experiment are presented in Table 4. It can be seen that reducing of the k parameter (reducing the amount of information to make a decision) not only does not lead to a degradation in the accuracy of the network, but also improves its quality.

Table 4. Neural network accuracy on the test dataset at different values of parameter k of observability criterion of constellations.

Criterion	F_1, min	F_1, median	F_1, max
At least 7 stars	0.623	0.915	0.986
At least 5 stars	0.554	0.884	0.980
At least 3 stars	0.688	0.912	0.984
At least 1 stars	0.862	0.954	0.992

The condition for a single star constellation to enter is the weakest requirement compared to the other options presented, so it is with this value of the parameter that experiments will be conducted later.

Also, when substituting neural network errors into the feature correlation graph, the correlation between the network error and the constellation's relative area was observed when at least one star from the constellation was observed. We will therefore adjust the weighting of the loss function to balance the relative area of the constellation instead of the number of instances in order to continue the task.

5.4 Constellation Recognition Solution

This section provides a final solution to the task of recognising constellations, taking into account the previous stages of work. We will also study the effect of noise generated by star sensors on the accuracy of the solution.

To train the neural network, we developed our own loss function based on the binary cross entropy weighted by relative constellation areas. The basic learning strategy was changed: for the first 7 epochs, the Adam optimizer with standard parameters was used, then, for 15 epochs, the SGD optimizer was used, with an initial learning rate of 0.05, which is then reduced by 2 times every 3 epochs. Moreover, starting from the 15th epoch SGD optimizer was used with the change of the loss function to a logarithmically weighted product of the constellation area and number of stars in it, binary cross entropy. For all the components of the presented training strategy, the mini-batch size is 256 samples.

The results of the training and the test after the above modifications are presented in Table 5, where experiment №1 is performed on "pure data", and experiment №2 on noisy data (the training and the test). It is necessary to state that introduced noise disturbances physically correspond to influences to which in a reality the star sensor is exposed, namely:

1. Noise with normal distribution law – corresponds to the noise of the image quantization.
2. Impulse noise distributed by binomial law – corresponds to impulse effects that lead to "false stars".
3. Closing a random part of the image is equivalent to obstructing the vision of the part of the stars that are actually in view.
4. Random rotation of the image related to the center can be regarded as rotation of the camera around the line of sight.
5. Random mirroring - image orientation errors in the star sensor.

An example of a starry sky image noise is shown in the Figs. 6 and 7.

Table 5. Accuracy of the modified solution on a test set of data. Experiment: №1 – "clear images", №2 – images with superimposed noise.

	F_1, min	F_1, median	F_1, max
Experiment №1	0.940	0.981	0.996
Experiment №2	0.81	0.927	0.971

5.5 Outcome Analysis

From the presented results it follows that the artificial neural network successfully approximates the mapping between the image of the starry sky and the space of constellations, and also has a generalizing ability and copes with the solution of the problem in conditions of a priori uncertainty, incomplete observability and stochastic disturbances.

It is necessary to notice that in experiment №2, with noisy images of the star sky, on an input of a neural network any "pure image" during training did not arrive. That is the neural network did not observe the $\Gamma_{\circ}^{\mathrm{N}}$ graph, but formed its representation in its latent space and successfully identified on the new images.

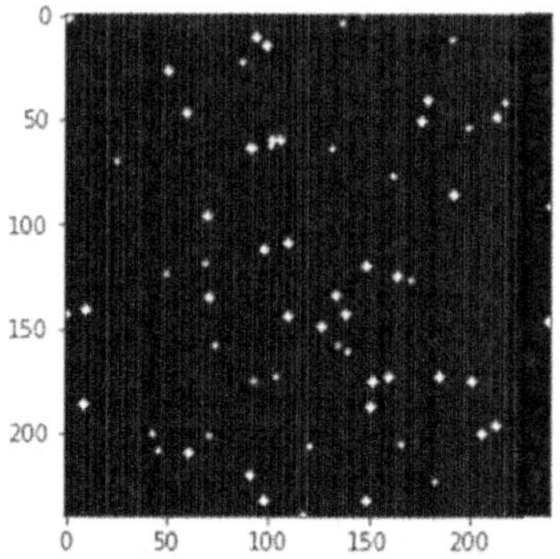

Fig. 6. The example of the image of the starry sky

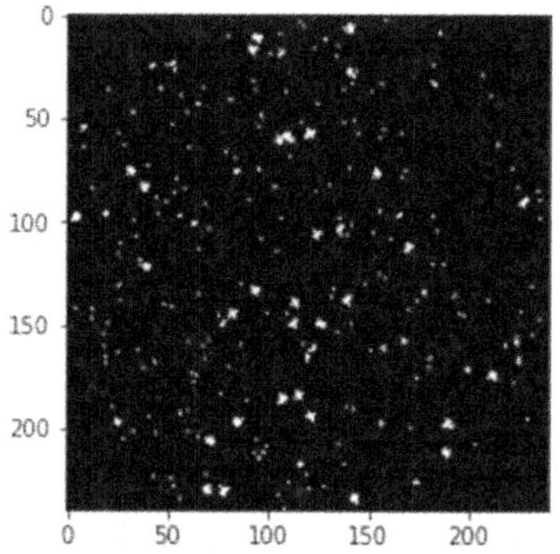

Fig. 7. Photo of the starry sky under the conditions of noise

6 Conclusion

In the presented work the possibility of effective application of deep neural networks to the solution of structural identification problem on graphs in conditions of a priori uncertainty, incomplete observability and stochastic disturbances was shown. The problem of identification of constellations on a photo of the night sky was considered as a test. The quality of solution 0.927 by metric F_1 is obtained. To achieve the result, the original ResNet (He et al. 2016) similar neural network architecture was synthesized (26 trainable layers, 415 193 configurable parameters), statistical analysis of structural characteristics of the dataset was carried out, and a special neural network training strategy was developed to form an adequate criterion of the constellation observability in the image. Besides, the study of noise influence on the quality and stability of the solution was carried out and it is shown that it has a pronounced adaptability.

In contrast to a number of other works (Spratling and Mortari 2009; Rijlaarsdam et al. 2020) on identification of stars and constellations, which require only and/or mainly empirical (manual) synthesis of informative features, methods of deep machine learning require significantly less development efforts, and, most importantly, they allow very flexible algorithm tuning in case of significant changes in the structure of input data, conditions of observability of objects, sets of recognized characteristics, etc.

Further research of this question assumes movement in two directions. Firstly, the study of theoretical aspects of the neural network approach to structural identification on graphs in conditions of a priori uncertainty, incomplete observability and stochastic disturbances. Secondly, the analysis of the obtained constellation identification solution and its integration into the astronomical navigation and orientation problems.

References

Newman, M.E.: Modularity and community structure in networks. Proc. Natl. Acad. Sci. **103**(23), 8577–8582 (2006)

Airoldi, E.M., Blei, D.M., Fienberg, S.E., Xing, E.P.: Mixed membership stochastic blockmodels. J. Mach. Learn. Res. **9**(Sept), 1981–2014 (2008)

Newman, M.E.: Spectral methods for community detection and graph partitioning. Phys. Rev. E **88**(4), 042822 (2013)

Du, N., Wu, B., Pei, X., Wang, B., Xu, L.: Community detection in large-scale social networks. In: Proceedings of the 9th WebKDD and 1st SNA-KDD 2007 Workshop on Web Mining and Social Network Analysis, pp. 16–25 (2007)

Chen, J., Saad, Y.: Dense subgraph extraction with application to community detection. IEEE Trans. Knowl. Data Eng. **24**(7), 1216–1230 (2010)

Li, P.-Z., Huang, L., Wang, C.-D., Lai, J.-H., Huang, D.: Community detection by motif-aware label propagation. ACM Trans. Knowl. Discov. Data (TKDD) **14**(2), 1–19 (2020)

Yang, J., McAuley, J., Leskovec, J.: Community detection in networks with node attributes. In: 2013 IEEE 13th International Conference on Data Mining, pp. 1151–1156. IEEE (2013)

Yan, B., Gregory, S.: Detecting community structure in networks using edge prediction methods. J. Stat. Mech. Theory Exp. **2012**(09), P09008 (2012)

Yan, B., Gregory, S.: Finding missing edges and communities in incomplete networks. J. Phys. A Math. Theoret. **44**(49), 495102 (2011)

Zhou, M.: Infinite edge partition models for overlapping community detection and link prediction. In: Artificial Intelligence and Statistics, pp. 1135–1143 (2015)

Goodfellow, I., Bengio, Y., Courville, A.: Deep Learning, MIT Press, Cambridge (2016)

Lin, L., He, Z., Peeta, S.: Predicting station-level hourly demand in a large-scale bike-sharing network: a graph convolutional neural network approach. Transp. Res. Part C Emerg. Technol. **97**, 258–276 (2018)

Zügner, D., Akbarnejad, A., Günnemann, S.: Adversarial attacks on neural networks for graph data. In: Proceedings of the 24th ACM SIGKDD International Conference on Knowledge Discovery and Data Mining, pp. 2847–2856 (2018)

Ji, S., Wang, J., Liu, X.: Constellation detection. In: Final Project for Spring 2014–2015. Stanford Press, Palo Alto (2015)

He, K., Zhang, X., Ren, S., Sun, J.: Deep residual learning for image recognition. In: Proceedings of the IEEE Conference on Computer Vision and Pattern Recognition, pp. 770–778 (2016)

Cybenko, G.: Approximation by superpositions of a sigmoidal function. Math. Control Sig. Syst. **2**(4), 303–314 (1989)

Spratling, B.B., Mortari, D.: A survey on star identification algorithms. Algorithms **2**(1), 93–107 (2009)

Rijlaarsdam, D., Yous, H., Byrne, J., Oddenino, D., Furano, G., Moloney, D.: A survey of lost-in-space star identification algorithms since 2009. Sensors **20**(9), 2579 (2020)

Residual Life Time of the Gnedenko Extreme - Value Distributions, Asymptotic Behavior and Applications

Vladimir Rusev(✉) and Alexander Skorikov

Gubkin University, Leninskiy Prospekt, 65, Moscow, Russia
{rusev.v,skorikov.a}@gubkin.ru

Abstract. Asymptotic estimates parameters residual lifetime of the Weibull-Gnedenko and Gompertz distributions for long-term operation of the object are obtained. It is found that domain of attraction of exponential distribution includes residual lifetime of the Gnedenko extreme distributions.

Keywords: Weibull - Gnedenko distribution · Mean residual life · Asymptotic behavior · Domain of attraction

1 Introduction and Background

The remaining operating time of the system and its characteristics are becoming a popular tool for solving equipment maintenance tasks. In reliability theory, Mean Residual Life - MRL as a function of time may be a more relevant characteristic of aging processes than the uptime function or failure rate (hazard function). Average residual operating time (mean residual life) is an important characteristic of aging processes in applications of reliability theory. Its theoretical properties were considered by Cox in 1962. An overview of the theory and applications of average residual operating time is available in the book Chin-Die Lai, Min Xie [1]. The average remaining operating time sums up the distribution of the remaining resource over time, while the failure rate is a characteristic of immediate failure. Muth E. in [2] has established also that the average residual operating time is more informative and useful than the failure rate. The study of the limit distributions of the residual operating time goes back to the classical work of B. V. Gnedenko [3]. B. V. Gnedenko found the only possible limiting distributions: the Weibull-Gnedenkol and Gompertz distributions, which are often used as distribution models in reliability theory and insurance theory.

The study of the limit distributions of the residual life time of such extreme distributions is a topical issue, both for the general theory and for applications. De Haan L. in [4] and Meilijson I. in [5] proposed to use MRL to describe the domain of attraction of limit distributions. The description of the domain of attraction of the exponential distribution in terms of the convergence of the moments was obtained by Balkema A. A., De Haan L. see [6]. Banjevic D. in

A. N. Shiryaev et al. (Eds.): ICSM-5 2020, PROMS 371, pp. 292–305, 2021.
https://doi.org/10.1007/978-3-030-83266-7_22

[7] noticed that under some general conditions on the failure rate, which include the Weibull - Gnedenko distribution with the shape parameter >1, the limit distribution is exponential.

1.1 Definitions and Basic Properties

Let T be a nonnegative random variable with probability distribution F, absolutely continuous so that its density function $f(t)$, and its hazard function $\lambda(t)$ exist. Consider the conditional random variable $X_t = (T - t\,|T > t)$, which is called the residual life time (remaining useful life). The mathematical expectation of random variable X_t, i.e. the function of the average residual time before failures (MRL) is defined as

$$\mu(t) = E(T - t\,|T > t) = \frac{\int\limits_t^{+\infty} P(x)dx}{P(t)},$$

where $P(x) = 1 - F(x)$ is the survival function or reliability function. The following relations shows the equivalence of the mean residual life function $\mu(t)$, the hazard function $\lambda(t)$ and the reliability function $P(t)$:

$$\lambda(t) = \frac{1 + \mu'(t)}{\mu(t)}; \quad P(t) = \frac{\mu(0)}{\mu(t)} \cdot e^{-\int\limits_0^t \frac{dz}{\mu(z)}}.$$

Calabria and Pulcini in [8] derived the relationship

$$\lim_{t\to\infty} \mu(t) = \lim_{t\to\infty} \frac{1}{\lambda(t)},$$

provided the latter limit exists and is finite.

The function $\lambda(t)$ has an obvious visual meaning, but the statistical estimation of $\lambda(t)$ is very unstable. On the contrary, the statistical properties of the estimated averages $\mu(t)$ are much more stable than the characteristics $\lambda(t)$.

2 Analytical Representation MRL and Residual Variance for the Weibull-Gnedenko Model

Let T denotes a random variable equal to the element uptime and satisfying the two-parameter Weibull-Gnedenko distribution law with the distribution function:

$$F(t; \alpha, \beta) = 1 - e^{-(\alpha t)^\beta}, \quad t \ge 0$$

with $\alpha > 0, \beta > 0$ and expected value

$$ET = T_0 = \int\limits_0^{+\infty} e^{-(\alpha t)^\beta} dt = \alpha^{-1} \cdot \Gamma\left(1 + \frac{1}{\beta}\right), \tag{1}$$

where $\Gamma(\cdot)$ is the Euler gamma function.

The analytical representation $\mu(t)$ in case of the Weibull-Gnedenko distribution law looks as follows:

$$\mu(t) = \frac{\int\limits_t^{+\infty} (1 - F(x))\,dx}{1 - F(t)} = e^{(\alpha t)^\beta} \cdot \int\limits_t^{+\infty} e^{-(\alpha x)^\beta} dx$$

$$= e^{(\alpha t)^\beta} \cdot \left(\int\limits_0^{+\infty} e^{-(\alpha x)^\beta} dx - \int\limits_0^t e^{-(\alpha x)^\beta} dx \right)$$

$$= e^{(\alpha t)^\beta} \cdot \left[\alpha^{-1} \cdot \Gamma\left(1 + \frac{1}{\beta}\right) - \int\limits_0^t e^{-(\alpha x)^\beta} dx \right].$$

The analytical representation of the MRL can also be obtained via the Kummer-Pochhammer function. Using formula 8.351(2), see [9] we get the following expression:

$$\int\limits_0^t e^{-(\alpha x)^\beta} dx = t \cdot e^{-(\alpha t)^\beta} \cdot {}_1F_1\left(1;\ \frac{1}{\beta} + 1;\ (\alpha t)^\beta\right), \tag{2}$$

where ${}_1F_1(\rho;\ \gamma;\ x)$ is the standard notation for a degenerate hypergeometric function of the 1st kind or the Kummer-Pochhammer function from the class of special functions. Consequently,

$$\mu(t) = e^{(\alpha t)^\beta} \cdot \left[\alpha^{-1} \cdot \Gamma\left(1 + \frac{1}{\beta}\right) - t \cdot e^{-(\alpha t)^\beta} \cdot {}_1F_1\left(1;\ \frac{1}{\beta} + 1;\ (\alpha t)^\beta\right)\right]. \tag{3}$$

Using the representation of the Kummer-Pochhammer function as a series, we obtain

$$\mu(t) = T_0 \cdot \sum_{k=0}^{+\infty} \frac{(\alpha t)^{\beta k}}{k!} \left(1 - \frac{\alpha \cdot t \cdot k!}{\Gamma\left(k + 1 + \frac{1}{\beta}\right)}\right), \tag{4}$$

where $(a)_k$ the Pochhammer symbol:

$$(a)_k = \frac{\Gamma(a + k)}{\Gamma(a)}.$$

We use the incomplete gamma function $\gamma(a,\ z) = \int\limits_o^z y^{a-1} e^{-y} dy$. By formula 3.381 (8) in [9] we get

$$\mu(t) = e^{(\alpha t)^\beta} \cdot \frac{1}{\alpha}\left[\Gamma\left(1 + \frac{1}{\beta}\right) - \frac{1}{\beta} \cdot \gamma\left(\frac{1}{\beta},\ (\alpha t)^\beta\right)\right]. \tag{5}$$

We transform (5) using the properties of the Euler gamma function:

$$\mu(t) = e^{(\alpha t)^{\beta}} \cdot \frac{1}{\alpha\beta}\left[\Gamma\left(\frac{1}{\beta}, (\alpha t)^{\beta}\right)\right], \tag{6}$$

through the incomplete gamma function $\Gamma(a, z) = \int\limits_{z}^{+\infty} y^{a-1}e^{-y}dy$ by 8.350(2) in [9].

Note that there is a generalization (5) for a more general exponentiated Weibull distribution [10].

Numerical Analysis. For numerical analysis, we set without loss of generality the scale parameter $\alpha = 1$. Let's apply the mathematical package Wolfram Mathematica using the functions: incomplete gamma functions and Kummer's hypergeometric function, respectively. There is an oscillation of the values calculated using the Wolfram Mathematica, according to formulas (3), (5) for the value $\beta = 2$ shape parameter.

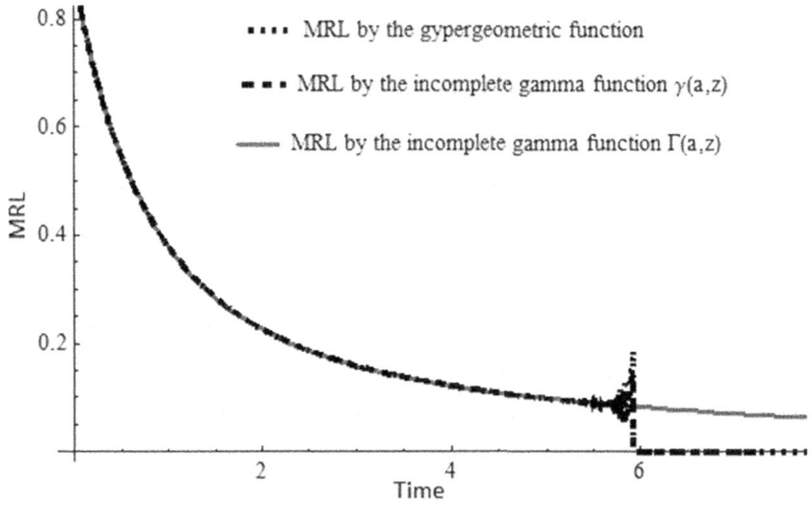

Fig. 1. Mean residual life graphics for different analytical representations

Using the formula

$$\sigma^2(t) = E(X_t^2) - \mu^2(t) = \frac{2}{P(t)} \int\limits_{t}^{+\infty} P(x)\mu(x)\, dx - \mu^2(t),$$

we derive an analytical representation for the residual variance of the Weibull - Gnedenko distribution.

$$\sigma^2(t) = 2e^{(\alpha t)^\beta} \int_t^{+\infty} e^{-(\alpha x)^\beta} e^{(\alpha x)^\beta} \int_x^{+\infty} e^{-(\alpha z)^\beta} dz\, dx - \mu^2(t)$$

$$= 2e^{(\alpha t)^\beta} \int_t^{+\infty} dx \int_x^{+\infty} e^{-(\alpha z)^\beta} dz - \mu^2(t)$$

$$= 2e^{(\alpha t)^\beta} \int_t^{+\infty} dz \int_t^{z} e^{-(\alpha z)^\beta} dx - \mu^2(t)$$

$$= 2e^{(\alpha t)^\beta} \int_t^{+\infty} e^{-(\alpha z)^\beta} (z-t) dz - \mu^2(t)$$

$$= 2e^{(\alpha t)^\beta} \left(\int_t^{+\infty} e^{-(\alpha z)^\beta} z dz - t \int_t^{+\infty} e^{-(\alpha z)^\beta} dz \right) - \mu^2(t).$$

Since

$$\int_t^{+\infty} e^{-(\alpha x)^\beta} dx = \frac{\mu(t)}{e^{(\alpha t)^\beta}}, \qquad \int_t^{+\infty} z^\beta e^{-(\alpha z)^\beta} dz = \frac{t+\mu(t)}{\alpha^\beta \beta e^{(\alpha t)^\beta}},$$

then

$$\sigma^2(t) = 2e^{(\alpha t)^\beta} \left(\int_t^{+\infty} e^{-(\alpha z)^\beta} z dz - t \frac{\mu(t)}{e^{(\alpha t)^\beta}} \right) - \mu^2(t)$$

$$= 2e^{(\alpha t)^\beta} \int_t^{+\infty} e^{-(\alpha z)^\beta} z dz - 2t\mu(t) - \mu^2(t),$$

and by the formula

$$\int_t^{+\infty} e^{-(\alpha z)^\beta} z dz = \frac{1}{\alpha^2 \beta} \Gamma\left(\frac{2}{\beta}, (\alpha t)^\beta \right)$$

so that

$$\sigma^2(t) = 2e^{(\alpha t)^\beta} \frac{1}{\alpha^2 \beta} \Gamma\left(\frac{2}{\beta}, (\alpha t)^\beta \right) - 2t\mu(t) - \mu^2(t). \tag{7}$$

3 Asymptotic Expansion for the MRL and Residual Variance

In this section explicit asymptotic expressions for the mean, variance of residual lifetime Weibull - Gnedenko and Gompertz distributions are found. Residual lifetime distribution is discussed in terms of its mean, variance and behavior of those quantities as t tends to infinity.

Theorem 1. *Let T be a random variable with the two-parameter Weibull-Gnedenko distribution, then $\forall\alpha > 0,\ \forall\beta > 0$*

$$\mu(t) = \frac{t}{\beta(\alpha t)^{\beta}}\left(1 + \frac{1-\beta}{\beta(\alpha t)^{\beta}} + \frac{(1-\beta)(1-2\beta)}{\beta^2(\alpha t)^{2\beta}} + O\left(\frac{1}{t^{3\beta}}\right)\right), \quad t \to +\infty, \quad (8)$$

$$\sigma^2(t) = \frac{1}{\beta^2\alpha^{2\beta}t^{2(\beta-1)}}\left(1 + \frac{4(1-\beta)}{\beta(\alpha t)^{\beta}} + \frac{(1-\beta)(11-17\beta)}{\beta^2(\alpha t)^{2\beta}} + O\left(\frac{1}{t^{3\beta}}\right)\right), t \to +\infty. \quad (9)$$

Proof. We use asymptotic expansions 8.357 in [9] and formula 6. Then

$$\Gamma\left(\frac{1}{\beta}, (\alpha t)^{\beta}\right) = \frac{(\alpha t)^{1-\beta}}{e^{(\alpha t)^{\beta}}}[1 + \frac{1-\beta}{\beta}\cdot\frac{1}{(\alpha t)^{\beta}} + \frac{1-\beta}{\beta}\cdot\frac{1-2\beta}{\beta}\cdot\frac{1}{(\alpha t)^{2\beta}}$$
$$+ \frac{1-\beta}{\beta}\cdot\frac{1-2\beta}{\beta}\cdot\frac{1-3\beta}{\beta}\cdot\frac{1}{(\alpha t)^{3\beta}} + O\left(\frac{1}{t^{4\beta}}\right)],$$

$$\mu(t) = e^{(\alpha t)^{\beta}}\frac{1}{\alpha\beta}\frac{(\alpha t)^{1-\beta}}{e^{(\alpha t)^{\beta}}}[1 + \frac{1-\beta}{\beta}\cdot\frac{1}{(\alpha t)^{\beta}} + \frac{1-\beta}{\beta}\cdot\frac{1-2\beta}{\beta}\cdot\frac{1}{(\alpha t)^{2\beta}}$$
$$+ \frac{1-\beta}{\beta}\cdot\frac{1-2\beta}{\beta}\cdot\frac{1-3\beta}{\beta}\cdot\frac{1}{(\alpha t)^{3\beta}} + O\left(\frac{1}{t^{4\beta}}\right)].$$

Let $\xi = \frac{1}{(\alpha t)^{\beta}}, \quad \rho = \frac{1}{\beta}$. Then

$$\mu(t) = \rho t\xi\left[1 + (\rho-1)\xi + (\rho-1)(\rho-2)\xi^2 + (\rho-1)(\rho-2)(\rho-3)\xi^3 + O(\xi^4)\right],$$

hence,

$$\begin{aligned}\mu^2(t) &= (\rho t\xi)^2\left[1 + (\rho-1)\xi + (\rho-1)(\rho-2)\xi^2 + O(\xi^3)\right]^2\\ &= (\rho t\xi)^2\left[1 + 2(\rho-1)\xi + (\rho-1)(3\rho-5)\xi^2 + O(\xi^3)\right]\\ &= \rho^2t^2\xi^2 + 2\rho^2(\rho-1)t^2\xi^3 + \rho^2(\rho-1)(3\rho-5)t^2\xi^4 + O(\xi^5).\end{aligned}$$

Then

$$\begin{aligned}\mu^2(t) + 2t\mu(t) &= 2\rho t^2\xi + \rho(3\rho-2)t^2\xi^2 + 4\rho(\rho-1)^2t^2\xi^3\\ &+ \rho(\rho-1)(5\rho^2-15\rho+12)t^2\xi^4 + O(\xi^5).\end{aligned}$$

Since

$$2e^{(\alpha t)^\beta}\frac{1}{\alpha^2\beta}\Gamma\left(\frac{2}{\beta},(\alpha t)^\beta\right)$$

$$= 2\rho t^2\xi\left[1+(2\rho-1)\xi+(2\rho-1)(2\rho-2)\xi^2+(2\rho-1)(2\rho-2)(2\rho-3)\xi^3+O(\xi^4)\right]$$

$$= 2\rho t^2\xi+2\rho(2\rho-1)t^2\xi^2+2\rho(2\rho-1)(2\rho-2)t^2\xi^3+2\rho(2\rho-1)(2\rho-2)(2\rho-3)t^2\xi^4+O(\xi^5),$$

then from (7) follows

$$\begin{aligned}\sigma^2(t) &= \rho^2t^2\xi^2+4\rho^2(\rho-1)t^2\xi^3+\rho^2(\rho-1)(11\rho-17)t^2\xi^4+O(\xi^5)\\ &= \rho^2t^2\xi^2\left[1+4(\rho-1)\xi+(\rho-1)(11\rho-17)\xi^2+O(\xi^3)\right],\end{aligned}$$

or

$$\sigma^2(t)=\frac{1}{\beta^2\alpha^{2\beta}t^{2(\beta-1)}}\left(1+\frac{4(1-\beta)}{\beta(\alpha t)^\beta}+\frac{(1-\beta)(11-17\beta)}{\beta^2(\alpha t)^{2\beta}}+O\left(\frac{1}{t^{3\beta}}\right)\right).$$

Remark. This result makes more precise asymptotic expansions for the MRL and residual variance in [11], but the second and third terms of the asymptotic expansion (9) for the variance have differences with [11].

Theorem 2. *Let T be a random variable with the Gompertz distribution, then*

$$\mu(t)=e^{-t}\left(1-e^{-t}+2e^{-2t}-6e^{-3t}+24e^{-4t}+o(e^{-4t})\right),\quad t\to+\infty, \tag{10}$$

$$\sigma^2(t)=e^{-2t}\left(1-4e^{-t}+17e^{-2t}-84e^{-3t}+o(e^{-3t})\right)\quad t\to+\infty. \tag{11}$$

Proof. It is known

$$\mu(t)=\frac{\int_t^{+\infty} e^{-e^x}dx}{e^{-e^t}},\quad \sigma^2(t)=\frac{2}{e^{-e^t}}\int_t^{+\infty} e^{-e^x}\mu(x)dx-\mu^2(t).$$

By integration by parts, the integral is obtained as follows:

$$\int_t^{+\infty} e^{-e^x}dx=\int_{e^t}^{+\infty}\frac{e^{-\xi}}{\xi}d\xi=-\int_{e^t}^{+\infty}\frac{1}{\xi}d\left(e^{-\xi}\right)=e^{-e^t}e^{-t}-\int_{e^t}^{+\infty}e^{-\xi}\cdot\xi^{-2}d\xi.$$

We assume that $I_n=\int_{e^t}^{+\infty}\frac{e^{-\xi}}{\xi^n}d\xi$, $n=1,2,\ldots$. Then similarly,

$$I_n=\int_{e^t}^{+\infty}\frac{e^{-\xi}}{\xi^n}d\xi=-\int_{e^t}^{+\infty}\frac{de^{-\xi}}{\xi^n}=e^{-e^t}e^{-nt}-nI_{n+1}.$$

By applying the recurrence relation several times, it is found that

$$I_1 = e^{-e^t}e^{-t} - (e^{-e^t}e^{-2t} - 2(e^{-e^t}e^{-3t} - 3(e^{-e^t}e^{-4t} - 4(e^{-e^t}e^{-5t} - 5I_6))))$$

$$= e^{-e^t}e^{-t} - e^{-e^t}e^{-2t} + 2e^{-e^t}e^{-3t} - 2\cdot 3e^{-e^t}e^{-4t} + 2\cdot 3\cdot 4e^{-e^t}e^{-5t} - 2\cdot 3\cdot 4\cdot 5I_6$$

$$= e^{-e^t}e^{-t}\left(1 - e^{-t} + 2e^{-2t} - 6e^{-3t} + 24e^{-4t} + o(e^{-4t})\right).$$

The general case:

$$\int_t^{+\infty} e^{-e^x}dx = e^{-e^t}e^{-t}\sum_{n=0}^{+\infty}(-1)^n n! e^{-nt}.$$

Hence

$$\mu(t) = e^{-t}\left[1 - e^{-t} + 2e^{-2t} - 6e^{-3t} + 24e^{-4t} + o(e^{-4t})\right]$$

$$= e^{-t} - e^{-2t} + 2e^{-3t} - 6e^{-4t} + 24e^{-5t} + o(e^{-5t}), \quad (t \to +\infty).$$

Then

$$\mu^2(t) = e^{-2t}\left[1 - e^{-t} + 2e^{-2t} - 6e^{-3t} + 24e^{-4t} + o(e^{-4t})\right]^2$$

$$= e^{-2t} - 2e^{-3t} + 5e^{-4t} - 16e^{-5t} + o(e^{-5t}), \quad (t \to +\infty).$$

$$\int_t^{+\infty} e^{-e^x}\mu(x)dx = \int_t^{+\infty} e^{-e^x}e^{-x}\left(1 - e^{-x} + 2e^{-2x} - 6e^{-3x} + 24e^{-4x} + o(e^{-4x})\right)dx$$

$$= \int_t^{+\infty} e^{-e^x}\left[e^{-x} - e^{-2x} + 2e^{-3x} - 6e^{-4x} + 24e^{-5x} + o(e^{-5x})\right]dx$$

$$= \int_t^{+\infty} e^{-e^x}e^{-x}dx - \int_t^{+\infty} e^{-e^x}e^{-2x}dx + 2\int_t^{+\infty} e^{-e^x}e^{-3x}dx - 6\int_t^{+\infty} e^{-e^x}e^{-4x}dx + \cdots.$$

Note that

$$\int_t^{+\infty} e^{-e^x}e^{-nx}dx = \int_{e^t}^{+\infty}\frac{e^{-\xi}}{\xi^n}\frac{d\xi}{\xi} = \int_{e^t}^{+\infty}\frac{e^{-\xi}}{\xi^{n+1}}d\xi = I_{n+1}.$$

We find

$$\int_t^{+\infty} e^{-e^x}\mu(x)dx = I_2 - I_3 + 2I_4 - 6I_5 + 24I_6 + ...,$$

applying the recurrence relation several times

$$I_2 = e^{-e^t}e^{-2t} - 2(e^{-e^t}e^{-3t} - 3(e^{-e^t}e^{-4t} - 4(e^{-e^t}e^{-5t} + ...)))$$

$$= e^{-e^t}(e^{-2t} - 2e^{-3t} + 6e^{-4t} - 24e^{-5t} + ...),$$

$$I_3 = e^{-e^t}e^{-3t} - 3(e^{-e^t}e^{-4t} - 4(e^{-e^t}e^{-5t} + ...)) = e^{-e^t}(e^{-3t} - 3e^{-4t} + 12e^{-5t} + ...),$$

$$I_4 = e^{-e^t}e^{-4t} - 4(e^{-e^t}e^{-5t} + ...) = e^{-e^t}(e^{-4t} - 4e^{-5t} + ...),$$

$$I_5 = e^{-e^t}e^{-5t} - 5(e^{-e^t}e^{-6t} + ...) = e^{-e^t}(e^{-5t} + ...).$$

Hence

$$\int_t^{+\infty} e^{-e^x}\mu(x)dx = e^{-e^t}(e^{-2t} - 2e^{-3t} + 6e^{-4t} - 24e^{-5t} + ...)$$

$$-e^{-e^t}(e^{-3t} - 3e^{-4t} + 12e^{-5t} + ...) + 2e^{-e^t}(e^{-4t} - 4e^{-5t} + ...) - 6e^{-e^t}(e^{-5t} + ...) + ...$$

$$= e^{-e^t}(e^{-2t} - 3e^{-3t} + 11e^{-4t} - 50e^{-5t} + ...).$$

We derive asymptotic expansion for the residual variance:

$$\sigma^2(t) = \frac{2}{e^{-e^t}} \cdot e^{-e^t}(e^{-2t} - 3e^{-3t} + 11e^{-4t} - 50e^{-5t} + o(e^{-5t}))$$

$$-(e^{-2t} - 2e^{-3t} + 5e^{-4t} - 16e^{-5t} + o(e^{-5t}))$$

$$= e^{-2t} - 4e^{-3t} + 17e^{-4t} - 84e^{-5t} + o(e^{-5t}) = e^{-2t}(1 - 4e^{-t} + 17e^{-2t} - 84e^{-3t} + o(e^{-3t})).$$

Theorems 1 and 2 make it possible to find the asymptotic expansions of the coefficient of variation.

Theorem 3. *Let $\mu(t)$ be residual mean life, $\sigma^2(t)$ is the residual variance function and $c_v(t) = \frac{\sigma(t)}{\mu(t)}$ is the coefficient of variation of residual life time $X_t = (T - t\,|T > t)$.*

1) If T be a random variable with the two-parameter Weibull-Gnedenko distribution, then $\forall\alpha > 0,\ \ \forall\beta > 0$

$$c_v(t) = 1 + \frac{1-\beta}{\beta(\alpha t)^\beta} + o\left(\frac{1}{t^\beta}\right) \quad t \to +\infty. \tag{12}$$

2) If T be a random variable with the Gompertz distribution, then

$$c_v(t) = 1 - e^{-t} + o(e^{-t}), \quad t \to +\infty. \tag{13}$$

Proof. By putting $\xi = \frac{1}{(\alpha t)^\beta},\quad \xi \to 0,\quad (t \to +\infty)$. By analytical representation MRL and residual variance for the Weibull-Gnedenko distribution

$$\mu(t) = e^{(\alpha t)^\beta} \cdot \frac{1}{\alpha\beta} \cdot \Gamma\left(\frac{1}{\beta}, (\alpha t)^\beta\right) = \frac{1}{\beta} \cdot t \cdot \xi\left(1 + \frac{1-\beta}{\beta}\xi + \frac{1-\beta}{\beta} \cdot \frac{1-2\beta}{\beta}\xi^2 + O(\xi^3)\right),$$

$$\sigma^2(t) = 2e^{(\alpha t)^\beta} \cdot \frac{1}{\alpha^2\beta} \cdot \Gamma\left(\frac{2}{\beta}, (\alpha t)^\beta\right) - 2\mu(t) \cdot t - \mu^2(t),$$

we have

$$(c_v(t))^2 = \frac{\sigma^2(t)}{\mu^2(t)} = 2\left(\frac{e^{(\alpha t)^\beta}}{\alpha^2\beta} \cdot \frac{\Gamma\left(\frac{2}{\beta},(\alpha t)^\beta\right)}{\mu^2(t)} - \frac{t}{\mu(t)}\right) - 1$$

$$= 2 \cdot \frac{\left(\frac{1}{\beta}t^2\xi\left(1+\frac{2-\beta}{\beta}\xi+\frac{2-\beta}{\beta}\cdot\frac{2-2\beta}{\beta}\xi^2+O(\xi^3)\right)-t\mu(t)\right)}{\mu^2(t)} - 1.$$

And after simplification

$$(c_v(t))^2 = 2\cdot\frac{1+\frac{3(1-\beta)}{\beta}\xi+o(\xi)}{\left(1+\frac{1-\beta}{\beta}\xi+\frac{1-\beta}{\beta}\cdot\frac{1-2\beta}{\beta}\xi^2+o(\xi^2)\right)^2}-1 = 2\cdot\frac{1+\frac{3(1-\beta)}{\beta}\xi+o(\xi)}{1+\frac{2(1-\beta)}{\beta}\xi+o(\xi)}-1$$

$$= \frac{2\left(1+\frac{2(1-\beta)}{\beta}\xi+\frac{1-\beta}{\beta}\xi+o(\xi)\right)}{1+\frac{2(1-\beta)}{\beta}\xi+o(\xi)} - 1 = 2\left(1+\frac{\frac{1-\beta}{\beta}\xi+o(\xi)}{1+\frac{2(1-\beta)}{\beta}\xi+o(\xi)}\right)-1$$

$$= 1+2\cdot\frac{\frac{1-\beta}{\beta}\xi+o(\xi)}{1+2\frac{1-\beta}{\beta}\xi+o(\xi)}, \quad \xi\to 0.$$

Hence

$$c_v(t) = \left(1+\frac{2\frac{1-\beta}{\beta}\xi+o(\xi)}{1+2\frac{1-\beta}{\beta}\xi+o(\xi)}\right)^{\frac{1}{2}} = 1+\frac{1}{2}\cdot\frac{2\frac{1-\beta}{\beta}\xi+o(\xi)}{1+2\frac{1-\beta}{\beta}\xi+o(\xi)}$$

$$= 1+\frac{1-\beta}{\beta}\cdot\frac{\xi+o(\xi)}{1+2\frac{1-\beta}{\beta}\xi+o(\xi)} = 1+\frac{1-\beta}{\beta}(\xi+o(\xi)), \quad \xi\to 0,$$

$$c_v(t) = 1+\frac{1-\beta}{\beta}\cdot\frac{1}{(\alpha t)^\beta}+o\left(\frac{1}{t^\beta}\right), \quad t\to+\infty.$$

To prove 2) we use next formulas from Theorem 2:

$$\mu(t) = e^{-t}\left(1-e^{-t}+2e^{-2t}+o(e^{-2t})\right), \quad t\to+\infty,$$

$$\sigma^2(t) = \frac{2}{e^{-e^t}}\int_t^{+\infty} e^{-e^x}\mu(x)dx - \mu^2(t)$$

$$= 2\left(e^{-2t}-3e^{-3t}+o(e^{-3t})\right)-\mu^2(t), \quad t\to+\infty$$

Then

$$(c_v(t))^2 = \frac{\sigma^2(t)}{\mu^2(t)} = 2\cdot\frac{e^{-2t}-3e^{-3t}+o(e^{-3t})}{\left(e^{-t}\left(1-e^{-t}+2e^{-2t}+o(e^{-2t})\right)\right)^2}-1$$

$$= 2 \cdot \frac{e^{-2t}\left(1 - 3e^{-t} + o(e^{-t})\right)}{e^{-2t}\left(1 - e^{-t} + 2e^{-2t} + o(e^{-2t})\right)^2} - 1 = 2 \cdot \frac{1 - 3e^{-t} - e^{-t} + o(e^{-t})}{1 - 2e^{-t} + o(e^{-t})} - 1$$

$$= 2 \cdot \frac{1 - 2e^{-t} - e^{-t} + o(e^{-t})}{1 - 2e^{-t} + o(e^{-t})} - 1 = 2 \cdot \left(1 - \frac{e^{-t} + o(e^{-t})}{1 - 2e^{-t} + o(e^{-t})}\right) - 1$$

$$= 1 - \frac{2e^{-t} + o(e^{-t})}{1 - 2e^{-t} + o(e^{t})} = 1 - 2e^{-t} + o(e^{-t}), \quad t \to +\infty.$$

So

$$c_v(t) = \left(1 - 2e^{-t} + o(e^{-t})\right)^{\frac{1}{2}} = 1 - 2e^{-t} \cdot \frac{1}{2} + o(e^{-t}) = 1 - e^{-t} + o(e^{-t}), t \to +\infty.$$

Corollary 1. *The exponential distribution is the limiting distribution for the residual operating time for random variable T that have a two-parameter Weibull-Gnedenko distribution with $\forall \alpha > 0$, $\forall \beta > 0$ or Gompertz distribution, i.e.*

$$\lim_{t \to \infty} P\left\{\frac{T - t}{\mu(t)} \le x \,|T > t\right\} = 1 - e^{-x}.$$

Proof. The proof of the theorem follows from the obtained asymptotic formulas and the criterion (Theorem 8. Corollary) see [6].

3.1 The General Case. The Distributions with Heavy Tails

We are interested now in behavior of $c_v(t) = \frac{\sigma(t)}{\mu(t)}$ for general distributions.

Theorem 4. *If an absolutely continuous distribution $F(x)$ on $[0, \infty)$ has density $f(x)$ and hazard function $\lambda(t)$, then*

$$\lim_{t \to +\infty} (c_v(t))^2 = \lim_{t \to +\infty} \left(\lambda(t) \cdot \frac{\int_t^{+\infty} dx \int_x^{+\infty} (1 - F(\xi))\, d\xi}{\int_t^{+\infty} (1 - F(x))\, dx}\right), \tag{14}$$

$$\lim_{t \to +\infty} (c_v(t))^2 = \lim_{t \to +\infty} \left(\lambda(t) \cdot \frac{\int_t^{+\infty} (1 - F(\xi))\,(\xi - t)\, d\xi}{\int_t^{+\infty} (1 - F(x))\, dx}\right). \tag{15}$$

Proof. It is known that

$$\lambda(t) = \frac{f(t)}{1 - F(t)}, \tag{16}$$

$$\mu(t) = \frac{\int_t^{+\infty} (1 - F(x))\, dx}{1 - F(t)}, \tag{17}$$

$$\sigma^2(t) = \frac{2}{1 - F(t)} \int_t^{+\infty} (1 - F(x)) \cdot \mu(x) dx - \mu^2(t), \tag{18}$$

hence

$$(c_v(t))^2 = \frac{\sigma^2(t)}{\mu^2(t)} = 2 \cdot \frac{\int_t^{+\infty} (1 - F(t))\, \mu(x) dx}{(1 - F(t)) \cdot \mu^2(t)} - 1 = 2 \cdot \frac{(1 - F(t)) \cdot \int_t^{+\infty} (1 - F(x))\, \mu(x) dx}{\left(\int_t^{+\infty} (1 - F(x))\, dx\right)^2} - 1,$$

$$\lim_{t \to +\infty} [c_v(t)]^2 = 2 \cdot \lim_{t \to +\infty} \frac{(1 - F(t)) \cdot \int_t^{+\infty} (1 - F(x))\, \mu(x) dx}{\left(\int_t^{+\infty} (1 - F(x))\, dx\right)^2} - 1$$

By applying L'Hospital's rule and combining (17) one derive the relationship

$$= 2 \lim_{t \to +\infty} \frac{F'(t) \int_t^{+\infty} (1 - F(x))\, \mu(x) dx + (1 - F(t))^2 \cdot \mu(t)}{2 \cdot (1 - F(t)) \int_t^{+\infty} (1 - F(x))\, dx} - 1$$

$$= \lim_{t \to +\infty} \frac{F'(t) \int_t^{+\infty} (1 - F(x))\, \mu(x) dx + (1 - F(t)) \int_t^{+\infty} (1 - F(x))\, dx}{(1 - F(t)) \int_t^{+\infty} (1 - F(x))\, dx} - 1.$$

So

$$\lim_{t \to +\infty} (c_v(t))^2 = \lim_{t \to +\infty} \left(\frac{f(t)}{1 - F(t)} \cdot \frac{\int_t^{+\infty} (1 - F(x))\, \mu(x) dx}{\int_t^{+\infty} (1 - F(x))\, dx} + 1 \right) - 1$$

$$= \lim_{t \to +\infty} \left(\lambda(t) \cdot \frac{\int_t^{+\infty} (1 - F(x))\, \mu(x) dx}{\int_t^{+\infty} (1 - F(x))\, dx} \right).$$

Then and from (17) we get (14). If we transform the integral

$$\int_t^{+\infty} dx \int_x^{+\infty} (1 - F(\xi))\, d\xi = \int_t^{+\infty} d\xi \int_t^{\xi} (1 - F(\xi))\, dx = \int_t^{+\infty} (\xi - t)\, (1 - F(\xi))\, d\xi,$$

then from (14) we get (15).

Example. The Distributions with Heavy Tails. Let T denotes a random variable satisfying the two-parameter Pareto distribution law with the reliability function:

$$P(t) = \left(\frac{a}{t}\right)^b, \quad t \geq a > 0, \quad b > 2$$

with hazard function $\lambda(t) = \frac{b}{t}$. We derive for Pareto distribution

$$\int_t^{+\infty} (1 - F(x))\, dx = \int_t^{+\infty} \left(\frac{a}{x}\right)^b dx = \frac{a^b}{b-1} t^{1-b}, \quad b \neq 1,$$

$$\int_t^{+\infty} dx \int_x^{+\infty} (1 - F(\xi))\, d\xi = \int_t^{+\infty} \frac{a^b}{b-1} x^{1-b} dx = \frac{a^b}{(b-1)(b-2)} t^{2-b}, \quad b \neq 2.$$

From (14) it follows

$$\lim_{t \to +\infty} (c_v(t))^2 = \lim_{t \to +\infty} \frac{\frac{b}{t} \cdot \frac{a^b t^{2-b}}{(b-1)(b-2)}}{\frac{a^b}{b-1} t^{1-b}} = \frac{b}{b-2} \neq 1.$$

Following Theorem 8. Corollary in [6], the limiting distribution of residual life time for Pareto distribution can not to be exponential distribution, i.e. the domain of attraction of exponential distribution doesn't include Pareto distribution.

Practical Applications. As an application, an assessment of the reliability of the "well - pump" system in terms of MRL is proposed. The paper [12] showcases a case study of modelling "well - pump" system failure using Weibull - Gnedenko distribution on life-failure-data samples collated from oil producing region. MRL of submersible equipment elements was obtained. The reliability of submersible equipment of oil wells with various failures that led to the lifting of pumps is estimated.

References

1. Lai, C.-D., Xie, M.: Stochastic Ageing and Dependence Reliability. Springer, New York (2006). https://doi.org/10.1007/0-387-34232-X
2. Muth, E.: Reliability models with positive memory derived from the mean residual life function. In: Tsokos, C.P., Shimi, I.M. (eds.) The Theory and Applications of Reliability, vol. 2. Academic Press, New York (1977)
3. Gnedenko, B.V.: Sur la distribution limite du terme maximum d une seria aleatoire. Ann. Math **44**, 423–453 (1943)
4. De Haan, L.: On regular variation and its application to weak convergence of sample extremes. Math. Centre Tracts, vol. 32, Amsterdam (1970)
5. Meilijson, I.: Limiting properties on mean residual lifetime function. Ann. Math. Stat. **43**(1), 354–357 (1972)

6. Balkema, A.A., De Haan, L.: Residual life time at great age. Ann. Probab. **2**(5), 792–804 (1974)
7. Banjevic, D.: Remaining useful life in theory and practice. Metrika (2009). https://doi.org/10.1007/s00184-008-0220-5
8. Calabria, R., Pulcini, G.: On the asymptotic behaviour of the mean residual life function. Reliab. Eng. **19**, 165–170 (1987)
9. Gradshteyn, I.R., Ryzhik, I.M.: Tables of Integrals, Series, and Products. Elsevier Inc., Oxford (2007)
10. Nassar, M.M., Eissa, F.H.: On the exponentiated Weibull distribution. Commun. Stat. Theory Methods **32**(7), 1317–1336 (2003)
11. Siddiqui, M.M., Caglar, M.: Residual lifetime distribution and its applications. Microelectron. Reliab. **34**, 211–227 (1994)
12. Dengaev, A.V., Rusev, V.N., Skorikov, A.V.: The mean residual life (MRL) of the Gnedenko-Weibull distribution. Estimates of residual life time of pump submersible equipment. In: Proceedings of Gubkin Russian State University of Oil and Gas, No. 1/298, pp. 25–37 (2020)

The Application of a Neural Network and Elements of Regression Analysis in the Development of a Methodology for Effective Foreign Exchange Trading

Elena Alymova[1(✉)] and Oleg Kudryavtsev[1,2]

[1] Russian Customs Academy, Rostov branch, Rostov-on-Don, Russian Federation
koe@donrta.ru
[2] Southern Federal university, Rostov-on-Don, Russian Federation

Abstract. The paper presents a combined approach of using machine learning methods to select an effective trading strategy on the currency exchange. The presented approach uses the calculation of the linear regression angle coefficient by log return indicators and determination of the currency pair quotes trend in the next period based on the calculated coefficient sign. The multilayer feed-forward neural network predicts the angular coefficient value in the next 10-min period for the current 20-min period. The research contains practical experiments that estimate the ratio of effective strategies to non-effective ones based on the linear regression coefficients predicted values.

Keywords: Deep learning · Machine learning · Neural network · Financial time series prediction · Trading strategy · Linear regression coefficients · Logarithmic return

1 Introduction

Predicting the behavior of financial time series is an essential part of investment activity. A rational choice of trading strategy at the currency exchange significantly increases the return on investment.

Recent studies of many authors address the applicability and effectiveness of machine learning methods to predict financial time series behavior. The paper [1] presents a solution to the problem of predicting the trend of the Shanghai Stock Exchange composite index at the close of the trading period based on Deep Learning technologies using an LSTM neural network. Researches [2] and [3] are also devoted to studying the possibilities to use the LSTM neural network for modeling financial time series behavior. In the study [4], the support vector regression predicts the financial time series and generates reasonable prediction uncertainty estimates to tackle flexible real-world financial time series prediction problems effectively. The possibility of predicting the realized volatility of cryptocurrency quotes using a neural network is considered in [5]. The

A. N. Shiryaev et al. (Eds.): ICSM-5 2020, PROMS 371, pp. 306–317, 2021.
https://doi.org/10.1007/978-3-030-83266-7_23

study [6] conducted the effectiveness comparison of predicting the financial time series behavior in several ways. The forecasting accuracy was measured using the LSTM neural network and the integrable autoregressive model. The second model under consideration was the moving average model (ARIMA). The authors concluded that the developing of a model of time series behavior based on a neural network gives a more accurate forecast result.

Despite the large number of studies carried out in this field of research, the problem of predicting the behavior of a financial time series remains relevant and necessary in the practice of traders in the foreign exchange market.

The current research is devoted to studying effectiveness of combining machine learning methods in solving the problem of predicting the behavior of cryptocurrency quotes. The study's goal is to develop a methodology for a reasonable choice of a trading strategy on the currency exchange in the next 10-min period, based on the data of the current 20-min trading period. This research continues the study [7], in which a model for classifying the behavior of a financial time series based on indicators of the logarithmic return of the BTC/USD currency pair implements in the form of a neural network.

2 Preparing Data for Developing a Trading Strategy Based on Linear Regression Indicators of Logarithmic Returns

The real-time historical trading data of the BTC/USD currency pair with a minute interval is using as initial information. The price at the moment of opening (OPEN), the price at the moment of closing (CLOSE), the highest (HIGH) and lowest (LOW) prices in the current period, as well as the amount of currency sold (VOLUME) in the period, are known for each one minute.

In the current study, trading strategies are developed based on the close price (CLOSE). In the works devoted to predicting the financial time series behavior, indicators of logarithmic returns are often used instead of price themselves [8], [9] and [10]. The logarithmic returns are close to zero and do not change dramatically from period to period, which gives the best result when training a neural network. Notice, the logarithmic return expresses the price change's dependence in the current period on the price in the previous period.

Based on the initial data, we calculate the logarithmic return indicator (the natural logarithm of the trading closing price ratio at the next minute to the trading closing price at the current minute) using the formula (1) where i is the number of the current minute.

$$LOGRET_i = LN\left(\frac{CLOSE_i}{CLOSE_{i-1}}\right) \tag{1}$$

The research assumes that the logarithmic return indicators during the current 20-min period determine the logarithmic return indicators in the next 10-min period.

To prove this assumption, we build a model for predicting the value of the linear regression coefficient of logarithmic returns in the next 10-min period based on the known indicators of logarithmic returns in the previous 20-min period. The linear regression equation used in the presented work has the form (2):

$$Y = A + B \cdot X, \tag{2}$$

where

- X is an independent variable (indicators of logarithmic returns),
- A is Y-intercept (level of Y when X is 0),
- B is a linear regression coefficient (slope).

OPEN	HIGH	LOW	CLOSE	VOLUME	PERIOD	LOG RETURN	SPAN
966.61	966.61	966.61	966.61	1			
966.62	966.63	966.62	966.63	1	1	0.0000206907	
967.05	967.06	967.05	967.06	10	2	0.0004447455	
967.37	967.40	967.37	967.39	9	3	0.0003411823	
967.39	967.39	967.39	967.39	0	4	0.0000000000	
967.55	967.79	967.55	967.79	2	5	0.0004133982	
967.46	967.46	967.45	967.45	0	6	-0.0003513776	
967.34	967.34	966.77	966.77	0	7	-0.0007031258	
965.79	965.79	965.79	965.79	0	8	-0.0010141989	
965.79	965.79	965.79	965.79	0	9	0.0000000000	
965.73	965.73	965.03	965.03	0	10	-0.0007872303	1 - 10
965.03	965.03	965.03	965.03	0	11	0.0000000000	2 - 11
965.03	965.03	965.02	965.02	1	12	-0.0000103624	3 - 12
965.01	965.36	965.00	965.36	11	13	0.0003522623	4 - 13
965.30	965.30	965.05	965.05	1	14	-0.0003211753	5 - 14
965.05	965.83	965.05	965.83	10	15	0.0008079218	6 - 15
965.11	965.11	965.11	965.11	0	16	-0.0007457508	7 - 16

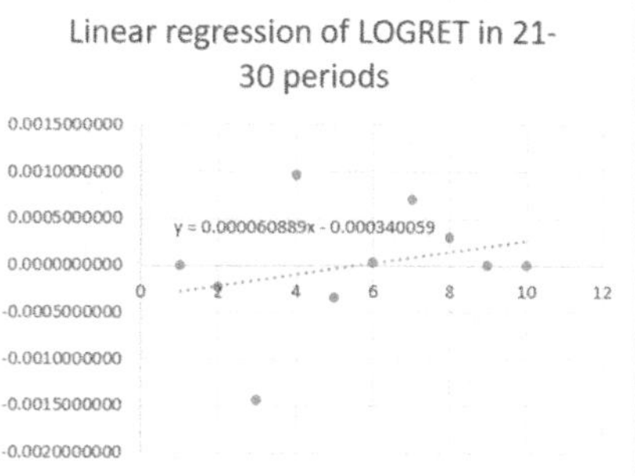

Fig. 1. Logarithmic returns changes chart in 10 min with a plotted trendline

The greater the absolute value of the coefficient B, the more noticeable the line slope and the more pronounced the quotes change. If the coefficient B value is positive, then the numerical series's values are expected to be increased. Otherwise, the values in the numerical series are expected to be decreased.

A trend line is plotted for every twenty minutes in the initial data on the column of calculated logarithmic returns (LOG RETURN). Figure 1 shows the result of plotting the logarithmic yield changes in ten minutes (values in the LOG RETURN column from 21 to 30 periods) and plotting a trend line for a series of logarithmic returns.

In the ten minutes from 21 to 30 min, the linear regression coefficient B is positive. Therefore, the logarithmic return values may increase in this period. There is a chance that a higher price will appear during the next 9-min period than in the first minute of the analyzed period. The highest price (HIGH column) values in the next 10-min period on historical data prove this assumption.

Figure 2 shows the data on price changes in the analyzed 2-min period. The column of the highest prices from 21 to 30 min (column HIGH) contains a price

OPEN	HIGH	LOW	CLOSE	VOLUME	PERIOD	LOG RETURN
965.11	965.11	965.11	965.11	0	17	0.0000000000
965.12	966.48	965.12	966.47	10	18	0.0014081739
966.46	966.46	966.46	966.46	0	19	-0.0000103470
966.41	967.05	966.41	967.05	10	20	0.0006102891
967.04	967.04	967.04	**967.04**	1	21	-0.0000103408
966.81	966.81	966.81	966.81	0	22	-0.0002378675
965.43	965.43	965.42	965.42	0	23	-0.0014387524
966.80	966.80	966.35	966.35	3	24	0.0009628476
966.81	**967.68**	966.01	966.01	5	25	-0.0003519013
966.63	966.73	966.03	966.03	2	26	0.0000207035
966.71	966.71	966.71	966.71	0	27	0.0007036643
966.71	967.00	966.71	967.00	3	28	0.0002999416
967.00	967.00	967.00	967.00	5	29	0.0000000000
967.00	967.00	967.00	967.00	1	30	0.0000000000

Fig. 2. Price values behavior in 10 min with a positive linear regression coefficient B for the observed period

965.10	965.10	965.10	965.10	1	78	0.0000000000	69 - 78
965.10	965.10	965.00	965.00	4	79	-0.0001036216	70 - 79
965.00	965.00	965.00	965.00	0	80	0.0000000000	71 - 80
965.01	965.01	965.01	**965.01**	0	81	0.0000103626	72 - 81
965.00	965.00	965.00	965.00	7	82	-0.0000103626	73 - 82
965.00	965.00	965.00	965.00	1	83	0.0000000000	74 - 83
965.00	965.00	965.00	965.00	0	84	0.0000000000	75 - 84
965.01	965.01	963.86	963.86	29	85	-0.0011820455	76 - 85
963.87	963.87	963.87	963.87	0	86	0.0000103749	77 - 86
963.87	963.87	963.87	963.87	0	87	0.0000000000	78 - 87
963.87	963.87	963.86	963.86	3	88	-0.0000103749	79 - 88
963.23	963.40	963.23	963.40	0	89	-0.0004773617	80 - 89
963.16	963.16	**963.16**	963.16	0	90	-0.0002491487	81 - 90
963.36	963.36	963.36	963.36	0	91	0.0002076283	82 - 91
963.36	963.36	963.36	963.36	0	92	0.0000000000	83 - 92

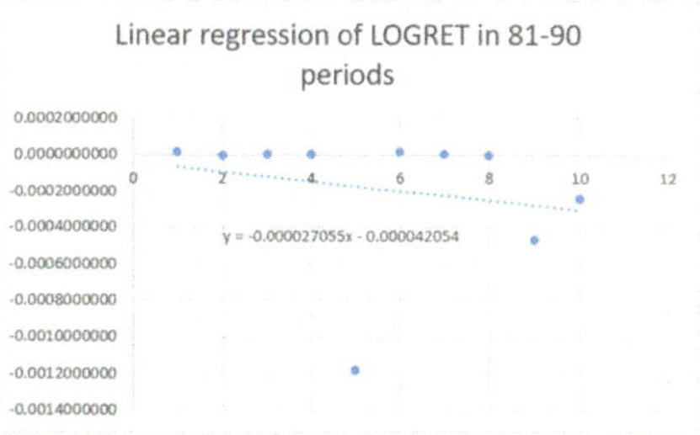

Fig. 3. Price values behavior in 10 min with a negative linear regression coefficient B for the observed period

higher in value than the price at the close of the first minute of the analyzed 10-min period.

The linear regression coefficient B of the logarithmic return trend line in the observed 10-min period turns out to be negative. Thus, we should expect a decrease in trading indicators in this period.

Figure 3 shows an example of when the coefficient B of the trend line equation is negative. Simultaneously, the lowest price column (LOW column) in the observed 10-min period contains a value that will be lower than the price at the close of the first minute of the target 10-min period.

There were added two columns to the original data table: a column for calculating the linear regression coefficient and a column for calculating the logarithmic return.

In the presented work, the linear regression coefficient for the logarithmic returns values is calculated using the LINEST function of the MS Excel spreadsheet, as shown in Fig. 4.

	E	H	I	J	K
1	CLOSE	LOG RETURN	SPAN	LIN REG COEF B	B SIGN
20	966.47	=LN(E20/E19)	9 - 18	=LINEST(H11:H20,,1,1)	
21	966.46	=LN(E21/E20)	10 - 19	=LINEST(H12:H21,,1,1)	
22	967.05	=LN(E22/E21)	11 - 20	=LINEST(H13:H22,,1,1)	=IF(J32>=0, 1,0)
23	967.04	=LN(E23/E22)	12 - 21	=LINEST(H14:H23,,1,1)	=IF(J33>=0, 1,0)
24	966.81	=LN(E24/E23)	13 - 22	=LINEST(H15:H24,,1,1)	=IF(J34>=0, 1,0)
25	965.42	=LN(E25/E24)	14 - 23	=LINEST(H16:H25,,1,1)	=IF(J35>=0, 1,0)
26	966.35	=LN(E26/E25)	15 - 24	=LINEST(H17:H26,,1,1)	=IF(J36>=0, 1,0)
27	966.01	=LN(E27/E26)	16 - 25	=LINEST(H18:H27,,1,1)	=IF(J37>=0, 1,0)
28	966.03	=LN(E28/E27)	17 - 26	=LINEST(H19:H28,,1,1)	=IF(J38>=0, 1,0)
29	966.71	=LN(E29/E28)	18 - 27	=LINEST(H20:H29,,1,1)	=IF(J39>=0, 1,0)
30	967	=LN(E30/E29)	19 - 28	=LINEST(H21:H30,,1,1)	=IF(J40>=0, 1,0)
31	967	=LN(E31/E30)	20 - 29	=LINEST(H22:H31,,1,1)	=IF(J41>=0, 1,0)
32	967	=LN(E32/E31)	21 - 30	**=LINEST(H23:H32,,1,1)**	=IF(J42>=0, 1,0)

Fig. 4. Formulas for calculating indicators of logarithmic returns and linear regression coefficients

Thus, as the initial data for building a model for predicting changes in BTC/USD quotes, the indicators of logarithmic returns, calculated at prices at the time of the closing of minute periods, and linear regression coefficients, calculated according to indicators of logarithmic returns in the next 10-min periods, are used. We use the linear regression coefficients calculated from historical data to train the prediction model and check its adequacy to the simulated process of the observed currency pair's behavior.

The neural network trains using a vector $\overline{\mathrm{X}} = \{X_1, X_2, \ldots, X_m\}$ of independent variables and a vector $\overline{\mathrm{Y}} = \{Y_1, Y_2, \ldots, Y_m\}$ of dependent variables (m is the number of 10-min intervals in the initial data for which the trend line coefficients were calculated). Both vectors defined as follows:

- $X_n = \{LOGRET_{n+1}, LOGRET_{n+2}, \ldots, LOGRET_{n+20}\}$;
- $Y_n = \{LINREGB_{n+21,n+30}\}$,

where

- $LOGRET_i$ – logarithmic return at the ith minute ($i \geq 1$);
- $LINREGB_{j,k}$—coefficient B of the linear regression equation based on logarithmic returns $LOGRET_j, \ldots, LOGRET_k$, ($k > j$);
- n—number of the observed period ($n \geq 0$).

3 An Approach for Choosing a Trading Strategy Based on Linear Regression of the Logarithmic Return

Predicting the direction of the observed currency pair quotes trend in the next 10-min period is carried out for the current 20-min period using the sign of the coefficient B in the linear regression equation.

Figure 5 and Fig. 6 show the calculations' results to choose a trading strategy based on historical data. When choosing a trading strategy, it is assumed that the results of trading in the previous 20-min period are known, the logarithmic returns of the closing price (LOG RETURN column) are calculated, and the linear regression coefficient is predicted for the logarithmic returns in the next 10-min period (column LIN REG COEF B).

OPEN	HIGH	LOW	CLOSE	VOLUME	PERIOD	LOG RETURN	SPAN	LIN REG COEF B	B SIGN
965.12	966.48	965.12	966.47	10	18	0.0014081739	9 - 18	0.000098405	
966.46	966.46	966.46	966.46	0	19	-0.0000103470	10 - 19	0.000089309	
966.41	967.05	966.41	967.05	10	20	0.0006102891	11 - 20	0.000061710	1
967.04	967.04	967.04	**967.04**	1	21	-0.0000103408	12 - 21	0.000035800	1
966.81	966.81	966.81	966.81	0	22	-0.0002378675	13 - 22	-0.000003085	1
965.43	965.43	965.42	965.42	0	23	-0.0014387524	14 - 23	-0.000080541	0
966.80	966.80	966.35	966.35	3	24	0.0009628476	15 - 24	-0.000050187	0
966.81	**967.68**	966.01	966.01	5	25	-0.0003519013	16 - 25	-0.000031838	0
966.63	966.73	966.03	966.03	2	26	0.0000207035	17 - 26	-0.000082684	0
966.71	966.71	966.71	966.71	0	27	0.0007036643	18 - 27	-0.000055852	0
966.71	967.00	966.71	967.00	3	28	0.0002999416	19 - 28	0.000034309	1
967.00	967.00	967.00	967.00	5	29	0.0000000000	20 - 29	0.000026974	1
967.00	967.00	967.00	967.00	1	30	0.0000000000	21 - 30	**0.000060889**	1

Fig. 5. A trading strategy determination with a positive linear regression coefficient B

If the linear regression coefficient B, obtained by plotting a trend line for ten logarithmic returns, is positive, then it is assumed that logarithmic returns will grow in these ten minutes. In this case, the strategy of selling the currency at the closing price of the first minute of the current 10-min period turns out to be efficient to buy the currency in the next nine minutes at the first higher price.

In Fig. 5, the linear regression coefficient B for the next 10-min period is determined to be positive at the end of the 20-min period (PERIOD = 20) (the SIGN column B value is one). The historical trading data for the next ten minutes shows that there will be a higher price in the next nine minutes than the one at which the purchase was made at the first minute. Thus, buying currency in the first minute to sell the currency in the next nine minutes at the first higher price that comes across is effective.

If the linear regression coefficient calculated from the logarithmic returns in the observed 10-min period is negative, then it can be supposed that the logarithmic returns will fall in these ten minutes (the SIGN column B value is zero).

Figure 6 shows a situation when the trend line slope coefficient of logarithmic returns from 81 to 90 min (PERIOD = 90) is negative. Among the lowest prices in these ten minutes, there is a lower price than the one at which we sold at the first minute. Thus, selling the currency in the first minute to buy the currency in the next nine minutes at the first lower price that comes across is effective.

OPEN	HIGH	LOW	CLOSE	VOLUME	PERIOD	LOG RETURN	SPAN	LIN REG COEF B	B SIGN
965.00	965.00	965.00	965.00	0	80	0.0000000000	71 - 80	-0.000002010	0
965.01	965.01	965.01	**965.01**	0	81	0.0000103626	72 - 81	0.000004834	1
965.00	965.00	965.00	965.00	7	82	-0.0000103626	73 - 82	0.000010422	1
965.00	965.00	965.00	965.00	1	83	0.0000000000	74 - 83	0.000016701	1
965.00	965.00	965.00	965.00	0	84	0.0000000000	75 - 84	0.000022979	1
965.01	965.01	963.86	963.86	29	85	-0.0011820455	76 - 85	-0.000062152	1
963.87	963.87	963.87	963.87	0	86	0.0000103749	77 - 86	-0.000046567	1
963.87	963.87	963.87	963.87	0	87	0.0000000000	78 - 87	-0.000031109	1
963.87	963.87	963.86	963.86	3	88	-0.0000103749	79 - 88	-0.000016217	1
963.23	963.40	963.23	963.40	0	89	-0.0004773617	80 - 89	-0.000033579	0
963.16	963.16	**963.16**	963.16	0	90	-0.0002491487	81 - 90	-0.000027055	0

Fig. 6. A trading strategy determination with a negative linear regression coefficient B

The analysis of the logarithmic returns and the sign of the linear regression B coefficient historical data allows making the following conclusion: if B >= 0, then the price increase of the cryptocurrency is predicted (signal to buy), otherwise the price of the cryptocurrency is predicted to fall (signal to sell).

Thus, the conditions for deciding to buy or sell in the next 10-min period are determined in the following way:

- strategy 1: if at the end of the current 20-min period the coefficient $B \geq 0$, then at the first minute of the next 10-min period, we buy the cryptocurrency at the current price and sell at the first higher price in the next nine minutes;
- strategy –1: if at the end of the current 20-min period the coefficient $B < 0$, then at the first minute of the next 10-min period, we sell the cryptocurrency at the current price and buy at the first lower price in the next nine minutes.

4 Including a Waiting Strategy in Trading to Improve the Reliability of the Strategies Choice

The chosen strategy may be ineffective if the price at the beginning of the next 10-min period is the highest for a buy signal or the lowest for a sell signal.

Figure 7 shows the signal's failure to sell the currency at the first minute and then to buy it at a higher price during the next 9-min period. We see that the maximal price presents at the first minute of the observed period. Thus, there is no possibility to sell and then buy the currency at a profit.

1	OPEN	HIGH	LOW	CLOSE	VOLUME	PERIOD	LOG RETURN	SPAN	LIN REG COEF B	B SIGN
42	967.21	967.21	967.21	967.21	0	40	-0.0000413552	31 - 40	0.000000689	1
43	967.21	**967.21**	967.00	**967.00**	2	41	-0.0002171429	32 - 41	-0.000013787	1
44	967.00	967.00	967.00	967.00	0	42	0.0000000000	33 - 42	-0.000011029	1
45	966.90	966.90	966.10	966.11	8	43	-0.0009207961	34 - 43	-0.000063511	1
46	966.11	966.11	966.10	966.10	3	44	-0.0000103508	35 - 44	-0.000052914	1
47	966.01	966.01	966.01	966.01	0	45	-0.0000931624	36 - 45	-0.000046709	1
48	966.02	966.02	966.02	966.02	0	46	0.0000103518	37 - 46	-0.000019942	1
49	966.11	966.11	966.00	966.00	10	47	-0.0000207037	38 - 47	-0.000006274	1
50	966.03	966.03	965.50	965.50	3	48	-0.0005177323	39 - 48	-0.000019466	0
51	965.89	965.89	965.89	965.89	0	49	0.0004038542	40 - 49	0.000027333	1
52	965.51	965.89	965.50	965.50	8	50	-0.0004038542	41 - 50	**0.000019602**	1

Fig. 7. An example of a failure of the signal to sell based on a positive linear regression coefficient

We check the efficiency of the strategies by the following criteria:

- if the coefficient $B \geq 0$ at the first minute of the next 10-min period, then the opening price should be less than one of the highest minute quotes in the next 10-min period. In this case, we consider the strategy 1 to be efficient;
- if the coefficient $B < 0$ at the first minute of the next 10-min period, then the closing price should be higher than one of the lowest minute quotes in the next 10-min period. In this case, we consider the strategy –1 to be efficient.

Checking the effectiveness of trading strategies based on the linear regression angular coefficient sign is implemented as follows. For the column of maximal prices (HIGH), the highest price is calculated within every ten minutes (column MAX NEXT 10 PERIOD); for the column of minimal prices (LOW), the lowest price is calculated within every ten minutes (MIN NEXT 10 PERIOD). If for a signal to sell (B SIGN = 1), the opening price at the first minute is less than the maximum of all price values in the current 10-min period, we put the value 1 into the WORKING STRATEGY column. With a buy signal (B SIGN = 0), the closing price at the first minute must be less than the minimum price values in the next ten minutes. In that case, WORKING STRATEGY takes the value 1. In all other cases, WORKING STRATEGY takes on the value 0, which indicates the failure of the chosen strategy.

Checking the formulated criteria for historical data shows one failure for every five successful strategies. It means that it is possible to make an unprofitable decision to buy or sell cryptocurrency in every sixth case.

We use the strategy 0 (a wait strategy) to minimize the number of unprofitable decisions. In this case, the trader performs no actions to buy or sell cryptocurrency.

The waiting strategy introduction is based on the coefficient B's absolute value in the linear regression equation. Figure 8 demonstrates the case when the chosen strategy turns out to be ineffective, while the coefficient B modulo was equal to 0.000019602.

OPEN	HIGH	LOW	CLOSE	VOLUME	PERIOD	LOG RETURN	SPAN	LIN REG COEF B	B SIGN	MAX NEXT 10 PERIOD	MIN NEXT 10 PERIOD	WORKING STRATEGY
967.21	967.21	967.21	967.21	0	40	-0.0000413552	31 - 40	0.000000689	1	967.21	965.50	0
967.21	**967.21**	967.00	**967.00**	2	41	-0.0002171429	32 - 41	-0.000013787	1	967.00	965.50	0
967.00	967.00	967.00	967.00	0	42	0.0000000000	33 - 42	-0.000011029	1	966.90	965.50	0
966.90	966.90	966.10	966.11	8	43	-0.0009207961	34 - 43	-0.000063511	1	966.11	965.50	0
966.11	966.11	966.10	966.10	3	44	-0.0000103508	35 - 44	-0.000052914	1	966.11	965.50	1
966.01	966.01	966.01	966.01	0	45	-0.0000931624	36 - 45	-0.000046709	1	966.11	965.50	1
966.02	966.02	966.02	966.02	0	46	0.0000103518	37 - 46	-0.000019942	1	966.11	965.50	0
966.11	966.11	966.00	966.00	10	47	-0.0000207037	38 - 47	-0.000006274	1	966.04	965.50	1
966.03	966.03	965.50	965.50	3	48	-0.0005177323	39 - 48	-0.000019466	0	966.04	965.50	1
965.89	965.89	965.89	965.89	0	49	0.0004038542	40 - 49	0.000027333	1	966.04	965.50	1
965.51	965.89	965.50	965.50	8	50	-0.0004038542	41 - 50	**0.000019602**	1	966.21	965.50	1

Fig. 8. Results of testing the selling strategy effectiveness with a positive linear regression coefficient

We choose the strategy 0 if the value of the coefficient B in the first minute of the next 10-min period is insignificant. It means that the coefficient B modulo is less than the specified parameter h. In the current research, the value selection of the parameter h was performed empirically on historical data and is equal to 0.0003, which made it possible to increase the number of profitable strategies. The wrong decision to buy or sell cryptocurrency is making in every eighth case when strategy 0 is using as a third alternative of behavior on the trade market.

5 Implementation and Results of Training a Neural Network to Predict Linear Regression Coefficients

Within current research, the neural network is implementing and training on the Rapid Miner analytical platform. Figure 9 demonstrates the process chain of training and testing the linear regression coefficients predictive model. The Deep Learning block implements a multilayer feed-forward artificial neural network trained with stochastic gradient descent using back-propagation.

The neural network uses four hidden layers. Each odd layer contains 100 neurons, and each even layer contains 50 neurons. Each hidden layer uses a linear rectification function (ReLU) to activate neurons. These neural network parameters are chosen empirically to obtain greater forecasting accuracy. The mean square prediction error (RMSE) with given parameters is 0.000086186.

Figure 10 demonstrates the comparative chart of actual and predicted linear regression coefficients. The chart shows that the constructed regression model based on the log-return trend slopes is consistent with the observed process.

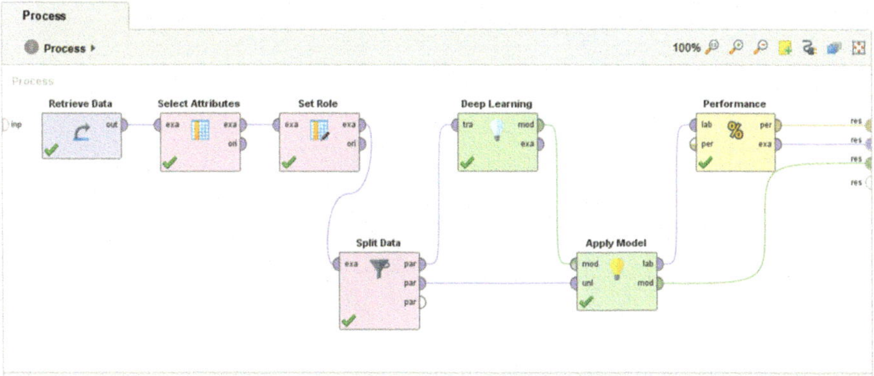

Fig. 9. The process of training and testing the forecasting model in Rapid Miner

6 Evaluating the Effectiveness of Using a Trading Strategy Based on Linear Regression Indicators

The neural network predicts the values of the linear regression coefficients for 1000 min periods (16.5 h of trading). The effectiveness of the proposed trading methodology was compared to historical and predicted values.

In the current research, the correct prediction of the linear regression coefficient sign is the goal. For 1000 predicted values, 768 values have a sign corresponding to the value calculated from historical data. Thus, the coincidence of the actual and forecast values by sign is about 77.

The methodology's effectiveness for choosing a trading strategy decreases due to a decrease in the sign determination accuracy of the linear regression coefficient in the forecasting values. For every three successful strategies on the forecasting values of coefficient B, there is one ineffective one. It means that it is possible to make an unprofitable decision to buy or sell cryptocurrency in every fourth case.

Checking the effectiveness of introducing strategy 0 (waiting) on the predicted values showed that the indicator of significance h of the linear regression coefficient modulo, which was selected empirically on historical data, should be revised due to the low forecasting accuracy of the linear regression coefficient value.

The choice of the significance indicator h should take into account the estimates of the model's accuracy. In this work, the significance indicator h for the linear regression coefficients' predicted values is 0.00005, which reduces the number of both ineffective and effective strategies.

Nevertheless, the introduction of the 0 (waiting) strategy made it possible to somewhat increase trading efficiency according to the proposed method. With the introduction of strategy 0 on forecasting values for every four successful strategies, there is one ineffective one. That is, in every fifth case, the chosen strategy turns out to be unprofitable.

Fig. 10. Comparison of actual and predicted values of linear regression coefficients

7 Conclusion

The current paper proposes a methodology for choosing a strategy for trading on the currency exchange in the next 10-min period, based on the current 20-min trading period. The proposed approach identifies three trading activities: buying at the beginning of the next period to sell at the first higher price (strategy 1), sell at the beginning of the next period to buy at the first lower price (strategy –1). The third strategy is inaction (strategy 0), when the linear regression coefficient's value at the beginning of the next 10-min period is insignificant for the chosen criterion.

As the experiments on historical data show, every sixth decision in choosing a strategy leads to losses without introducing strategy 0, and when strategy 0 is applying, every eighth decision is unprofitable.

The feed-forward neural network is using to predict future values of the linear regression coefficients. Neural network parameters were selected empirically to improve forecasting accuracy.

The neural network model predicts the values of linear regression coefficients and has an accuracy of 77% in predicting the sign, which affects the overall profitability of trading according to the proposed method. Nevertheless, the experiments on historical and forecasting data prove the proposed trading methodology's effectiveness on the currency exchange. When entering the strategy of waiting on the predicted linear regression coefficient values, there is one ineffective one for every four successful strategies, which generally implies a break-even trade according to the methodology proposed in the current research.

Acknowledgments. This work was supported by the Russian Foundation for Basic Research under Grant No 18-01-00910.

References

1. Yan, H., Ouyang, H.: Financial time series prediction based on deep learning. Wireless Pers. Commun. **102**(2), 683–700 (2018)
2. Kondratyeva, T.N.: Forecasting the trend of financial time series using LSTM neural network. Eurasian Sci. J. **9**(4), 61–67 (2017)
3. Labusov, M.V.: Application of long short-term memory neural networks to modeling financial time series. Innov. Invest. **4**, 167–171 (2020)
4. Law, T., Shawe-Taylor, J.: Practical Bayesian support vector regression for financial time series prediction and market condition change detection. Quant. Finan. **17**(9), 1403–1416 (2017)
5. Shintate, T., Pichl, L.: Trend prediction classification for high-frequency bitcoin time series with deep learning. J. Risk Finan. Manage. **12**(1), 17–33 (2019)
6. Alzheev, A.V., Kochkarov, R.A.: Comparative analysis of ARIMA and LSTM predictive models: evidence from Russians stocks. Finan.: Theory Pract. **4**,(1), 14–23 (2020)
7. Alymova, E.V., Kudryavtsev, O.E.: Neural networks usage for financial time series prediction. In: Abstracts of Talks Given at the 4th International Conference on Stochastic Methods In: Theory of Probability and Its Applications, vol. 65, no. 1, pp. 122–123 (2020)
8. Miura, R., Pichl, L., Kaizoji, T.: Artificial neural networks for realized volatility prediction in cryptocurrency time series. In: Lu, H., Tang, H., Wang, Z. (eds.) Advances in Neural Networks – ISNN 2019. ISNN 2019. Lecture Notes in Computer Science, vol. 11554. Springer, Cham (2019). https://doi.org/10.1007/978-3-030-22796-8_18
9. Kodama, O., Pichl, L., Kaizoji, T.: Regime change and trend prediction for Bitcoin time series data. CBU Int. Conf. Proc. **5**, 384–388 (2017)
10. Arévalo, A., et al.: High-frequency trading strategy based on deep neural networks. In: Huang, DS., Han, K., Hussain, A. (eds.) Intelligent Computing Methodologies. ICIC 2016. Lecture Notes in Computer Science, vol. 9773. Springer, Cham (2016). https://doi.org/10.1007/978-3-319-42297-8_40

Statistical Analysis of Generalized Jackson Network with Unreliable Servers via Strong Approximation

Elena Bashtova and Elena Lenena(✉)

Lomonosov Moscow State University, Moscow, Russia
{elena.bashtova,elena.lenena}@math.msu.ru

Abstract. We consider a Jackson network with regenerative input flows in which every server is subject to a random environment influence generating breakdowns and repairs. They occur in accordance with two independent sequences of i.i.d. random variables. We are using a theorem on the strong approximation of the vector of queue lengths by a reflected Brownian motion in positive orthant to establish a consistent estimates construction and present an approach to recognize bottlenecks.

Keywords: Jackson network · Strong approximation · Heavy traffic · Unreliable systems · Reflected Brownian motion · Asymptotic covariance matrix estimation

1 Introduction

Jackson networks are among the most fundamental and complicated objects in the queueing theory. Such networks were introduced in Jackson in 1957 [12]. This type of queueing systems models have long been used for a wide range of applications in transportation, production, distributed computing systems, team workflows, social networks etc., so results concerning their asymptotic behavior present a practical interest. Within the wide set of problems concerning Jackson networks specifically interesting research directions include the evaluation of the limit distribution and its product forms in the case when this limit distribution exists and consideration of the systems with heavy traffic. These studies provide a motivation to estimate limit traffic cases and bottleneck conditions.

Heavy traffic and overloaded system cases are analytically non-trivial and at the same time the most crucial, as overloading may lead to breakdowns, production shutdown, losses from downtime, repair costs and other overheads.

So, a lot of papers were devoted to the heavy traffic limit analysis. Harrison [9] considered tandem queueing systems and proved a heavy traffic limit theorem for the stationary distribution of the sojourn times. His limit was also given as a complicated function of multidimensional Brownian motion. Harrison later again considered tandem queueing systems, and introduced reflected Brownian motion on the non-negative orthant as the diffusion limit. Further investigation

A. N. Shiryaev et al. (Eds.): ICSM-5 2020, PROMS 371, pp. 318–328, 2021.
https://doi.org/10.1007/978-3-030-83266-7_24

in the areas of heavy traffic limit theorems and diffusion approximations is surveyed in Whitt [21] and Lemoine [14]. Additionally, Reiman [17] presented heavy traffic limit theorems for the queue length and sojourn time processes associated with open queueing networks. To sum it up, limit theorems state that properly normalized sequences of queue length and sojourn time converge to a reflected Brownian motion in an orthant.

In this paper we employ a more powerful limit theorem, namely strong invariance principle or strong Gaussian approximation. This huge exploration area saw pioneering investigation in the classical paper by Strassen [20]. This was followed by many beautiful results, with the most complete overview of which for the i.i.d. case has been done in Zaytsev [22]. Markov Chains result was presented in Merlevede, Rio[15], that studied strong approximation for additive functionals of geometrically ergodic Markov chain. These results typically allow to replace a complex random system for which some almost sure convergence is needed by one given by (possibly transformed) Gaussian family. Concerning Jackson networks, strong (almost sure) approximation with infinite time horizons and reliable server networks were considered by Chen and Yao [5].

It is natural to generalize results to networks with unreliable servers. There has been a growing literature on queues with an unreliable server. To emphasize the significance of queueing models with unreliable servers in applications we will refer to some crucial works [7,8,13,18]. Particularly one should mention Gaver's [8] research in which the input flow is a compound Poisson process, the interruptions appear at random in the sense that if the system is currently free of interruptions, then the time until the next interruption is exponentially distributed. Gaver introduced completion time that is the generalization of the service time. This notion made it possible to apply results for $M|G|1$ to investigate a system subjected to interruptions, i.e. with an unreliable server. So, total basic results were expressed in terms of the distribution function (d.f.) of the completion time. The systems with unreliable servers and general input flow was considered in Afanasyeva, Bashtova [1]. An interesting interpretation of unreliable servers emerged while modeling of unregulated crossroads. The simplest model of a vehicle crossing problem in probabilistic terms was considered in [10]. There is a one-lane road S_1 which is intersected on one side by a single-lane secondary road S_2. A car waiting on the secondary road S_2 will turn right only if there is at least distance J between the intersection and the first car on S_1. We may consider the crossroads with respect to cars arriving on the secondary road S_2 as a queueing system with an unreliable server. The server is in working state when there are no cars within a distance J of the crossroads on the road S_1 and it becomes out of order when the first car appears on this interval. By the nature of the system we have to suppose that the breakdown of the server can occur at any time, even when the server is free. Note that a queueing system with an unreliable server may also be considered in the stochastic analysis of crossroads with traffic lights.

Estimation of the asymptotic variance (or covariance matrix in the vector-valued case) in complex random systems is a well-known problem. Corresponding estimates provide a natural way to establish limit theorems with random normalization and thereby to testing the hypotheses on the unknown mean. We apply the local averaging techniques initiated by Peligrad and Shao [16], known to be applicable to many stationary dependent families of random variables and having good asymptotic properties.

One research direction which is especially interesting in practice is the bottleneck recognition. The aim of this statistical inference problem is to detect the buffers in which the queue length will systematically approach infinity, as well as those for which it is stochastically bounded. This class of real problems usually relies on the following assumptions: for each station one should know the general input flow (cumulated input streams from outside of the system and streams being transferred from other stations) and queue length vector. Unreliable server concept implies that the breakout and repair times also being observable. As for the routing matrix, we assume that it is an attribute of the system and either is known or is estimated on another data.

Under these assumptions we give strongly consistent estimates of the traffic coefficients, drift coefficients and the covariance matrix. Using these estimates as building blocks, we provide an algorithm for the identification of strict bottlenecks and non-bottlenecks.

2 System Description

The queuing network $\mathcal{N}$ we study has K single server stations, and each of them has an associated infinite capacity waiting room. At least one station has an arrival stream from outside the network, and the vector $A(\cdot)$ of arrival streams is assumed to be a multi-dimensional regenerative flow. Recall

Definition 1. *A multi-dimensional coordinatewise càdlàg stochastic process $A(t)$ is called regenerative one if there exists an increasing sequence of random variables $\{\theta_i, i \geq 0\}$, $\theta_0 = 0$ such that the sequence*

$$\{\kappa_i\}_{i=1}^{\infty} = \{A(\theta_{i-1} + t) - A(\theta_{i-1}), \theta_i - \theta_{i-1}, t \in [0, \theta_i - \theta_{i-1})\}_{i=1}^{\infty}$$

consists of independent identically distributed random elements. The random variable θ_j is said to be the jth regeneration point of A, and $\tau_j = \theta_j - \theta_{j-1}$ (where $\theta_0 = 0$) to be the jth regeneration period.

From Smith [19] it is known that we can define the asymptotic intensity vector $\lambda = \lim\limits_{t\to\infty} \frac{A(t)}{t}$ a.s., and the asymptotic covariance matrix $V = \lim\limits_{t\to\infty} \frac{\mathsf{Var} A(t)}{t}$.

Each station's service times $\{\eta_j^i\}_{i=1}^{\infty}$, $j = 1, \ldots, K$, are mutually independent sequences of i.i.d. random variables with mean $1/\mu_j$ and variance σ_j^2. After being served at station k, the customer is routed to station j $(j = 1, \ldots, K)$ with probability p_{kj}. The routing matrix $P = \|p_{kj}\|_{k,j=1}^{K}$ is assumed to have spectral radius strictly less than unity, i.e. there is always a positive probability than a served customer leaves the system immediately. For $i = 1, 2, \ldots$ and $k = 1, \ldots K$, define φ_k^i to be a random variable equal to $j = 1, \ldots, K$ whenever ith customer served on station k is routed to station j, and $\varphi_k^i = 0$ if this customer exits the network. Routing vectors are defined by $\phi_k^i = e_{\varphi_k^i}$, where e_j is the K-vector whose jth component is 1 and other components are 0, if $j = 1, \ldots, K$, and $e_0 = 0$. Customers routing happens independently and immediately.

In our model the service on every channel is influenced by a random environment which causes breakdowns of the server (falling into OFF state from ON state) at random moments. The repair of the server also takes a random time. We suppose that consecutive time intervals of states ON and OFF form two independent sequences of i.i.d. random variables and, for jth server, denote them by $\{u_j^i\}_{i=1}^{\infty}$ and $\{v_j^i\}_{i=1}^{\infty}$ respectively.

Let for $j = 1, \ldots, K$

$$a_j = \mathsf{E}u_j^1,\ b_j = \mathsf{E}v_j^1,\ \alpha_j = a_j(b_j + a_j)^{-1},\ s_j^2 = \mathsf{Var}u_j^1, d_j^2 = \mathsf{Var}v_j^i.$$

We suppose that the service that was interrupted by the breakdown is continued upon repair from the point at which it was interrupted.

Let $Q(t) = (Q_1(t), \ldots, Q_K(t))$ be the vector of number of customers in each channel at time t. Next we need vector-valued busy time processes $B(t) = (B_1(t), \ldots, B_K(t))$. The jth component of $B(t)$, where $j = 1, \ldots, K$, indicates the amount of time up to t the server at station j is busy (i.e. in state ON and serving jobs). Then,

$$Q(t) = A(t) + \sum_{j=1}^{K} L_j(B_j(t)),$$

where

$$L_j(u) = \sum_{i=1}^{S_j(u)} (\phi_j^i - e_j), \quad 1 \leq k \leq K.$$

Next, we consider a system of equations

$$\gamma_j = \lambda_j + \sum_{i=1}^{k} (\gamma_i \wedge \alpha_i \mu_i) p_{ij}, \quad j = 1, \ldots, K. \tag{1}$$

In the Jackson networks theory, the systems like above play a significant role and are known as the traffic equations. Due to Theorem 7.3 in Chen and Yao [5] our traffic equation (1) has a unique solution $\gamma = (\gamma_1, \ldots, \gamma_K)$. Therefore we may define a jth station traffic coefficient

$$\rho_j = \frac{\gamma_j}{\alpha_j \mu_j},$$

$j = 1, \ldots, K$.

Buffer j is called a nonbottleneck if $\rho_j < 1$, a bottleneck if $\rho_j \geq 1$, a balanced bottleneck if $\rho_j = 1$, and a strict bottleneck if $\rho_j > 1$.

3 Reflected Brownian Motion in Orthant and Strong Approximation

Reflected Brownian motion (RBM) was initially investigated in Harrison, Reiman (1981) [10]. A lot of RBM's useful properties were derived in Chen, Yao [5]. Unfortunately, in multidimensional case there is no explicit form of marginal distributions as it takes place for RBM on real line. However it is possible to model multidimensional RBM as was shown in Blanchet, Murthy (2018) [3].

Consider a pair of K-dimensional processes $Z = \{Z(t); t \geq 0\}$ and $Y = \{Y(t); t \geq 0\}$ which jointly satisfy the following conditions:

$$Z(t) = W(t) + Y(t)(I - P), t \geq 0,$$

where $W = \{W(t); t \geq 0\}$ is a K-dimensional Brownian motion with covariance matrix Γ, drift vector b and $W(0) \in \mathbb{R}_+^K$; $Z(t)$ takes values in $\mathbb{R}_+^K$, $t \geq 0$; $Y_j(.)$ is continuous and nondecreasing with $Y_j(0) = 0$; and $Y_j(.)$ increases only at those times t where $Z_j(t) = 0$, $j = 1, \ldots K$. It was shown in [10] that for any given Brownian motion W there exists a unique pair of processes Y and Z satisfying conditions above.

In the language of [10] and [5], Z is a reflected Brownian motion on $\mathbb{R}_+^K$ with drift b, covariance matrix Γ, and reflection matrix $(I - P)$.

In this paper instead of approximation by Wiener process we are using an approximation by a Reflected Brownian motion. Such a problem for a system with reliable server was investigated in Chen, Yao [5], further results for a system with unreliable server were presented in Bashtova, Lenena [2].

Definition 2. *We say that a vector-valued random process* $\zeta = \{\zeta_t, t \geq 0\}$ *admits a* r*-strong approximation by some process* $\overline{\zeta} = \{\overline{\zeta}_t, t \geq 0\}$, *if there exists a probability space* $(\Omega, \mathcal{F}, \mathsf{P})$ *on which one can define both* ζ *and* $\overline{\zeta}$ *in such a way that*

$$\sup_{0 \leq u \leq t} \|\zeta_u - \overline{\zeta}_u\| = o(t^{1/r}), \quad a.s., \ when\ t \to \infty.$$

The following theorem was proved in Bashtova, Lenena [2].

Theorem 1. *Let* $\mathsf{E}\tau_i^p < \infty$, $\mathsf{E}\|A(\theta_{i+1}) - A(\theta_i)\|^p < \infty$, $\mathsf{E}(\eta_j^i)^p < \infty$ *for* $i \in \mathbb{N}$ *and any* $j = 1, \ldots, K$. *Then* Q *admits* p'*-strong approximation by a reflected Brownian motion on* $\mathbb{R}_+^K$ *with drift* $\lambda - \alpha\mu(I - P)$, *reflected matrix* $(I - P)$, *covariance matrix* $\Gamma = ||\Gamma_{kl}||_{k,l=1}^K$,

$$\Gamma_{kl} = V_{kl} + \sum_{j=1}^{K} (\gamma_j \wedge \alpha_j \mu_j) p_{jk}(\delta_{kl} - p_{jl}) + (\sigma_j^2 \mu_j^3 \alpha_j + \mu_j^2 D_j)(\rho_j \wedge 1)(p_{jk} - \delta_{jk})(p_{jl} - \delta_{jl}),$$

where

$$D_j = \frac{a_j^2 d_j^2 + b_j^2 s_j^2}{(a_j + b_j)^3}$$

and $p' = p$ *for* $p < 4$ *and* p' *is any number less than* 4 *for* $p \geq 4$.

4 Parameters Estimation

For the statistic inference we assume, for each station $k = 1, \dots, K$, the following data to be known: a general input flow (both input streams: from outside of the system and transferred from another station), queue length, breakdown and repair times. Moreover, the routing matrix is also considered to be known by the structure of the system or to be estimated based on the other considerations. Realistically, for systems with unreliable servers it is doubtful that the two flows (outside input flow and inter-system transferring input flow) are known separately. As every station is operating as a single-server queuing system, we know input flow, queue length, breakdowns and repair time, so that service time is known too.

Service time calculation logic is the following: it equals either the time between the customer arrival to the empty system and the first following drop in $Q(t)$, or the difference between the time from one drop of the process $Q(t)$ to the next one during the busy period and the lengths of the intervals breakdowns within a given period of time.

Thus we can produce consistent estimates

$$\widehat{\gamma}_k = \frac{X_k(t)}{t}; \quad \widehat{\mu_k} = \left(\frac{1}{l_k(t)} \sum_{i=1}^{l_k(t)} \eta_k^{(i)} \right)^{-1} \quad \widehat{\alpha}_k = \frac{C_k(t)}{t}$$

where $C_k(t)$ and $l_k(t)$ equal the total time kth server being in the state ON before t and the number of customers served before t respectively, $\mathrm{X}(t)$ is total input flow (both internal and external).

Accordingly, for kth traffic coefficient we have

$$\widehat{\rho_k} = \frac{\widehat{\gamma}_k}{\widehat{\alpha}_k \widehat{\mu}_k}$$

One can now estimate the intensity of incoming flow and the drift vector of the approximating Brownian motion, making use of the traffic equation:

$$\widehat{\lambda_k} = \widehat{\gamma}_k - \sum_{i=1}^{K} (\widehat{\gamma}_i \wedge \widehat{\alpha}_i \widehat{\mu}_i) p_{ik}; \quad \widehat{\theta} = \widehat{\lambda} - \widehat{\alpha}\widehat{\mu}(I - P).$$

Let us tackle covariance matrix Γ estimation. To this end we introduce the random sums

$$\widehat{\Gamma}_k^Q = t^{\frac{3\alpha}{2}-1} \sum_{j=1}^{[t^{1-\alpha}]} \left| \overline{Q}_k^{(j)} - \overline{\overline{Q}}_k \right|,$$

$$\widehat{\Gamma}_k^Z = t^{\frac{3\alpha}{2}-1} \sum_{j=1}^{[t^{1-\alpha}]} \left| \overline{Z}_k^{(j)} - \overline{\overline{Z}}_k \right|, \quad \widehat{\Gamma}_k^W = t^{\frac{3\alpha}{2}-1} \sum_{j=1}^{[t^{1-\alpha}]} \left| \overline{W}_k^{(j)} - \overline{\overline{W}}_k \right|, \quad k = 1, \dots, K,$$

where $\alpha \in (0, 1)$ and for any process

$$\zeta(t) = \{(\zeta_1(t), \dots, \zeta_k(t)), t \geq 0\}$$

one has set

$$\overline{\zeta}_k^{(j)} = \frac{\zeta_k(jt^\alpha) - \zeta_k((j-1)t^\alpha)}{t^\alpha}; \qquad \overline{\overline{\zeta}}_k = \frac{\zeta_k(t)}{t}.$$

Define also

$$\widehat{\Gamma}^Q_{k,m} = t^{\frac{3\alpha}{2}-1} \sum_{j=1}^{[t^{1-\alpha}]} \left| \overline{Q}_k^{(j)} + \overline{Q}_m^{(j)} - \overline{\overline{Q}}_k - \overline{\overline{Q}}_m \right|$$

for $k, m = 1, \ldots, K$.

Theorem 2. *If $\rho_k \geq 1$ for all $k = 1, \ldots, K$, $\alpha > \frac{2}{p'}$, then for any $k, m = 1, \ldots, K$ one has*

$$\widehat{\Gamma}^Q_k \xrightarrow[t\to\infty]{a.s.} \sqrt{\frac{2}{\pi}\Gamma_{kk}},$$

$$\widehat{\Gamma}^Q_{k,m} \xrightarrow[t\to\infty]{a.s.} \sqrt{\frac{2}{\pi}(\Gamma_{kk} + \Gamma_{mm} + 2\Gamma_{km})}.$$

Proof. We prove the first statement only (second one being proved analogously), and its proof relies on four lemmas below.

Lemma 1. *For any $\alpha \in (0,1)$ and $k = 1, \ldots, K$ one has*

$$\widehat{\Gamma}^W_k \xrightarrow[t\to\infty]{a.s.} \sqrt{\frac{2}{\pi}\Gamma_{kk}}.$$

Proof. First, we note that we can assume $\mathsf{E}W_t = 0$, $t \geq 0$; in which case one has

$$t^{\frac{3\alpha}{2}-1} \left| \sum_{j=1}^{[t^{1-\alpha}]} \left| W_k^{(j)} - \overline{\overline{W}} \right| - \sum_{j=1}^{[t^{1-\alpha}]} \left| W_k^{(j)} \right| \right| \leq \frac{|W_t|}{t^{1-\alpha/2}} \xrightarrow[t\to\infty]{\text{a.s.}} 0$$

due to the law of the iterated logarithm. Therefore, we have to prove that

$$t^{\frac{3\alpha}{2}-1} \sum_{j=1}^{[t^{1-\alpha}]} \left| W_k^{(j)} \right| \xrightarrow[t\to\infty]{\text{a.s.}} \sqrt{\frac{2}{\pi}\Gamma_{kk}}. \tag{2}$$

As for a given t, random variables $\beta_j = t^{\frac{\alpha}{2}} \left| W_k^{(j)} \right|$, $j = 1, \ldots, [t^{1-\alpha}]$ are independent and distributed as $|\mathcal{N}(0, \Gamma_{kk})|$, convergence (2) follows from Theorem 2 of Hu, Móricz and Taylor [11].

Let Ψ be the mapping of the trajectory of a Wiener process to the trajectory of Y appearing in the definition of reflected Brownian motion (see § 7.2 in Chen, Yao [5]).

Lemma 2. *Let $\theta = (\theta_1, \ldots, \theta_k)$ be a nonnegative vector, and $B = \{B_t, t \geq 0\}$ be a K-dimensional Wiener process with zero drift and an arbitrary covariance matrix Γ; $B^\theta(t) = B_t + \theta_t$. Then*

$$\mathsf{E}\Psi(B(t))_k \geq \mathsf{E}\Psi(B^\theta(t))_k, \quad t \geq 0, \quad k = 1, \ldots, K.$$

Proof. Let us build $Y = \Psi(B)$ and $Y^\theta = \Psi(B^\theta)$ as in the proof of Theorem 7.2 in Chen, Yao [5]. Namely, we set iteratively $Y_{(0)} = Y^\theta_{(0)} \equiv 0$ and

$$Y_{(n+1)}(t) = \sup_{0\le u\le t}\left[-B(u) + (I-P^T)^{-1}Y_{(n)}(t)\right]^+,$$

$$Y^\theta_{(n+1)}(t) = \sup_{0\le u\le t}\left[-B(u) + (I-P^T)^{-1}Y^\theta_{(n)}(t)\right]^+, \ n\in\mathbb{N}.$$

Then clearly $Y_{(n)}(t) \ge Y^\theta_{(n)}(t)$ coordinate-wisely, hence the same is true for the limit processes.

Lemma 3. *For any $\alpha\in(0,1)$ and $k=1,\dots,K$*

$$\left|\widehat{\Gamma}^Z_k - \widehat{\Gamma}^W_k\right| \xrightarrow[t\to\infty]{a.s.} 0.$$

Proof. The definition of reflected Brownian motion in orthant leads us to

$$\left|\widehat{\Gamma}^Z_k - \widehat{\Gamma}^W_k\right| \le t^{\frac{3\alpha}{2}-1}\sum_{j=1}^{[t^{1-\alpha}]}\left|\left[(I-P)\left(\overline{Y}^{(j)} - \overline{\overline{Y}}\right)\right]_k\right|.$$

Furthermore,

$$\left|\left[(I-P)\left(\overline{Y}^{(j)} - \overline{\overline{Y}}\right)\right]_k\right| \le \sum_{m=1}^{K}\left(\left|\overline{Y}^{(j)}_m\right| + \left|\overline{\overline{Y}}_m\right|\right),$$

so by monotonicity of Y

$$\left|\widehat{\Gamma}^Z_k - \widehat{\Gamma}^W_k\right| \le t^{\frac{3\alpha}{2}-1}\sum_{k=1}^{K}\left(\frac{Y_k(t)}{t^\alpha} + \frac{Y_k(t)}{t}t^{1-\alpha}\right) = \sum_{k=1}^{K}\frac{2Y_k(t)}{t^{1-\alpha/2}}.$$

As by Theorem's condition the drift of W is coordinate-wise nonnegative, due to Lemma 2 we can consider only the case of zero drift (corresponding to all the buffers being balanced bottlenecks). As proved in § 7.2 in Chen, Yao [5], the mapping $Y = \Psi(W)$ is a Lipschitz continuous one with respect to the uniform topology (the Lipschitz constant being determined by the matrix P). Thus for some $C = C(P) > 0$

$$\left|\widehat{\Gamma}^Z_k - \widehat{\Gamma}^W_k\right| \le \frac{C\max\limits_{k=1,\dots,K}\sup\limits_{0\le u\le t}|W_k(u)|}{t^{1-\alpha/2}},$$

which tends to 0 almost surely due to the law of the iterated logarithm.

Lemma 4. *For any $\alpha > \frac{2}{p'}$, and $k=1,\dots,K$*

$$\left|\widehat{\Gamma}^Q_k - \widehat{\Gamma}^Z_k\right| \xrightarrow[t\to\infty]{a.s.} 0.$$

Proof.

$$\left|\widehat{\Gamma}_k^Z - \widehat{\Gamma}_k^W\right| \le t^{\frac{3\alpha}{2}-1} \sum_{j=1}^{[t^{1-\alpha}]} \left|\left|\overline{Z}_k^{(j)} - \overline{\overline{Z}}_k\right| - \left|\overline{Q}_k^{(j)} - \overline{\overline{Q}}_k\right|\right|$$

$$\le t^{\frac{3\alpha}{2}-1} \sum_{j=1}^{[t^{1-\alpha}]} \left|\overline{Q}_k^{(j)} - \overline{Z}_k^{(j)}\right| + t^{\frac{\alpha}{2}-1} \left|Q_k(t) - Z_k(t)\right|$$

Therefore applying Theorem 1, one can see that

$$\left|\widehat{\Gamma}_k^Q - \widehat{\Gamma}_k^Z\right| = \overline{o}(t^{\frac{1}{p'}-\frac{\alpha}{2}}), \quad t \to \infty.$$

Theorem 2 now follows from Lemmas 1, 3 and 4.

Remark 1. One can consider a more general parametric family of local averaging estimates, for $s \in [1, 2]$ letting

$$\hat{\Gamma}_k^Q = \frac{t^{\alpha s/2}}{t^{1-\alpha}} \sum_{j=1}^{[t^{1-\alpha}]} \left|\overline{Q_k}^{(j)} - \overline{\overline{Q}}\right|^s$$

(and defining similar quantities for W and Z). While we are studying in detail the case $s = 1$ as being the most robust one, other cases attract interest as well [4,6].

Remark 2. Under any condition on the traffic coefficients vector ρ, the statistics $\hat{\Gamma}_{kk}$ $(k = 1, \ldots, K)$ obeys the relations

$$\frac{\widehat{\Gamma}_{kk}}{\sqrt{t}} \to 0, \quad \sqrt{t}\widehat{\Gamma}_{kk} \to \infty.$$

5 Application to Bottlenecks Recognition

Theorems 1 and 2 give us an opportunity to decide which of the nodes are bottlenecks and which are not, based on observing the vector of queue lengths.

In the paper Blanchet, Murthy (2018) [3] an algorithm for modelling of a multidimensional RBM with zero drift and identity covariance matrix is proposed. This algorithm in fact is valid for modelling RBM with arbitrary covariance matrix, allowing to approximate the marginal distribution of RBM at a given point by empirical distributions.

Now consider a statistics $\frac{Q(t)}{\sqrt{t}}$. If all the nodes are balanced bottlenecks (what we will consider as a null hypothesis), then Theorem 1 implies that it is asymptotically distributed like $\frac{Z(t)}{\sqrt{t}}$, which has the same distribution as $Z(1)$ (by Property 3 on p. 165 in Chen, Yao [5]). Consequently, it is reasonable to expect the values of this statistics to lie (coordinate-wisely) within 2.5% and 97.5% quantiles of the said distribution, which depends on the estimated covariance matrix Γ.

Our Theorems 1–2 and Remark 2 imply that the violation of the lower quantile bound in kth coordinate supports the evidence that $\rho_k < 1$. Indeed, the strict bottleneck case means a positive drift in the kth coordinate of the approximating reflected Brownian motion, which would over-weigh the possible rate at which $\widehat{\Gamma}_{kk}$ tends to infinity.

Similarly, a strong intersection of the upper quantile boundary substantiates the existence of positive drift. Thus our proposed algorithm of bottlenecks detection is to construct the statistics $\frac{Q(t)}{\sqrt{t}}$ and $\widehat{\Gamma}$ and mark the buffers having substantial lower (respectively upper) violations in the coordinate-wise comparison procedure as non-bottlenecks (respectively strict bottlenecks).

Acknowledgments. The reported study was funded by RFBR according to the research project 20-01-00487 at the Department of Mechanics and Mathematics, Lomonosov Moscow State University.

References

1. Afanasyeva, L.G., Bashtova, E.E.: Coupling method for asymptotic analysis of queues with regenerative input and unreliable server. Queue. Syst. **76**(2), 125–147 (2013). https://doi.org/10.1007/s11134-013-9370-x
2. Bashtova, E., Lenena, E.: Jackson network in a random environment: strong approximation. Far Eastern Math. J. **20**(2), 144–149 (2020)
3. Blanchet, J., Murthy, K.: Exact simulation of multidimensional reflected Brownian motion. J. Appl. Probab. **55**(1), 137–156 (2018)
4. Bulinski, A., Vronski, M.: Statistical variant of the central limit theorem for associated random fields. Fundam. Prikl. Mat. **2**, 999–1018 (1996)
5. Chen, H., Yao, D.D.: Fundamentals of Queueing Networks. Springer, New York (2001). https://doi.org/10.1007/978-1-4757-5301-1
6. Damerdji, H.: Strong consistency of the variance estimator in steady-state simulation output analysis. Math. Oper. Res. **19**(2), 494–512 (1994)
7. Djellab, N.V.: On the $M|G|1$ retrial queue subjected to breakdowns. RAIRO - Oper. Res. **36**, 299–310 (2002)
8. Gaver, D.P.: A waiting line with interrupted service, including priorities. J. R. Stat. Soc. (B) **24**(1), 73–90 (1962)
9. Harrison, J.M.: The heavy traffic approximation for single server queues in series. J. Appl. Probab. **10**(3), 613–629 (1973)
10. Harrison, J.M., Reiman, M.I.: Reflected Brownian motion on an orthant. Ann. Probab. **9**(2), 302–308 (1981)
11. Hu, T.-C., Móricz, F., Taylor, R.L.: Strong laws of large numbers for arrays of rowwise independent random variables. Acta Math. Hung. **54**, 153–162 (1989)
12. Jackson, J.R.: Networks of waiting lines. Oper. Res. **5**(4), 518–521 (1957)
13. Kalimulina, E.: Analysis of unreliable Jackson-type queueing networks with dynamic routing. SSRN: https://ssrn.com/abstract=2881956
14. Lemoine, A.J.: State of the art - networks of queues: a survey of weak convergence results. Manag. Sci. **24**(11), 1175–1193 (1978)
15. Merlevede, R., Rio, E.: Strong approximation for additive functionals of geometrically ergodic Markov chains, hal-01044508 (2014)

16. Peligrad, M., Shao, Q.: Self-normalized central limit theorem for sums of weakly dependent random variables. J. of Theor. Probab. **7**(2), 309–338 (1994)
17. Reiman, M.I.: Open queueing networks in heavy traffic. Math. Oper. Res. **9**(3), 441–458 (1984)
18. Sherman, N., Kharoufen, J., Abramson, M.: An $M|G|1$ retrial queue with unreliable server for streaming multimedia applications. Prob. Eng. Inf. Sci. **23**, 281–304 (2009)
19. Smith, W.L.: Regenerative stochastic processes. Proc. Royal Soc. London Ser. A **232**(1188), 6–31 (1955)
20. Strassen, V.: An invariance principle for the law of the iterated logarithm. Z. Wahrsch. Verw. Geb. **3**, 211–226 (1964)
21. Whitt, W.: Heavy traffic limit theorems for queues: a survey. In: Clarke, A.B. (ed.) Mathematical Methods in Queueing Theory. Lecture Notes in Economics and Mathematical Systems (Operations Research), vol. 98. Springer, Berlin, Heidelberg (1974). https://doi.org/10.1007/978-3-642-80838-8_15
22. Zaitsev, A.: The accuracy of strong Gaussian approximation for sums of independent random vectors. Russ. Math. Surv. **68**(4), 721–761 (2013)

Applications of Vacation Queues with Close-Down Time to Maintenance of Residential Buildings

E. A. Korol and G. A. Afanasyev(✉)

Moscow State University of Civil Engineering, 129337 26, Yaroslavskoye Shosse, Moscow, Russia
kafedraGKK@mgsu.ru

Abstract. The paper is devoted to applications of the probability theory to the maintenance of the engineering systems of residential buildings. FIrstly, we consider a single-server queueing system with vacations and close-down times that operates in the following manner. When the server returns from a vacation it observes the following rule. If there is at least one customer in the system, the server commences service. If the server finds the system empty a close-down period begins. If no customers have arrived during this period the server commences a vacation. Otherwise the server begins service of the first arrived customer at the instant of this arrival. The input flow is supposed to be a Poisson one outside of the vacation period and the flow of arrivals during vacation period has a single jump at the end of this period. Under general assumptions with respect to distributions of the service time, vacation and close-down periods the distribution and the mean of the number of customers at the system in the stationary regime are obtained. The proposed system is considered as the mathematical model for estimation of the number of emergency calls $q(t)$ which the service team has to respond to at time t. Another function for servicing housing is the scheduling of preventive inspections and repairs. Based on the renewal theory we estimate the number of completed preventive inspections and repairs during a time interval $(0, t)$ for large t.

Keywords: Queueing systems · Vacations · Close-down times · Residential buildings

1 Introduction

A vacation queueing system is one in which a server may become unavailable for a random period of the time from a primary service center. The time away from the primary service center is called a vacation. There are various types of behavior of the server on the vacation period. In classical models the server completely stops service or is switched off when he is on a vacation. Many new vacation queueing systems have been proposed in literature. For example, Servi

A. N. Shiryaev et al. (Eds.): ICSM-5 2020, PROMS 371, pp. 329–340, 2021.
https://doi.org/10.1007/978-3-030-83266-7_25

and Finn [1] introduced the working vacation scheme, in which the server works at a different rate rather than completely stopping service during a vacation. A vacation can be the result of many factors. In particular, it can be a deliberate action taken to utilize the server in a secondary service center when there are no customers present at the primary service center. Namely, this situation occurs in the model for technical operation of residential buildings. We assume that the service team can begin a scheduled preventive inspection and repair only once all breakdowns are reduced. If this time period starts the service of new arrived sudden calls are shutdowned. Therefore this period is called vacation. Queueing systems with server vacations have attracted the attention of many researchers since the idea was first discussed in the paper of Levy and Yechiali [2]. Several excellent surveys on these vacation models have been done by Doshi [3,4] and the books by Tacagi [5] and Zhang [6] are devoted to the subject.

Many new vacation queueing systems have been proposed in literature. Among them queueing systems with close-down times [7] or timeout [8]. The proposed paper is devoted to vacation systems of this kind. The main novelty of the paper is the assumption that the input flow during a vacation has a unique jump at the end of the vacation period. There are no other essential conditions with respect to this flow. We obtain the stationary distribution and the mean of the number of customers at the system. It also makes it possible to find the lower and upper bounds for the mean of the number of customers at the system under enough general assumptions.

2 Model Description and Main Results

We consider a vacation single-server queueing system $M|G|\infty$ with close-down times [7]. This system operates in the following manner. Customers are served by the single server until the system becomes empty. If the server finds no customers in the system at a customer departure, it enters the close-down phase ζ. If a customer arrives at the system before the close-down phase ζ expires the server immediately goes back to the busy phase. If no customers arrive during the close-down period, the server goes to the vacation period η. When the server returns from a vacation, it observes the following rule. If there is at least one customer in the system the server commences service. If the server finds the system empty a close-down phase begins and the situation described above takes place. The close-down period, the vacation and the service times are assumed to be generally distributed with distribution functions $F(x)$, $G(x)$ and $B(x)$ with LSTs (Laplace-Stiltjes transforms) $f(s)$, $g(s)$, $\beta(s)$, respectively, and corresponding means by $\overline{\zeta}$, $\overline{\eta}$ and b. The input flow is supposed to be a Poisson one with rate λ outside the vacation period.

Let η_n be the nth vacation period, T_n the moment of the nth vacation start and $\tau_n = T_{n+1} - T_n$ $(n = 1, 2, \ldots)$. Define the random process $Y_n(t)$ as the number of customers which are present in the system at time $T_n + t$. The sequence $\{Y_n(t), t \leq \eta_n\}_{n=1}^{\infty}$ consists of identically distributed independent random elements. We do not assume that $Y_n(t)$ is a Poisson process with rate λ. It allows us to study many new vacation queueing models mentioned above.

Define the function

$$V(z,t) = Ez^{Y(t)}\mathbb{I}(\eta > t) = \sum_{j=0}^{\infty} z^j P(Y(t) = j, \eta > t),$$

$$G(z,s) = EZ^{Y(\eta)}e^{s\eta},$$

$$G(t) = P(\eta \le t)$$

$$C(z) = Ez^{Y(\eta)} = G(z,0) = \sum_{j=0}^{\infty} c_j z^j,$$

$$(|z| \le 1, Res \ge 0).$$

Here and later we omit the index n when it does not involve difficulties in the understanding.

3 Stability Theorem

We study the process $q(t)$ that is the number of customers in the system at the instant t assuming that sample paths are right continuous functions.

Condition 1. $\overline{Y}_1 = EY(\eta) < \infty, \quad \overline{\eta} = E\eta < \infty.$

Theorem 1. *Let Condition 1 be fulfilled and* $\rho = \lambda b$.
If $\rho < 1$ *then* $q(t)$ *is a stable process, i.e. for any initial state there exists*

$$\lim_{t\to\infty} Ez^{q(t)} = \pi(z)$$

and $\pi(1) = 1$, $(|z| \le 1)$.
If $\rho \ge 1$ *then*

$$q(t) \xrightarrow[t\to\infty]{P} \infty.$$

Proof. The second statement is the simple corollary results for the classical model $M|G|1|\infty$ without vacations, i.e. $\eta = 0$. It is enough to remark that for the number of customers $q_0(t)$ being in this system at time t we have the stochastic inequality

$$q_0(t) \le q(t) \quad (t \ge 0).$$

Of course we assume identical initial conditions for the both systems. Since $q(t) \xrightarrow[t\to\infty]{P} \infty$ if $\rho \ge 1$ (see e.g. [9]), the process $q(t)$ is unstable one in this case.

Consider the case $\rho < 1$. Let as note that $q(t)$ is a regenerative process and as points of regeneration we take the sequence $\{\theta_n\}_{n=1}^{\infty}$ such that

$$\theta_n = \inf\{t > \theta_{n-1} \;:\; q(t-0) > 0,\; q(t) = 0\},\; \theta_0 = 0. \tag{1}$$

According to Smith's Theorem [10] $q(t)$ is a stable process if the mean of the regeneration period $\hat{\varkappa}_n = \theta_{n+1} - \theta_n$ is finite, i.e. $E\varkappa_n < \infty$. For the system $M|G|1|\infty$ let β_{k0} be the busy period which starts when there are $k-1$ customers in the queue ($k = 1, 2, ...$). It is well known (see e.g. [11]) that $E\beta_{10} = b(1-\rho)^{-1}$ and therefore $E\beta_{k0} = kb(1-\rho)^{-1}$.

Let ζ_n be the first close-down period, a_n - the time of the next customer arrival after the moment θ_n. Then in distribution the following equality takes place

$$\varkappa_n = (a_n + \beta_{10})\mathbb{I}(\zeta_n \geq a_n) + \sum_{k=0}^{n}(\zeta_n + \eta_n + \beta_{k0})\mathbb{I}(Y_n(\eta_n) = k, \zeta_n < a_n) + \\ + \tilde{\varkappa}_n \mathbb{I}(Y_n(\eta_n) = 0, \zeta_n < a_n). \quad (2)$$

Here $\mathbb{I}(A)$ is an indicator function and in distribution $\varkappa_n = \tilde{\varkappa}_n$.

In view of independence random variables a_n, ζ_n, β_{k0} $(k = 1, 2, ...)$, $\tilde{\varkappa}_n$ and $(Y_n(\eta_n), \eta_n)$ we obtain from (2)

$$E\varkappa = E\min(a, \zeta) + E\eta P(\zeta < a) + E\beta_{10} P(\zeta \geq a) + \\ + E\beta_{10} EY(\eta) P(\zeta < a) + E\tilde{\varkappa} P(Y(\eta) = 0) P(\zeta < a). \quad (3)$$

Since

$$P(\zeta < a) = f(\lambda) = \int_0^\infty e^{-\lambda x} dF(x)$$

and

$$E\min(a, \zeta) = \lambda^{-1}(1 - f(\lambda)),$$

we have from (3)

$$E\varkappa = \frac{1 - f(\lambda) + \lambda f(\lambda)(\overline{\eta}(1-\rho) + b\overline{Y}_1)}{\lambda(1-\rho)(1 - c_0 f(\lambda))}, \quad (4)$$

where $\overline{\eta} = E\eta$, $\overline{Y}_1 = EY(\eta)$, $c_0 = P(Y(\eta) = 0)$.

Hence if $\rho < 1$ then $E\varkappa < \infty$ and in accordance with Smith's theorem $q(t)$ is a stable process. □

Under some additional assumptions we may obtain the stationary distribution for $q(t)$. Here we consider the model S assuming that the following condition is fulfilled.

Condition 2. *The sample paths of Y have a unique jump at the point η.*

This condition means that all customers arriving on a vacation period come at the end of this period. Now we give the main result.

4 Limit Theorem and Corollaries

Theorem 2. *Let Conditions 1 and 2 be fulfilled and $\rho < 1$. Then*

$$\lim_{t\to\infty} Ez^{q(t)} = \pi(z) = \pi_0 + \\ + (1-\pi_0)\frac{(1-\rho)(1-c_0 f(\lambda))}{1 - f(\lambda)(1-\overline{Y}_1)} \frac{z(1-P_1(z))}{\beta(\lambda - \lambda z) - z} \frac{1 - \beta(\lambda - \lambda z)}{\rho(1-z)}, \quad (5)$$

where

$$\pi_0 = \frac{(1-\rho)(1-f(\lambda)+\lambda f(\lambda)\overline{\eta})}{1-f(\lambda)(1-\lambda(1-\rho)\overline{\eta}-\rho\overline{Y}_1)} \tag{6}$$

and

$$P_1(z) = \frac{(1-f(\lambda))z + f(\lambda)(C(z)-c_0)}{1-f(\lambda)c_0}, \tag{7}$$

$$c_0 = P(Y(\eta)=0).$$

Proof. First we prove (6). Without loss of generality we assume that $\theta_0 = 0$ is a point of regeneration for the process $q(t)$. In view of the condition 2 the regeneration period consists of the interval v_n in which there are no customers in the system (free period) and the busy period u_n.

According to Smith's theorem [10] there exists

$$\lim_{t\to\infty} P(q(t)=0) = \pi_0 = \frac{Ev}{E\varkappa}, \tag{8}$$

Since $E\varkappa$ is defined by formula (4) and $Ev = E\varkappa - Eu$ we need to find Eu.

Let ξ_n be the number of customers in the system at the moment when the busy period starts on the n-th regeneration period, i.e. at time $\theta_n + v_n$.

Analogously formula (2) with the proceeding notation we have the following equality

$$\begin{aligned} P(\xi_n = 1) &= P(a_n \le \zeta_n) + P(a_n > \zeta_n, Y_n(\eta_n) = 1) + \\ &+ P(a_n > \zeta_n, Y_n(\eta_n) = 0, \tilde{\xi}_n = 1) = P_1, \end{aligned} \tag{9}$$

where $\tilde{\xi}_n = \xi_n$ in distribution and does not depend on $a_n, \zeta_n, Y_n(\eta_n), \eta_n$.

Analogously for $j > 1$

$$\begin{aligned} P(\xi_n = j) &= P(a_n > \zeta_n, Y_n(\eta_n) = j) \\ &+ P(a_n > \zeta_n, Y_n(\eta_n) = 0, \tilde{\xi}_n = j) = P_j. \end{aligned} \tag{10}$$

Since

$$P(a > \zeta, Y(\eta) = 0, \tilde{\xi} = j) = c_0 f(\lambda) P_j \quad (j = 1, 2, \ldots)$$

we have from (9) and (10)

$$P_1(z) = Ez^{\xi} = \sum_{j=1}^{\infty} z^j P_j = \frac{f(\lambda)(C(z)-z) + zc_0 f(\lambda)}{1-c_0 f(\lambda)}. \tag{11}$$

Therefore

$$E\xi = P_1'(1) = \frac{1-f(\lambda)+f(\lambda)\overline{Y}_1}{1-c_0 f(\lambda)}. \tag{12}$$

Since the mean of the busy period in the system $M|G|1|\infty$ with unique customer at the beginning is equal to $b(1-\rho)^{-1}$ [11], from (12) we get $Eu = E\xi b(1-\rho)^{-1}$. Now (6) follows from (4) and (8).

To prove (5) we introduce two auxiliary systems $\hat{S}$ and $\tilde{S}$ by identification points $\theta_n + v_n$ and θ_{n+1} for $\hat{S}$ and points θ_n and $\theta_n + v_n$ for $\tilde{S}$.

Then $\hat{S}$ and $\tilde{S}$ describe the behavior of S on the close-down and vacation periods (if there is) and on the busy period respectively. In view of Condition 2 the number of customers $\hat{q}(t) = 0$ in the system $\hat{S}$ for any t. The system $\tilde{S}$ has a Poisson input flow with rate λ and in addition at the end of the n-th busy period ξ_n customers arrive at the system ($\xi > 0$). Let $\tilde{q}(t)$ be the number of customers in the system $\tilde{S}$ and $\tilde{P}(z) = \lim_{t\to\infty} Ez^{\tilde{q}(t)}$.

Putting $\hat{p} = \frac{Ev}{E\varkappa}$ and $\tilde{p} = 1 - \hat{p} = \frac{Eu}{E\varkappa}$ we obtain from renewal Theorem

$$\pi(z) = \hat{p} + (1 - \hat{p})\tilde{P}(z) = \pi_0 + (1 - \pi_0)\tilde{P}(z). \tag{13}$$

To obtain $\tilde{P}(z)$ we introduce the embedded process $\tilde{q}_n$, putting $\tilde{q}_n = \tilde{q}(t_n)$. Here $\{t_n\}_{n=1}^{\infty}$ are sequential moments of service completion times in the system $\tilde{S}$. Since $\tilde{q}(t) > 0$ for any t the sequence $\{t_n\}_{n=1}^{\infty}$ is a renewal process and $P(t_{k+1} - t_k \le x) = B(x)$. Therefore $\{\tilde{q}_n\}_{n=1}^{\infty}$ is a Markov chain with state space $\{1, 2, ...\}$.

Lemma 1. *Under Conditions 1 and 2 and $\rho < 1$ there exists the limit*

$$\lim_{n\to\infty} Ez^{\tilde{q}_n} = \tilde{P}^*(z)$$

and

$$\tilde{P}^*(z) = \frac{(1-\rho)(1 - c_0 f(\lambda))}{1 - f(\lambda)(1 - Y_1)} \cdot \frac{z(1 - P_1(z))}{\mathcal{K}(z) - z}, \tag{14}$$

were

$$\mathcal{K}(z) = \beta(\lambda - \lambda z) = \sum_{j=1}^{\infty} k_j z^j$$

and $P_1(z)$ is defined by (11).

Proof. Transient probabilities $\{\tilde{P}_{ij}\}$ for Markov chain $\tilde{q}_n$ can be written as follows

$$\tilde{P}_{1j} = k_j + k_0 P_j \quad (j = 1, 2, ...)$$

and for $i > 1$

$$\tilde{P}_{ij} = \begin{cases} k_{j-i+1} & if\ j \ge i - 1, \\ 0 & if\ j < i - 1. \end{cases}$$

Therefore stationary probabilities

$$\tilde{P}_j^* = \lim_{n\to\infty} P(\tilde{q}_n = j) \quad (j = 1, 2, ...)$$

satisfy the system of equations

$$\tilde{P}_j^* = k_0 P_j \tilde{P}_1^* + k_j \tilde{P}_1^* + k_{j-1}\tilde{P}_2^* + ... + k_0 \tilde{P}_{j+1}^* \quad for\ j = 1, 2, ...$$

For the generating function $\tilde{P}(z) = \sum_{j=1}^{\infty} z^j \tilde{P}_j^*$ these equations give

$$\tilde{P}^*(z) = k_0 \tilde{P}_1^* \frac{z(1 - P_1(z))}{\mathcal{K}(z) - z}. \tag{15}$$

The normalization condition $\tilde{P}^*(1) = 1$ gives

$$k_0 \tilde{P}_1^* = \frac{1-\rho}{P_1'(1)}.$$

Therefore (14) follows from this equality, (11) and (15). □

Now we have to express $\tilde{P}(z)$ by means of $\tilde{P}^*(z)$. We use results from renewal theory. Let

$$n(t) = \min\{k \; : \; t_k < t\}$$

and $\gamma_t = t - n(t)$. According to the renewal theorem [10] there exists

$$\lim_{t\to\infty} P(\gamma_t \le x) = \frac{1}{b}\int_0^x (1 - B(y))dy. \tag{16}$$

If $\xi(t)$ the number of customers arrived at the system $\tilde{S}$ during interval $(t_{n(t)}, t_{n(t)} + \gamma_t)$, then

$$P(\xi(t) = j) = \int_0^\infty \frac{(\lambda y)^j}{j!} e^{-\lambda y} dP(\gamma_t \le y).$$

In view of (16) there exists

$$\lim_{t\to\infty} P(\xi(t) = j) = \frac{1}{b}\int_0^\infty \frac{(\lambda y)^j}{j!} e^{-\lambda y}[1 - B(y)]dy = \delta_j.$$

and the generating function

$$\delta(z) = \sum_{j=1}^{\infty} z^j \delta_j =$$

$$= \frac{1}{b}\int_0^\infty e^{-\lambda(1-z)y}[1 - B(y)]dy = \frac{1 - \mathcal{K}(z)}{\rho(1-z)}. \tag{17}$$

Since $\tilde{P}(z) = \tilde{P}^*(z)\delta(z)$ from (15) and (17) we have

$$\tilde{P}(z) = \frac{(1-\rho)(1 - c_0 f(\lambda))}{1 - f(\lambda)(1 - Y_1)} \cdot \frac{z(1 - P(z))}{\beta(\lambda - \lambda z) - z} \cdot \frac{1 - \beta(\lambda - \lambda z)}{\rho(1-z)}. \tag{18}$$

Substituting this formula in (13) with regard equalities $\hat{P} = \pi_0$, $\hat{P}(z) = 1$ proves the theorem. □

Differentiating $\pi(z)$ with respect to z and assuming $z = 1$, from (5) we obtain the formula for the mathematical expectation $\overline{q}$ of the number of customers in the system in the stationary regime.

Corollary 1. *Let Conditions 1 and 2 be fulfilled,* $\rho < 1$ *and*

$$\overline{Y}_2 = EY^2(\eta) < \infty, \qquad b_2 < \infty. \tag{19}$$

Then

$$\overline{q} = \pi'(1) = (1 - \pi_0)\left(1 + \frac{f(\lambda)\overline{Y}_2}{1 - f(\lambda)(1 - \overline{Y}_1)} + \frac{\lambda^2 b_2}{2\rho(1-\rho)}\right). \tag{20}$$

If $f(\lambda) = 0$ we have a queueing system $M|G|1|\infty$ without vacations. From (20) we obtain a well-known result [11] .

$$q = \rho + \frac{\lambda^2 b_2}{2(1-\rho)}.$$

When Condition 2 is not fulfilled (20) gives the lower bound for the average number q of customers in the system. To obtain the upper bound we take the random variables $\tau_n = T_{n+1} - T_n$ $(n = 1, 2, \ldots)$, where $\{T_n\}_{n=1}^{\infty}$ are defined in Sect. 3.

Lemma 2. *Let Condition 1 be fulfilled and* $\rho < 1$ *then*

$$\tau = E\tau_n = \frac{1 - f(\lambda) + \lambda f(\lambda)(\overline{\eta}(1-\rho) + b\overline{Y}_1)}{\lambda f(\lambda)(1-\rho)}. \tag{21}$$

The proof of Lemma 2 is given in Appendix.

Condition 3. *Process* $\{Y_n(t), t \geq 0\}$ *and the duration of the nth vacation* η_n *are independent ones* $(n = 1, 2, \ldots)$.

Corollary 2. *Let conditions 1 and 3 be fulfilled. Then the average number* q *of customers in the system satisfies the inequalities*

$$\overline{q} \leq q \leq \overline{q} + \frac{1}{\tau}\int_0^{\infty} tEY(t)dG(t), \tag{22}$$

where $\overline{q}$ *and* τ *are defined by (20) and (21) respectively.*

Proof. Consider an auxiliary system S^* which is the same as S with a unique distinction: all customers arriving during vacation period come at the beginning of this period. Let q^* be the number of customers in the system S^* in the stationary regime. In just the same way as for the system S we obtain the following equality

$$Eq^* = \overline{q} + \frac{1}{\tau}\int_0^{\infty} tEY(t)dG(t).$$

Since $Eq \leq Eq^*$ the corollary is proved. □

As an example consider the case when $Y(t)$ is a Poisson process with rate λ. Then from formulas (20), (21) and (22) we have

$$\overline{q} = \rho + \frac{\lambda^2 b_2}{2(1-\rho)} + \frac{\rho f(\lambda)(\lambda^2 E\eta^2 + \lambda\overline{\eta})}{1 - f(\lambda) + f(\lambda)\lambda\overline{\eta}}$$

and

$$\overline{q} \leq q \leq \rho + \frac{\lambda^2 b_2}{2(1-\rho)} + \frac{\lambda f(\lambda)(\lambda E\eta^2 + \rho\overline{\eta})}{1 - f(\lambda) + f(\lambda)\lambda\overline{\eta}}.$$

5 Application to Maintenance of Residential Buildings

At this point we will show that the mathematical models we constructed can be used for organisation of the management company (MC) of housing and community services.

We assume that the whole area for which MC is responsible is split into some districts. In each one there is a team of technicians responsible for servicing the objects in the district. Each team has two main objectives.

1. Regular preventive services of the objects in the district.
2. Repairs of the faulty objects.

The latter may occur as a result of random breakdown of the equipment as a result of many random factors including old age of the object.

Let us note that in recent years there was a sharp increase in the interest to applications of the theory of probability, in particular, queueing theory, to analysis of the activities of the managing companies of residential properties.

These applications usually start from collection of statistical data with the objective of constructing statistical estimations of the parameters and functions defined in the mathematical model in use.This is a quite complex problem. It is based on various methods of mathematical statistics.

In order to describe the process $q(t)$ representing the number of breakdowns requiring urgent repairs at time t we use queueing systems with vacations, described in paragraph 2. The following assumptions are made.

- The flow of requests for emerging repairs is a Poisson process with rate λ.
- The times required for the emergency repairs are independent random variables with distribution function $B(x)$ and mean b.
- The team of the technicians can work on preventive services only if there are no calls for urgent services.

When the team is free of urgent calls the close down period starts and lasts time α. If there are no urgent calls during this period when this period ends the preventive service (vacation) starts. The length of the vacation η is a random variable with distribution function $G(x)$. If during the close down period the urgent call occurs then the period is interrupted and the repair of the call starts. During the vacation period η the urgent calls come in accordance to Poisson flow with rate λ.

Suppose that MC wants to organise the activity of the technicians' team in a such way that average number of urgent calls q would not exceed some level $\delta > 0$, and expected value of the number $n(T)$ of finished preventive services during time T would be not less than $N(T)$, i.e.

$$q \leq \delta, \qquad \overline{n}(T) \geq N(T). \tag{23}$$

To solve this problem, we need to express q and $\overline{n}(T)$ in terms of previously introduced parameters λ, α and functions G and B. For the average number of calls q the lower and upper bounds are given by (22). We have to take

$$\overline{\eta} = \int_0^\infty x dG(x), \qquad \overline{Y}_1 = \lambda \overline{\eta}, \qquad f(\lambda) = e^{-\lambda \alpha},$$

$$\overline{Y}_2 = \lambda^2 \int_0^\infty x^2 dG(x) + \lambda \overline{\eta},$$

$$b = \int_0^\infty x dB(x), \qquad b_2 = \int_0^\infty x^2 dB(x).$$

Then we calculate τ, π_0 and $\overline{q}$ by means of formulas (21), (6), (20) respectively. If

$$\overline{q} + \frac{\lambda^2(1-\rho)E\eta^2}{e^{\alpha\lambda} - 1 + \lambda\overline{\eta}} < \delta \tag{24}$$

the first condition in (23) may be considered realized.

A formula for the average number of completed scheduled repairs (or vacations) $\overline{n}(T)$ for sufficiently large T follows from the elementary renewal theorem [10]

$$\overline{n}(t) = \frac{t}{\tau}(1 + o(1)), \quad t \to \infty,$$

where τ is defined by (21). For the second inequality in (23) we set $N(T) = \gamma T$. Then this inequality is satisfied if

$$\tau \leq \gamma^{-1}. \tag{25}$$

If the system parameters λ, α, b, b_2, $E\eta^2$ satisfy the inequalities (24) and (25), then we can assume that the team satisfactorily copes with the tasks. If one of the conditions (24), (25) is not satisfied, managing actions should be taken.

6 Conclusion

In the present paper a vacation queueing system with close-down times is considered. Under enough general assumptions limit distribution for the number of customers in the system was obtained. The proposed model is an essential generalization of vacation systems described in literature. Nevertheless, one may think that our assumption to the process Y may be restrictive in applications. Therefore, we give the upper and lower bounds for the mean of the number of customers in the system and employ our results for organisation of the management company of housing and community services.

Acknowledgments. This work is supported by Russian Foundation for Basic research project 20-01-00-487.

7 Appendix

Proof. (of Lemma 2)

Denote by $\{\theta_n^{(1)}\}_{n=1}^{\infty}$ the sequence of close-down period starts. Let ζ_n be the duration of the n-th close-down period and $\tilde{a}_n$ the time of the first customer arrival after $\theta_n^{(1)}$. Assuming that $\theta_1^{(1)} = 0$ define the sequence

$$\nu(n) = \min\{k > \nu(n-1) : \zeta_k < \tilde{a}_k\}(\nu(1) = 1).$$

Then $T_n = \theta_{\nu(n)}^{(1)} - \zeta_{\nu(n)}$ and therefore

$$\tau_n = T_{n+1} - T_n = \theta_{\nu(n+1)}^{(1)} - \zeta_{\nu(n+1)} - \theta_{\nu(n)}^{(1)} + \zeta_{\nu(n)}. \tag{26}$$

From (26) we obtain the equality in distribution

$$\tau_n = \zeta_{\nu(n)} + \eta_n + \tau_0(Y_{\nu(n)}(\eta)) + \sum_{j=1}^{k(n)} \delta_j. \tag{27}$$

There $k(n) = \nu(n+1) - \nu(n) - 1$, $\tau_0(k)$ is the busy period in the system $M|G|1|\infty$ which starts when there are $k-1$ customers in a queue. Random variable δ_j in distribution has a form $(j = 1, 2, ..., k_n)$

$$\delta = a\mathbb{I}(a < \zeta) + \tau_0(1).$$

Since

$$\begin{aligned} E(a \mid a \leq \zeta) &= \frac{1}{\lambda} + \frac{f'(\lambda)}{1 - f(\lambda)}, \\ E(\zeta \mid \zeta < a) &= -\frac{f'(\lambda)}{f(\lambda)}, \\ E\tau_0(Y_{\nu(n)}(\eta)) &= b(1-\rho)^{-1}Y_1 \end{aligned} \tag{28}$$

and

$$Ek(n) = \frac{1 - f(\lambda)}{f(\lambda)},$$

taking mathematical expectation from (28) we obtain (21) with the help of (28). □

References

1. Servi, L.D., Finn, S.G.: m|M|1 queue with working vacations (M|M|1|WV), Perform. Eval. **50**, 41–52 (2002)
2. Levy, Y., Yechiali, U.: Utilization of idle time in an $M|G|1$ queueing system. Manage. Sci. **22**, 202–211 (1975)
3. Doshi, B.T.: Queueing systems with vacations, a survey. Queue. Syst. **1**, 29–66 (1986)

4. Doshi, B.T.: Single-server queues with vacations. In: Takagi, H. (ed.) Stochastic Analysis of Computer and communications Systems, pp. 217–265. Elsevier, Amsterdam (1990)
5. Takagi, H.: Queueing Analysis: A Foundation of Performance Analysis. Volume 1: Vacation and Priority Systems. Part 1, Elsevier Science Publishers B.V., Amsterdam (1990)
6. Tain, N., Zhang, G.: Vacation Queueing Models: Theory and Applications. Springer-Verlag, New York (2006). https://doi.org/10.1007/978-0-387-33723-4
7. Zhisheng, N., Tao, S., Takahashi, Y.: A Vacation quene with setup and close-down times and batch Markovian arrival Process. Perform. Eval. **54**, 225–248 (2003)
8. Ibe, O.C.: $M|G|1$ Vacation queueing systems with server timeout. Am. J. Oper. Res. **5**, 77–88 (2015)
9. Borovkov, A.A.: Stochastic Processes in Queueing Theory. Springer, Berlin (1976). https://doi.org/10.1007/978-1-4612-9866-3
10. Smith, W.L.: Renewal theory and its ramifications. J. Roj. Statist. Soc., Ser. B. **29**(2) 95–150 (1961)
11. Saaty, T.L.: Elements of Queueing Theory with Applications, vol. 520. Mc Graw Hill Book Company (1961)

Output Process of Retrial Queue with Two-Way Communication Under Low Rate of Retrials Limit Condition

Ivan L. Lapatin(✉) and Anatoly A. Nazarov

Tomsk State University, 36 Lenina ave., Tomsk, Russia

Abstract. In this paper, we consider retrial queue with MAP input and two-way communication. Upon arriving, an incoming call makes the server busy for an exponentially distributed time if it's idle at the moment. Otherwise, the incoming call goes to the orbit and repeat its request for service after random delay. In its idle time the server also makes an outgoing calls. We use asymptotic analysis method under low rate of retrials limit condition to derive characteristic function of the number of calls in the output flow of the system

Keywords: Output process · Retrial queue · Markovian arrival process · Asymptotic analysis method

1 Introduction

Mathematical modeling is effectively used in various spheres of modern human activity. The queuing theory [10] considers models of claim service nodes. The configuration of these models is very diverse, which allows to choose the one necessary for a specific applied problem. In this regard, the study of various characteristics of the proposed models makes it possible to effectively simulate the operation of various service nodes. In telecommunication systems, automated call-centers, computer networks, etc., one of the most important characteristics of great practical interest is the number of requests served.

The main results on the analytical study of the outflows of classical models were made in the second half of the 20th century such scientists as Burke [5], Reich [16], Mirasol [13]. The study of output proccess was continued in the future [4,6,9].

The paper proposes to consider the output proccess of the system with repeated calls [1] and two-way communication. Retrial phenomenon arises from various communication systems with random access [2,8]. Such a system can be interpreted as a node of a communication network with random multiple access, which in its free time from processing requests can request a self-test or another procedure that will continue for a random time. Retrial queues with two-way communication have been extensively studied recently [3,15,17].

A. N. Shiryaev et al. (Eds.): ICSM-5 2020, PROMS 371, pp. 341–352, 2021.
https://doi.org/10.1007/978-3-030-83266-7_26

Individual nodes form a communication network model in which the outgoing flow of one node is incoming for another, therefore the results of the study of outgoing flows of queuing systems are widely applicable for designing real data transmission systems and analyzing complex processes consisting of several stages.

We used the method of asymptotic analysis to find the approximation of the distribution of the number of serviced calls of the incoming flow for some time t under low rate of retrial condition.

The results of this article for the two-way communication model generalize the results of the article cite lapatin 2019 asymptotic, in which the device served only arrival calls. In the two-way communication model when the server is free it makes outgoing calls and serves such calls. This changes the distribution of the output process.

The remainder of the paper has follows structure. In Sect. 2, we describe the model in detail. Section 3 contains the derivation of the Kolmogorov equations and the transition to characteristic functions. In Sect. 4, we present the main results, which are formulated in three theorems. Section 5 presents main results.

2 Mathematical Model

We consider a single server retrial queue with two types of calls: incoming calls and outgoing calls. Incoming calls form a Markov Arrival Process (MAP). MAP is determined by matrixs $\mathbf{Q}$, $\mathbf{\Lambda}$ and $\mathbf{D}$.

The matrix $\mathbf{Q}$ of q_{ij} elements is the infinitesimal generator of underlying process of MAP $k(t)$. $k(t)$ is a continuous time Markov chain with finite set of states $k = 1, 2, \ldots, K$ Diagonal matrix $\mathbf{\Lambda}$ contains conditional arrival rates λ_k. Matrix $\mathbf{D}$ contains probabilities d_{ij} of that an event will come at the moment of changing the state of the Markov chain $k(t)$ from i to j.

Upon arrival a incomming call occupies the server if it is free. Duration of the service time of incomming calls is an exponentially distributed random variable with rate μ_1. If the incomming call arrival finds the server busy, this call joins the orbit. After a random exponentially distributed time with rate σ repeats his request for service. When the server is free it makes outgoing calls with rate α and serves such calls for an exponentially distributed time with parameter μ_2.

Denote random processes: $i(t)$ - the number of calls in the orbit at the time t; $k(t)$ - the state of the underlying process of the MAP at time t; $n(t)$ - the state of server at the moment t:

$$n(t) = \begin{cases} 0, & \text{server is free;} \\ 1, & \text{server is busy serving an incoming call;} \\ 2, & \text{server is busy serving an outgoing call,} \end{cases}$$

the process $m(t)$ is the number of served incomming calls at the moment t.

The problem is to find the probability characteristics of the number of serviced incomming calls in the system by the time t.

3 Kolmogorov System of Equations and Characteristic Function

Four-dimensional random process is determined by the probability distribution

$$P\{n(t) = n, k(t) = k, i(t) = i, m(t) = m\} = P_n(k, i, m, t). \tag{1}$$

For the probability distribution (1) of the Markovian process $\{n(t), k(t), i(t), m(t)\}$ we can write Kolmogorov system of differential equations

$$\frac{\partial P_0(k,i,m,t)}{\partial t} = -(\lambda_k + i\sigma + \alpha)P_0(k,i,m,t) + \mu_1 P_1(k,i,m-1,t)$$

$$+\mu_2 P_2(k,i,m,t) + \sum_{\nu=1}^{K} P_0(\nu,i,m,t)q_{\nu k}(1-d_{\nu k}),$$

$$\frac{\partial P_1(k,i,m,t)}{\partial t} = -(\lambda_k+\mu_1)P_1(k,i,m,t)+(i+1)\sigma P_0(k,i+1,m,t)+\lambda_k P_0(k,i,m,t)$$

$$+\lambda_k P_1(k,i-1,m,t) + \sum_{\nu=1}^{K} P_1(\nu,i,m,t)q_{\nu k}(1-d_{\nu k}) + \sum_{\nu=1,\nu\neq k}^{K} P_0(\nu,i,m,t)q_{\nu k}d_{\nu k}$$

$$+ \sum_{\nu=1,\nu\neq k}^{K} P_1(\nu,i-1,m,t)q_{\nu k}d_{\nu k},$$

$$\frac{\partial P_2(k,i,m,t)}{\partial t} = -(\lambda_k + \mu_2)P_2(k,i,m,t) + \alpha P_0(k,i,m,t) + \lambda_k P_2(k,i-1,m,t)$$

$$+\sum_{\nu=1}^{K} P_2(\nu,i,m,t)q_{\nu k}(1-d_{\nu k}) + \sum_{\nu=1,\nu\neq k}^{K} P_2(\nu,i-1,m,t)q_{\nu k}d_{\nu k}. \tag{2}$$

Let $H_n(k,u_1,u,t)$ denotes the partial characteristic functions

$$H_n(k,u_1,u,t) = \sum_{i=0}^{\infty}\sum_{m=0}^{\infty} e^{ju_1 i}e^{jum}P_n(k,i,m,t), \tag{3}$$

where $j = \sqrt{-1}$. System (2) for functions (3) can be rewritten

$$\frac{\partial H_0(k,u_1,u,t)}{\partial t} = -(\lambda_k + \alpha)H_0(k,u_1,u,t) + j\sigma\frac{\partial H_0(k,u_1,u,t)}{\partial u_1}$$

$$+\mu_1 e^{ju}H_1(k,u_1,u,t) + \mu_2 H_2(k,u_1,u,t) + \sum_{\nu=1}^{K} H_0(k,u_1,u,t)q_{\nu k}$$

$$-\sum_{\nu=1}^{K} H_0(k,u_1,u,t)q_{\nu k}d_{\nu k},$$

$$\frac{\partial H_1(k,u_1,u,t)}{\partial t} = -(\lambda_k+\mu_1)H_1(k,u_1,u,t) - j\sigma e^{-ju_1}\frac{\partial H_0(k,u_1,u,t)}{\partial u_1}$$
$$+\lambda_k H_0(k,u_1,u,t) + \lambda_k e^{ju}H_1(k,u_1,u,t)$$
$$+e^{ju}\sum_{\nu=1,\nu\neq k}^{K} H_1(\nu,u_1,u,t)q_{\nu k}d_{\nu k} + \sum_{\nu=1,\nu\neq k}^{K} H_0(\nu,u_1,u,t)q_{\nu k}d_{\nu k}$$
$$+\sum_{\nu=1}^{K} H_1(\nu,u_1,u,t)q_{\nu k}(1-d_{\nu k}),$$
$$\frac{\partial H_2(k,u_1,u,t)}{\partial t} = -(\lambda_k+\mu_2)H_2(k,u_1,u,t) + \alpha H_0(k,u_1,u,t)$$
$$+\lambda_k e^{ju}H_2(k,u_1,u,t) + \sum_{\nu=1}^{K} H_2(\nu,u_1,u,t)q_{\nu k}$$
$$+ e^{ju}\sum_{\nu=1,\nu\neq k}^{K} H_2(\nu,u_1,u,t)q_{\nu k}d_{\nu k}. \tag{4}$$

Denoting

$$\mathbf{H}_n(u_1,u,t) = \{H_n(1,u_1,u,t), H_n(2,u_1,u,t), .., H_n(K,u_1,u,t)\},$$

we rewrite the system in following form

$$\frac{\partial \mathbf{H}_0(u_1,u,t)}{\partial t} = \mathbf{H}_0(u_1,u,t)(\mathbf{Q}-\mathbf{B}-\alpha\mathbf{I}) + \mu_1 e^{ju}\mathbf{H}_1(u_1,u,t)$$
$$+\mu_2\mathbf{H}_2(u_1,u,t) + j\sigma\frac{\partial \mathbf{H}_0(u_1,u,t)}{\partial u_1},$$
$$\frac{\partial \mathbf{H}_1(u_1,u,t)}{\partial t} = \mathbf{H}_0(u_1,u,t)\mathbf{B} + \mathbf{H}_1(u_1,u,t)(\mathbf{Q}+(e^{ju_1}-1)\mathbf{B}-\mu_1\mathbf{I})$$
$$-j\sigma e^{-ju_1}\frac{\partial \mathbf{H}_0(u_1,u,t)}{\partial u_1},$$
$$\frac{\partial \mathbf{H}_2(u_1,u,t)}{\partial t} = \alpha\mathbf{H}_0(u_1,u,t) + \mathbf{H}_2(u_1,u,t)(\mathbf{Q}+(e^{ju}-1)\mathbf{B}-\mu_2\mathbf{I}) \tag{5}$$

where $\mathbf{B} = \mathbf{\Lambda} + \mathbf{Q} * \mathbf{D}$, here $*$ means the product of Hadamard, $\mathbf{I}$ is unit matrix of K dimension.

System of equations (5) we will solved using asymptotic analysis method under low rate of retrials condition ($\sigma \to 0$).

4 Asymptotic Solution

Denoting $\sigma = \varepsilon$ we introduce the following notations in the system (5)

$$u_1 = \varepsilon w, \quad \mathbf{H}_n(u_1, u, t) = \mathbf{F}_n(w, u, t, \varepsilon),$$

and write the system (6)

$$\frac{\partial \mathbf{F}_0(w,u,t,\varepsilon)}{\partial t} = \mathbf{F}_0(w,u,t,\varepsilon)(\mathbf{Q} - \mathbf{B} - \alpha\mathbf{I}) + \mu_1 e^{ju}\mathbf{F}_1(w,u,t,\varepsilon)$$

$$+\mu_2\mathbf{F}_2(w,u,t,\varepsilon) + j\frac{\partial \mathbf{F}_0(w,u,t,\varepsilon)}{\partial w},$$

$$\frac{\partial \mathbf{F}_1(w,u,t,\varepsilon)}{\partial t} = \mathbf{F}_0(w,u,t,\varepsilon)\mathbf{B} + \mathbf{F}_1(w,u,t,\varepsilon)(\mathbf{Q} + (e^{jw\varepsilon} - 1)\mathbf{B} - \mu_1\mathbf{I})$$

$$-je^{jw\varepsilon}\frac{\partial \mathbf{F}_0(w,u,t,\varepsilon)}{\partial w},$$

$$\frac{\partial \mathbf{F}_2(w,u,t,\varepsilon)}{\partial t} = \alpha\mathbf{F}_0(w,u,t,\varepsilon) + \mathbf{F}_2(w,u,t,\varepsilon)(\mathbf{Q} + (e^{jw\varepsilon} - 1)\mathbf{B} - \mu_2\mathbf{I}). \quad (6)$$

An asymptotic solution of the system of equations was carried out (6) .

Theorem 1. *Let $i(t)$ is the number of customers in $MAP/M/1$ retrial queue with two-way communication, then in the stationary regime we obtain*

$$\lim_{\varepsilon\to 0}\{\mathbf{F}_0(w,0,t,\varepsilon) + \mathbf{F}_1(w,0,t,\varepsilon) + \mathbf{F}_2(w,0,t,\varepsilon)\} = \lim_{\sigma\to 0} Me^{jw\sigma i(t)} = e^{jw\kappa}, \quad (7)$$

where κ is the positive root of the equation

$$\kappa\mathbf{R}_0(\kappa)\mathbf{e} = [\mathbf{R}_1(\kappa) + \mathbf{R}_2(\kappa)]\mathbf{B}\mathbf{e}. \quad (8)$$

Furthermore, vectors $\mathbf{R}_n(\kappa)$ are defined by

$$\begin{cases} \mathbf{R}_0(\kappa) = \mathbf{r}\left\{\mathbf{I} + [\mathbf{B} + \kappa\mathbf{I}](\mu_1\mathbf{I} - \mathbf{Q})^{-1} + \alpha(\mu_2\mathbf{I} - \mathbf{Q})^{-1}\right\}^{-1}, \\ \mathbf{R}_1(\kappa) = \mathbf{R}_0(\kappa)[\mathbf{B} + \kappa\mathbf{I}](\mu_1\mathbf{I} - \mathbf{Q})^{-1}, \\ \mathbf{R}_2(\kappa) = \alpha\mathbf{R}_0(\kappa)(\mu_2\mathbf{I} - \mathbf{Q})^{-1}. \end{cases} \quad (9)$$

$\mathbf{r}$ *is the stationary probability distribution of the underlying process $k(t)$.*

Proof. Consider the system (6) in the stationary mode and set $u = 0$. We obtain a system of equations for the probability distribution of the process $\{k(t), n(t), i(t)\}$.

Denoting

$$\mathbf{F}_n(w,\varepsilon) = \lim_{t\to\infty} \mathbf{F}_n(w,0,t,\varepsilon),$$

in order to get the following system:

$$\mathbf{F}_0(w,\varepsilon)(\mathbf{Q} - \mathbf{B} - \alpha\mathbf{I}) + j\mathbf{F}_0'(w,\varepsilon) + \mu_1\mathbf{F}_1(w,\varepsilon) + \mu_2\mathbf{F}_2(w,\varepsilon) = 0,$$

$$\mathbf{F}_1(w,\varepsilon)(\mathbf{Q}+(e^{jw\varepsilon}-1)\mathbf{B}-\mu_1\mathbf{I})+\mathbf{F}_0(w,\varepsilon)\mathbf{B}-je^{-jw\varepsilon}\mathbf{F}_0'(w,\varepsilon)=0,$$
$$\mathbf{F}_2(w,\varepsilon)(\mathbf{Q}+(e^{jw\varepsilon}-1)\mathbf{B}-\mu_2\mathbf{I})+\alpha\mathbf{F}_0(w,\varepsilon)=0. \tag{10}$$

Considering the limit as $\varepsilon\to 0$ in the system (10) yields

$$\mathbf{F}_0(w)(\mathbf{Q}-\mathbf{B}-\alpha\mathbf{I})+j\mathbf{F}_0'(w)+\mu_1\mathbf{F}_1(w)+\mu_2\mathbf{F}_2(w)=0,$$
$$\mathbf{F}_1(w)(\mathbf{Q}-\mu_1\mathbf{I})+\mathbf{F}_0(w)\mathbf{B}-j\mathbf{F}_0'(w)=0,$$
$$\mathbf{F}_2(w)(\mathbf{Q}-\mu_2\mathbf{I})+\alpha\mathbf{F}_0(w)=0, \tag{11}$$

where

$$\lim_{\varepsilon\to 0}\mathbf{F}_n(w,\varepsilon)=\mathbf{F}_n(w).$$

The key idea of our proof is to look for the solution of (11) in form of

$$\mathbf{F}_n(w)=\Phi(w)\mathbf{R}_n, \tag{12}$$

where $\mathbf{R}_n$ is the server state probability distribution. Substituting (12) into (11) yields

$$\mathbf{R}_0(\mathbf{Q}-\mathbf{B}-\alpha\mathbf{I})+j\frac{\Phi'(w)}{\Phi(w)}\mathbf{R}_0+\mu_1\mathbf{R}_1+\mu_2\mathbf{R}_2=0,$$
$$\mathbf{R}_1(\mathbf{Q}-\mu_1\mathbf{I})+\mathbf{R}_0\mathbf{B}-j\frac{\Phi'(w)}{\Phi(w)}\mathbf{R}_0=0,$$
$$\mathbf{R}_2(\mathbf{Q}-\mu_2\mathbf{I})+\alpha\mathbf{R}_0=0. \tag{13}$$

Because $\frac{\Phi'(w)}{\Phi(w)}$ does not depend on w, the scalar function $\Phi(w)$ is obtained in the following form

$$\Phi(w)=\exp jw\kappa_1,$$

We have $j\frac{\Phi'(w)}{\Phi(w)}=-\kappa$. Substituting this expression into the system (13) yields

$$\mathbf{R}_0(\mathbf{Q}-\mathbf{B}-\alpha\mathbf{I})-\kappa\mathbf{R}_0+\mu_1\mathbf{R}_1+\mu_2\mathbf{R}_2=0,$$
$$\mathbf{R}_1(\mathbf{Q}-\mu_1\mathbf{I})+\mathbf{r}_0\mathbf{B}+\kappa\mathbf{R}_0=0,$$
$$\mathbf{R}_2(\mathbf{Q}-\mu_2\mathbf{I})+\alpha\mathbf{R}_0=0. \tag{14}$$

Let us write the normalization condition for the stationary server state probability distribution

$$\mathbf{R}_0+\mathbf{R}_1+\mathbf{R}_2=\mathbf{r}.$$

From this equation and the last two equations in (14), we obtain

$$\begin{cases}\mathbf{R}_1=\mathbf{R}_0[\mathbf{B}+\kappa\mathbf{I}](\mu_1\mathbf{I}-\mathbf{Q})^{-1},\\ \mathbf{R}_2=\alpha\mathbf{R}_0(\mu_2\mathbf{I}-\mathbf{Q})^{-1},\\ \mathbf{R}_0+\mathbf{R}_1+\mathbf{R}_2=\mathbf{r}.\end{cases} \tag{15}$$

By summing equations in (10), we will get the following equation

$$[\mathbf{F}_0(w,\varepsilon)+\mathbf{F}_1(w,\varepsilon)+\mathbf{F}_2(w,\varepsilon)]\mathbf{Q}$$
$$+\mathbf{F}_1(w,\varepsilon)(e^{jw\varepsilon}-1)\mathbf{B}+\mathbf{F}_2(w,\varepsilon)(e^{jw\varepsilon}-1)\mathbf{B}+je^{-jw\varepsilon}(e^{jw\varepsilon}-1)\mathbf{F}_0'(w,\varepsilon)=0.$$

Multiplying this equation by a single column vector $\mathbf{e}$ yields

$$\{\mathbf{F}_1(w,\varepsilon)+\mathbf{F}_2(w,\varepsilon)\}\mathbf{Be}+je^{-jw\varepsilon}\mathbf{F}_0'(w,\varepsilon)\mathbf{e}=0.$$

Substituting the product (12) into this equation, we obtain

$$[\mathbf{R}_1+\mathbf{R}_2]\mathbf{Be}+j\frac{\Phi'(w)}{\Phi(w)}\mathbf{R}_0\mathbf{e}=0$$

and then

$$[\mathbf{R}_1+\mathbf{R}_2]\mathbf{\Lambda e}-\kappa\mathbf{R}_0\mathbf{e}=0. \tag{16}$$

From (16) we can find the expression of κ in terms of $\mathbf{R}_0$, $\mathbf{R}_1$ and $\mathbf{R}_2$. Furthermore, we rewrite (15) as follows:

$$\begin{cases}\mathbf{R}_0(\kappa)=\mathbf{r}\left\{\mathbf{I}+[\mathbf{B}+\kappa\mathbf{I}](\mu_1-\mathbf{Q})^{-1}+\alpha(\mu_2\mathbf{I}-\mathbf{Q})^{-1}\right\}^{-1},\\ \mathbf{R}_1(\kappa)=\mathbf{R}_0(\kappa)[\mathbf{B}+\kappa\mathbf{I}](\mu_1\mathbf{I}-\mathbf{Q})^{-1},\\ \mathbf{R}_2(\kappa)=\alpha\mathbf{R}_0(\kappa)(\mu_2\mathbf{I}-\mathbf{Q})^{-1}.\end{cases}$$

The equations for κ and $\mathbf{R}_n(\kappa)$ coincide with the equalities (8) and (9).
The theorem is proved.

Theorem 1, only defines the mean asymptotic value κ of the number of calls in an orbit and the server state probability distributionin $\mathbf{R}_n$ the limit situation where σ is close to zero. This result was obtainted in [14] for $MMPP/M/1$ retrial queue with two-way communications. We generalized this result for MAP input.

The results of Theorem 1 are auxiliary for our article. The main results are formulated in Theorem 2.

Theorem 2. *Let $m(t)$ is the number of served customers in $MAP/M/1$ retrial queue with two-way communications, then*

$$\lim_{\varepsilon\to 0}\{\mathbf{F}_0(0,u,t,\varepsilon)+\mathbf{F}_1(0,u,t,\varepsilon)+\mathbf{F}_2(0,u,t,\varepsilon)\}\mathbf{e}$$
$$=\lim_{\sigma\to 0}Me^{jum(t)}=\mathbf{R}e^{\mathbf{G}(u)t}\mathbf{ee}, \tag{17}$$

where the block matrix mathbfG(u) has dimension 3K times3K and has the following form

$$\mathbf{G}(u)=\begin{bmatrix}\mathbf{Q}-(\mathbf{B}+(\kappa+\alpha)\mathbf{I}) & \mathbf{B}+\kappa\mathbf{I} & \alpha\mathbf{I}\\ \mu_1 e^{ju}\mathbf{I} & \mathbf{Q}-\mu_1\mathbf{I} & 0\\ \mu_2\mathbf{I} & 0 & \mathbf{Q}-\mu_2\mathbf{I}\end{bmatrix},$$

vector $\mathbf{R} = \{\mathbf{R}_0, \mathbf{R}_1, \mathbf{R}_2\}$ *has* $3K$ *dimensions and its blocks* $\mathbf{R}_0$, $\mathbf{R}_1$, $\mathbf{R}_2$ *are two-dimensional probability distribution of the random process* $\{k(t), n(t)\}$, κ *- normalized mean number of calls in orbit,* $\mathbf{e}$ *and* $\mathbf{ee}$ *unit column vectors of dimensions* K *and* $3K$.

Proof. Let us make the limiting transition $\varepsilon \to 0$ in the system (6), denoting the following functions

$$\mathbf{F}_n(w,u,t) = \lim_{\varepsilon \to 0} \mathbf{F}_n(w,u,t,\varepsilon)$$

then we obtaine the system in the following form

$$\frac{\partial \mathbf{F}_0(w,u,t)}{\partial t} = \mathbf{F}_0(w,u,t)(\mathbf{Q} - \mathbf{B} - \alpha\mathbf{I}) + \mu_1 e^{ju}\mathbf{F}_1(w,u,t)$$

$$+\mu_2\mathbf{F}_2(w,u,t) + j\frac{\partial \mathbf{F}_0(w,u,t)}{\partial w},$$

$$\frac{\partial \mathbf{F}_1(w,u,t)}{\partial t} = \mathbf{F}_0(w,u,t)\mathbf{B} + \mathbf{F}_1(w,u,t)(\mathbf{Q} - \mu_1\mathbf{I})$$

$$-j\frac{\partial \mathbf{F}_0(w,u,t)}{\partial w},$$

$$\frac{\partial \mathbf{F}_2(w,u,t)}{\partial t} = \alpha\mathbf{F}_0(w,u,t) + \mathbf{F}_2(w,u,t)(\mathbf{Q} - \mu_2\mathbf{I}). \tag{18}$$

We will write a solution to the (18) system in the following form

$$\mathbf{F}_n(w,u,t) = \Phi(w)\mathbf{F}_n(u,t).$$

Then we rewrite this system as follows

$$\frac{\partial \mathbf{F}_0(u,t)}{\partial t} = \mathbf{F}_0(u,t)(\mathbf{Q} - \mathbf{B} - \alpha\mathbf{I}) + \mu_1 e^{ju}\mathbf{F}_1(u,t)$$

$$+\mu_2\mathbf{F}_2(u,t) + j\frac{\Phi'(w)}{\Phi(w)},$$

$$\frac{\partial \mathbf{F}_1(u,t)}{\partial t} = \mathbf{F}_0(u,t)\mathbf{B} + \mathbf{F}_1(u,t)(\mathbf{Q} - \mu_1\mathbf{I})$$

$$-j\frac{\Phi'(w)}{\Phi(w)},$$

$$\frac{\partial \mathbf{F}_2(u,t)}{\partial t} = \alpha\mathbf{F}_0(u,t) + \mathbf{F}_2(u,t)(\mathbf{Q} - \mu_2\mathbf{I}). \tag{19}$$

$\frac{\Phi'(w)}{\Phi(w)}$ does not depend on w, the scalar function $\Phi(w)$ is obtained in the following form

$$\Phi(w) = \exp jw\kappa,$$

then

$$\frac{\Phi'(w)}{\Phi(w)} = j\kappa.$$

Then the system (19) can be written as

$$\frac{\partial \mathbf{F}_0(u,t)}{\partial t} = \mathbf{F}_0(u,t)(\mathbf{Q} - \mathbf{B} - (\alpha+\kappa)\mathbf{I}) + \mu_1 e^{ju}\mathbf{F}_1(u,t)$$
$$+\mu_2\mathbf{F}_2(u,t),$$
$$\frac{\partial \mathbf{F}_1(u,t)}{\partial t} = \mathbf{F}_0(u,t)\mathbf{B} + \mathbf{F}_1(u,t)(\mathbf{Q} - (\mu_1+\kappa)\mathbf{I}),$$
$$\frac{\partial \mathbf{F}_2(u,t)}{\partial t} = \alpha\mathbf{F}_0(u,t) + \mathbf{F}_2(u,t)(\mathbf{Q} - \mu_2\mathbf{I}). \tag{20}$$

denote vector $\mathbf{FF}(u,t) = \{\mathbf{F}_0(u,t), \mathbf{F}_1(u,t), \mathbf{F}2(u,t)\}$, which has $3K$ dimention, and matrix $\mathbf{G}(u)$

$$\mathbf{G}(u) = \begin{bmatrix} \mathbf{Q} - (\mathbf{B} + (\kappa+\alpha)\mathbf{I}) & \mathbf{B} + \kappa\mathbf{I} & \alpha\mathbf{I} \\ \mu_1 e^{ju}\mathbf{I} & \mathbf{Q} - \mu_1\mathbf{I} & 0 \\ \mu_2\mathbf{I} & 0 & \mathbf{Q} - \mu_2\mathbf{I} \end{bmatrix}.$$

We rewrite system (20) in matrix form

$$\frac{\partial \mathbf{FF}(u,t)}{\partial t} = \mathbf{FF}(u,t)\mathbf{G}(u), \tag{21}$$

with the initial condition

$$\mathbf{FF}(u,0) = \mathbf{R}. \tag{22}$$

Vector $\mathbf{R} = \{\mathbf{R}_0, \mathbf{R}_1, \mathbf{R}_2\}$ has $3K$ dimensions and its blocks $\mathbf{R}_0$, $\mathbf{R}_1$, $\mathbf{R}_2$ are two-dimensional probability distribution of the random process $\{k(t), n(t)\}$.

Solution to the Cauchy problem (21), (22) has the form

$$\mathbf{FF}(u,t) = \mathbf{R}e^{\mathbf{G}(u)}.$$

From where we write the asymptotic characteristic function of the number of serviced claims

$$\lim_{\sigma\to 0} Me^{jum(t)} = \lim_{\varepsilon\to 0}\{\mathbf{F}_0(0,u,t,\varepsilon) + \mathbf{F}_1(0,u,t,\varepsilon)\}\mathbf{e}$$
$$= \mathbf{R}e^{\mathbf{G}(u)t}\mathbf{ee},$$

which coincides with (17), $\mathbf{e}$ and $\mathbf{ee}$ unit column vectors of dimensions K and $3K$. **The theorem is proved.**

The expression (17) has the same form as the formulas for the characteristic function of the number of events in the Markov arrival process [7,12]. Let us bring the matrix $\mathbf{G}(u)$ to the appropriate form and formulate Theorem 3.

Theorem 3. *Output process of retrial queue with two-way communication and MAP input under low rate of retrials condition ($\sigma \to 0$) is synchronous Markovian arrival process [12] defined by infinitesimal generator* $\mathbf{Q1}$ *of underlying Markov chain of* $3K$ *dimensions*

$$\mathbf{Q1} = \begin{bmatrix} \mathbf{Q} - (\mathbf{B} + (\kappa + \alpha)\mathbf{I}) & \mathbf{B} + \kappa\mathbf{I} & \alpha\mathbf{I} \\ \mu_1\mathbf{I} & \mathbf{Q} - \mu_1\mathbf{I} & 0 \\ \mu_2\mathbf{I} & 0 & \mathbf{Q} - \mu_2\mathbf{I} \end{bmatrix},$$

and probability matrix $\mathbf{D1}$ *of event occurrence in MAP at the moment of state changes of underlying Markov chain*

$$\mathbf{D1} = \begin{bmatrix} 0 & 0 & 0 \\ \mathbf{I} & 0 & 0 \\ 0 & 0 & 0 \end{bmatrix},$$

Proof. In the work [12], it is shown that the equation that determines the characteristic function of the probability distribution of the number of events occurring in the MAP flow for some time t has the form (17). In this case, the matrix $G(u)$ is defined in terms of the matrices defining the MAP flow as follows

$$\mathbf{G}(u) = \mathbf{Q1} + (e^{ju} - 1)[\mathbf{\Lambda}1 + \mathbf{Q1} * \mathbf{D1}].$$

Matrix $\mathbf{Q1}$ - matrix of infinitesimal characteristics of the control Markov chain; $\mathbf{\Lambda}1$ is the matrix of conditional intensities of the occurrence of events on the intervals of constancy of states of the control Markov chain; $\mathbf{D1}$ matrix of probabilities of occurrence of events when the state of the control Markov chain changes. Transform the matrix $\mathbf{G}(u)$ to this form

$$\mathbf{G}(u) = \begin{bmatrix} \mathbf{Q} - (\mathbf{B} + (\kappa + \alpha)\mathbf{I}) & \mathbf{B} + \kappa\mathbf{I} & \alpha\mathbf{I} \\ \mu_1 e^{ju}\mathbf{I} & \mathbf{Q} - \mu_1\mathbf{I} & \mathbf{0} \\ \mu_2\mathbf{I} & \mathbf{0} & \mathbf{Q} - \mu_2\mathbf{I} \end{bmatrix}$$

$$= \begin{bmatrix} \mathbf{Q} - (\mathbf{B} + (\kappa + \alpha)\mathbf{I}) & \mathbf{B} + \kappa\mathbf{I} & \alpha\mathbf{I} \\ ((e^{ju} - 1)\mu_1\mathbf{I} + \mu_1\mathbf{I}) & \mathbf{Q} - \mu_1\mathbf{I} & \mathbf{0} \\ \mu_2\mathbf{I} & \mathbf{0} & \mathbf{Q} - \mu_2\mathbf{I} \end{bmatrix}$$

$$= \begin{bmatrix} \mathbf{Q} - (\mathbf{B} + (\kappa + \alpha)\mathbf{I}) & \mathbf{B} + \kappa\mathbf{I} & \alpha\mathbf{I} \\ \mu_1\mathbf{I} & \mathbf{Q} - \mu_1\mathbf{I} & \mathbf{0} \\ \mu_2\mathbf{I} & \mathbf{0} & \mathbf{Q} - \mu_2\mathbf{I} \end{bmatrix}$$

$$+ (e^{ju} - 1)\left(\begin{bmatrix} \mathbf{0} & \mathbf{0} & \mathbf{0} \\ \mathbf{0} & \mathbf{0} & \mathbf{0} \\ \mathbf{0} & \mathbf{0} & \mathbf{0} \end{bmatrix} + \begin{bmatrix} \mathbf{0} & \mathbf{0} & \mathbf{0} \\ \mu_1\mathbf{I} & \mathbf{0} & \mathbf{0} \\ \mathbf{0} & \mathbf{0} & \mathbf{0} \end{bmatrix} \right)$$

That is, in our problem we got that

$$\mathbf{Q1} = \begin{bmatrix} \mathbf{Q} - (\mathbf{B} + (\kappa + \alpha)\mathbf{I}) & \mathbf{B} + \kappa\mathbf{I} & \alpha\mathbf{I} \\ \mu_1\mathbf{I} & \mathbf{Q} - \mu_1\mathbf{I} & \mathbf{0} \\ \mu_2\mathbf{I} & \mathbf{0} & \mathbf{Q} - \mu_2\mathbf{I} \end{bmatrix},$$

and probability matrix $\mathbf{D}$ of event occurrence in MAP at the moment of state changes of underlying Markov chain

$$\mathbf{D1} = \begin{bmatrix} \mathbf{0} & \mathbf{0} & \mathbf{0} \\ \mathbf{I} & \mathbf{0} & \mathbf{0} \\ \mathbf{0} & \mathbf{0} & \mathbf{0} \end{bmatrix}, \mathbf{Q1} * \mathbf{D1} = \begin{bmatrix} \mathbf{0} & \mathbf{0} & \mathbf{0} \\ \mu_1\mathbf{I} & \mathbf{0} & \mathbf{0} \\ \mathbf{0} & \mathbf{0} & \mathbf{0} \end{bmatrix}, \mathbf{\Lambda1} = \begin{bmatrix} \mathbf{0} & \mathbf{0} & \mathbf{0} \\ \mathbf{0} & \mathbf{0} & \mathbf{0} \\ \mathbf{0} & \mathbf{0} & \mathbf{0} \end{bmatrix}$$

where sign $*$ is Hadamard product.

We have expressed the matrix $\mathbf{G}(u)$ in terms of matrices defining some map flow. The matrix of conditional intensities $\mathbf{\Lambda1}$ in this process is equal to zero. Consequently, the output process of the system under consideration, when the asymptotic condition of large delay in the orbit is satisfied, belongs to the class of synchronous MAP. **The theorem is proved.**

5 Conclusion

In this paper, we have considered the output of $MAP/M/1$ retrial queue with two-way communication using asymptotic analysis method under low rate of retrials condition. We have formulated and proved Theorem 2, where we obtained an explicit formula (7) for the characteristic function of the number of served customers in the system. We have shown that the output process in the system is synchronous Markovian arrival process. This result was formulated and proved in Theorem 3.

References

1. Artalejo, J., Falin, G.: Standard and retrial queueing systems: a comparative analysis. Rev. Mat. Complut. **15**(1), 101–129 (2002)
2. Artalejo, J.R., Gómez-Corral, A.: Retrial queueing systems: a computational approach (2008). https://doi.org/10.1007/978-3-540-78725-9
3. Artalejo, J.R., Phung-Duc, T.: Markovian retrial queues with two way communication. J. Ind. Manag. Optim. **8**(4), 781–806 (2012)
4. Bean, N., Green, D., Taylor, P.: The output process of an mmpp/m/1 queue. J. Appl. Probab. **35**(4), 998–1002 (1998)
5. Burke, P.J.: The output of a queuing system. Oper. Res. **4**(6), 699–704 (1956)
6. Daley, D.: Queueing output processes. Adv. Appl. Probab. **8**(2), 395–415 (1976)
7. Dudin, A., Klimenok, V.: Corellated flow queueing systems. Minsk: BSU publ. 175 (2000)
8. Falin, G., Templeton, J.G.: Retrial Queues, vol. 75. CRC Press, Boca Raton (1997)
9. Ferrari, P.A., Fontes, L.R.G.: The net output process of a system with infinitely many queues. Ann. Appl. Probab. **4**(4), 1129–1144 (1994)
10. Kendall, D.G.: Stochastic processes occurring in the theory of queues and their analysis by the method of the imbedded Markov chain. The Annals of Mathematical Statistics, pp. 338–354 (1953)
11. Lapatin, I., Nazarov, A.: Asymptotic analysis of the output process in retrial queue with Markov-modulated poisson input under low rate of retrials condition. In: Vishnevskiy, V.M., Samouylov, K.E., Kozyrev, D.V. (eds.) DCCN 2019. CCIS, vol. 1141, pp. 315–324. Springer, Cham (2019). https://doi.org/10.1007/978-3-030-36625-4_25

12. Lopuhova, S.V.: Asymptotic and numerical methods of research of special flows of homogeneous events. Ph.D. thesis, Tomsk State University (2008)
13. Mirasol, N.M.: The output of an $m/g/\infty$ queuing system is poisson. Oper. Res. **11**(2), 282–284 (1963)
14. Nazarov, A., Phung-Duc, T., Paul, S.: Slow retrial asymptotics for a single server queue with two-way communication and Markov modulated poisson input. J. Syst. Sci. Syst. Eng. **28**(2), 181–193 (2019)
15. Phung-Duc, T., Rogiest, W., Takahashi, Y., Bruneel, H.: Retrial queues with balanced call blending: analysis of single-server and multiserver model. Ann. Oper. Res. **239**(2), 429–449 (2014). https://doi.org/10.1007/s10479-014-1598-2
16. Reich, E.: Waiting times when queues are in tandem. Ann. Math. Stat. **28**(3), 768–773 (1957)
17. Sakurai, H., Phung-Duc, T.: Two-way communication retrial queues with multiple types of outgoing calls. TOP **23**(2), 466–492 (2014). https://doi.org/10.1007/s11750-014-0349-5

Stochastic Lists of Multiple Resources

Konstantin Samouylov[1,2(✉)] and Valeriy Naumov[3]

[1] Department of Applied Informatics and Probability, Peoples' Friendship University of Russia (RUDN University), Miklukho-Maklaya St. 6, 117198 Moscow, Russian Federation
samuylov-ke@rudn.university
[2] Institute of Informatics Problems, Federal Research Center "Computer Science and Control" of the Russian Academy of Sciences, Vavilov Street 44-2, 119333 Moscow, Russian Federation
[3] Service Innovation Research Institute, Annankatu 8A, 00120 Helsinki, Finland
valeriy.naumov@pfu.fi

Abstract. In this article, we introduce the concept of stochastic lists and pseudo-lists and apply them to analyze a loss system with Poisson arrivals, exponential service times, and multiple positive and negative resources. For this system, we prove that the total volumes of resources in the stochastic list and pseudo-list have the same stationary distribution.

Keywords: Multi-resource loss system · Negative customers · Stochastic list

1 Introduction

Lists are widely used in programming [1] and when defining computer and communication systems [2]. In this paper we show how lists can be adopted in queueing theory to study multi-resource loss systems, in which resources of multiple types are allocated to customers for the total length of their service times and then released. In such systems, an arriving customer that finds the required amounts of resources unavailable is lost. Once service completed, a customer releases the exact amounts of resources that have been allocated to it upon arrival. Thus, for each customer in service we must "remember" the vector of its allocated resource amounts, which greatly complicates stochastic processes modeling the behavior of such systems.

The analysis of these systems can be facilitated by its simplification [3]. Simplified loss system is identical to the original one, except that the amounts of resources released upon a departure are assumed random and may differ from the amounts allocated to the departing customer upon its arrival. Given the totals of allocated resources and the number of customers in service, the amounts of resources released upon a departure are independent of the system's behavior prior to the departure instant and have an easily calculable CDF. Stochastic processes representing the behavior of such simplified systems are easier to study

A. N. Shiryaev et al. (Eds.): ICSM-5 2020, PROMS 371, pp. 353–357, 2021.
https://doi.org/10.1007/978-3-030-83266-7_27

since there is no need to remember the amounts of resources used by each customer: the totals of allocated resources suffice. In [4] it was shown that for a multi-resource loss system with Poisson arrivals and exponential service times, simplified and original systems have the same stationary distribution of the totals of allocated resources.

In this paper we formalize notion of simplified system by introducing stochastic lists and pseudo-lists. We will show that for the system studied in [4] lists and pseudo-lists yield the same stationary distribution of the totals of allocated resources. We allow the quantities of resources requested by customers to be positive or negative and observe that requests for a negative quantity of a resource increase the amount of the resource available to customers requesting positive quantities of it. The notion of positive and negative customers was introduced in [5,6]. Positive customers can represent resource requests whereas negative customers can increase the amount of available resources by canceling some requests. We study a system with a different behavior of negative customers. Here, instead of requests cancelation, negative customers temporary increase the amount of resources available to positive customers.

2 Stochastic Lists

Let $S \subset \mathbb{R}^M$ represent a nonempty measurable subset of the real M-space. We will refer to the elements of S^k as lists of length k and denote the set of all such lists by $\bar{S} = \sum_{k=0}^{L} S^k$, where L represents the maximum list length. Denote by $[\xi]$ the length of list $\xi \in \bar{S}$. The set S^0 is assumed to consist of a single list () of zero length. The operation of deleting the ith element from a list $\xi = (s_1, s_2, ..., s_k)$ results in the list $\mathrm{Del}_i(\xi) = (s_1, \ldots, s_{i-1}, s_{i+1}, \ldots, s_k)$, $1 \leq j \leq k$., while the operation of inserting $u \in S$ as the jth element yields the list $\mathrm{Ins}_j(\xi, u) = (s_1, \ldots, s_{j-1}, u, s_j, \ldots, s_k)$, $1 \leq j \leq k+1$..

Let $\xi(t) \in \bar{S}$, $t \geq 0$, be a right-continuous jump stochastic process with jumps at random times $0 < \tau_1 < \tau_2 < \ldots$. Let $\tau_0 = 0$ and denote $\xi_n = \xi(\tau_n)$,, $n \geq 0$. We call the process $\xi(t)$ a stochastic list if, for any $n \geq 1$, the values of the process before and after time τ_n are related as either

1) $\xi_n = \mathrm{Del}_j(\xi_{n-1})$ for some index j, $1 \leq j \leq [\xi_{n-1}]$, or
2) $\xi_n = \mathrm{Ins}_j(\xi_{n-1}, u)$ for some $u \in S$ and j, $1 \leq j \leq [\xi_n]$.

Put simply, a process $\zeta(t) \in \bar{S}$ is a stochastic list if each of its states is obtained from the previous one either through insertion or deletion of some element.

Let $\mathbf{v} \in \mathbb{R}^M$ be some nonnegative vector.

A stochastic list $\xi(t) = (\zeta_1(t), ..., \zeta_{n(t)}(t))$ is said to be a list of resources of capacity $\mathbf{v}$, if at any time t we have $\boldsymbol{\sigma}(t) = \zeta_1(t) + ... + \zeta_{n(t)}(t) \leq \mathbf{v}$. Vector $\boldsymbol{\sigma}(t)$ contains the totals of allocated resources at time t, while the difference $\mathbf{v} - \boldsymbol{\sigma}(t)$ represents a vector of idle resources. Since $\boldsymbol{\sigma}(t) \leq \mathbf{v}$ for all t, at insertion times τ_k we have $n(\tau_k - 0) < L$ and $\boldsymbol{\sigma}(\tau_k - 0) + \mathbf{r}_k \leq \mathbf{v}$, while at deletion times $n(\tau_k - 0) > 0$ and $\boldsymbol{\sigma}(\tau_k - 0) - \zeta_{\beta_k}(\tau_k - 0) \leq \mathbf{v}$.

If some element $r_k(i)$ of $\mathbf{r}_k$ is positive, then the amount of idle resources of the ith type decreases when $\mathbf{r}_k$ is inserted into the list and increases once $\mathbf{r}_k$ is deleted. If, on the contrary, $r_k(i) < 0$ then the amount of idle type i resources increases when inserting $\mathbf{r}_k$ and decreases upon its deletion. Thus, vectors with negative entries, while present in a list, increase the capacity of resources of the corresponding types.

The process $\boldsymbol{\sigma}(t)$ is similar to the stochastic storage processes studied in [7], with a major difference that only elements which are present in the list can be deleted. Consequently, only vectors that have been previously added to $\boldsymbol{\sigma}(t)$ are subtracted. This is why even if we are interested only in the totals of allocated resources, we still have to deal with all elements of a list of resources.

3 Stochastic Pseudo-Lists

To evaluate the totals of allocated resources, instead of $\xi(t)$, one can use a simpler jump process $(\kappa(t), \boldsymbol{\vartheta}(t))$, which we will refer to as the pseudo-list of resources of capacity $\mathbf{v}$. This process has an integer component $\kappa(t) \in \mathbb{N}$, which we will call the pseudo-list's length, and a vector component $\boldsymbol{\vartheta}(t) \leq \mathbf{v}$ called the pseudo-list's volume. At an insertion time its length $\kappa(t)$ increases by 1 and its volume increases by a vector $\mathbf{r}_k \in S$. At a deletion time $\kappa(t)$decreases by 1 and $\boldsymbol{\vartheta}(t)$ decreases by a vector $\boldsymbol{\delta}_k$.

The vectors $\boldsymbol{\delta}_k$ correspond to the elements $\zeta_{\beta_k}(b_k - 0)$ of the list of resources $\xi(t)$. Given the state $(n, \mathbf{y})$ of the pseudo-list before time b_k, vector $\boldsymbol{\delta}_k$ is independent of the pseudo-list's behavior prior to b_k and has the CDF

$$P\{\boldsymbol{\delta}_k \leq \mathbf{x} | \kappa(b_k - 0) = n, \boldsymbol{\vartheta}(b_k - 0) = \mathbf{y}\} = F_n(\mathbf{x}|\mathbf{y}).$$

$F_1(\mathbf{x}|\mathbf{y})$ is a CDF of a constant vector $\mathbf{y}$, and it follows from the definition of the conditional probability [8] that for $k \geq 2$ the function $F_k(\mathbf{x}|\mathbf{y})$ solves

$$\int\limits_{\mathbf{z}\leq\mathbf{y}} \int\limits_{\mathbf{u}\geq\mathbf{z}-\mathbf{x}} F_k(d\mathbf{u}|\mathbf{z})F^{(k)}(d\mathbf{z}) = \int\limits_{\mathbf{z}\leq\mathbf{x}} F(\mathbf{y}-\mathbf{z})F^{(k-1)}(d\mathbf{z}), \ \mathbf{x}, \mathbf{y} \in \mathbb{R}^M.$$

The pseudo-list of resources $(\kappa(t), \boldsymbol{\vartheta}(t))$ imitates the process $(n(t), \boldsymbol{\sigma}(t))$ in the list of resources $\xi(t) = (\zeta_1(t), ..., \zeta_{n(t)}(t))$, however, instead of previously inserted vectors $\mathbf{r}_k$, we subtract from $\boldsymbol{\vartheta}(t)$ the random vectors $\boldsymbol{\delta}_k$ distributed with the CDFs $F_n(\mathbf{x}|\mathbf{y})$.

4 Multi-resource Loss System

Consider a loss system with L servers, a Poisson arrival process of rate λ and service times exponentially distributed with parameter μ. Assume the system to possess M types of resources, each of which has limited capacity. Let each customer require a variable amount of each resource. A customer is lost if upon its arrival the idle service capacity of some resource is less than the amount required by the customer. Once service begins, the idle capacity of each resource decreases by the amount required by the newly arrived customer.

Denote by $v(m)$ the total service capacity of the type m resource, $\mathbf{v} = (v(1), \ldots, v(M))$, and by $\mathbf{r}_j = (r_j(1), \ldots, r_j(M))$ the vector of resource amounts required by the jth customer, $j = 1, 2, \ldots$. We assume vectors $\mathbf{r}_j$ independent of the arrival and service processes, mutually independent and identically distributed with CDF $F(\mathbf{x})$, $F(\mathbf{v}) > 0$.

The state of the system at time t can be described by a stochastic list $\xi(t) = (\zeta_1(t), \ldots, \zeta_{n(t)}(t))$, consisting of vectors $\zeta_i(t) \in S$ of resource amounts allocated to customers in service.

Theorem 1. *Limit distribution of the process* $\xi(t) = (\zeta_1(t), \ldots, \zeta_{n(t)}(t))$, $p_0 = \lim\limits_{t\to\infty} P\{n(t) = 0\}$, $P_k(\mathbf{x}_1, \ldots, \mathbf{x}_k) = \lim\limits_{t\to\infty} P\{\, n(t) = k, \zeta_1(t) \le \mathbf{x}_1, \ldots, \zeta_k(t) \le \mathbf{x}_k\}$, *is given by*

$$p_0 = \left(1 + \sum_{k=1}^{L} F^{(k)}(\mathbf{v})\frac{\rho^k}{k!}\right)^{-1},$$

$$P_k(\mathbf{x}_1, \ldots, \mathbf{x}_k) =$$

$$= p_0\frac{\rho^k}{k!} \int\limits_{\substack{\mathbf{y}_1 \le \mathbf{x}_1, \ldots, \mathbf{y}_k \le \mathbf{x}_k \\ \mathbf{y}_1 + \ldots + \mathbf{y}_k \le \mathbf{v}}} F(\mathbf{y}_1) \ldots F(\mathbf{y}_k), \mathbf{x}_1, \ldots, \mathbf{x}_k \in \mathbb{R}^M, 1 < k \le L,$$

where $\rho = \lambda/\mu$ *and* $F^{(k)}(\mathbf{x})$ *is the k-fold convolution of* $F(\mathbf{x})$.

Corollary. *Limit distribution of the process* $(n(t), \boldsymbol{\sigma}(t))$ *is given by*

$$\lim_{t\to\infty} P\{n(t) = k\ ,\ \boldsymbol{\sigma}(t) \le \mathbf{x}\} = p_0 F^{(k)}(\mathbf{x})\frac{\rho^k}{k!}, \mathbf{x} \in \mathbb{R}^M, 0 \le k \le L.$$

Theorem 2. *Limit distribution of the process* $(\kappa(t), \boldsymbol{\vartheta}(t))$ *is given by*

$$\lim_{t\to\infty} P\{\kappa(t) = k\ ,\ \boldsymbol{\vartheta}(t) \le \mathbf{x}\} = p_0 F^{(k)}(\mathbf{x})\frac{\rho^k}{k!}, \mathbf{x} \in \mathbb{R}^M, 0 \le k \le L.$$

Therefore, in this case an approximation of the process $(n(t), \boldsymbol{\sigma}(t))$ by the pseudo-list $(\kappa(t), \boldsymbol{\vartheta}(t))$ gives correct limit distribution.

5 Conclusion

Each random list $\xi(t) = (\zeta_1(t), ..., \zeta_{n(t)}(t))$ can be associated with a process of total volumes of allocated resources $X(t) = (n(t), \boldsymbol{\sigma}(t))$ and a pseudo-list $Y(t) = (\kappa(t), \boldsymbol{\vartheta}(t))$. We will call a list $\xi(t)$ simplifiable if $X(t)$ and $Y(t)$ have the same stationary distribution. The example above shows that the class of simplifiable lists is not empty. The question arises as to how wide this class is and would the list simplificability be a consequence of the product form of its stationary distribution. These questions are to be answered in subsequent studies.

Acknowledgments. This paper has been supported by the RUDN University Strategic Academic Leadership Program (K.S., validation, formal analysis, project administration). The reported study was partially funded by RFBR, project № 19-07-00933.

References

1. Knuth, D.E.: The Art of Computer Programming: Fundamental Algorithms, vol. 1. Addision Wesley. Reading, Mass (1997)
2. Le Boudec, J.Y.: Performance evaluation of computer and communication systems. In: Computer and communication sciences. Lausanne: EPFL Press London (2010)
3. Naumov, V.A., Samuylov, K.E.: On the modeling of queueing systems with multiple resources. Discrete Continuous Models Appl. Comput. Sci. **3**, 60–64 (2014)
4. Naumov, V.A., Samuilov, K.E., Samuilov, A.K.: On the total amount of resources occupied by serviced customers. Autom. Remote. Control. **77**(8), 1419–1427 (2016). https://doi.org/10.1134/S0005117916080087
5. Gelenbe, E.: Random neural networks with negative and positive signals and product form solution. Neural Comput. **1**(4), 502–510 (1989)
6. Gelenbe, E.: Product-form queueing networks with negative and positive customers. J. Appl. Prob. **28**(3), 656–663 (1991)
7. Prabhu, N.U.: Stochastic Storage Processes: Queues, Insurance Risk, Dams, and Data Communication, 15. Springer-Verlag, New York (1998)
8. Shiryaev, A.N.: Mathematical foundations of probability theory. In: Probability-1. GTM, vol. 95, pp. 159–371. Springer, New York (2016). https://doi.org/10.1007/978-0-387-72206-1_2

Author Index

A. N. Shiryaev et al. (Eds.): ICSM-5 2020, PROMS 371, pp. 359–360, 2021.
https://doi.org/10.1007/978-3-030-83266-7

Printed by Printforce, the Netherlands